ADVANCES IN SUGARBEET PRODUCTION: PRINCIPLES AND PRACTICES

[Sugarbeet Congress, Salt Lake City, 1969]

Advances in
SUGARBEET

PRODUCTION:
Principles and Practices

Edited by

Russell T. Johnson
John T. Alexander
George E. Rush
George R. Hawkes

THE IOWA STATE UNIVERSITY PRESS,
AMES, IOWA, U.S.A.

The chapters of this book are the result of a symposium sponsored by Chevron Chemical Company, December 8–10, 1969.

© 1971 The Iowa State University Press
Ames, Iowa 50010. All rights reserved

Composed and printed by
The Iowa State University Press

First edition, 1971

International Standard Book Number: 0-8138-1415-4
Library of Congress Catalog Card Number: 75-137094

Table of Contents

Preface vii

1. A Food Resource 3
   Thomas Theis
2. Environmental Factors 19
   R. S. Loomis, Albert Ulrich, and Norman Terry
3. Seedbed Preparation, Planting, and Thinning 49
   G. E. Nichol, L. M. Burtch, and D. J. Traveller
4. Weed Control 69
   E. F. Sullivan and B. B. Fischer
5. Nitrogen Nutrition 111
   F. Jackson Hills and Albert Ulrich
6. Phosphorus and Potassium Nutrition 137
   W. R. Schmehl and D. W. James
7. Secondary Nutrients and Micronutrients 171
   Frank G. Viets, Jr., and Lynn S. Robertson
8. Irrigation and Water Management 189
   Marvin E. Jensen and Leonard J. Erie
9. Diseases and Their Control 223
   C. W. Bennett and L. D. Leach
10. Insects and Mites and Their Control 287
    W. H. Lange
11. Nematodes and Their Control 335
    Jack Altman and Ivan Thomason
12. Factors Affecting Quality 371
    J. T. Alexander
13. Harvesting and Delivery 383
    Stewart Bass and P. B. Smith
14. Variety Development 401
    J. S. McFarlane

15. Seed Production 437
    Sam C. Campbell and A. A. Mast
16. Economics of Production 451
    Bion Tolman and Charles Whipple

Index 461

Preface

 SUGARBEET growing for sugar in Europe became a successful industry in the early 1800s. The three things required for success were (1) the development of a satisfactory process for recovering the sugar from the roots; (2) the development of varieties with a sufficiently high sugar content to make recovery of sugar attractive; and (3) the economic push created for the domestic industry in France by the British blockade of cane-sugar imports to France during the Napoleonic wars.

In America, it was 1870 before a successful beet-sugar operation was established. Many unsuccessful attempts had been made earlier in many parts of this country, but most had failed due to lack of adequate, reliable supplies of the raw product—sugarbeets. Once established, the sugarbeet became an agricultural mainstay and the industry has now spread to some 26 states. The history of the development of this industry is a chronicle of successes in research seldom achieved with an agricultural crop.

Many of the early unsuccessful attempts to produce sugarbeets in America were failures, due to disease and insect problems. The conquest of diseases by the development of resistant varieties or other control methods has been responsible for the very existence of sugarbeets as a crop in many areas. Similarly, this is true with insects in some areas. While much has been accomplished in disease and insect control, these two factors remain as a continuing threat to the industry in most beet-growing areas, requiring continued awareness and research. The ultimate goal in sugarbeet production is the development of a crop yielding high tonnage and sugar content that can economically be processed into sugar. This is best acquired by producing a healthy beet crop provided with optimum amounts of nutrients. Some nutrients that are required for growth, if still available to the plant late in the season, are deleterious to a desired sugar content. Improper amounts or timing in fertilizer application can result in a crop of lower value. For this reason, an attempt should be made to obtain maximum crop value by making optimum application of nutrients. While much remains to be learned about the singular and multiple effects of many nutrients in the development of the sugarbeet crop, fundamental knowledge is available that if properly applied will permit an improvement in quality and crop value in essentially every beet-growing area in the United States.

Variety development has advanced through several stages. The original varieties used in this country were open-pollinated varieties imported directly from Europe. Successive improvements have led to varieties better adapted to the specific conditions of the different growing areas, to varieties with resistance to some diseases, to varieties with monogerm seed, and to the currently used hybrids. While these accomplishments are impressive, the search continues for new and improved sources of disease and insect resistance and for hybrids that will produce more high-quality beets per acre.

Weed control in sugarbeets has always been recognized as a requirement for successful production. Until relatively recent years, however, the only available means of accomplishing this were by mechanical cultivation or hand weeding or both. Starting with a small number of chemical herbicides that found relatively restricted use in beet fields, more and more chemicals have become available which are effective against a wider range of weed species and a wider range of conditions under which they can be applied. No one chemical appears to be a cure-all for weed problems. Even the best of them should be considered as a supplement to, not a substitute for, mechanical cultivation. Improved chemical weed control has made possible the final stand establishment in beet fields with less dependence on hand labor. In some beet-growing areas, the combination of effective herbicides, together with timely cultural operations with proper mechanical devices, has almost eliminated the requirement of hand labor in the production of a satisfactory beet crop.

Economically, the sugarbeet has a different status in different areas of the United States. It ranges from areas in which the rotations on many farms are built around the sugarbeet crop, on one extreme, to areas where increasing land values, competing crops, and high production costs have threatened the continued existence of sugarbeets. In this latter case, sugarbeet acreage is more dependent upon the economic status of competing land uses than upon the economic level of sugarbeets alone. These differences are likely to continue to exist in the various parts of the country.

With this background, the Sugarbeet Congress was convened on December 8, 9, and 10, 1969, in Salt Lake City, Utah, to discuss the various agricultural principles involved in the production of sugarbeets. Authorities on all facets of agricultural production were brought together for this three-day conference to present and discuss the topics that comprise this book. In these discussions, attempts were made to rely heavily on the principles involved in problem solving, and rely less intensively on the specifics in the hope that by understanding the principles involved the readers of this book could interpret those principles to fit specific needs. It is hoped the contents of this book will find valuable use among many beet growers, university extension workers, sugar company agricultural representatives, agricultural service personnel, and students at colleges and universities.

PREFACE

The editors of this book wish to express their appreciation and thank the Ortho Division of Chevron Chemical Company for financing this seminar-type congress. We also wish to thank the various participants for the efforts to which they have gone in screening the available material on each of the subjects to present that most applicable in a manner that could be understood, and it is hoped, used to produce better sugarbeet crops.

R. T. JOHNSON
J. T. ALEXANDER
G. E. RUSH
G. R. HAWKES

ADVANCES IN SUGARBEET PRODUCTION: PRINCIPLES AND PRACTICES

1
A Food Resource

THOMAS THEIS (deceased)
Crops Research Division, ARS, USDA
Beltsville, Maryland

HISTORY	4
SUGAR CONTROLS	6
THE INDUSTRY TODAY	9
THE INDUSTRY TOMORROW	12
OUTLOOK	16

Dr. Theis was killed in an automobile accident, June 1970.

THE SUGARBEET is vital to man as a source of a high energy, pure food. Along with sugarcane, it provides the 70 million tons of sugar consumed annually in the world. The two crops, sugarcane and sugarbeets, contribute 55 and 45 percent of the supply, respectively.

The sugarbeet as a food source occupies a unique position in the plant kingdom. Taxonomically, it is a small portion of the 300,000 species of higher plants on our planet. Nevertheless, the plant is one of a dozen species that feeds the world's population (13). During his evolution, man experimented with at least 3,000 species of plants for food. He concentrates now on approximately 150 of these. The most efficient of the group includes the sugarbeet, which is ranked for its food value along with such crops as rice, corn, wheat, potatoes, beans, and others.

Why is the sugarbeet so valuable? For one thing, it is efficient. Few other plants equal its ability to convert sunlight into stored energy. For example, in 1959 an average yield of 17,036 pounds of sugar per acre was obtained from a 21-acre plot in California (15). It is a tough plant. Seedlings will withstand mechanical abuse, cold, heat, drouth, and defoliation with aplomb. The recovery of a badly battered sugarbeet field after a hailstorm is a remarkable sight.

The sugarbeet is adaptable. It grows in northern states at 7,000-foot altitudes in arid, sedimentary, mountainous valleys. Compare this with beet crops produced in the semitropic, below-sea-level California Imperial Valley where soils are alkaline, the crop is irrigated, and temperatures may reach 120° F. An entirely different ecologic area of production is in the Red River Valley of the north-central states where production is expanding. Soils are prairie, a limited rainfall is the moisture source, and the frost-free period is short.

The sugarbeet is compatible with modern agricultural technology. The crop is highly mechanized. The development of monogerm varieties broke the last obstacle to complete mechanization of production. Beets are well suited to rotations. The deep-rooted nature of the plant and weed-free culture make it a useful crop in grain-pasture-beet cycles, for example. The by-products have value. Beet tops are used for animal feed as is the pelleted mixture of pulp and molasses remaining after the extraction process.

The sugarbeet has an earned and deserved place as an essential world food crop and key component of the nation's agricultural economy.

HISTORY

Man's demand for a sweet food is universal. This was as true during primitive times as it is now. Observe, for example, practices of the bushmen of southwest Africa whose precarious primitive existence includes honey in their diet (18). In one form or another, man satisfies his "sweettooth." The historical documentation of his means to do this is thorough (4, 7).

Sugarcane was the first source of sugar. The crop was grown in tropical areas hundreds of years before sugar from beets became available. Brandes (4) noted that clumps of "chewing cane" were common in New Guinea villages, a cultural trait that probably can be traced to primitive times. Sugarcane was known in India and in the Orient long before the Christian era. The Arabs, Venetian merchants, Portuguese, and others were instrumental in developing and spreading the crop.

The sugarbeet has a different history. As man experimented with his variable plant materials, many claims were made for the beet. Its medicinal value was reported by the Greek physician Hippocrates. The sweetness of the beet root was reported in the 1500s, along with other claims about its usefulness as a hair tonic and additive to sour wine. The beet was recognized as a plant with valuable properties; these became more apparent in the 1700s.

Proof that the beet stored extractable sucrose was provided in 1747 by Andreas Marggraf, an exceptional German chemist. Marggraf studied the "white mangold," "sugar-root," and the "red mangold." He extracted plant material with alcohol and proved that the crystals that formed had the same physical and chemical characteristics as sugar from sugarcane. He said, "This sweet salt, sugar, may be made as well from our plants as from sugarcane."

Marggraf's achievement was translated 40 years later into practical use by his student Franz Karl Achard. His work won him acclaim as father of the sugarbeet industry.

Achard's achievement was not without its difficulties. The sugarbeets with which he worked were composed of many types. They came from material commonly grown for cattle feed. The various sizes and shapes of the beets ranged in colors from red to white. Achard discovered that the white-skinned and fleshed types with conical shape that grew below ground had the sweetest juice. These guidelines led to the development of the White Silesian beet, ancestress of all the sugarbeets in the world.

Achard worked diligently and published widely. His work came to the attention of Frederick William III, King of Prussia, who provided funds for the first beet sugar factory erected at Cunern, Silesia. It was apparent after operations began in 1802 that sugar could be produced economically from beets. The production of sugar from beets spread throughout the European continent. It became a substantial industry by the middle 1800s.

The sugarbeet industry in the United States is a success-failure story (6). For every advance there was a retreat, fortunately of lesser scale. The successful enterprise of today is built on dogged determination, enthusiasm, visions of profit, demands of a developing nation, and enterprising risk-taking entrepreneurship.

The successful production of sugar from beets in Europe sparked bright hopes and ambitions in the United States for a home sugarbeet industry. The Beet Sugar Society of Philadelphia—the title alone should have assured success—made the first effort. It was abortive because they lacked technical skills. The first beet sugar produced in the United States came from a factory at Northampton, Massachusetts. Thirteen-hundred pounds were produced in 1838, but the plant closed. The beet sugar company at White Pigeon, Michigan, also failed.

An account of sugarbeet production in the United States would not be complete without mention of interest in the crop (2) by leaders in the Mormon Church. A missionary, John Taylor, studied the sugarbeet process in France. When he returned to Utah, he founded the Deseret Manufacturing Company. They purchased 1,200 pounds of sugarbeet seed from France, and also bought a mill. From New Orleans they floated it in barges up the Mississippi and Missouri rivers. By an incredible feat they hauled it by wagon and oxen to Utah. The venture was unsuccessful. Again, technical skills were lacking.

Promoters were having a difficult time in other states as well. Fourteen sugarbeet factories were constructed between 1838 and 1879. All failed.

E. H. Dyer deserves credit for establishing the sugarbeet industry in the United States. Despite a record of failures in this country, there were still many who were interested in producing the crop. Dr. Lewis S. Ware was one of these disciples. He preached the potential of sugarbeet production. Dyer was one of those who heard, listened, and acted. He and associates formed the Standard Sugar Refining Company and bought a bankrupt sugarbeet company at Alvarado, California. They revamped it and put the operation in the black in 1879. Dyer proved that sugar from sugarbeets could be produced economically in the United States. The Alvarado factory site is commemorated by the California Historical Society with a plaque for the first factory constructed that successfully produced sugar. It is dated 1870.

SUGAR CONTROLS

The subsequent growth and current status of the domestic sugarbeet industry have been shaped by federal controls. An understanding of the basic legislation, known as the Sugar Act, is essential. There are many clear renditions of this, one of them in the "Beet Sugar Story" (3). The present dimension, location, profitability, and future of the sugarbeet industry are predicated in the act and its amendments. Why do we have it? How does it work?

The first law of our country concerning sugar was passed during the first administration of George Washington by the First Congress. The raw-sugar tariff of one cent a pound, enacted in 1789, was to provide money for the treasury. It was an important source of revenue, providing about 20 percent of all import duties. The tariff became more important in a different sense in later years.

The sugar tariff was in effect for about 100 years, acting as a dominant protective force for a struggling, new, domestic industry. The early cane growers of Louisiana had some protection from world competition. Reciprocal trade treaties in 1876 with the Kingdom of Hawaii excluded Hawaiian sugar from the tariff. Dyer surely must have considered this legislation as he risked his capital in California.

The year 1890 was crucial, especially for some portions of the domestic industry. Congress repealed the raw-sugar tariff, reduced the tariff on refined sugar, and created a bounty of two cents per pound for domestic sugar. The results were violent and chaotic in Hawaii, since they were excluded from tariff protection and bounty. This contributed to the downfall of an empire and the formation of a Republic. Four years later Congress again changed the law. The nation returned to a sugar tariff that included protection for Hawaiian sugar; it remained as the law of the land until 1934.

The Jones-Costigan Act of 1934 is the basis of our current Sugar Act. When enacted it provided stability to a troubled domestic industry. The events that preceded this law caused the economy of the domestic sugar industry to fluctuate wildly. The sugar tariff was not adequate legislation to insure a stable domestic sugar supply.

Various events took place prior to the Jones-Costigan Act that altered the domestic industry. The annexation of the Republic of Hawaii converted this source of sugar from foreign to domestic. Puerto Rico and Philippine Island imports became domestic after the Spanish-American War. The same war provided Cuba with her independence. In a duty preference exchange, Cuba obtained preferential rights for her sugar in the United States. The investment of capital from the United States and the industrious Cuban people helped double Cuban production. These combined forces essentially drove other foreign sugar out of our domestic market.

Meanwhile, sugar supplies in the United States vacillated widely as did the price. Conditions were right for market manipulations too. And they were manipulated, causing prices to skyrocket to 27 cents per pound and to dash down just as severely. Clearly, something had to be done. The Jones-Costigan Act was the result.

The most recent legislation governing the domestic sugar industry is the "Sugar Act Amendments of 1965" (14). Amendments to the law have changed the original Jones-Costigan Act. The principal purposes, however, have not been altered. They are:

1. To assure American consumers of an adequate supply of sugar at reasonable prices.
2. To encourage foreign trade.
3. To provide a healthy economic climate for a competitive domestic sugar industry.

The means by which these goals are achieved are remarkably fair. The law and its enforcement are a credit to the nation.

The estimate of annual sugar requirements made by the United States Department of Agriculture is basic to the sugar industry. Each year the secretary of agriculture is enjoined by law to determine the estimated sugar demand. This is done according to prescribed procedures. He must arrive at a figure that is not too high or prices will drop to unsatisfactory levels. If the estimate is too low, prices will rise unreasonably.

A very important feature of our Sugar Act is the fact that the secretary does not establish or control sugar prices. He controls supply. Our competitive system determines the price. Sugar companies compete vigorously in the market.

Marketing quotas are determined by Congress. They establish who will provide the sugar. One goal of the Sugar Act is to stimulate foreign trade. About 40 percent of our national consumption is allocated to foreign production. The domestic beet-sugar-producing area has a quota amounting to about 30 percent of United States sugar requirements.

Marketing allotments are determined by the secretary. They determine how the domestic quota shall be allocated. They permit the secretary to control the supply of sugar by areas. The allotments are used when the supply of sugar for an area threatens to exceed the quota substantially.

Proportionate shares set acreage allotments for individual farms. They are determined after a national sugar acreage is established. This is a complicated activity that considers new growers, size of farm, past history, crop potential, and all the other factors that affect the potential of an area to produce.

Conditional payments are payments to the beet producers by the government, as the term implies, conditional upon the producer having complied with the terms and requirements of the Sugar Act.

The excise tax—a more than adequate source of funds for conditional payments and administration of the act—are obtained for the treasury by taking a share from the sale of refined sugar. This fund is disbursed in a manner that not only assures compliance with the Sugar Act but by paying the small producer at a higher rate enables him to compete with larger sugarbeet enterprises.

A few comments should be made about international controls and sugar prices since they bear on the United States Sugar Act. Critics of the act say that consumers in this country pay dearly for their sugar. They compare domestic prices with world prices on the "free market." The critics are uninformed.

Each nation in the world is concerned about self-sufficiency, strategic materials, and balance of payments. They are concerned about sugar. Sugar is one of the most regulated commodities in the world. The types of national sugar legislation in the world differ, but most countries have some form of control. Import countries seek assured supplies at stable, reasonable prices. Export countries seek assured markets and acceptable profit. The United States Sugar Act is not unusual legislation nor does it penalize consumers.

Sugar in the international market is under the purview of members of the International Sugar Agreement. About 70 percent of the world's sugar supply is consumed in the areas in which it is grown. Twenty percent is reserved for markets for which the growers have prearranged agreements or some form of preference.

The remaining 10 percent is "world market" or "free market." This surplus sugar is sold at a price that has no relationship to the cost of production. It is a dumping price. "Free market" sugar would not be a reliable source for the United States.

THE INDUSTRY TODAY

Under the aegis of the Sugar Act, a vigorous national sugarbeet industry evolved. It produces about 30 percent of the 11 million tons of sugar consumed in the United States annually. The industry is of age as a stable, efficient, key component of the agricultural economy.

Sugarbeets are grown from Maine to California and from North Dakota to Texas. Twenty-nine states produced sugarbeets in 1969. The selection of these areas is a result of trials, tribulations, failures, and successes. The process of seeking the most efficient center for production, processing, and marketing goes on today. Three states, California, Colorado, and Idaho, must be ideal sites for they produce about one-half of the nation's annual requirement.

The sparkle of sugar in California may be obscured by the glitter of Hollywood and the glamour of aerospace, but they do not diminish its importance. California is the leading sugarbeet-producing state. The multimillion dollar industry is one of the leading segments in an agriculturally oriented state.

The early trials in California were the origin of large companies that currently are major producers of sugar from the beet. Dyer's achievement, of course, gave the state fame as the birthplace of the industry. Claus Spreckels, another California resident and founder of the Spreckels Sugar Company, experimented with beets and became a dominant figure in the refining industry. The Oxnard brothers founded the American Crystal Sugar Company by their activities at Chino in Riverside County. The founders of the Union Sugar Company built a plant in Santa Barbara County in 1898. Today, sugarbeet companies extend their activities all over the nation; for some of them, however, their history remains rooted in the exploits of their California pioneers.

The sugarbeet is a western industry. The situation is anomalous in terms of history, since capital, population centers, and interest were available in the East. While it is true that much of the sugarbeet acreage migrated West, it is also true that sugarbeets can be and are grown in the humid East.

For example, the fiasco at White Pigeon, Michigan, did not deter the development of a sugarbeet industry in that state. There are now five factories in Michigan; they have a daily capacity of 10,900 tons of beets. Michigan produced 1,256,000 tons of beets in 1967 (1), and 1,708,000 in 1968. This is a significant amount of national production.

The history of the sugarbeet industry in Michigan parallels that of other states and portrays some of the problems. The history is ably presented by Gutleben (9). The White Pigeon effort dampened enthusiasm for 60 years. Nevertheless, over the years, 26 beet sugar factories were built. Kalamazoo, Lansing, Bay City, Menominee, Charlevoix, and towns and cities with English, French, and Indian names, all were sites of sugarbeet factories. For many reasons, including poor financing, poor crops, lack of technical knowledge, and poor economics, most of the 26 failed. Nevertheless, a large, strong beet industry evolved in Michigan.

Efforts are being made to extend the range of beets to the eastern coast. A marketing quota was granted to Maine Sugar Industries, Inc. Farmers are readjusting their potato culture to include sugarbeets. A marketing quota was also granted to a factory in New York. Montezuma, New York, is now the site of the New York Sugar Industries, Inc., sugar factory. New York farmers are shifting their vegetable, dairy, grain, and bean options to make room for sugar.

Sugarbeet culture in Arizona and Oregon deserves special mention. A portion of the acreage is planted in late summer, overwintered, and harvested the following summer with cutting bars. The end product is seed. The history of our domestic sugarbeet seed industry is interesting and is well covered in Chapter 15.

Thirteen major sugar companies are the core of making sugar from sugarbeets. Those not mentioned previously are The Amalgamated Sugar Company, Buckeye Sugars, Inc., The Great Western Sugar Company, Michigan Sugar Company, Monitor Sugar Company, Northern Ohio Sugar Company, and the Utah-Idaho Sugar Company.

Sugarbeet companies differ in size; there are giants such as Great Western whose factories can process 39,000 tons of beets in a day; there are new, small companies such as Maine Sugar Industries, Inc., whose one factory will handle 4,000 tons per day.

The national sugarbeet crop is processed in 58 plants. This appears to be a small number to handle a large, bulky, perishable crop. The number, however, is misleading. The size and operating procedures are important considerations. Sugarbeet processing plants are constantly growing in size and efficiency. Automation is standard. Computers are taking over more of the management procedures. A modern sugarbeet factory compares with the best of technology in United States industry.

Who produces the national sugarbeet crop? A small segment of the nation's work force is responsible. Furthermore, the number of people becomes smaller each year. In 1937 sugarbeets were grown on 47,100 farms. Thirty years later, this was reduced to 18,300 farms (17). Conversely, the output per person is higher. In 1937 the larger work force tilled 816,000 acres, producing 8,770,000 tons of beets. In 1968 the smaller work force tilled 1,410,000 acres, producing 25,363,000 tons of beets. This is possible, of course, because of improved technology and the large capital investment in machines to mechanize the crop.

Farmers arrange to produce beets by contracting with sugarbeet companies. The process of arriving at a mutually acceptable contract is complex. The contracts are detailed, specifying acreage, varieties, quality, price, and other factors. Once an agreement has been reached, the grower and processor are closely joined in a common effort seeking profit.

Growers are well organized. The Farmers and Manufacturers Beet Sugar Association provides a forum for concerted action by growers in the eastern part of the country. The expansion of production in the midwestern Red River Valley was coincidental with the formation of the Red River Valley Sugarbeet Growers Association. The California Beet Growers Association, Ltd., is a strong organization. The National Sugarbeet Growers Federation has a large active membership. These and other groups through individual and collective meetings arrive at a consensus on grower reaction to national sugarbeet issues.

How efficient is the sugarbeet industry? By most standards of measurement, it ranks alongside leading crops that make the United States the most efficient agricultural economy in the world.

The increased yield of beets and sugar per acre is a phenomenal success. During the period 1913–37 less than 12 tons per acre was the national average. The average yield in 1968 was 18.1 tons per acre. The achievements making this possible, such as improved varieties, nutrition, cultural practices, pest control, and others will be discussed in detail in the chapters that follow.

The mechanization of all aspects for producing and harvesting sugarbeets was vital to the current efficiency of the industry. When the muscle of men and animals powered the farm, beets in the ground were loosened by a lifter. Laborers pulled them from the soil and heaped them in a pile after slicing off the crown and leaves. Today, farmers choose from a number of different machines for one to do the work. The type they select depends in part on how they wish to handle the sugarbeet tops. Many of the machines pull, top, and load in one operation as they pass through the field.

The sight of long lines of migrant laborers blocking and thinning sugarbeets is now historical, a striking example of increased farm efficiency. Sugarbeets formerly were planted with multigerm seed. The rows contained a large number of seedlings. Short-handled hoes and strong backs were required to reduce plant populations for maximum yields. The development of monogerm varieties having seeds that yield one plant per seed ball, mechanical thinners, and selective herbicides have permitted machines to replace men. We are rapidly approaching the time when all of our sugarbeet fields will be space planted—one seed every eight inches, for example. The seedlings, plants, and roots will be untouched by hand throughout the entire growing, harvesting, and processing cycle. Hand labor for producing beets will be history.

The efficiency of sugarbeet growers is based on research. Vigorous programs are underway by state, federal, and industry scientists. For example, the State Agricultural Experiment Station in California has a long history of research on the crop. The Agricultural Experiment Station in Maine is devoting its talents to solve production problems of the new sugarbeet venture in that state. These activities by state experiment stations are cooperative with federal agencies and the industry. The USDA, in cooperation with state stations and the industry, maintains six specialized research installations to serve the major areas of production. Associated federal research is also conducted at other locations.

Sugarbeet companies are concerned with all phases of research. Through the Beet Sugar Development Foundation, they disperse funds to state and federal scientists to further studies on production and processing of the crop. The individual companies also have their own research laboratories and staffs.

The industry has its own large-scale research programs. Producer groups such as the California Beet Growers Association, Ltd., contribute funds to public agencies for research. They also provide farm facilities. Others such as the Red River Valley Sugarbeet Growers Association have their own staff and also contribute funds. All growers associations in one manner or another foster science.

Sugarbeet research enables growers, processors, and merchandisers to provide consumers with a dependable supply of pure food at a reasonable price. Technology has changed drastically since 1910–14, when the raw sugar price was 4.9 cents per pound. The benefits of research enabled the industry to absorb a steep rise in costs without an equivalent rise in price for the product. Despite inflation since 1910, raw sugar sold for 7.5 cents per pound in 1968. This rise of 86 percent is much less than for other farm products or nonfood items.

The efficient sugarbeet industry provides American consumers with a pound of sugar for a fraction of their work hours. Sugar is cheaper in the United States than any country in the world, when you use the earning power of an average United States worker as a measure. It takes 2.3 minutes of work to earn a pound of sugar. In other countries, comparable figures are Sweden 4.1, Great Britain 4.9, Germany 6.9, Italy 14.0, Japan 16.9, and France 18.9. The benefits of sugarbeet research accrue to the producer, processor, merchandiser, and the consumer.

THE INDUSTRY TOMORROW

The future of the sugarbeet industry depends on the Sugar Act, consumption trends, and the industry's ability to solve its problems. These factors are interdependent.

The Sugar Act has been and is good legislation. The provisions of the law, however, are rooted in compromises of differing expert opinion. This is evident at hearings preceding renewal of the law. For example, as foreign quotas are considered, how much emphasis shall be placed on the international balance of payments? What will be the intensity and nature of federal control? Will there be limitations on size of payments to individual producers? The changing strength of the forces that bear on Sugar Act legislation will be a powerful factor in shaping the future of the sugarbeet industry.

Past history is reliable proof that a dependable large demand by American consumers for sugar will continue. The per capita annual consumption has been quite constant at 97 pounds per person. During a period when personal incomes were rising and the population increased nearly 8 percent, sugar consumption increased a little more than 7 percent. Our affluent society is obviously able to sate its hunger for sugar. Though we represent a relatively small portion of the world's people, we consume a seventh of the total annual supply. The sugarbeet industry, all things being equal, will probably expand as population in the United States increases.

The future of the industry is being shaped today. Production, processing, and marketing problems exist. How they are solved today is basic to tomorrow.

The producer has more than his share of industry problems. For example, at a rapidly accelerated rate, he is forced to depend more and more on chemical weed control. If these chemicals are ineffective, the failure to control weeds is tantamount to crop failure, or more certainly profit failure.

Disease and insect control for sugarbeets is now vastly superior to previous years. When the crop commenced in this country, pests were a dominant limiting factor. However, despite our achievements in control, losses still occur. The most recent report, "Losses in Agriculture" (10), lists the following for sugarbeets in percent loss: diseases, 16; insects, 12; and nematodes, 4.

The major diseases that infect beets are old friends, or more properly, enemies. Some of them can be controlled with chemicals; for example, leaf spot. Producers, however, are still plagued by the profit-loss dilemma. When are sprays essential to insure a good crop and how often should they be applied for maximum profit? We are not certain. Curly top, a disastrous pathogen in the early 1900s, has not disappeared. Its mutated virulent strains pose a serious threat to production. We have gained on virus yellows, but the tax it levies in the form of reduced yields is still being paid.

The control of insect pests of sugarbeets must be considered in an entirely new light. Pollution of the environment by use of persistent pesticides has become a critical national issue. Concern about the spread of these materials throughout our biotic environment is the basis for stringent controls. It is no longer adequate to develop an effective, cheap chemical for wireworm control, for example. Acceptability is now based on its pollution effects and pesticide residues in foods and feeds.

To further complicate the problem, sugarbeets absorb soil pesticide residues resulting from previous crops. Tons of beet pulp scheduled for feeding lots have failed to meet the maximum tolerance levels for pesticide residues because of this trait. Restrictive controls on pesticides are forcing agriculture to change its technology.

The sugarbeet cyst nematode is one of the most serious pests of the crop. Crop rotation is a practical control measure; failure to do so results in a serious nematode infestation and a poor sugarbeet crop. Control of nematodes by rotation is difficult because nematodes can live on some common weeds. The kinds of crops and the types of weed-control practices used in the rotation, therefore, are important to subsequent sugarbeet crops.

Efforts to develop nematode-resistant commercial varieties have not been successful. Some wild forms of sugarbeets are immune, but the genetic factors that govern this are not easily transferred to commerical beets. Nematode control continues to be a serious production problem.

Sugarbeet processors have equally perplexing difficulties. An important one that affects most of the producing regions is the loss of sugar that occurs when beets are stored in piles. Some background will help explain this situation.

The sugarbeet is a temperate-zone crop. During the spring and summer, it produces a leaf rosette and a large storage organ, the root. Sugar is stored in the root when night temperatures are cool and the days are bright, as in the fall. Producers know their profits increase with every day that beets remain in the ground during such periods. Processors know that high-quality beets are essential for an efficient factory. But, beets must be out of the ground before a permanent winter freeze. Therein lies the dilemma.

Campaigns, the harvest-processing activity, may begin in many areas about October 1. The harvest will be completed by the second week of November. The factory campaign, however, will extend through February. Beets are stored in mountainous piles, awaiting processing. During this time, they lose sugar worth about 8.5 cents per pound at the rate of one-half pound per day per ton of beets. The monetary loss is shared by the grower and the processor in most instances, and in the scheme of things by the consumer too. The consumer shares the efficiencies and inefficiencies of our national sugarbeet industry.

The processor has another insidious problem, one that is more recent in origin; namely, the declining yield of sucrose per ton of beets. We are enthusiastic about the leap in field productivity of beets. By known standards of measurement, quality has not been sacrificed as the geneticist changed the beet and the agronomist changed cultural practices to increase yields. But something happened. We are getting less sugar per ton of processed beets today than we did previously.

The problem is complex. Sugar chemists may find one answer as they determine the nature and activity of nonsucrose constituents of the sugarbeet. These ingredients, as they wend their way through the processing cycle to the molasses vats, carry with them a portion of otherwise crystallizable sucrose.

Modern technology may contribute to the declining yield of sucrose per ton of processed beets. Producers, caught in a cost-profit squeeze and labor shortage, turn to larger, more efficient machines. The old one-row harvester did a tailor-made job of topping, digging, and cleaning beets. The machines that dig six rows at a time are not as fastidious and quality is lowered.

Beet sugar is sold in a competitive market. To compete, merchandisers must have a dependable supply in adequate amounts; also it must be available at a competitive price. The problems of meeting marketing requests parallel those of processing and production.

The weather is an important factor that affects the supply of sugar. The year 1969 is a good example. It started out well. Contracts were let for one of the largest crops in history. The growing season was generally favorable. At the start of the harvest season, the national crop yield was estimated at 3.8 million tons of sugar, raw value. However, an early mid-October freeze in Colorado, Nebraska, Wyoming, Kansas, Texas, and Montana changed the situation. As much as 30 inches of snow fell. Temperatures dropped to record low levels. This caused sugar percentage and extraction rates to drop to unprecedented lows. Two successive weekend storms lowered yield estimates 0.3 million tons. Farming is a risky business and productivity is as variable as the weather.

Competitive crops affect the production of sugarbeets. A $20,000,000 processing plant located in an area where beets can be produced provides no assurance that raw materials will be forthcoming. The crop must be profitable to the farmer. Generally speaking, growers respond favorably to sugarbeets. The cash return is dependable; the crop is more likely to be a blue-chip investment, not speculative. But, there are a multitude of reasons why producer interest may vacillate; and as it vacillates, so does the annual production of beet sugar. It behooves sugar companies and the public to share with growers the hazards of production; farmers, too, must make a profit. A beet sugar factory cannot process corn, or wheat, or beans; the sugarbeet must be a profitable venture to producers or they use their option of growing other crops.

Competitive synthetic sweeteners are a concern to the sugarbeet industry. Some of these products are receiving front-page attention—the cyclamates, for example. Concern about their effects on health of consumers brings to public attention their widespread use. The packaged diet sweeteners are a relatively new but common addition to sugar bowls. Saccharin has been spared the public attention of cyclamates. It is used commonly, however, as a sugar substitute.

Competitive agricultural products are a concern to the sugarbeet industry. The potential for making sugar from sweet sorghum opens the possibility that the crop may become one of these. Sorghum is commonly thought of as a grain or forage crop. Some members of this plant family, however, will store sugar in the stalks. The sweet character of these varieties was identified in the 1500s. Throughout the world and at different periods, efforts were made to extract sugar from sorghum (7). One of the first projects of the USDA in 1878 was to make sugar in this manner (11). The day may be closer than we imagine.

Sugar can be made from sweet sorghum. A major problem, development of disease-resistant varieties, has been solved (5). Conventional methods to process sweet sorghum for sugar production are not applicable. Starch and a high aconitic acid content interfered with sugar recovery. New techniques, however, are available that permit the extraction of sucrose from sorghum without undue difficulty (12, 14). Sweet sorghum bears close scrutiny as a source of sugar.

Corn is being considered as a crop to produce a competitive sweetener. The fact that the stalks of some varieties of maize have a sweet juice has been known for a long time. Cortes, for example, described the use of corn for producing sweet food products by the Indians in Mexico. Currently, attempts are being made to develop the sugar potential of the crop. Special selections of corn—male steriles—that produce no seed are grown for their sweet stalks. The juice is extracted and processed. The end product is a liquid sugar comprised of a mixture of sucrose and other sugars in solution. Purportedly, this product is for the liquid sugar market in the United States.

The hydrolyzed starch from the corn grain, which produces dextrose, is available in crystalline form or as corn syrup, and has long been a substitute for sucrose in some uses.

Sugarcane is a most important competitive crop. True, the production of sugar from cane is governed by marketing quotas. But, the competitive price is not controlled. Every advance in the efficiency of producing the crop is reflected in a shift of competitive standings in the marketplace. The development of hybrid beets that yielded 15 percent more tons per acre was viewed with concern by the cane industry. The development of a sugarcane variety in Florida that yields 47 percent more sugar per acre is not cheered by the beet growers. The cane and beet industries compete not only in the marketplace, but also in the search for improved technology.

OUTLOOK

The domestic sugarbeet industry views the future with confidence. The continuing flow of dollars into capital investment is partial evidence. Cumulatively, growers invest millions of dollars in new equipment to improve efficiency. Cumulatively, corporations invest large sums in modernization of factories and related facilities. There is confidence that legislative controls, demand for sugar, and problem solving will insure a profitable future.

The Sugar Act has been too successful to be discarded easily or altered drastically. It has assured an abundant supply of sugar at a fair price to consumers. A thriving domestic sugar industry coexists with importation of foreign sugar from friendly countries. The social gains from improved wages and working conditions for farm labor are a tribute to the act. Furthermore, the program not only provides for administrative costs but it put more than a half billion dollars of surplus into the treasury.

The demand for sugar in the United States is stable. One can expect that per capita consumption will continue at its present level. It is true, we are a diet-conscious nation; noncaloric synthetic sweeteners, for example, are used as sugar substitutes. It is also true that sugar's share of the market for the major sweeteners—sugar, corn, and noncaloric—declined from 87 percent of the total in 1957 to 81 percent in 1965. Nevertheless, the annual 97 pounds per capita consumption of sugar remains the same. In other words, per capita consumption of all sweeteners as a group increased, due to expanded consumption of nonsugar sweeteners. As consumers, we demand sugar in one form or another in our diet.

The sugarbeet industry has an exceptional record of problem solving. In the West beets are grown profitably in areas that were abandoned before curly top control measures became available. The loss of low-priced foreign labor in California and elsewhere was offset by monogerm varieties, mechanical thinners, and herbicides. A generous constant support to basic and applied research is the backbone of problem solving. The sugarbeet industry deserves its dividends from research.

The prospects for continued problem solving are bright. Nematode-imposed rotations that are a burden may be eased through recent achievements in genetic research. The biological control of insect pests, a useful tool in averting pesticide residue problems, is having remarkable success. Virus yellows, an international disease, is finally yielding to the research of USDA scientists.

Possibly beets may be stored in silos or plastic tents for a year with minimal quality decline by using special gas mixtures. Weed control problems must eventually yield to inventive minds such as those that developed 2,4-D and selective grass killers. An entirely new concept of pest control may evolve from studies on the absorption of chemicals by leaves, translocation, and root exudation. The prospects are excellent that current problems will be solved.

Crops that compete with sugarbeets have their problems, too. Their competitive status with beets fluctuates from year to year. The economics of some of them—winter vegetables, for example—is volatile. Profits may be high some years—as high as the losses in other years. Labor and production problems and rising costs of operation plague off-shore suppliers of domestic cane sugar. Corn and sorghum are yet to be proved as a source of a competitive product. Synthetic sweeteners are under scrutiny. There is general concern about food additives and pesticide residues, because of their effect on health.

The outlook for the crop is excellent, because the sugarbeet is a remarkable plant, one of nature's important gifts to man. The world is indebted to Marggraf who developed the pattern of how to utilize sugarbeets; and to Achard who used the pattern to develop a practical economic means of making sugar from the crop. In the United States, we are indebted to Dyer, who established the crop here, and to the efficient domestic sugarbeet industry for annually producing an important high-energy, pure food.

REFERENCES

1. Agricultural Statistics, 1969. USDA, U.S. Govt. Printing Office, Washington, D.C. 1969.
2. Arrington, L. J. Beet Sugar in the West. Univ. Wash. Press, Seattle. 1966.
3. Beet Sugar Story, The. U.S. Beet Sugar Assoc., Washington, D.C. 1959.
4. Brandes, E. W., G. B. Sartoris, and G. H. Coons. Sugar Plants, USDA Yearbook Separate 1556, U.S. Govt. Printing Office, Washington, D.C. 1939.
5. Coleman, O. H., W. R. Cowley, D. M. Broadhead, K. C. Freeman, and I. E. Stokes. Rio: A new disease-resistant variety of sweet sorghum for southern United States. Crops Research, USDA, ARS 34–72, Washington, D.C. 1965.
6. Coons, G. H., F. V. Owen, and D. Stewart. Improvement of the sugar beet in the United States. Advan. Agron., Academic Press, Inc., New York. 1955.
7. Deerr, N. The History of Sugar, vols. 1 and 2. Chapman and Hall, Ltd., London. 1949–50.
8. Elcock, H. A. and J. C. Overpeck. Methods of producing sugarbeet seed in New Mexico. N. Mex. Agr. Exp. Sta. Bull. 207, State College. 1933.
9. Gutleben, D. The Sugar Tramp. Bay Cities Duplicating Co., San Francisco, Calif. 1955.
10. Losses in Agriculture. USDA, ARS, Agr. Handbook 291. U.S. Govt. Printing Office. 1965.
11. Moore, E. G. The Agricultural Research Service. Frederick A. Praeger, New York. 1967.
12. Poe, W. E. and B. F. Barentine. Colorimetric determination of aconitic acid in sorgo. J. Agr., Ford Chemical 6 (16): 983–84. 1968.
13. Plant Sciences Now and in the Coming Decade. Nat. Acad. Sci., Nat. Res. Council, Washington, D.C. 1966.
14. Smith, B. A. Sweet sorghum, obstacles overcome. Agr. Res. 18 (3): 18. 1969.
15. Spreckels Sugar Beet Bulletin. The Honor Roll for 1969, vol. 24, p. 12. 1960.
16. Sugar Act Amendments of 1965. Public Law 89–331, 89th Congr., H. R. 11135, Washington, D.C. 1965.
17. Sugar Statistics and Related Data, vol. 2, rev., Statistical Bull. 244, USDA Agr. Stabilization and Conservation Serv., Washington, D.C. 1969.
18. Thomas, Elizabeth M. The Harmless People. Alfred A. Knopf, New York. 1959.

2

Environmental Factors

R. S. LOOMIS
University of California
Davis

ALBERT ULRICH
University of California
Berkeley

NORMAN TERRY
University of California
Berkeley

GROWTH AND DEVELOPMENT	20
ENVIRONMENT-PLANT RELATIONSHIPS	24
Primary Effects of Temperature	24
Night Temperature	25
Root Temperature	25
Day Temperature	26
Antecedent Climate Effects	27
Day-length Effects	29
Low Night Temperature and Nitrogen Deficiency	30
Sunlight Illumination, Nitrogen Nutrition, and Growth	33
Genetic Variations in Adaptation	35

SOIL RELATIONSHIPS	38
ENVIRONMENT-COMMUNITY RELATIONSHIPS	38
Microclimate Processes in Beet Fields	38
Transpiration in the Soil-Plant-Air-Water System	40
Production Processes	42

AGRICULTURE may be characterized as a series of systems for environmental management. The choice of crop and variety, timing of cultural events, space relationships, and inputs of fertilizer, pesticides, and water all represent techniques for controlling certain features of the environment. The objective is to exploit to the greatest extent the potential for crop production of the local combination of soil, water, and sunlight resources.

There are two distinct aspects to the environmental relations of any crop. The first area concerns the individual plant and how its growth and development are affected by climate and soil variables; that is, the direct effects of environment. Weather and growth are closely coupled. The environment controls photosynthesis and the rates of growth and development, and thus at any point in time the actual size of the plant and its capacity to exploit the environment, through the extent of its photosynthetic cover, are products of such feedback mechanisms.

The second area considers groups of plants in communities, their competitive relationships, and their relation as a unit to various environmental processes, including radiation exchange and aerodynamic transfers. These processes, together with the crop canopy and the physiological activities of the individual leaves in transpiration and photosynthesis, establish profiles of light, temperature, water vapor, and CO_2 within and above the crop foliage framework.

The production processes of sugarbeet plants and communities in relation to the influences of the major components of climate (day and night temperature, day length, and light intensity) and other environmental factors such as carbon dioxide concentration, soil moisture, and wind velocity are subjects of the present chapter. We plan to approach our subject by first reviewing some principal features of growth and development.

GROWTH AND DEVELOPMENT

Under favorable temperatures, and with ample nutrients and moisture, sugarbeet seeds germinate rapidly and the cotyledons usually emerge within 4 to 5 days after planting (Fig. 2.1). During emergence, the supply of energy from carbohydrates stored in the embryo in the seed is sufficient to force

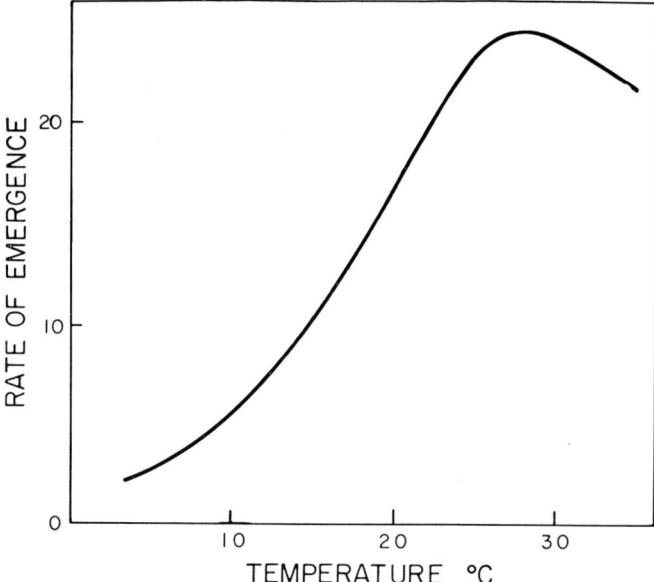

Fig. 2.1. Rate at which sugarbeet seedlings emerge from soils as a function of temperature. (After Leach, 3.)

the emergence of cotyledons above the soil surface through growth of the hypocotyl and to form the primary rootlets. In the presence of light, the cotyledons and newly formed leaves expand rapidly. These are usually displayed horizontally to the soil surface, thus capturing a maximum of solar energy for photosynthesis. As the cotyledons and leaves unfold, the taproot grows downward rapidly (Fig. 2.2). By the time the first leaf is fully developed the taproot may reach 12 inches or more, providing the soil is not too compact or water-logged. Damage to the main root during this time from soil compaction, fertilizer placement, or other causes may result later in the formation of forked or sprangled storage roots.

The seedling evolves gradually into a vegetative plant with a large rosette top and a storage root (Fig. 2.2). As the leaves become more numerous and crowded, the newly expanding leaves remain in a more vertical pattern, thereby maintaining a suitable exposure to light for photosynthesis.

The rate and pattern of growth of the sugarbeet plant is dependent principally upon internal supplies of sugar produced in photosynthesis. Maintenance of basic metabolic processes has first priority over these supplies of sugar, followed by top and fibrous root growth (Fig. 2.3). Only when sugar is in excess of these needs does appreciable growth of the storage root take place. Normally, this occurs when the tops have nearly reached their maximum size relative to climate. Concurrently with rapid storage-root growth, the sucrose concentration in the root reaches a steady

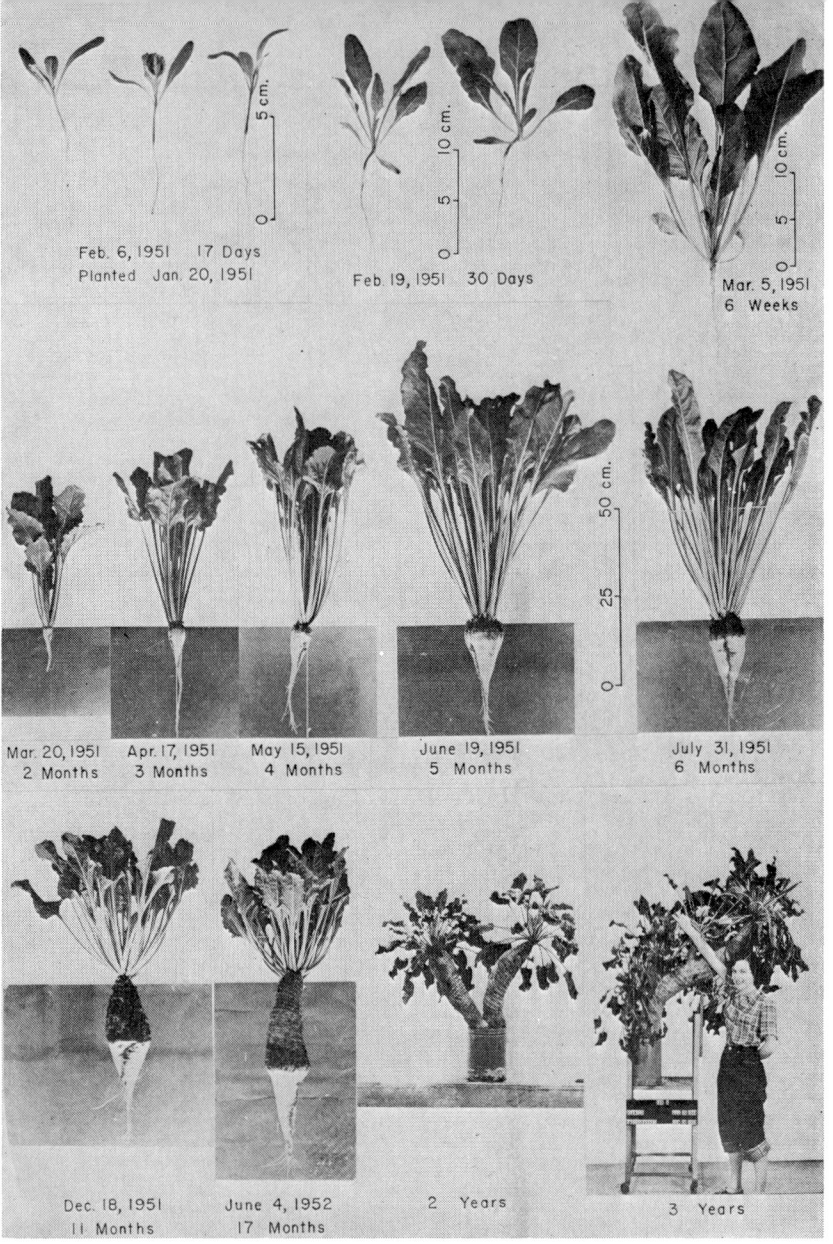

FIG. 2.2. Growth and development of sugarbeet plants in a controlled temperature greenhouse from the seedling stage to the age of 3 years. Growth for the first 2 months was confined primarily to tops and to fibrous roots. Top size increased for more than 3 months and then remained relatively constant as old leaves began to die as rapidly as new leaves were produced. Storage root development began slowly at the age of 2 months, increased in rate at 3 months and thereafter proceeded at a rapid pace for 5 to 6 months. Beyond that time, crown growth constituted most of the growth of the storage tissues. No storage-root growth took place after the first year. Greenhouse temperatures were kept at 23° C (73° F) from 8 a.m. to 4 p.m. and at 17° C (63° F) from 4 p.m. to 8 a.m. (Ulrich, 9.)

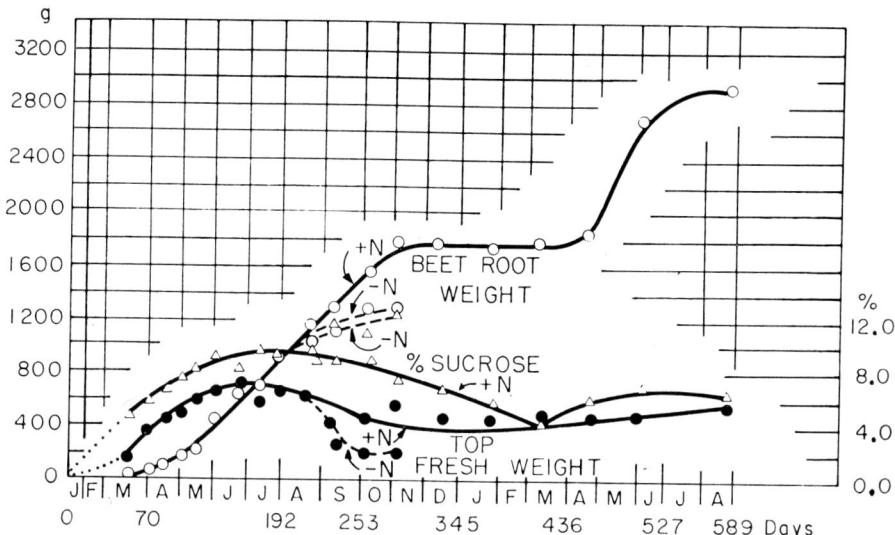

FIG. 2.3. Influence of time and nitrogen upon beet root weight, sucrose concentration, and top fresh weight of sugarbeet plants. With ample supplies of water and nutrients (solid lines, +N), sugarbeet plants failed to ripen or sugar-up during the 83-week growth period. Ripening, however, was induced by watering comparable plants with a nitrogen-free culture solution starting July 31, 1951 (broken lines, —N). (After Ulrich, 9.)

state value, depending to some extent on variety and plant spacing, but principally on climate and nitrogen nutrition. The rate of storage-root growth then varies with the supply of surplus sugar, and in most environments continues at a fairly constant rate. Top growth also remains relatively constant, with new leaves forming at a uniform rate. The new leaves increase in size, age slowly, and then gradually die. As a result of this cyclic leaf growth, the amount of living leaves remains fairly constant but the total dry matter of the tops, living and dead, increases at a relatively uniform rate.

The sucrose concentration in storage roots remains constant and relatively low until sugar utilization in growth decreases, due to some change in environment. Such changes as low night temperatures, nitrogen deficiency, and possibly other phenomena restrict the growth of roots and tops, and concurrently the sucrose concentration in the storage root increases. This increase in sucrose concentration often compensates for losses in root growth as far as sucrose yield is concerned. Such a sequence of environmental events normally occurs during the fall of the year, when night temperatures decrease, available nitrogen becomes depleted, and warm sunny days still prevail for effective photosynthesis and sugar storage.

ENVIRONMENT-PLANT RELATIONSHIPS

PRIMARY EFFECTS OF TEMPERATURE

Climate has three primary effects on the growth and development of the sugarbeet plant: one effect is primarily transitory in nature, such as may be observed in photosynthesis, respiration, or transpiration; another is mainly additive in character as observed through total vegetative growth; and a third is concerned with developmental events as observed in the time-delayed reaction of flowering following an extended period of low temperature.

The response of plants to changes in temperature when all other growth factors are not limiting is a summation of all of the chemical and biological processes that are influenced by temperature from the start of germination to the time of harvest. The transitory and additive effects of temperature on the growth and development of sugarbeets are best understood from studies done in controlled temperature environments, such as

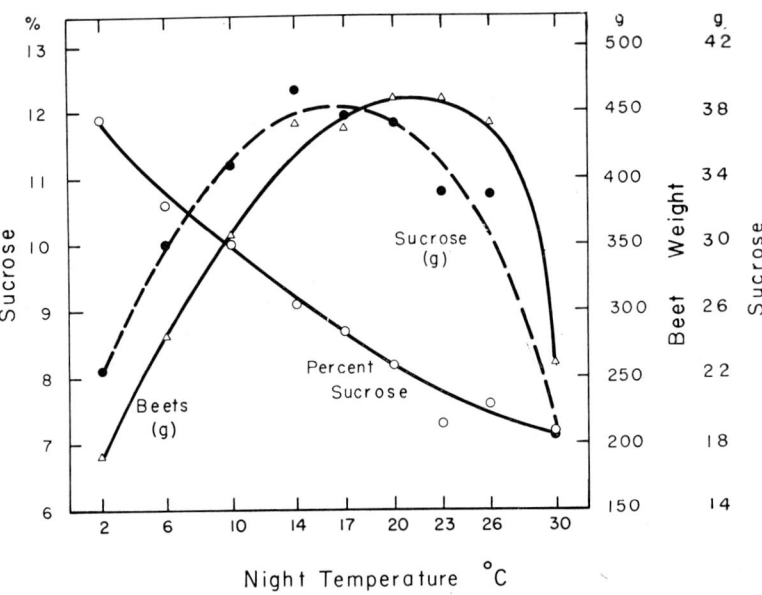

FIG. 2.4. Relation of night temperature to percent sucrose, beet root weight, and sucrose produced. The plants during the day periods were kept in greenhouse sunlight from 8 a.m. to 4 p.m. and for the night periods in dark rooms from 4 p.m. to 8 a.m. Day temperatures were comparable [20°, 23°, and 26° C (68°, 73°, and 79° F)] and nutrient supplies, including water, were ample at all times. The maximum sucrose concentration, 11.9%, occurred at 4° C (37° F) and the minimum sucrose concentration, 7.2%, occurred at 30° C (86° F). The optimum temperature for storage-root growth is approximately 20° C (68° F) and for sugar production, 15° C (59° F). (After Ulrich, 10)

ENVIRONMENTAL FACTORS

are now available at a few locations in phytotrons with sunlight illumination or with combinations of fluorescent and incandescent illumination.

NIGHT TEMPERATURE

With adequate nutrition and moisture, low night temperature increases sucrose concentration, mainly a transitory phenomenon, and decreases top and root size (Fig. 2.4). Maximum sucrose production occurs at a night temperature of about 15° C (59° F), and storage-root size is maximum at about 20° C (68° F). When night temperature increases to 30° C (86° F), sucrose yield decreases greatly, partly because of the low sucrose concentration of the storage root, but mostly because of the large decrease in root size that takes place at this high night temperature.

ROOT TEMPERATURE

Contrary to expectation, decreasing root temperature from 25° to 10° C and with air temperatures fluctuating from 20° to 30° C, the sucrose concentration of the storage root not only fails to increase but it actually

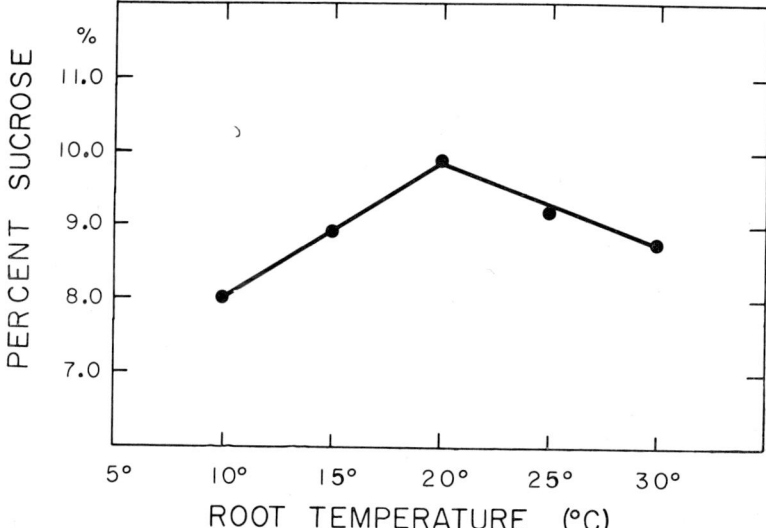

Fig. 2.5. Low root temperature under conditions of ample nutrition and water supply failed to increase sucrose concentrations of storage roots as anticipated from the increases observed for low night temperature (Fig. 2.4). Minimum temperature of the tops during the day or night was 17° C; maximum temperature during the day was 30° C. Sucrose values for the plants harvested September 2 and September 30 are pooled, since the interaction of root temperature and date of harvest was not significant. (Unpublished data of Albert Ulrich.)

decreases from the maximum value of 10 percent at 20° C to a low value of 8 percent at 10° C (Fig. 2.5). Root size for the September 30 harvest (Fig. 2.6) increased more than threefold from a value of about 370 grams per pot at 10° C to 1,250 grams per pot at 25° C. Root size decreased to 1,000 grams per pot at 30° C. Sugar produced followed root size primarily.

DAY TEMPERATURE

Day temperature effects over a limited range of 20° to 26° C have a similar effect on sucrose production as night temperature. The highest sucrose concentrations occur with 20° to 23° C and these values decrease significantly at 26° C. Root size increased at 23° and 26° C over that at 20° C, with the maximum sucrose production observed at 23° C (Table 2.1).

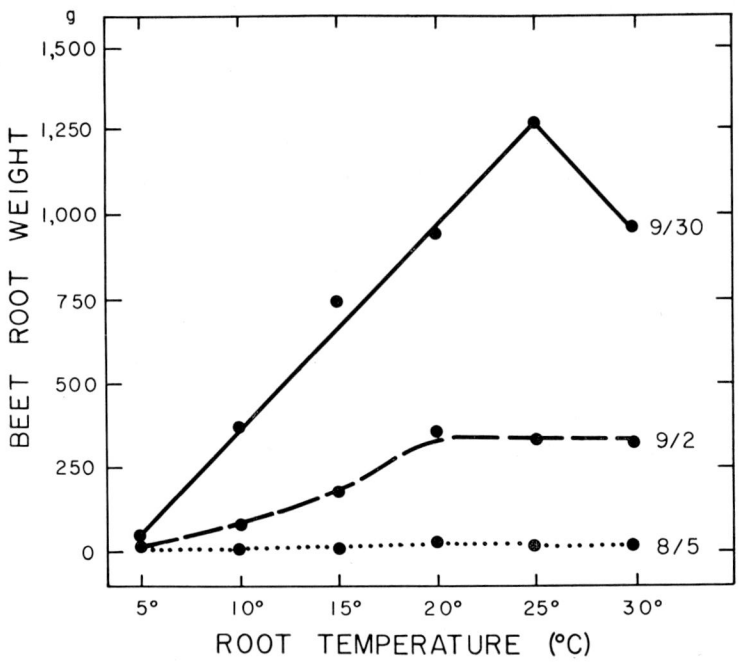

FIG. 2.6. Root weight for the September 30 harvest increased dramatically with root temperature up to 25° C, followed by a marked decrease at 30° C. Root weight for the September 2 harvest shows a small increase with temperature up to 20° C, followed by no change at 25° and 30° C. On August 5 root weight did not differ significantly with root temperature. (Unpublished data of Albert Ulrich.)

TABLE 2.1. Day temperature effects upon sugarbeet plants

	Beet Root			Tops	
Day Temperature, °C	Fresh weight, g	Sucrose %	Sucrose g	Fresh weight, g	Dry weight, g
20	307	9.20	27.3	584	85.2
23	392	9.16	34.5	564	89.5
26	413	8.52	33.8	544	91.6
Significant difference*	84	0.49	n.s.	n.s.	n.s.
Coefficient of variability	48.1	11.6	46.8	40.3	28.4

NOTE: Mean values for 36 beets grown four each at night temperatures of 2°, 6°, 10°, 14°, 17°, 20°, 23°, 26°, and 30° C.
* Significant difference at the 5% level; n.s. indicates not significant (8).

ANTECEDENT CLIMATE EFFECTS

Whereas root size and total top growth are determined primarily by the total climate experienced from germination to harvest, sucrose concentration and the size and shape of leaves are determined by the climate prevailing just prior to harvest. Under conditions of ample nutrition and adequate moisture supply, an early season cold spell of 8 weeks decreased root size by 23 percent from a value of 435 grams per pot to 335 grams per pot as shown in Table 2.2. In this same climate study, an early season hot spell had no significant effect on subsequent root size at harvest. The late season hot and cold periods of nearly 11 weeks duration reduced root size in comparison with a late season warm climate. The losses from the hot or cold spells were approximately 30 percent. The poorest combination of early and late season climates was an early season cold spell followed by a late season hot spell; this combination yielded only 274 grams of roots, or nearly 50 percent less than the 522 grams produced in the continuously warm climate. An entire season of warm weather or a combination of early season hot weather, followed by late season warm weather, gave the best root growth. An entire season of either cold or hot weather was greatly inferior to an entire season of warm weather.

TABLE 2.2. Beet root weights in grams per pot

Initial Climate May 14–July 9 (56 days)	Final Climate July 9–Sept. 22 (75 days)			Initial Climate Mean
	Cold 63°–54° F	Mild 73°–63° F	Hot 86°–72° F	
Cold	300	432	274	335
Mild	395	522	388	435
Hot	356	510	347	404
Final climate mean	350	488	337	

NOTE: Significant differences within climates are 98 and 134, and between climates 56 and 77, at the 5% and 1% levels, respectively (11).

TABLE 2.3. Fresh weight of tops in grams per pot

Initial Climate May 14–July 9 (56 days)	Final Climate July 9–Sept. 22 (75 days)			Initial Climate Mean
	Cold 63°–54° F	Mild 73°–63° F	Hot 86°–72° F	
Cold	752	656	356	588
Mild	594	642	329	522
Hot	456	531	284	424
Final climate mean	600	610	323	

NOTE: Significant differences within climates are 115 and 158, and between climates 67 and 91 at the 5% and 1% levels, respectively (11).

A cold climate is favorable to top size whether the cold weather occurs early or late in the growing season (Table 2.3). Conversely, hot weather reduces top size whether the hot weather occurs early or late in the growing season. The largest tops occur for plants continuously in a cold climate and the smallest tops occur for plants continuously in a hot climate. The life expectancy of a sugarbeet leaf for plants in a hot climate is 44 days, which increases to 58 days in a warm climate and to 67 days in a cold climate.

Climate also affects the outward appearance and age of leaves. In a cold climate, the blades and petioles of the leaves are dark green. The blades are generally broad and rounded in shape and the petioles are relatively short and thick. Conversely, the blades of plants in a hot climate are pale green and the petioles pale yellow to almost white. The blades of these leaves are long and narrow, with the outer edges of many blades curled upward. The corresponding petioles are long, brittle, and narrow in structure. In a warm climate, the leaves are intermediate in appearance to those in a cold and a hot climate, and are similar to those observed in midseason in fields where the plants are growing vigorously. A change in climate is followed by a change in kind of leaf formed. The new leaves formed assume the appearance of the leaves characteristic of the climate regardless of earlier antecedent climates. In this way an experienced observer can read the climate under which each leaf is formed.

Sucrose concentration, much like leaf growth, reflects the climate immediately preceding the time the storage root is sampled. Only the late season climate influences the sucrose concentration of the storage root. As illustrated in Table 2.4, average values of 11.6 percent, 8.8 percent, and 7.8 percent were observed for the cold, warm, and hot climates, respectively. The early season climate had no significant effect on the final sucrose concentration as may be seen in the early season climate means of 9.2 percent, 9.6 percent, and 9.5 percent (Table 2.4). The sucrose yield (Table 2.5) reflects both the early and late season climates, just as was observed and described for root size (Table 2.2).

ENVIRONMENTAL FACTORS

TABLE 2.4. Percent sucrose

Initial Climate May 14–July 9 (56 days)	Final Climate July 9–Sept. 22 (75 days)			Initial Climate Mean
	Cold 63°–54° F	Mild 73°–63° F	Hot 86°–72° F	
Cold	11.45	8.72	7.49	9.22
Mild	12.00	8.70	7.98	9.56
Hot	11.40	9.08	8.07	9.52
Final climate mean	11.62	8.84	7.85	

NOTE: Significant differences within climates are 0.80 and 1.09 and between climates 0.46 and 0.63 at the 5% and 1% levels, respectively (11).

DAY-LENGTH EFFECTS

Day length seems to influence sugarbeets almost entirely through its effects on photosynthesis. In growth chambers, day length can be varied while keeping irradiance constant. Under these conditions, increasing day length results in an increase in photosynthates produced by the plant, which might appear as an increase in sucrose concentration in the storage root or as an increase in the size of roots or tops. As illustrated in Table 2.6, the principal difference between 8-hour days and normal days of 10 to 14 hours was a doubling of root size, while top size and sucrose percentage increased slightly. Thus, the photosynthates seemed to be partitioned in the normal way in that supplies of sucrose not needed in top growth were used in root growth, while the concentration of sugar in the root remained nearly constant. The effect is similar to that obtained by increasing irradiation while holding day length constant.

Under natural conditions, the interpretation of day-length effects is more complex. Long days occur in early summer when radiation is also greatest. At high latitudes, the days are quite long but the sun is lower in the sky than in the lower latitudes, and thus solar radiation is not as intense. As a result, despite the longer day length, daily radiation totals may be less. However, as we shall see, photosynthesis in plant communi-

TABLE 2.5. Sucrose in grams per pot

Initial Climate May 14–July 9 (56 days)	Final Climate July 9–Sept. 22 (75 days)			Initial Climate Mean
	Cold 63°–54° F	Mild 73°–63° F	Hot 86°–72° F	
Cold	34.4	38.2	20.5	31.0
Mild	47.3	46.0	31.6	41.6
Hot	41.4	46.2	28.4	38.7
Final climate mean	41.0	43.5	26.8	

NOTE: Significant differences within the table at the 5% and 1% levels are equal to 9.7 and 13.3 and between climate means 5.6 and 7.7, respectively (11).

TABLE 2.6. Day-length effects upon sugarbeet plants grown in sunlight

Day		Night		Light Source	Beet Root			Tops	
Temp. °C	Hr.*	Temp. °C	Hr.*		Fresh weight, g	Sucrose %	Sucrose g	Fresh weight, g	Dry weight, g
20	8	14	16	Nat.	443	8.9	40.3	745	108
20	10–14	14	14–10	Nat.	913	9.9	90.6	810	136
23	8	17	16	Nat.	474	8.5	40.7	838	112
23	10–14	17	14–10	Nat.	927	9.5	86.3	824	114
26	8	20	16	Nat.	463	7.1	32.2	663	121
26	10–14	20	14–10	Nat.	794	8.5	67.8	892	115
Observed F-value†					2.52	3.86	3.38	0.31	0.30
Significant difference‡					438	1.4	41.1	n.s.	n.s.
Mean (24 plants)					669	8.7	59.6	795	118
Coefficient of variability					44.1	11.2	46.5	36.1	30.4

NOTE: Means of 4 plants grown one per pot.
* Natural day length in Pasadena during the growth period of the plants increased gradually from 10 hours in January to 14 hours in May. The air temperatures of these plants were maintained at the day temperatures from 8 a.m. to 4 p.m. and at the night temperatures from 4 p.m. to 8 a.m.
† Required F-values at the 5% and 1% levels are 2.13 and 2.92, respectively (8).
‡ At the 5% point; n.s. indicates not significant.

ties is a curvilinear function of radiation level. Thus, under long days of moderate brightness, as much sugar production may be obtained as under short days with greater total radiation.

Day length also influences flowering and reproduction. A period of cold temperatures (several weeks or more near 4° C) is sufficient to induce flowering. Long days then tend to accelerate the subsequent development of flowers and fruits. This long-day effect also seems interpretable as being due to increased photosynthesis. Thus, for most varieties of sugarbeet, day length does not serve as a regulating influence for changes in phases of development as is common in the flowering, leaf fall, and dormancy of many other species.

LOW NIGHT TEMPERATURE AND NITROGEN DEFICIENCY

Sugarbeets "ripen," under conditions of ample nutrition and water supply, when night temperatures decrease from 17° C (63° F) to below 10° C (50° F). Under these conditions, the storage roots increase in sucrose concentration from around 8 to 9 percent to a value of about 12 percent. A much larger increase in sucrose concentration occurs when the beet plants become deficient in nitrogen for a period of 4 to 6 weeks prior to harvest. This increase in sucrose concentration occurs even under conditions of relatively high night temperature; for example, from a value of about 9 percent to about 15 percent. However, when the night temperature decreases to 10° C (50° F) or less shortly after the plants become deficient in nitrogen, the sucrose values increase to an average value of about 18 percent, with some roots as high as 20 percent sucrose (Fig. 2.7). Beet root weight also responds dramatically to a deficiency in nitrogen and a decrease in night temperature (Fig. 2.8). The storage roots for the high-nitrogen plants at a night temperature of 17° C weighed 770 grams on August 12, 1,120 grams on September 9, and 1,110 grams on October 7. Deficient plants were not affected significantly by night temperature as seen for September 9, although on October 7, the largest roots occurred at 17° C. Low night temperatures decreased root size significantly for high-nitrogen plants.

Storage roots of nitrogen-deficient plants sometimes contain more sugar than nondeficient plants, regardless of night temperature (Fig. 2.9). Thus,

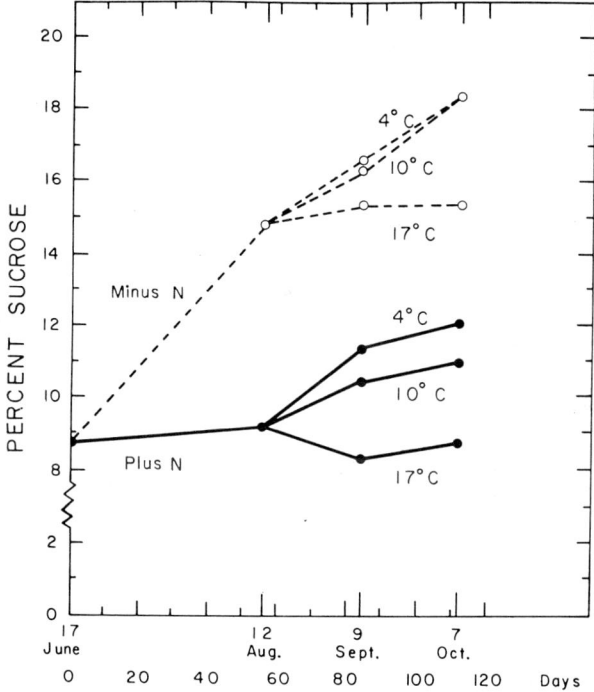

Fig. 2.7. Influence of nitrogen nutrition and night temperature upon the sucrose concentration of sugarbeet roots. At low night temperature [4° and 10° C (37° and 50° F)] and nitrogen depletion (—N), the sucrose concentrations for the October 7 harvest averaged 18.3%, with individual roots as high as 20.0%. At all times and for all night temperatures, the sugar produced by the nitrogen-deficient beet roots consistently exceeded that of the high-nitrogen beet roots, although the differences were not always significant statistically. (Ulrich, 10.)

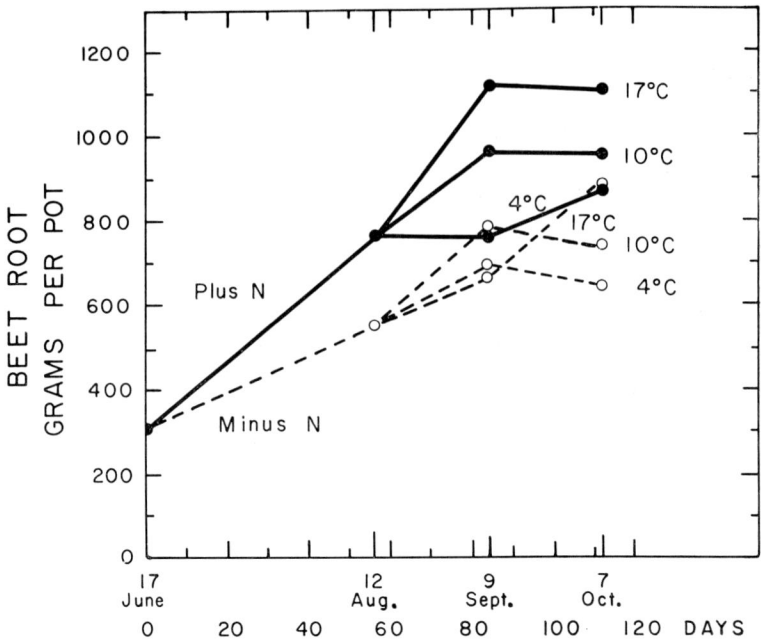

Fig. 2.8. Influence of nitrogen nutrition and night temperature upon the storage-root growth of sugarbeet plants. (See also Fig. 2.7.) (Ulrich, 10.)

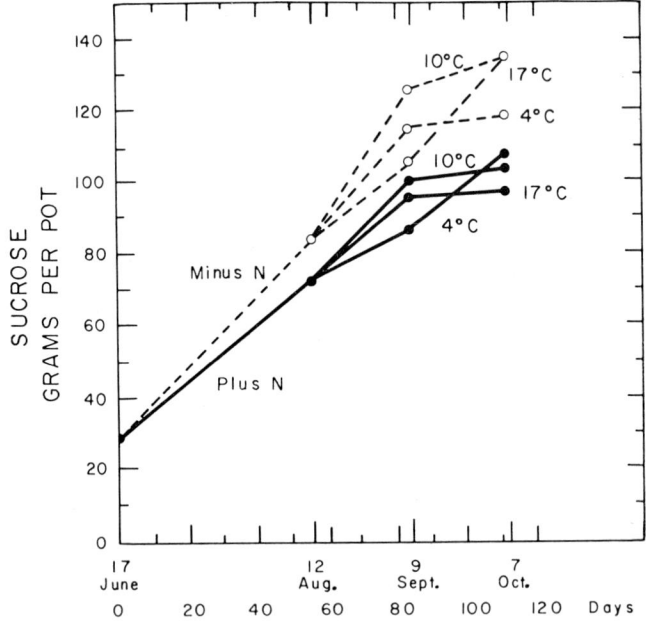

Fig. 2.9. Influence of nitrogen nutrition and night temperature upon the sucrose stored within the storage root of sugarbeet plants. (See also Fig. 2.7.) (Ulrich, 10.)

large increases in sucrose concentration of storage roots of the nitrogen-deficient plants sometimes more than make up for losses in root size caused by nitrogen deficiency. Plants at an intermediate night temperature of 10° C produce as much as or more sucrose than plants at 4° or 17° C, whether they are adequately or inadequately supplied with nitrogen for a period of time prior to harvest.

Top size is drastically reduced under conditions of nitrogen deficiency but this reduction in top size is not affected by decreases in night temperature (Fig. 2.10). Low night temperature, however, decreases top size of plants adequately supplied with nitrogen and this is associated with an increase in sucrose concentration of the storage roots (Fig. 2.7).

Under natural conditions, nitrogen deficiency usually will not be as sharply defined or as severe as illustrated by these nutrient culture experiments. This is because the soil may continue to supply marginal amounts of nitrogen. In such cases, the optimal length of the deficiency period will be greater than 4 to 6 weeks.

SUNLIGHT ILLUMINATION, NITROGEN NUTRITION, AND GROWTH

As we have seen, the amount of sunlight strongly influences the amount of growth. As a consequence, the amount of nitrogen that can be used most effectively by sugarbeet plants is also strongly affected. This is illustrated in Figure 2.11 with results from a pot experiment. Dry weight of tops in-

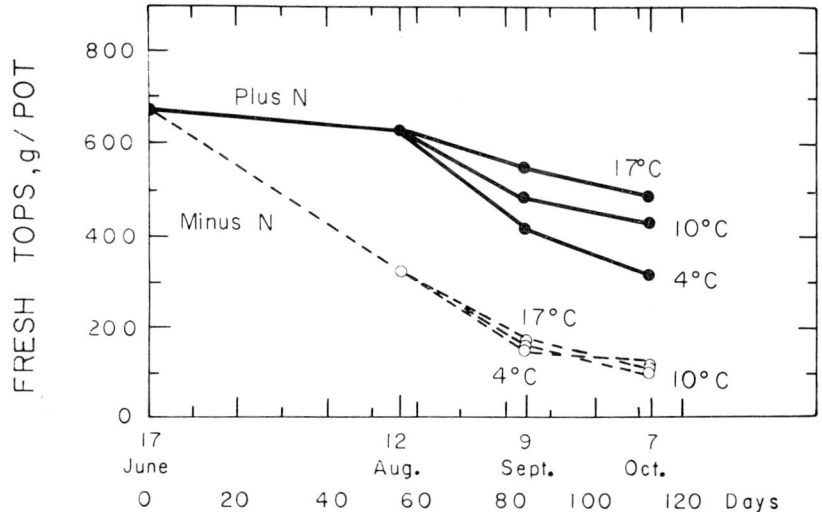

FIG. 2.10. Influence of nitrogen nutrition and night temperature upon the fresh weight of tops of sugarbeet plants. (See also Fig. 2.7.) (Ulrich, 10.)

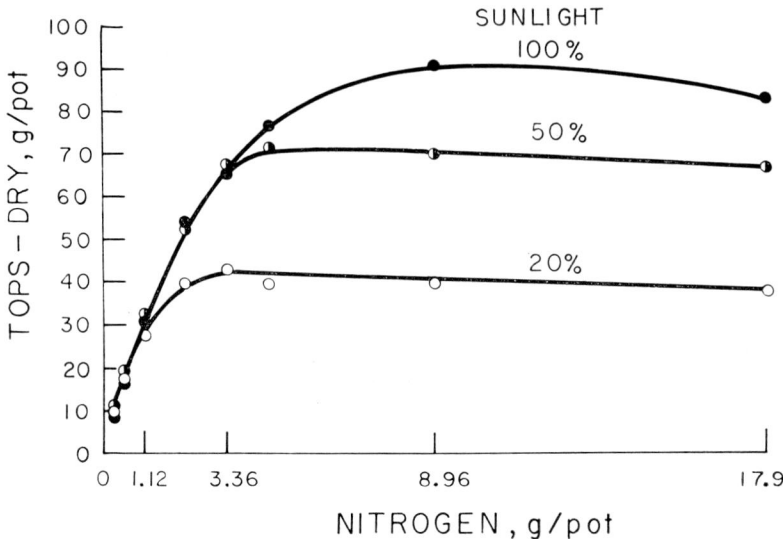

Fig. 2.11. Relation of dry weight of tops to sunlight and nitrogen nutrition for sugarbeets raised in culture solutions. (Ulrich, 13.)

creased with each addition of nitrogen with the 20 percent full sunlight illumination, until 2.2 grams of N had been added per pot. No further increases in top dry matter occurred with more nitrogen unless there was an increase in light intensity to 50 percent and 100 percent full sunlight. Maximum top dry weight was attained at 50 percent sunlight with 3.4 to 4.5 grams of N per pot, and with about 9.0 grams of N per pot at full sunlight illumination.

Increases in top dry matter with increases in sunlight illumination and N addition follow to a large extent the basic concept expressed by the law of the minimum, in which growth is limited by the factor present in the least amount. Storage-root growth, in contrast to top size, increases with each increase in sunlight illumination for a given nitrogen addition when nitrogen limits storage-root growth (Fig. 2.12). Nitrogen becomes far more effective in terms of storage-root growth as radiation increases. For example, the addition of 2.2 grams of N per pot produced a maximum storage-root growth of 190 grams per pot at 20 percent sunlight, but this is increased to 250 and 340 grams per pot at 50 and 100 percent sunlight, respectively. A similar interaction of nitrogen and radiation was observed with the 0.56 and 1.12 grams of N additions.

Radiation level has only a small effect on the sucrose concentration of nitrogen-deficient sugarbeet plants even when the illumination is as low as 20 percent of full sunlight, providing the plants have been deficient in nitrogen for a relatively long time (Fig. 2.13). These plants contained 14.2 percent sucrose, which compares very favorably with the 14.2 percent

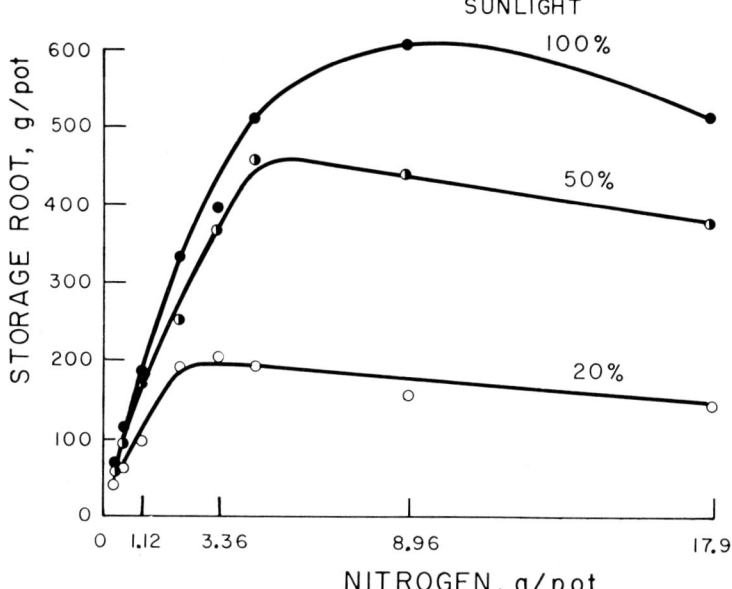

Fig. 2.12. Relation of the root size (fresh weight) to sunlight and nitrogen nutrition for sugarbeets raised in culture solutions. (Ulrich, 13.)

and 13.8 percent observed in half and full sunlight illumination, respectively. With adequate nitrogen nutrition, the average sucrose concentration decreased to a value of about 5 percent for all three sunlight illuminations, including that of only 20 percent of full sunlight illumination.

The combined effect of radiation level and nitrogen nutrition on the amount of sugar produced by the sugarbeet plants is illustrated in Figure 2.14. The results indicate that more nitrogen can be used effectively for sugar production as light intensity is increased to that of full sunlight illumination. This nitrogen response pattern for sugar production differs greatly from that observed for dry weight top growth, which follows the law of the minimum, and for storage-root growth, which follows an interaction growth pattern as more nitrogen or sunlight is provided.

GENETIC VARIATIONS IN ADAPTATION

A final point to be made here is that differences exist among genetic lines of sugarbeets in how they respond to environmental conditions. Compared to many other crop species, sugarbeets have a narrow genetic base; that is, there is a small range of variation in adaptation. Fortunately, we are concerned only with vegetative growth characteristics and not with the

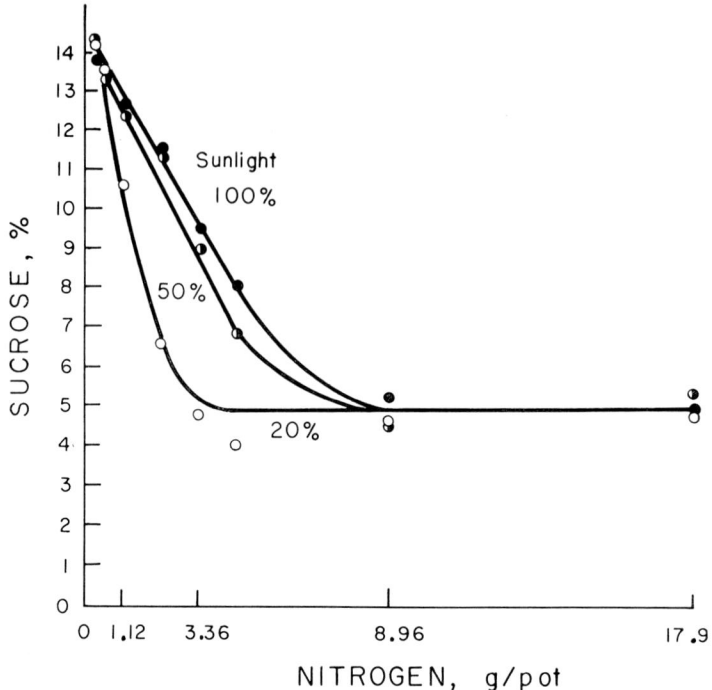

FIG. 2.13. Relation of sucrose percent of beet roots to sunlight and nitrogen nutrition for sugarbeets raised in culture solutions. (Ulrich, 13.)

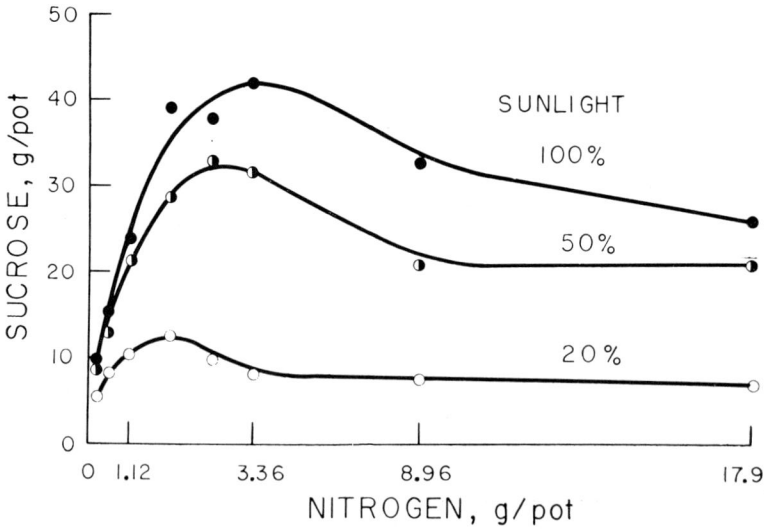

FIG. 2.14. Relation of sucrose produced to sunlight and nitrogen nutrition for sugarbeets raised in culture solution. (Ulrich, 13.)

complexities in flowering and reproduction. As a result, a single new superior genotype frequently is superior over a quite broad range of environments and our principal concerns in varietal improvement have been with other features such as resistance to bolting or disease. However, illustrated in Table 2.7 is one case in which an appreciable amount of cool-weather tolerance was lost during the selection for curly top virus resistance. This may have occurred because the selections and testing for

TABLE 2.7. Effects of climate on root weight, sucrose percentage, and sucrose weight on yield- and sugar-type curly top resistant (U.S. 22/3, 35/2) and nonresistant (E, ZZ) sugarbeet varieties (means of 8 values, expressed on a per pot basis)

Simulated Climate	Temperature		Yield-type		Sugar-type		Climate Mean
	8 AM to 4 PM	4 PM to 8 AM	E	U.S. 22/3	U.S. 35/2	ZZ	
			Beet roots, g				
Cold	17° C	12° C	497 **	253 **	247 **	49	261
Cool	20° C	14° C	621	637	558 **	199	504
Warm	26° C	20° C	466 *	480 *	381 *	225	388
Hot	30° C	22° C	304	303	213	121	235
Variety mean	...	...	472	418	350	149	...
			Sucrose, %				
Cold	17° C	12° C	11.3	10.8	11.0 **	13.7 **	11.71
Cool	20° C	14° C	10.3	10.0	10.9	11.9	10.76
Warm	26° C	20° C	9.2	9.1	9.8 **	11.6	9.94
Hot	30° C	22° C	8.4	9.4	9.4 **	12.0	9.81
Variety mean	...	...	9.81	9.83	10.27 **	12.31	...
			Sucrose, g				
Cold	17° C	12° C	56.8 *	27.8 *	27.2 **	6.8 **	29.7
Cool	20° C	14° C	64.7	64.3	61.8 **	23.6	53.6
Warm	26° C	20° C	44.3 *	44.1	37.8 *	26.3	38.1
Hot	30° C	22° C	25.8	28.9	21.1	14.6	22.6
Variety mean	...	...	47.9	41.3	37.0	17.8	36.0
			Fresh tops, g				
Cold	17° C	12° C	861	842	922 **	161	696
Cool	20° C	14° C	813 *	834	890 **	518	764
Warm	26° C	20° C	479 *	617 *	595 * **	403	523
Hot	30° C	22° C	316	365	302	294	319
Variety mean	...	...	617	664	677	344	...

NOTE: The symbols * and ** indicate significant differences between the adjacent means at the 5% and 1% levels, respectively, calculated from the t-test for paired means (12).

disease resistance were made under hot climate conditions and this masked the poor ability of the resistant variety to grow in cool climates. The results shown here were obtained in a disease-free controlled-environment phytotron—a practice which could prove highly profitable in future efforts toward sugarbeet improvement.

SOIL RELATIONSHIPS

The soil environment serves as the primary source of plant nutrients, water, oxygen (for root growth), and to a small extent, as a source of carbon dioxide for photosynthesis. The roots in their intimate contact with the soil anchor the plant to a fixed position in the field. The soil environment reflects the macroclimate, as manifested by the air temperature and prevailing moisture regime. Bare soil temperatures rise gradually with the onset of spring, reach a maximum in late summer or early fall, and thereafter decline to a minimum in late winter. Surface soil temperatures, in contrast with temperatures at lower depths, fluctuate more and are related to the daily changes in radiation and air temperature. Surface temperatures of bare soils usually exceed air temperature during the daytime, being warmest in late afternoon. However, when covered by a canopy of leaves, soil temperatures fluctuate less and are cooler than air during the daytime and warmer at night.

Soil texture determines to a large extent moisture-holding capacity and soil drainage, and these in turn affect soil aeration. Light sandy soils tend to have a small moisture capacity and to be well drained and aerated, whereas silty to clay soils tend to have a greater moisture capacity and to be poorly drained and aerated. Sandy soils also have a low cation-exchange capacity and thus tend to be low in fertility, and usually require more fertilization than silty or clayey soils. The heavier soils are generally higher in fertility, but because of poor soil structure they require the incorporation of organic matter from trashy crop residues to improve soil aeration and drainage. Nitrification of soil organic matter is also an important source of nitrogen for the crop, and therefore at times, it may be impossible to develop a low-nitrogen status in sugarbeets late in the growing season. A deep, well-drained loamy soil is generally the most satisfactory for sugarbeets. Soils with pH values of less than 5.0 to 5.5 are generally not favorable for sugarbeet growth unless limed to correct the acidity and to improve soil structure. Microelement deficiencies often occur on acid, poorly aerated soils, or following a period of low soil temperature or excessive soil moisture.

ENVIRONMENT-COMMUNITY RELATIONSHIPS

MICROCLIMATE PROCESSES IN BEET FIELDS

The microweather in sugarbeet fields is determined by the prevailing patterns of air movement and temperature (the regional weather) and by the radiation balance within the vegetation. The shortwave solar radiation received by the crop surface is composed of direct radiation from the sun, diffuse radiation from the sky, and reflected radiation from the soil surface. The amount varies with latitude, time of day and year, cloud cover, and with the clarity of the air. On a clear summer day in North America, one may expect to receive up to 1.0 cal of energy cm^{-2} min^{-1} or more, and between 500 and 700 cal cm^{-2} day^{-1}, the larger values being observed in arid regions with clear skies. Under average clear sky conditions, about 45 percent of the incoming radiant energy is in the visible region of the spectrum, that is, 0.4 to 0.7μ wavelengths, and thus is capable of being used in photosynthesis. Most of the rest of the incoming energy is middle infrared, 0.7 to 3.0μ wavelengths. Plant leaves strongly absorb the visible light, but the infrared, particularly 0.7 to 1.0μ wavelengths, is more readily transmitted and reflected. Thus, the leaf, and hence, the crop are fairly well designed, being able to absorb and use in photosynthesis the spectrum in which nearly one-half the solar energy is received and to restrict absorption of unneeded energy of other wavelengths.

About 10 to 20 percent of the incoming solar energy is reflected by a crop plant community. Additional amounts of long-wave thermal radiation are exchanged between the crop and its surroundings, including the atmosphere. Everything in the environment emits long-wave thermal radiation at a rate which increases with the fourth power of temperature. Thus, when the plant system is heated by incoming radiation, outgoing thermal radiation increases. Water vapor and clouds in the atmosphere serve as a barrier to thermal radiation cooling. This is particularly important at night and accounts for the cool nights common in arid regions—and also is the reason that frost is more apt to occur with clear skies than with cloudy skies.

A portion of each of the shortwave radiant fluxes exchanged by a crop is converted to other forms of energy. The amount of radiant energy which is converted is termed net radiation (R_N) and an equation may be written to describe its fate:

$$R_N = E + Q + S + P$$

where energy used in evaporation of water *(E)*, changes in sensible heat of the air *(Q)*, and soil *(S)*, as well as photosynthesis *(P)* are considered. Sometimes, wind advection to and from neighboring fields must also be considered. This is important in arid regions when hot, dry winds from desert lands bring in considerable energy and thus increase evaporation losses.

Ignoring complications due to wind advection and disregarding *P* which is usually small, the fate of net radiation which may amount to 200 or more cal cm^{-2} day^{-1} is described by: $R_N = E + Q + S$. Early in the growth of an annual plant, cover is low and much of the incoming radiation reaches the soil surface. The water of this surface which is free to evaporate is quickly exhausted (energy lost as *E*) and then *S* and *Q* become large, that is, soil and air temperatures rise. Cooling of the soil surface then occurs through transfer of heat to air and through outgoing thermal radiation. Later in the season with greater cover, a greater proportion of the incoming radiation is dissipated through evaporation of water from the leaves (transpiration). As long as water supply is adequate, E will be the largest term in the equation (up to 50 to 75 percent of the heat loss) and evaporative cooling will have a strong influence on temperatures.

To illustrate the importance of these net-radiation processes, temperatures may reach 60° C (140° F) in the surface of an unshaded dry soil, while on the same day, a moist soil under a sugarbeet canopy may never exceed 25° C. The air temperature (about 35° C in the above case) and soil and plant temperatures obviously may differ greatly.

Thus, the radiation environment, and hence, temperature are dependent upon cover, reflectivities, and absorptivities of leaves and soil, and on water supply. Thermal emissivities vary for leaves and soil and this also affects the radiation balance. Sugarbeet is similar to most other crops in these respects except perhaps in its slowness to achieve full cover (due in part to the wide spacing plans used in commercial production).

One other important microclimate process, eddy transfer, helps determine the crop environment (4). In this process, wind movements interact with the roughness and drag of the crop surface to set up eddies of air. Since this air contains heat, CO_2, and water, and is moving at some velocity, the eddies during daytime serve to transfer CO_2 and momentum (observed in part as leaf movement) to the crop, and equally important, water vapor and heat are transferred away. The direction of transport depends upon concentration gradients. For example, at night CO_2 and heat are usually transferred away, while water may be deposited as dew. Eddy transfers can be visualized from the profiles of CO_2 and water vapor gradients which develop between the crop and the atmosphere, as illustrated for CO_2 in Figure 2.15. Although the eddy transfer process becomes greater as wind speed increases, it is quite rapid at surprisingly low wind speeds of 1 to 2 mph. For this reason, attempts at CO_2 fertilization by releasing CO_2 gas within the crop are largely ineffective.

TRANSPIRATION IN THE SOIL-PLANT-AIR-WATER SYSTEM

Because of its importance as a microclimate process, transpiration will be considered briefly. Details are given in later chapters.

The soil-plant-atmosphere forms a continuous system for the movement of water. Liquid water moves into the roots and through the plant vascular system to the mesophyll tissue of the leaf where evaporation occurs. The water vapor then diffuses out through the leaf's stomatal pores into the surrounding air. This flow of water from the leaves into the surrounding air is termed transpiration.

Water movement in transpiration follows basic physical laws. If we take pure water as a reference, then its potential to diffuse is greater than

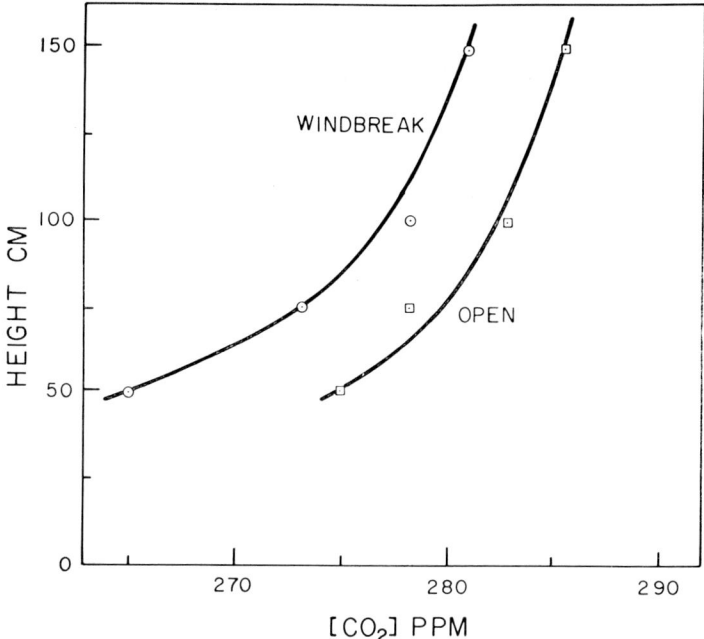

Fig. 2.15. CO_2 profiles in midafternoon of sugarbeet crops protected and unprotected by porous windbreaks of corn plants at 15-m intervals. Gradients in CO_2 concentrations (CO_2) occur between atmosphere and crop and are greater in the quieter air of the protected plot. Such curves usually follow the logarithmic profile of wind with height. (After K. W. Brown, 1.)

where the water is diluted by solutes, held by insoluble soil or plant matrixes, or under tension.

For convenience, we will speak of the energy content in terms of atmospheres of equivalent osmotic solution. The energy content of water in a moist but well-drained soil (moisture equivalent) will be only a small fraction of an atmosphere less than pure water. As plants take up the capillary water, soil-water potential drops curvilinearly, because a larger and larger proportion of the remaining water is held strongly by adsorptive forces of soil colloids, or is affected by the increasing concentration of solutes. However, the solute component really becomes significant only with saline soils. Soil-water potential decreases very rapidly with further drying after the capillary water is depleted (about -15 atm). Most plants can survive on soils well beyond the -15 atm point, but vegetative growth, for practical crop production, may be very slow or absent.

Solutes accumulated by the plant serve to keep leaf mesophyll tissues at a lower water potential than the soil moisture so that water is absorbed by the roots and flows to the leaves along this gradient. As the soil dries, there is less capillary conductivity and the resistance to water movement steepens considerably. Within the plant, there is a high resistance to liquid water transport across the root cortex. Thus, when transpiration demand is high and quite low water potentials occur in the leaves, negative pressures may occur in the stem vascular tissues.

The resistance in the vascular tissues to water flowing from the roots to the leaves is low. In the leaves, evaporation takes place mainly in the cell walls lining the substomatal cavities, and the water vapor diffuses to the atmosphere mainly through the stomata. Little water moves through the cuticle, since cuticle resistance is perhaps 10 times greater than for open stomates. The movement of water from the leaves to the atmosphere is determined by the gradient in water-vapor pressures between the intercellular space system of the leaf and the atmosphere. Since the air inside the leaf is probably nearly saturated with water vapor, even at 93 percent relative humidity outside the leaf, the diffusion gradient may still be the equivalent of about 100 atmospheres of water potential (compared with 5 to 20 between soil and leaf). A tremendous gradient exists, therefore, between the plant and the atmosphere.

The boundary layer of still air adjacent to the leaf surface presents a further significant resistance to water loss. The thickness of this layer is influenced strongly by wind velocity, being rather small for most leaves whenever wind exceeds 1 to 2 mph. Finally, one may speak of a canopy resistance for the eddy transfer between crop surface and atmosphere. As we noted above, even with poor ventilation this resistance is relatively small. However, one can reduce ventilation of a crop by introducing wind breaks, and this has been shown in Nebraska to give small but measurable improvements in yield through improvements in water balance—the greater resistance of the becalmed air slowed transpiration, thus preventing restrictions on CO_2 uptake from stomatal closure due to stress for water. Our conclusion is that the stomates and the root-soil interface are the main regulatory sites in the transpiration stream.

PRODUCTION PROCESSES

Photosynthesis is driven by light energy from the 0.4 to 0.7μ wavelength region absorbed by chlorophyll. Thus, only a portion of the shortwave solar radiation is used. The nature of the physiological processes is also limiting, and with sunlight, the theoretical maximum efficiency of photosynthesis (energy converted $\times$ 100/energy received) is about 12 percent of the visible light, or about 5.3 percent of the total solar radiation received (5). A useful generalization is that about $14\mu g$ of dry organic material might be produced for each calorie of incoming radiant energy. Thus, the upper limit for crop growth with 500 cal cm^{-2} day^{-1} would be about 70 g m^{-2} day^{-1} or over 600 lb of dry matter acre^{-1} day^{-1}. Under intensive agriculture, sugarbeets may achieve 50 percent of this potential growth rate or about 30 to 40 g m^{-2} day^{-1}. In the following, we will consider (1) photosynthesis in leaves and (2) the behavior of foliage canopies.

CO_2 Assimiliation by Individual Leaves. In photosynthesis, carbon dioxide is converted to organic products in green leaves, using energy obtained from the sun. The overall reaction, greatly simplified and generalized, is

$$CO_2 + H_2O + \text{light energy} \rightarrow [CH_2O] + O_2$$

where the carbohydrate [CH$_2$O] contains much greater chemical bond energy than CO_2. In addition to simple carbohydrates, amino acids and other organic materials may be produced. The captured light energy may also be used in nitrate reduction. Generally, however, the rate of CO_2 assimilation adequately characterizes the photosynthetic capabilities of green leaves.

CO_2 exchange rates for sugarbeet leaves are illustrated in Figure 2.16 as a function of varying light flux and CO_2 concentration. The maximum rates of CO_2 uptake observed in normal air, 35 to 45 mg CO_2 dm^{-2} hr^{-1} (and the temperature optimum, near 35° C), are much higher than reported earlier. The light response curve shows saturation at one-half to two-thirds of the full sunlight of a bright day. This saturation phenomenon is related to limitations in CO_2 transfer since it can be eliminated by elevating the CO_2 level (7).

The net rate of CO_2 uptake shown here is the gross rate of CO_2 fixation less the rate of CO_2 evolved through respiration. In the dark, respiration may be 2 to 5 mg CO_2 dm^{-2} hr^{-1} or about 5 to 10 percent of maximum photosynthesis, but in the light, it is estimated that photorespiration may be as much as 10 to 30 percent of photosynthesis. The physiological role of photorespiration is not yet clear.

These photosynthetic characteristics place sugarbeet in a group with other sun-loving dicotyledonous species, such as sunflower and cotton, having the Benson-Calvin carbon fixation mechanism, but marks it as quite different from corn, sorghum, and pigweed (Hatch and Slack mechanism) which have higher rates, no light respiration, and which do not saturate in full sun. We do not know whether different strains of sugarbeet vary significantly in the shape of their light-response curves and in their maximum rates of photosynthesis. Such variations would have important effects on production.

Photosynthesis in Sugarbeet Communities. When sugarbeet leaves are arranged in foliage canopies, the illumination of individual leaves is influenced markedly by angle of display and mutual shading. Generally, leaves inclined to the direction of the sun's rays use the light more efficiently than leaves displayed at right angles to the sun's rays. This is because the

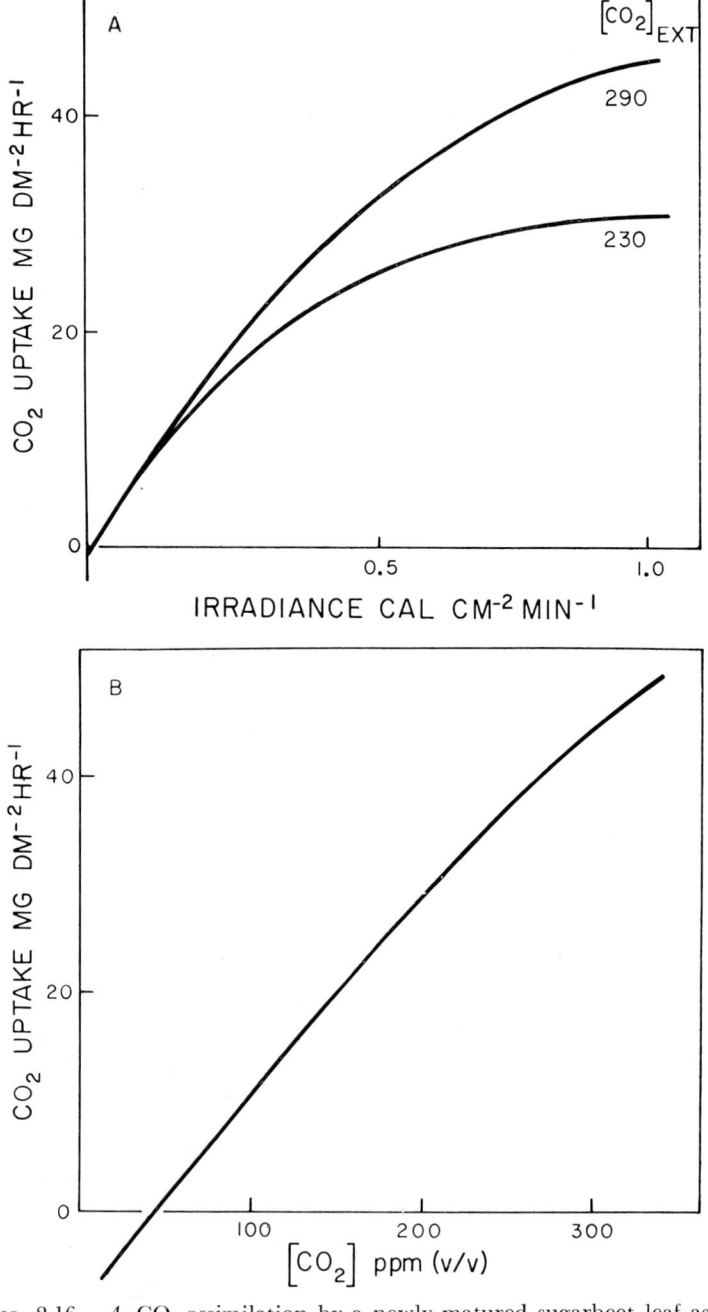

Fig. 2.16. *A.* CO_2 assimilation by a newly matured sugarbeet leaf as a function of light intensity at two external CO_2 levels 230 and 290 ppm (v/v). Ambient air fluctuates around 320 ppm but under intense photosynthesis (CO_2) in the beet field may drop to 290 ppm or less. Single, attached leaves were enclosed in carefully controlled and monitored turbulent air in a plexiglass chamber. (Previously unpublished data of A. E. Hall.) *B.* CO_2 assimilation in intense light as a function of CO_2 concentration external to the leaf. The response is near linear even beyond 300 ppm CO_2, thus higher concentrations than this would yield greater photosynthesis. Net CO_2 exchange is zero near 40 ppm; at less than this value, light respiration exceeds photosynthesis and the leaves leak CO_2. (Previously unpublished data of A. E. Hall.)

dim illumination gives a higher yield per unit light received (Fig. 2.16). Thus the density of leaves and their arrangement in the canopy affect productivity.

Two generalizations about canopy architecture can be made. First, percent cover is critical, especially during early season; as cover varies from 0 to 100 percent, production rate per unit land area (C) increases linearly with increases in percent cover. In commercial practice, wide rows and low population densities delay the achievement of full cover and reduce total production. Second, if leaf density is expressed in leaf area index units (L) and C is measured at various levels of L, a diminishing returns response of C is noted with increasing levels of L.

D. J. Watson, who first employed L in canopy descriptions, concluded that there must be some optimum value of L beyond which C would decline, because the additional leaves would contribute more to respiration than to P. However, Watson never was able to demonstrate an optimum for sugarbeet—C increased as L increased up to 8 or 9 units of leaf area per unit ground area. In our experiments with higher L values there was a plateau response, and C did not decrease from its maximum. Apparently, the heavily shaded leaves and other respiring tissues adjust to lower respiration rates when respiratory substrates are in short supply due to intense mutual shading. Such large values of L are not found at commercial planting densities where one attempts to maximize root rather than total plant production. While an optimum L for maximum rate of dry matter production does not exist, this is not the case for root and sugar production. Leaf growth has priority over the new photosynthates and at the high plant densities required for high L, each plant may be so limited for space that most of its new photosynthate goes to leaf growth with little surplus for root growth and sugar accumulation. Optimum L values for root and sugar production for various environments have not been determined.

Graphical representations of sugarbeet canopies are given in Figure 2.17. In this case, with plants uniformly spaced in a triangular pattern, $L = 3$ to 4 gave 95 percent interception of solar radiation and near maximal production rate. Planting in rows, particularly the wide-spaced, single rows now commonly used, gives a clumped leaf pattern and $L = 4$ or 5 is sometimes required before all of the ground is shaded. There has been a great deal of discussion about optimum leaf angles. Some workers have argued that rosette plants, like the sugarbeet, ought to have an ideal pattern—upper leaves are near vertical and lower leaves are near horizontal. As seen in Figure 2.17, many of the upper leaves are actually near horizontal.

The best information on what constitutes an optimum canopy has come from computer simulation models. Results calculated with Duncan's model (2) are illustrated in Figure 2.18. The illumination of each leaf by direct and diffuse sunlight as well as by reflected and transmitted light is considered in estimating its photosynthesis. In the three cases shown, all leaves are assumed to be at either 0°, 45°, or 90° elevation to the ground. At small L, horizontal leaves give best cover and highest production; and

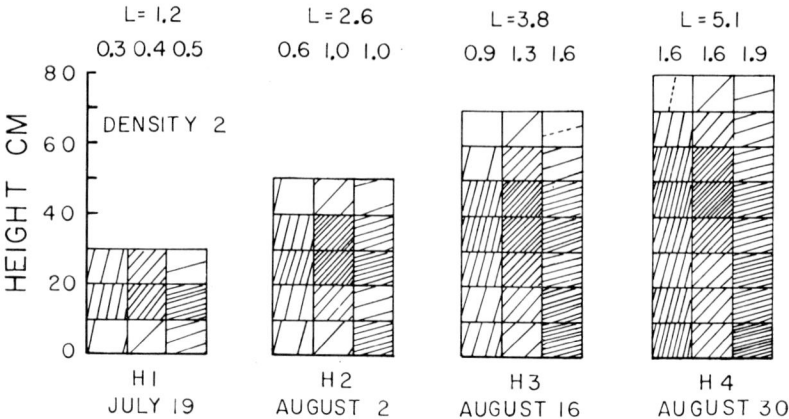

Fig. 2.17. Canopy architecture in the field for a sugarbeet population of 30,000 plants per acre at various times during the season. The crop was planted in late May at Davis, California, and grown with adequate nutrition and water. Leaf area was measured for each 10-cm (4-inch) stratum, beginning at the soil surface. Within each stratum, the leaf area found within three angle classes (angle of the leaf relative to the ground) was determined. The number of lines in each box is proportional to leaf area and the angle of the lines is the mean angle of the class. Total and partial leaf area indices (L) are given (L is m^2 leaf per m^2 ground).

the three leaf angles give equal production near $L = 2$ to 3. The model considers the leaves to be randomly distributed in the horizontal plane—in the field the leaves would be clustered on individual plants and the curves would be shifted to the right so that the crossover point would be nearer to $L = 3$ to 4. Strongly erect leaves are a distinct advantage only at large L. Leaf distributions of the type shown in Figure 2.17 behave similarly to the 45°-angle class.

Such simulations show clearly that canopies of $L = 3$ to 6 (commercial range) are more efficient when top leaves are erect and bottom leaves horizontal (like sugarbeet) than in the reverse case. However, they also show that with the leaf distributions seen with sugarbeet (Fig. 2.17), the arrangement would have to be changed drastically (to more leaves and with a much greater proportion of vertical leaves) to have much affect on production (6). Perhaps the most significant conclusion to be drawn from Figure 2.18 is that for sugarbeet-type canopies (45°), C varies little as L increases beyond the amount required for full cover. Thus, we should focus our attention on the length of time required to achieve full cover, and on maintaining full cover for as long as possible.

The most practical change in canopy architecture suggested by such studies is to seek a planting arrangement giving a more uniform distribution of plants in order to minimize time to full cover (that is, avoid wide

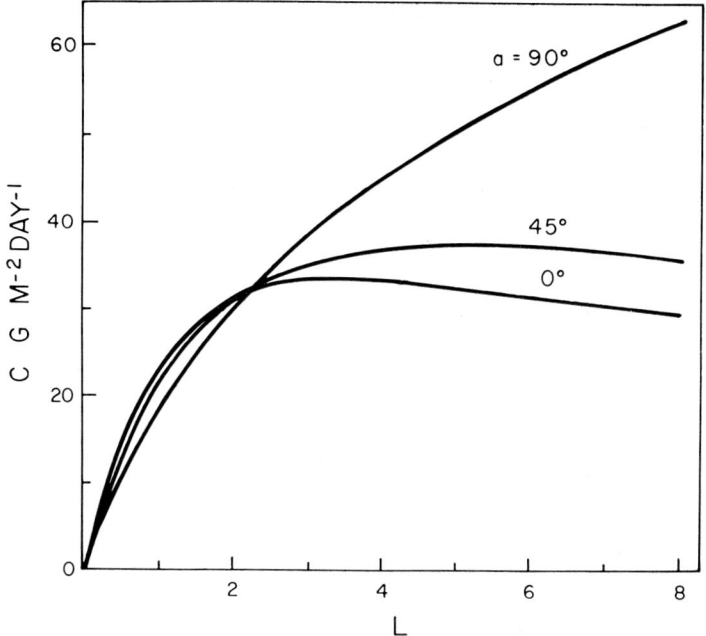

FIG. 2.18. Simulated crop growth rates (total net production of leaves, roots, and stems expressed as dry matter) for three hypothetical canopies with all leaves displayed at 0°, 45°, or 90° to the ground, as a function of leaf area index, L. The 45° canopy behaves in a fashion similar to a sugarbeet canopy. Light conditions were those of Davis, California, for a July 1 day. The photosynthetic capabilities of the individual leaves were taken as similar to the CO_2 level illustrated in Figure 2.16A for 290 ppm.

rows). Beyond this we should look to plants with a high photosynthetic capacity (this markedly affects the curves shown in Fig. 2.18) which also partition a large proportion of their photosynthates to root production and sugar accumulation.

REFERENCES

1. Brown, K. Influence of windbreaks on sugar beet production. Univ. Nebr. Hort. Prog. Rept. 71. Lincoln. 1969.
2. Duncan, W. G., R. S. Loomis, W. A. Williams, and R. Hagan. A model for simulating photosynthesis in plant communities. Hilgardia 38:181–205. 1967.
3. Leach, L. D. Growth rates of host and pathogen as factors determining the severity of preemergence damping-off. J. Agr. Res. 75:161–79. 1947.
4. Lemon, E. Gaseous exchange in crop stands. In J. D. Eastin, F. A. Haskins, C. Y. Sullivan, and C. H. M. van Bavel, eds., Physiological Aspects of Crop Yield, pp. 117–37. Am. Soc. Agron., Madison, Wis. 1969.

5. Loomis, R. S. and W. A. Williams. Maximum crop productivity: an estimate. Crop Sci. 3:67–72. 1963.
6. ———. Productivity and the morphology of crop stands: patterns with leaves. *In* J. D. Eastin, F. A. Haskins, C. Y. Sullivan, and C. H. M. van Bavel, eds., Physiological Aspects of Crop Yield, pp. 27–47. Am. Soc. Agron., Madison, Wis. 1969.
7. Nevins, D. J. and R. S. Loomis. Nitrogen nutrition and photosynthesis in sugar beet *(Beta vulgaris* L.). Crop Sci. 10:21–25. 1970.
8. Ulrich, A. The influence of temperature and light factors on the growth and development of sugar beets in controlled climatic environments. Agron. J. 44:66–73. 1952.
9. ———. Growth and development of sugar beet plants at two nitrogen levels in a controlled temperature greenhouse. Proc. Am. Soc. Sugar Beet Technologists 8:325–38. 1954.
10. ———. Influence of night temperature and nitrogen nutrition on the growth, sucrose accumulation and leaf minerals of sugar beet plants. Plant Physiol. 30:250–57. 1955.
11. ———. The influence of antecedent climates upon the subsequent growth and development of the sugar beet plant. J. Am. Soc. Sugar Beet Technologists 9:97–109. 1956.
12. ———. Variety climate interactions of sugar beet varieties in simulated climates. Proc. Am. Soc. Sugar Beet Technologists 11:376–87. 1961.
13. ———. Light, temperature, and nitrogen nutrition studies of sugar beets. Proc. 15th Ann. Calif. Fert. Conf. 15:85–86A-I. 1967.

3

Seedbed Preparation, Planting, and Thinning

G. E. NICHOL
Monitor Sugar Company
Bay City, Michigan

L. M. BURTCH
Spreckels Sugar Company
Mendota, California

D. J. TRAVELLER
The Amalgamated Sugar Company
Burley, Idaho

FIELD SELECTION	50
SEEDBED PREPARATION—EARLY STAGES	51
Plowing	52
Deep Tillage	52
Leveling	53
SEEDBED PREPARATION—FINAL STAGES	54
SEEDBED PREPARATION—HUMID AREAS	55
PLANTER SELECTION	56
PLANTER MAINTENANCE	57
PLANTING DATES	58
PLANTING	59
STAND REDUCTION	60
PLANT POPULATION	63

THE PURPOSE of seedbed preparation is to provide a soil environment which will encourage maximum field emergence and subsequent growth under existing climatic conditions.

When does seedbed preparation begin? The first management decision in growing a sugarbeet crop concerns the selection of a field. The history of a field has a great deal to do with the kind and amount of tillage necessary for seedbed preparation. In some instances, good soil management over a period of years has provided soil conditions which will require a minimum effort to achieve a good seedbed. Conversely, a history of poor soil management can leave soil in a condition where it is impossible to prepare an acceptable seedbed. Good seedbeds are, therefore, the end result of continued good soil and crop management plus the application of timely seedbed preparation operations.

The measure of what constitutes a good seedbed has changed materially during the past twenty years. Standard advice in the past was to "prepare a firm, fine, well-compacted seedbed." When the transition was made from horse power to tractor power, this ceased to be good advice, since it became too easy to overtill the soil. Overtillage results in poor soil structure and poor plant growth. Because of this, the principle of minimum tillage has been adapted to the sugarbeet crop. The best systems of today place the seed in a firm, fine-structured seedbed. The area between the rows is loose enough to permit air exchange and water percolation, but rough enough to resist wind erosion, and smooth enough to permit precise cultivation.

This combination is possible because of modern high-capacity tractors which permit tillage operations to be performed when soil conditions are near optimum. To use this equipment effectively requires a skilled operator. No fixed rules can be followed in preparing a seedbed. Every farm, every field, and every season pose a unique set of problems which can be solved only by the experience and judgment of the beet grower.

FIELD SELECTION

Field selection is often given too little consideration by the average sugarbeet grower. Generally, the importance of such factors as weather and economics is weighed heavily, while too little attention is given to the many other facets of field selection that can adversely affect production.

The basic consideration should be crop rotation, since this is the most practical means of avoiding the buildup of harmful disease and insect problems. Unfortunately, this phase of sugarbeet culture was neglected during the industry's early development period. The practice of planting sugarbeets for several consecutive years in the same field resulted in a buildup of soil-borne insects and diseases to the point that yields were seriously depressed. Out of necessity, then, crop rotation practices evolved. It is now common to plant sugarbeets only once in a cropping sequence which may range in length from three to eight years. The length of a rotation is often determined by the presence and severity of nematodes and other soil-infesting pests.

Not all rotation crops are compatible. For example, crops which are hosts to common diseases and insects should be avoided or carefully separated in the rotation sequence. This is an important factor and is discussed in more detail in Chapters 9 through 11.

Sound crop rotation practices, however, have many other advantages that are well documented. One of the most important of these is the maintenance and improvement of soil structure which is favorable to rapid seedling emergence and root development. Poor soil structure is a hostile medium for plant growth. It increases production costs and decreases net returns.

Anticipated weed population is another important consideration, for sugarbeet seedlings are not strong competitors with weeds. Consequently, most successful growers will attempt to plant in a rotation sequence which tends to minimize weed and crop residue problems.

The presence of toxic chemical residues is also becoming a rotation factor as more chemicals are applied to more crops in the rotation. Herbicides applied to agricultural crops are by nature specific in their action and sometimes persist in the soil for extended periods. The sugarbeet grower, therefore, needs to be aware of the possible existence of toxic herbicide residues which may result from herbicides having been applied to the crops preceding sugarbeets. The trifluralin and triazine compounds are examples of products used safely on some rotation crops but which can be highly toxic to sugarbeets.

Basic fertility in the form of crop residues or fertilizer carry-over should always be considered in field selection. This is especially true in areas where a low sucrose concentration tends to be a problem at harvest. In some irrigated areas, growers are encouraged to avoid following crops such as potatoes, melons, and other heavily fertilized vegetable crops because of the adverse influence of excessive nitrogen on sucrose concentration and beet quality.

In summary, field selection is the first, and often the most important decision made by the grower.

SEEDBED PREPARATION—EARLY STAGES

The primary goals of early seedbed preparation are:
1. To manage crop residues effectively.
2. To change or improve the soil structure to meet the needs of the forthcoming sugarbeet crop.
3. To eliminate growing weeds.

PLOWING

Plowing is perhaps the oldest and commonest practice used to achieve the above goals. Effective plowing will cover crop residue and existing weeds with freshly turned soil. In addition, the "slice" of overturned soil is loosened, creating a condition which is conducive to air exchange and water percolation. Plowing depth varies with soil type and condition, and with the type of farm equipment used. Normally, the depth of plowing will range from 6 to 14 inches.

While plowing can be accomplished any time soil and weather conditions permit, the normal time for this operation is either in the spring or fall. Fall plowing has the advantage of getting a head start in the decomposition of residue and organic matter. It also helps to reduce the spring work load. In addition, in low rainfall areas, fall working of the soil helps to conserve winter moisture and enables earlier planting in the spring. Spring plowing is sometimes necessary when weather conditions prohibit fall work, or when soil and climatic conditions are such that it is necessary to leave crop residues over winter as an aid in preventing soil erosion.

When soil is spring plowed, care should be exercised to insure that the ground is not worked too wet. Traveling on wet soil with the heavy equipment normally employed by farmers today can result in soil compaction and less desirable soil structure. The resulting seedbed is a poor environment for the emergence and growth of sugarbeet seedlings. This potential problem can be minimized through the use of low-pressure flotation or dual tires on wheel tractors, or by the use of track-layer tractors.

Either spring or fall plowing is often preceded by application of fertilizer to the soil surface. This practice is considered desirable, since the plowing operation helps to incorporate the fertilizer into the soil and place it in a location where it will be available to the beet crop.

DEEP TILLAGE

While plowing helps to create a favorable seedbed environment, it does not necessarily help to improve the status of the subsoil or root zone. Since sugarbeets have a deep taproot, the root zone should also receive serious consideration in soil preparation. Several successive years of plowing to the same depth will often create a "plow-sole," or hard semi-impervious layer of soil at the bottom of the plow furrow. This layer of soil will resist movement of water and air as well as penetration of plant roots.

Chiseling is a deep-tillage operation wherein the subsoil is loosened without overturning the surface soil. This practice is employed in some areas once every three to five years for breaking subsoil hardpans, thereby providing for better root penetration. In other areas it is an annual practice. (See Fig. 3.1.)

The advantages of chiseling, in addition to breaking up a hardpan,

SEEDBED PREPARATION, PLANTING, AND THINNING 53

FIG. 3.1. Deep chiseling or subsoiling to break up compacted soil layers. It may substitute for plowing in some western areas.

are to allow for deep air and water percolation and to leave crop residues on the soil surface. This in turn will aid in the prevention of wind and water erosion by minimizing freshly exposed surface soil and allowing for water penetration rather than runoff. Chiseling also avoids returning leached salts to the soil surface and reduces the amount of tillage required to prepare the seedbed for planting.

Another less common deep-tillage practice is that of deep plowing. This differs from conventional plowing in the size and depth of the plow furrow or slice. From two to three feet of soil may be overturned in a deep-plowing operation.

LEVELING

After most early tillage operations, it becomes necessary to relevel the land before planting a crop of sugarbeets. The amount of leveling necessary depends in large measure upon the irrigation practices and cultivation procedures that will follow in the production of the crop.

The most critical leveling requirements are when gravity or furrow irrigation is to be used. Under this system, effective water management necessitates land planing to remove small variations that would cause high and dry spots, as well as areas where flooding or ponding might occur.

Where sprinkler or subirrigation is employed, uniform land leveling is less critical. Under all conditions, however, the land must be level enough to insure uniform and precise placement of seed during planting. This is also an advantage for subsequent cultivating and thinning operations. (See Fig. 3.2.)

SEEDBED PREPARATION—FINAL STAGES

The final stages of seedbed preparation are designed to eliminate germinating weed seeds and to create a surface soil environment conducive to precision planting and rapid germination and emergence of the freshly planted sugarbeet seed.

One common practice is to "bed" the ground. This consists of listing or shaping the soil into ridges or beds, on top of which the seed will be planted. Several types of equipment are used to accomplish this task. Sleds with bed shapers can be used either to form beds or to form beds and to plant. In this procedure, planters can be attached to the sled, and bedding and planting can be combined into one operation. The sled is later used as a carrier for cultivating tools. Another common method of "bedding" is simply to use shovels to throw soil up into a rough bed. Prior to planting, these beds are harrowed and shaped to leave a smooth, desirable working surface for the planter. Often fertilizer is injected or incorporated into the beds while they are being formed. Bedding is either a spring or fall operation. Generally, the rough-type beds are preferred in the fall, since a fine-textured bed will often set up hard from winter mois-

FIG. 3.2. Land leveling is necessary for regions where furrow irrigation is practiced.

ture, whereas a rough bed will stay mellow and loose. Rough beds are also more resistant to winter wind erosion.

There are several advantages to bedding. In general, bedding will help to form more definite furrows, thereby making early season irrigations easier to control. Bedding also leaves furrows in the soil, making planting and subsequent cultivations easier to perform because of having definite marks to follow with the tractor wheels. Another advantage is the conservation of spring soil moisture. This is especially true when fall bedding is used. Bedded ground will absorb the rays and warmth from the sun more readily.

When bedding is not employed, the land is usually harrowed or disced and harrowed ahead of planting. These operations are for breaking clods, killing small germinating weeds, and establishing a mulch on the soil surface. Preplant herbicides are normally applied during final soil preparation when not included as part of the planting operation.

In many of the semiarid areas of the intermountain West, early spring planting is completed several weeks before supplemental irrigation water becomes available. Under these conditions, seedbed preparation is kept to a minimum, since natural soil moisture is relied upon for early-season germination and growth.

It should be remembered that in all phases of seedbed preparation the type of operation to be performed will be determined by the existing soil condition and the goal to be achieved. By proper management, soil preparation can be minimized to the extent that little, or in some instances, no tillage is required for an adequate seedbed. One critical factor to keep in mind is that timeliness of operations is all-important. Working soil either too wet or too dry can easily destroy rather than enhance soil structure. With the high-capacity and high-speed farm equipment available today, large acreages can be covered in a short period of time. Sound judgment in farm management is important in knowing when and how to act and in being prepared to accomplish the job when conditions are at an optimum.

The overall goal in seedbed preparation is to provide a physical soil environment in which every planted seed will be assured of germination and every germinated seedling can continue for a full season of maximum uninterrupted growth. (See Fig. 3.3.)

SEEDBED PREPARATION—HUMID AREAS

In humid areas, particular care must be taken to avoid excessive spring tillage of wet soils. This has led to the adoption of the principle of minimum tillage. Since early planting is desirable, most soils are plowed or deep-tilled in the fall. Spring seedbed preparation generally consists of a once-over operation, using a dual-wheel tractor pulling a spring tooth harrow and spike tooth harrow. This breaks down the coarse clods and

Fig. 3.3. Planting with a shaper-equipped sled is a very important step toward the complete mechanization of the sugarbeet crop.

levels the soil. Planting follows this operation closely, usually on the same day. Where soils are spring plowed, a "clod-buster" is generally pulled behind the plow. Subsequent tillage is done before the soil dries out, and consists of only enough tillage to firm and level the soil. This can often be done in one operation. Again, planting is done as quickly as possible after the plowing and tillage operations are started.

PLANTER SELECTION

The precise planting of sugarbeet seed becomes more important year by year. The cost and availability of hand labor make it necessary for growers to mechanize in lieu of using the traditional hand-labor systems of the past. Whether planting to stand, planting for random mechanical thinning, or for electric-eye thinning, dependable planter performance and dependable seedling emergence is essential.

Planting units fall into two broad categories:
1. Those designed specifically for planting sugarbeet seed or small-sugarbeet seed.
2. Those capable of planting a broad spectrum of seeds, including sugarbeet seed.

Most of the major implement manufacturers in the United States build and market multipurpose seeding units of the type included in 2 above. These units meter seed through horizontal seed plates having proper thickness and with cells of correct size to accommodate the seed being planted. Furrow openers are generally the double-disc type, which permit operation under a wide range of conditions. These planters have generally proved to be satisfactory if the seed fits the seed plates being used, and if the planter is operated at a reasonable speed by a competent operator.

The type of planter included in category 1 above is designed to plant sugarbeet seed and other small seeds requiring precision planting. Some are manufactured in the United States, but several foreign makes are also available in this country. These planters use either belts or vertical rotors for seed distribution. They usually require that seed be graded very precisely to fit their particular cell size. These planters, operated properly, result in very accurate distribution of seed.

Either type of seeding unit may be used in conjunction with other operations. These operations include:
1. Building ridges or beds.
2. Incorporating herbicides or insecticides.
3. Applying fertilizer in bands.
4. Metering of insecticides.
5. Spraying preemergence herbicides.

Usually not more than two of the above operations are combined with seeding, since each additional operation provides more opportunities for problems to develop and tends to slow the seeding operation.

PLANTER MAINTENANCE

Timeliness of operations is always important in handling the sugarbeet crop. The planting season is particularly critical in that respect. Therefore, it is necessary to have the planter thoroughly checked and adjusted before the planting season arrives. Probably the best time to begin planter maintenance is at the end of any planting season. Problems which have occurred are fresh in mind and there is adequate time to replace worn or broken parts. Prior to the planting season, a thorough lubrication and check should be made to ascertain that all planter parts are in good working order. A check list, including the following items, could be used to good advantage:
1. Check spacing between seeding units to make certain that rows will be spaced at a uniform width.
2. Check length of markers so that guess rows will be spaced correctly.
3. Check depth bands to ascertain that seed will be planted at proper depth.
4. Check freedom of operation of disc openers.
5. Check dirt scrapers, covering blades, and covering chains, if used.
6. Check seed tube to verify that it is in place and free of obstructions.
7. Check seed cans for the following:
    a. Proper hopper bottom.

b. Proper spring tension on hopper bottom.
c. Proper filler ring.
d. Proper seed plate—clean and rust-free.
e. Proper cutoff.
f. Proper seed knocker.
g. Proper drive sprocket combination to obtain desired seed spacing.
h. Proper press wheel.
8. Lubricate all working parts.
9. Secure seed which fits seed plates to be used.

PLANTING DATES

Sugarbeets are considered as a crop most suited to the temperate climate typical of the latitudes ranging between the 40° and 45° parallels. In practice, however, sugarbeets are grown successfully under a much broader range of climatic conditions. Commercial sugarbeet acreages are now profitably grown from the southern tip of California, where air temperatures exceed 100° F for much of the growing season, to southwestern Canada, where long days compensate partially for short growing seasons. The range of adaptability present in North America is made possible because sugarbeets have the ability to adapt to a broad range of planting dates.

The commonest planting season in North America is in early spring, after much of the danger of severe frost has passed. The actual dates vary within local climatic limits, but generally plantings are completed between March 1 and June 15. This range permits a growing season which extends approximately 120 to 200 days, depending on local weather restrictions and harvest requirements. Dates of planting by areas are summarized in Table 3.1.

Since the sugarbeet grown for sugar production does not have a limiting growth cycle except as dictated by weather, growers usually prefer to plant as early as weather permits. This is especially true of the more

TABLE 3.1. Planting dates for the major beet-growing areas of the United States

Area	J	F	M	A	M	J ... S	O	N	D
Michigan, Ohio			x	X	x				
Red River Valley				X	X	x			
Eastern Slope			x	X	x				
Utah, Idaho				X	X	x			
Oregon, Washington		x	X	X					
Texas		X	X						
Arizona			x	X			X	x	
California (Coastal)	X	x							x
California (Imperial Valley)						X	x		
California (San Joaquin Valley)	X	x	x				x	X	x
California (Sacramento Valley)					x	X	x		

NOTE: X indicates main planting season.

northern latitudes where a delayed spring planting usually results in a lower yield at harvest.

In the southern latitudes and lower elevation areas in California and Arizona, mild year-round temperatures make stockpiling of beets impossible. Planting dates are a very important tool in regulating the beet supply available to factories. Thus, planting dates are scheduled for all calendar months except July and August. The actual growing season, therefore, ranges from 180 days in most fall harvest districts to over 300 days in districts where overwintering is practiced.

The Imperial Valley and central Arizona areas normally plant their entire acreage during September and October for harvest the following spring and early summer. Temperatures are usually above 100° F in September, and extreme care is required in order to obtain stands under the adverse temperature and salinity conditions.

Planting in the coastal valleys of California is carried out as weather permits from December until early March for harvest starting in late August and terminating in November. Early planting is especially prudent to take advantage of a longer growing season and to minimize the threat of the sugarbeet nematode.

The Central Valley of California is composed of two large river valleys which meet and form the San Francisco Bay. These valleys, the Sacramento in the north and San Joaquin in the south, produce the majority of California's sugarbeets. Maximum temperatures range from the low 30s in midwinter to slightly above 100° in midsummer. Rainfall occurs only during the fall, winter, and spring and ranges from 6 to 22 inches annually. Sugarbeets are grown on a 12-month basis, with planting date serving as the principal method of scheduling the growing season. Currently, sugarbeets are planted in the San Joaquin Valley from October through February for harvest in July and August. Since the harvest date is fixed by factory-slicing capacity rather than climate, growers prefer to plant as early as possible in each geographic section in order to produce the highest yield from their restricted growing season.

The northern sections of the Central Valley plant the majority of their acreage in the late spring after the threat of an aphid-borne virus disease has passed. Since the principal host for this disease complex is the sugarbeet, commercial plantings must be carefully scheduled so that successive crops do not overlap in the same area during the aphid flight season. Normally, spring harvest terminates in early May and the aphid flights reach a low level at about the same time. Harvest of these plantings is scheduled for the late fall and generally continues until rains interfere. The remaining acreage is overwintered and harvested in the spring.

PLANTING

As mentioned earlier, the optimum time for planting can be critical in most areas. However, if the planter has been properly prepared and checked, the planting operation can be done quickly and with few mechanical delays.

Some adjustments may be needed at the time planting begins. Seeding units are designed to operate best with the tool bar at a prescribed height and with the units working on a level plane. These requirements are spelled out in the operator's manual, but final adjustments must be made under field conditions. After these adjustments are complete, spring pressure on seeding units should be checked. It can then be decided if covering blades or covering chains should be used.

Speed of planting is a critical factor. Most operators are not aware of the large number of seeds which each seeding unit must handle at normal planting speeds. To illustrate, Table 3.2 shows the number of seeds which each seeding unit meters *per minute* at various combinations of ground speed × seed spacing.

Thus, planter plates are often turning too fast for dependable cell fill, resulting in erratic stands of beets. Excessive speed can also result in bunching and uneven covering of seed.

Depth of seeding should be checked at the time planting begins. This depth will vary, depending on conditions and personal preference. Since beet seed is small, it should be kept shallow (approximately ¾ inches) but with all seed covered. Seed metering must be checked frequently as planting progresses to make sure that all units are seeding properly. Missed rows are difficult to replant and result in reduced yields.

It was mentioned earlier that the seeding operation is often combined with other operations such as fertilizer placement and preemergence spraying. Because these operations are so important to the overall success of the crop, many beet growers assign an extra man to this job to make certain that everything works as it should. Constant checking is the key to their success.

STAND REDUCTION

To achieve optimum beet growth and maximum sugar production per acre, a uniform spacing between plants within the row is desirable.

Two main procedures are used to arrive at the desired plant spacing. The least common of these methods is planting to a stand. (See Fig. 3.4.)

TABLE 3.2. Seeds per unit per minute

Forward Speed	Seed Spacing (inches)			
	2	3	4	6
2.5 mph	1320	880	660	440
3.0 mph	1584	1056	792	528
4.0 mph	2112	1408	1056	704
5.0 mph	2640	1760	1320	880

Fig. 3.4. Emerging sugarbeet seedlings in a field which has been planted to a stand.

This consists of planting only the number of seeds which are needed for the desired final acre population. While this procedure, in theory, solves the spacing problem, it has not met general acceptance by growers. Many hazards await the small, early-planted beet seed. When the unpredictable conditions of wind, freezing temperatures, seedling diseases, insect damage, and beating spring rains combine forces, the mortality rate of young seedlings is sometimes high. Due to these hazards, planting to a stand often results in an uneven and inadequate plant population which may require replanting. Interest in this approach, however, is increasing in some areas where emergence conditions are generally favorable.

The second and commoner practice of establishing an adequate number of beet plants per acre is to plant extra seed and to remove the extra plants after the beets are well established. A requirement for successful mechanical thinning is to have the area in the row smooth and free of clods. Either a roller or cultipacker is used to accomplish this.

Machine thinning of beets can be divided into two types: (1) random stand reduction blockers, and (2) electronic or electric-eye thinning.

Random stand reduction is successfully accomplished with a number of different machines. Some go across the row and cross-block or remove at random the plants in front of the knives or cutting mechanism. The number of plants removed is determined by the size or position of the cutting device or the angle of travel against the row. Another type of random thinner travels up and down the row. Here again the number of plants removed is determined by the size, shape, and number of cutting heads being used.

The advantage of random thinners lies in that once properly mounted and adjusted they are easy and inexpensive to operate and allow the grower to be independent. All of these machines, however, have some limitations. The main limitation is that the random removal of beets is based on chance or mathematical probability. As a result, in a normal sugarbeet field, skips or thin spots are made larger, and thick spots are left too thick. The machines can be adjusted to leave only an average number of beets in a given number of feet, with thick and thin spots making up this average. Therefore, this type of machine under normal conditions is not capable of leaving a uniform spacing between plants. Nevertheless, with this equipment the careful operator can accomplish a satisfactory job of stand reduction under a broad range of field conditions.

A new approach to machine thinning of beets has occurred with the development of electronic or electric-eye thinners. The disadvantage of random thinners is minimized with electronic thinners, since these machines can be pre-set to leave a desired spacing between beets. This minimizes the bunching of beets where plants are thick and the widening of gaps where they are thin.

While electronic thinners sound sophisticated and perhaps appear complicated, they are basically quite simple to operate and relatively trouble free. (See Fig. 3.5.)

The requirements for accurate thinning with electric-eye thinners are as follows:

1. Good weed control. As yet the machines available cannot distinguish between weeds and beets. Therefore, good weed control is essential for proper functioning of the machine.

Fig. 3.5. Electronic row crop thinners are used successsfully when combined with precision-planted monogerm seed and chemical weed control.

2. Proper plant spacing and plant size. There must be a minimum of 2 to 2½ inches between plants since this is approximately the size of the block left by the thinners. Plants must be tall enough to activate the sensing mechanism but not so large as to have overhanging leaves which will trigger the cutting knife.
3. A smooth working surface in the row. Clods or other foreign matter coming in contact with the eye or sensing mechanism will activate the knife the same as a beet plant. Uniform soil conditions also aid in maintaining proper cutting depth.

The supply of labor available to hand-thin sugarbeets is decreasing in quality and quantity and increasing in cost each year. Nevertheless, most sugarbeet growers in areas where labor supplies are available still use hand thinning in spite of its limitations. More precise seed spacing and effective weed control are enabling hand labor to cover more acreage per unit of time. However, the monetary advantages of these production improvements are in large part going to the laborer rather than the grower.

The future now appears bright for the complete spring mechanization of the sugarbeet crop. Through proper employment of herbicides, precision planters, monogerm seed, and mechanical and electronic thinners, the beet grower is now fully capable of growing profitable crops of sugarbeets without hand labor.

PLANT POPULATION

What is the optimum harvest stand of sugarbeets? is a difficult question to answer. From a strictly theoretical standpoint, the square pattern obtained from a 12-inch spacing in both directions should be ideal. This arrangement would offer each plant an equal opportunity to reach its maximum potential. In actual practice, however, this plant spacing is impractical because of higher priority items such as cultivation, irrigation, and harvest operations. Climatic variability and its influence on the growing season is another very important consideration.

Because of the wide variety of factors which affect plant population, numerous field experiments have been conducted to determine the optimum plant population for the production of maximum yields. These experiments have employed variations in spacing between rows as well as spacings within the row. The results from most of these experiments have been similar even when conducted in areas having broad differences in climate. The data have generally shown that the highest production of both roots and gross sugar per acre is obtained from stands ranging from 21,000 to 36,000 plants per acre. High fertility conditions and long growing seasons tend to minimize the differences between plant populations. These observations indicate that sugarbeets have the ability to compensate for wide population variations in the row, providing that large gaps are not present. Since plant population is determined by two intervals—distance between rows and spacing within the row—these points will be discussed individually as well as in combination.

The most important variable in plant population is row width, since this factor can be selected by the grower in advance of land preparation. Generally, row spacings in North America vary between 20 and 32 inches on single rows. In the arid West where furrow irrigation is practiced, row width combinations are more restricted because of the need to maintain an adequate furrow for irrigation. In these areas, growers customarily plant on ridges or beds which vary in width from 20 to 44 inches. Beds spaced less than 32 inches apart always have a single row which is centered on top of the bed. However, some growers elect to list broader beds (40 to 44 inches wide) and plant two rows on top of the bed. This two-row planting pattern can vary from 12 to 16 inches between rows, with 14 inches as the common interval. The width of the irrigation furrow ranges from 26 inches across the furrow to a maximum of 32 inches. The double row thus is a compromise which leaves an adequate distance between beds for tractor tires and irrigation but produces the same number of rows per acre as the 20- to 22-inch uniform single-row system.

There are valid reasons in addition to personal preference in a grower's selection of row spacing. As indicated earlier, comparative row-spacing experiments have consistently favored closely spaced rows over wide rows. In commercial practice, the 20-inch single row or its 40-inch double row equivalent have been preferred by the majority of growers. Recent trends in chemical weed control and increased emphasis on mechanical cultural practices have favored wider single rows. This trend has been accentuated as modern tractors have increased in size and power. In today's economy, time, skilled labor, and other costs force growers to use large multirow equipment requiring large tractors and carriers with wider tires. Nevertheless, the relationship of close row spacing and higher gross returns encourages growers to keep row width to its practical minimum, with the result that several row spacings currently satisfy the requirements of most sugarbeet growers. These are single rows spaced 22, 24, or 30 inches apart and the 40-inch double-row bed with the two rows spaced 14 inches apart. Some of the special features which are associated with narrow 20- to 24-inch single- or double-row beds are as follows:

1. Higher per acre populations are possible and small skips or gaps in the stand can be tolerated.
2. Yields of roots, gross sugar, and extractable sugar per acre tend to be higher, especially in areas where the growing season is restricted.
3. The harvest of a greater percentage of smaller roots is invariably accompanied by a higher sucrose concentration.
4. Furrow irrigation can be more difficult, since the furrows are narrower and less clearly defined.
5. Cultural operations are generally slower and more expensive to accomplish. This is especially true of mechanical and chemical weed control as well as harvest operations.

6. More rows per acre increase the costs of thinning and weeding labor.
7. The presence of high salinity conditions or other factors which may affect germination and emergence of seedlings often favors the double-row bed or narrow single rows.

The final decision regarding a choice of row spacing may be dictated by factors other than listed above. Since most sugarbeet growers grow additional crops, the decision of row spacing may be determined by the row width of other crops grown on the same farm. In areas where fall bedding is practiced, having several crops utilizing the same bed width provides flexibility in the cropping program. This is especially useful as a hedge against weather or contracting restrictions.

The relationship which row spacing and distance between beets has on acre populations of sugarbeets is apparent from Table 3.3.

This table shows how rapidly the spacing within the row is affected by the width of row. Since the maximum yield potential is desired by all growers, the above relationship should be carefully considered. Growers should also be aware that in the field strict mathematical relationships are not achieved. That is, compensation within the row is limited, and therefore, too heavy populations within the row can restrict root growth to the point that heavy harvest losses can result from the small beets passing through the harvesting and factory-receiving facilities. Growers should, therefore, watch their stand reduction or thinning operations carefully in order to achieve the maximum yield potential from a field.

Generally, two stand determinations, percent of row unoccupied (skips of more than 24 inches), and 4-inch spaces containing 3 or more plants quite consistently show a close relationship to yield (1). Total beets per unit of row length (usually 100 feet) is, therefore, an unreliable measure of an adequate stand. Hence, it is difficult to establish reliable ground rules for the per acre populations which are essential before maximum yields can be achieved. Researchers are in general agreement, however, that growers

TABLE 3.3. Effect of spacing interval (rows and within row) on the number of plants per acre

Beet Population per Acre*	Row Width (inches)				
	20	22	24	28	30
	(beets per 100 feet of row)				
18,000	70	77	84	98	104
21,000	80	88	96	112	121
26,000	100	109	119	139	149
31,500	120	133	145	169	181
42,000	160	177	193	225	241
52,000	200	219	239	279	298
Percent more plants required than 20-inch rows	...	10	20	40	50

* To nearest 500 beets.

can leave from 6- to 12-inch intervals between individual beets before the yield potential is seriously reduced.

One of the most comprehensive plant population investigations has been conducted at the Mesa, Arizona, experiment station by University of Arizona personnel. This experiment effectively summarizes the relationship between plant population and yield factors for three-row widths and three-plant spacings (2). (See Table 3.4.)

The first significant conclusion is that harvest date or length of growing season is a primary factor and that plant population variables do not effectively compensate for a short growing season. Within the restrictions of a short growing season, however, yields were reduced by row widths greater than 24 inches. It was found also that plant spacings of 5 inches or 15 inches reduced yields for all row widths. The close spacing resulted in a large percentage of unmarketable small beets, whereas the wide spacing resulted in inefficient use of the land. Yield losses resulting from the widest single row were not serious, provided that a plant population of about 21,000 plants per acre was achieved (10-inch spacings). Less yield reduction was noticed with the closer row spacings, but again yields were depressed when the population was less than 21,000 or more than 31,000 plants. Sucrose percentage was improved when the population was above 31,000 plants, however.

Similar relationships are shown for the late harvest date (long growing season for the area) but the percent increase obtained by row width or in-row spacing was reduced. The wide row, however, produced consistently lower yields than the narrow rows. For this harvest date, maximum yields were again reached with populations ranging from 21,000 to 31,000 plants per acre. The longer growing season enabled the wider-spaced beets to partially compensate for the relatively low plant population. The high populations achieved from close spacings (5 inches) resulted in many small unmarketable roots at harvest time. The sucrose percentage was again

TABLE 3.4. Effect of row width, plant spacing, and harvest date on yield and sucrose concentration of sugarbeets

Row Width	Plant Population		Harvest Date			
			April 20		July 13	
	Plant spacing	Plants per acre	Root yield	Sucrose	Root yield	Sucrose
	inches		*tons/acre*	*percent*	*tons/acre*	*percent*
40-inch double row beds	5	62,700	16.8	13.4	28.5	14.2
	10	31,400	18.9	13.2	31.8	14.6
	15	21,000	18.1	12.8	32.3	14.4
24-inch single row beds	5	52,000	16.9	13.4	30.6	14.8
	10	26,000	18.4	12.6	33.1	14.1
	15	17,000	17.6	12.2	27.9	14.2
30-inch single row beds	5	41,800	15.7	13.0	25.0	14.2
	10	21,000	17.8	12.6	27.6	13.6
	15	14,000	16.4	12.4	26.1	13.6

increased with higher plant populations per acre, but the differences were less significant than for the short growing season.

In conclusion, experiences of beet growers and agronomists throughout North America point toward an optimum plant population ranging between 21,000 and 31,000 plants per acre regardless of the row width. Row width is an important consideration which must usually be decided by factors other than yield.

The successful beet grower is advised to select the beet population per acre which will produce the maximum yield possible for his row spacing.

REFERENCES

1. Hills, F. J. and D. Ririe. An evaluation of mechanical thinning of sugarbeets in California. J. Am. Soc. Sugar Beet Technologists 9:337–44. 1958.
2. Nelson, J. M. Effect of row width, plant spacing, nitrogen rate, and time of harvest on yield and sucrose content of sugarbeets. J. Am. Soc. Sugar Beet Technologists 15:509–16. 1969.

4

Weed Control

E. F. SULLIVAN
Great Western Agricultural Experiment Station
Longmont, Colorado

B. B. FISCHER
University of California
Fresno

Weed Problems	71
Processing Costs	72
Concept of Zero Labor	72
Weed Control Practices	72
Plowing and Delayed Seedbed Tillage	72
Manual Hoeing and Weeding	73
Cultivation	73
Mowing and Flaming	74
Direct Biological Control of Weeds	74
Selective Herbicides	74
Herbicide Combinations	75
Weed Infestation	75
Weed Distribution	76
Weed Biology	80
Competitive Emergence	81
Herbicide Activity and Selectivity	81
Organic Chemical Groups	81
Formulation	84
Herbicide Placement	84
Plant Morphology and Physiology	85

PRINCIPLES OF HERBICIDE USE 86
    Herbicide Classification 86
    Formulation and Rate Expressions . 86
    Rate Calculations 87
    Applicator Calibrations 88

FACTORS AFFECTING SOIL-APPLIED
    HERBICIDES 90
    Tilth and Moisture 90
    Preemergence Application . . . 91
    Preplant Application 91
    Line or Side Injection 92
    Postplanting Application 92
    Herbicide Adsorption 92
    Herbicide Response to Soil Type . . 93

FACTORS AFFECTING FOLIAR-APPLIED
    HERBICIDES 93
    Age and Growth Rate 93
    Structural Features and Leaf
        Permeability 94
    Weather 95

HERBICIDE COMBINATIONS 95
    Split Applications 98
    Insecticide or Fertilizer Mixtures . . 99

TROUBLESHOOTING 100
    Misapplication 100
    Misjudgment 100
    Control Symptoms 101

HERBICIDE DISSIPATION 102
    Herbicide Residual 103
    Residual Detection 103

HERBICIDE REGISTRATION 104
    Toxicological Properties 104

EFFECTIVE AGRONOMIC PRACTICES . . . 105
    System Management 105

OUTLOOK 105
    Discovery Opportunities Ahead . . 106

IN NORTH AMERICA, with the exception of July and August, sugarbeets are planted every month of the year. Weed genera and species vary greatly from one geographical and management area to another and with season and planting time. Weeds in sugarbeet fields reduce average returns per acre more than other pests, with estimates of 10 to 15 percent of the crop annually lost to weed competition and control. Recurring infestations of annual weeds resulting from lack of effective residual control systems have prevented spring mechanization of the crop. Fortunately, the introduction of selective herbicides and other developments, including system management involving herbicidal-cultural control, offers the grower the opportunity to produce sugarbeets with minimal hand labor.

WEED PROBLEMS

Besides cost of control, weeds compete with sugarbeets to lower production, to serve as hosts for other pests, and to hinder crop harvesting. The degree of weed competition for growth factors and space varies with species, time, and management practices. They are strong competitors; therefore, their constant presence suppresses crop growth and yield.

Weeds compete for soil nutrients, water, light, carbon dioxide, and space for growth and survival. Often water and nutrient needs of weeds surpass those of crop plants. If weeds cause any one of these factors to become limiting, crop retardation results.

Competitive ability for growth differs among individual plants and between weed species and sugarbeets. It is greater among species than individuals of the same species (57). Brimhall et al. (10) showed that green foxtail was less competitive than rough pigweed, and Dawson (15) found that barnyardgrass was less competitive than lambsquarters on sugarbeets. Mixed weed populations are generally more competitive than simple infestations.

Generally, weeds that emerge early and establish quickly delay crop development during the initial 12 weeks of growth (15). Weeds that emerge later are controlled by crop competition in a uniform, vigorous row crop. Species that remain and provide dense shade, such as lambsquarters, kochia, and pigweed, reduce crop yield more than species with narrow leaves like the grasses, or those with prostrate growth like purslane and puncturevine. Apparently, some weed species, velvet leaf, kochia, and quackgrass secrete toxic substances during germination and root growth that inhibit crop seed germination and establishment. Competition from weeds is less detrimental when crop seed germination and stand establishment occur rapidly.

Weeds harbor diseases and insects that can infest sugarbeets. Weed distribution and species presence contribute to the spread of the sugarbeet virus complex. Many weed species serve as hosts for western yellows disease. Curly top is transmitted to sugarbeets by the beet leafhopper from host weeds Russian thistle and salt bush. Root knot and cyst-forming nematodes that infest sugarbeets overwinter on mustard, pigweed, purslane, dock, and several weed members of the potato family. Pest control is aided greatly by weed absence in and near sugarbeet fields. High weed populations may increase root rot and seedling damping-off because of close relationship of the crop to weed refuse harboring disease organisms.

PROCESSING COSTS

Soil and weeds hauled to the processing plant increase cleaning and labor costs before beet processing. In the field, weeds sometimes slow the harvest and lower the quality and quantity of sugarbeet ensilage and browse forage. Grasses with their extensive root system are more troublesome than taprooted, broadleaf weeds at harvest. Soil adhering to the fibrous root is elevated by the pinch wheels of the harvester and conveyed with the beet roots. Weed bunches and trash in beet piles become a focal point for decay.

CONCEPT OF ZERO LABOR

Smith (50) aptly observed in 1950 the advance design for zero labor or complete spring mechanization of sugarbeets. He stated:

Crops that are kept clean ahead of the beet crop or other row crops make easier and faster labor saving machinery. . . . The use of sprays in grain, corn and other crops is important. . . . The use of weed burners along ditches, fences and waste areas. . . . Prepare a good seed bed. . . . It is necessary to have a good stand and uniform pattern of beets to obtain maximum results with mechanical thinning. . . . The proper rotation and chemical weed control are certainly important to full realization of machines, particularly to spring mechanization.

This design, with little modification, remains objective today.

WEED CONTROL PRACTICES

Deliberate control of weeds in sugarbeets before and after thinning consists of using physical, cultural, and chemical techniques alone or in combination periodically during the growing season. These management treatments are complemented in practice by natural agents of weed destruction, both biological and environmental.

PLOWING AND DELAYED SEEDBED TILLAGE

Tillage practices allowing germination and eradication of weeds have proved effective, particularly under long growing seasons and in humid areas. Early seedbed preparation and refitting offer temporary weed control and reduce weed populations during the crop emergence period. However, repeated seedbed tillage may dissipate needed soil moisture. Often, conserving the soil moisture in the seedbed is more beneficial to crop stand than the cultivation. In nonirrigated areas delayed crop planting may miss early germinating precipitation and increase the chance for seedling disease damage during warmer periods of spring. The growing season is shortened, with oftentimes lowered yields.

MANUAL HOEING AND WEEDING

Hoeing, supplemented by hand weeding, has been the principal method of weed control since the seventeenth century. Close cultivation, coupled with hoe thinning and weeding the crop, has produced quite effective early weed control within and between the rows. This system of weed control relied on removal of plants by physical means, for example, severing the root system, pulling the whole plant from the soil, and smothering small weeds by a soil cover.

In dense weed stands and during wet weather, hand labor is often less effective than chemical weeding within the row. Observations suggest that in general more weeds are likely to escape hand labor than herbicides. Also, under the present socioeconomy, hand weeding is more costly and wasteful of energy and resource; thus, it will be used in the future only to remove some hard-to-kill weeds when few in number.

CULTIVATION

Machine cultivation, although effective and economical for removing weeds from between the rows, prunes crop roots if the tools are adjusted too close to the row. Following crop emergence, close cultivation can result in considerable savings by reducing the area the hand or selective thinner must cover. Cultivation to be effective must start with seedling weeds. Tools chosen will be influenced by soil type, field condition, size of the crop, and weed maturity. Tine weeders, rotary hoes, disc weeders, spider weeders, knife and sinner weeders, and the duck foot, bull tong, and sweep cultivator are among tools used to remove small weeds from seedbeds prior to and after thinning.

Sled planters and later cultivators mounted on the same sled enable close precision cultivation. Following beet and weed emergence, cross cultivation with a flex action harrow has been successful in many production areas. Sectioned rolling cultivators, effective when operated at 5 to 6 miles per hour, are versatile machines that can be operated whether beets are flat or bed planted.

Speciality harrows and cultivators are frequently used to supplement weed control obtained from herbicides. Subsequent tillage improves weed control more on some herbicide treatments than on others (58). The cultivation uncovers for subsequent germination weed seed from within the seedbed; therefore, in many instances, more weed growth results than is controlled from a single cultivation. Davis (14) concluded in 1941 that it was doubtful that between-the-row cultivation in sugarbeets had any direct effect on soil moisture conservation or for that matter soil conditioning and crop yield on most soils except removal of weed competition existing at the time.

MOWING AND FLAMING

Mowing is sometimes used in emergencies to eradicate weed growth above the standing crop, thereby reducing competition; to improve harvesting mechanics; or to save a field for thinning labor. Flaming of weeds prior to emergence of sugarbeets as reported by McBirney (38) in 1948 may be a suitable practice on weedy fields. Flame cultivation has damaged and retarded emerged beets, resulting in reduced yield and percent sugar. Additional flaming studies are needed on sugarbeets. Weed burners are successfully employed to control weeds along irrigation ditches, fences, and waste places near sugarbeet fields.

DIRECT BIOLOGICAL CONTROL OF WEEDS

Weed control from insects and pathogens appears remote at this time. Several annual weeds infesting sugarbeet fields are closely related to the crop, and the potential of cross-invasion by vectors exists. Weeds succumb to natural agents, including desiccation, freezing, and flooding, more rapidly after being disturbed by physical and chemical measures employed by man. Crop rotation and crop competition are effective means for weed control in sugarbeets. In a Colorado study, sugarbeets following beans had lower weed-seed populations than sugarbeets following barley or corn (17).

SELECTIVE HERBICIDES

Inorganic and organic chemicals were evaluated on sugarbeets beginning in the 1930s. Robbins (46) of California reported on the applicability of sulphuric acid, carbon bisulphide, acid arsenicals, and the chlorates. The first reasonably successful herbicide was sodium chloride applied either preemergence (before crop or weeds emerge) or postemergence (after weeds and/or crop emerge). Other salts and acids were screened and discarded during the early years, 1935 to 1945. The hydrocarbons, diesel fuel and other aromatic oils, had limited use as contact herbicides on sugarbeets. Perhaps, the effectiveness revealed by the first true herbicide, 2,4-D, was the event which stimulated the search for organic candidates on sugarbeets. Among the organics evaluated during the late forties and early fifties, IPC, TCA, dalapon, and endothall were selective on sugarbeets, both preemergence and postemergence. An advance in herbicide placement in sugarbeets occurred when the Coloradans Deming (16) and Nelson (41) reported promising weed control results from IPC applied preemergence and soil incorporated. This preplanting technique has since improved performance of many herbicides, especially in semiarid regions.

Rapid advances were made in screening selective herbicides on sugarbeets during the late fifties and the sixties. During that time, EPTC showed preplanting promise on certain soil types in the Midwest, and diallate, pebulate, pyrazon, Endothal 283 and 273, among others, were screened and released on sugarbeets. Because of commercial performance of these, augmented by TCA and dalapon, the years 1963 to 1969 are called the years of transition from dependence on physical control methods to chemical weeding. In this period, many compounds were evaluated, and among these, cycloate, benzadox, phenmedipham, nitralin, and CP-52223 had effectiveness.

HERBICIDE COMBINATIONS

Reeve (44) and Cormany (12) reported in 1950 of promising preemergence control from endothall + TCA, and later work proved the effectiveness of the postemergence mixture, endothall + dalapon. Wide-scale introduction after 1963 of the preplanting combination, pebulate + diallate, in irrigated regions and the preemergence mixture, pyrazon + TCA, in humid areas enhanced chemical weeding performance. Postemergence broad-spectrum control was advanced by pyrazon + dalapon concurrently evaluated in Ontario and Colorado (3, 55). Cycloate + diallate, pyrazon + CP-52223, phenmedipham + lenacil, and pyrazon + phenmedipham, and other mixtures appeared promising in many tests during the latter half of the sixties. Split-application of herbicides, preemergence followed by postemergence, including mechanical stand reduction, was proposed by Grigsby (27). These systems from an uncertain beginning had progressed to nearly complete weed control by 1969. Several investigations showed that cycloate + pyrazon + dalapon, cycloate + EPTC or trifluralin, pyrazon + TCA + phenmedipham, and cycloate + phenmedipham were promising for extensive commercial use.

WEED INFESTATION

Recent surveys indicate that approximately 130 weed species infest sugarbeet fields in North America. Depending on region and locality, only about 50 species are considered major pests, while 30 species appear frequently enough to be troublesome minor pests (Table 4.1). These undesirable plants are natives of America or they were introduced from Europe and Asia mainly as contaminants in crop seed or in ship refuse and ballast.

Weeds are classified as winter or summer annual, biennial, or perennial plants. All four types invade sugarbeet fields, although seed-propagated annual weeds are most numerous and widespread. Perennial weeds are difficult to eradicate because propagation occurs by seeds, creeping stems and roots, and bulbs, tubers, and rootstocks. Creeping perennials such as field bindweed are very aggressive. Other common perennials are Canada thistle, sow thistle, common milkweed, leafy spurge, curly dock, sour dock, groundcherry, nutgrass, quackgrass, and johnsongrass. Among annual weeds, barnyardgrass, the foxtails, kochia, lambsquarters, the mustards, the nightshades, the pigweeds, shepherdspurse, smartweed, sunflower, ragweed, wild buckwheat, and wild oats are most prevalent. Common mallow, velvet leaf, marshelder, prickly lettuce, puncturevine, Russian thistle, sandbur, ladysthumb, crabgrass, common burdock, marestail, and prostrate pigweed appear potentially troublesome, depending on region. Crop volunteers, corn, barley, sorghum, and sweet clover may be local problems. Weeds in the composite, grass, mustard, buckwheat, goosefoot, and potato families predominate in simple and complex infestations in sugarbeet fields (Table 4.1). The nutgrasses, mallows, sunflower, redmaids, and Russian thistle may escape control because of herbicide tolerance. Nevertheless, herbicides or combination treatments are availaible which control many annual weeds in sugarbeet fields.

WEED DISTRIBUTION

Weed dispersal is promoted by wind, water, animals, and man through commerce, agriculture, and settlement agents and activities. Once established near cultivated land, weeds encroach on plowed land at the edge and disperse within the field from waste places, manure, and flowing water. After field culture, those species that persist have germination, growth habits, and life cycles compatible yet competitive with crops and crop cultural practices. Soil tillage spreads the persistent species, while natural species selection and habitat adaptation may produce lines or agricultural subspecies which resist eradication more completely than the parent species. Some weed ecotypes have acquired natural herbicide tolerance as shown by wild oat lines resistant to diallate and barban (30). Initial species invasion; climate; soil reaction and drainage; soil fertility, particularly nitrogen nutrition; and weed species competition, including self inhibition and its effect on population density, are other ecological factors which affect species survival and infestation. Weed seed populations in fertile soils are generally higher than in soils of low fertility (17). The successful use of sanitary measures, seed laws, quarantines, soil management, and chemical-tillage

TABLE 4.1. Most common weeds infesting sugarbeet fields in United States from recent survey

Family Name	Binomial and Common Name	Origin and Life-span
Amaranthaceae (Pigweed)	†* *Amaranthus retroflexus* (rough pigweed)	IA
	* *A. graecizans* (prostrate pigweed)	NA
Polygonaceae (Buckwheat)	†* *Polygonum convolvulus* (wild buckwheat)	IA
	†* *P. pensylvanicum* (smartweed)	NA
	* *P. persicaria* (ladysthumb)	IA
	†* *Rumex acetosa* (sour dock)	IP
	* *R. crispus* (curly dock)	**IP**
Boraginaceae (Borage)	*Amsinckia intermedia* (fiddleneck)	NA
Zygophyllacea (Caltrop)	* *Tribulus terrestris* (puncturevine)	IA
Compositae (Composite)	†* *Ambrosia artemisiifolia* (common ragweed)	NA
	A. trifida (giant ragweed)	NA
	* *Arctium minus* (common burdock)	IB
	†* *Cirsium arvense* (Canada thistle)	IP
	* *Erigeron canadensis* (marestail)	NA
	* *Helianthus annuus* (sunflower)	NA
	Iva axillaris (mouse-ear poverty weed)	NP
	* *I. xanthifolia* (marshelder)	NA
	* *Lactuca scariola* (prickly lettuce)	IA
	Senecio vulgaris (groundsel)	IA
	†* *Sonchus oleraceus* (annual sowthistle)	IA
	* *S. arvensis* (perennial sowthistle)	IP
	* *Xanthium pensylvanicum* (common cocklebur)	NA
Chenopodiaceae (Goosefoot)	†* *Chenopodium album* (lambsquarters)	IA
	C. glaucum (oakleaf goosefoot)	IA
	C. leptophyllum (slimleaf lambsquarters)	IA
	†* *Kochia scoparia* (kochia)	IA
	* *Salsola kali* (Russian thistle)	IA

TABLE 4.1. *(cont.)*

Family Name	Binomial and Common Name	Origin and Life-span
Gramineae (Grass)	†* *Agropyron repens* (quackgrass)	IP
	†* *Avena fatua* (wild oat)	IA
	A. sativa (tame oat)	IA
	* *Cenchrus pauciflorus* (sandbur)	NA
	Cynodon dactylon (bermudagrass)	IP
	†* *Digitaria sanguinalis* (large crabgrass)	IA
	†* *Echinochloa crusgalli* (barnyardgrass)	IA
	Hordeum sp. (tame barley)	IA
	Lolium multiflorum (Italian rye grass)	IA
	* *L. perenne* (perennial rye grass)	IP
	Phalaris arundinacea (canarygrass)	IP
	* *Poa annua* (annual bluegrass)	IA
	†* *Setaria faberii* (giant foxtail)	IA
	†* *S. glauca* (yellow foxtail)	IA
	†* *S. viridis* (green foxtail)	IA
	* *Sorghum bicolor* (wild cane)	IA
	†* *S. halepense* (johnsongrass)	IP
	* *S. vulgare* (tame sorghum)	IA
	Triticum sp. (tame wheat)	IA
	Zea mays (maize)	IA
Malvaceae (Mallow)	†* *Abutilon theophrasti* (velvet leaf)	IA
	Hibiscus trionum (Venice mallow)	IA
	* *Malva neglecta* (common mallow)	IA
Asclepiadaceae (Milkweed)	*Asclepias speciosa* (showy milkweed)	NP
	* *A. syriaca* (common milkweed)	NP
Convolvulaceae (Morning glory)	†* *Convolvulus arvensis* (field bindweed)	NP
	C. sepium (hedge bindweed)	NP
Cruciferae (Mustard)	†* *Barbarea vulgaris* (yellow rocket)	IP

TABLE 4.1. *(cont.)*

Family Name	Binomial and Common Name	Origin and Life-span
Cruciferae *(cont.)*	*Brassica campestris* (wild turnip)	IA
	B. juncea (Indian mustard)	IA
	†* *B. kaber* (wild mustard)	IA
	B. nigra (black mustard)	IA
	†* *Capsella bursa-pastoris* (shepherdspurse)	IA
	Descurainia sophia (tansymustard)	NA
	* *Thlaspi arvense* (fanweed)	IA
Urticaceae (Nettle)	*Urtica urens* (burning nettle)	IA
Caryophyllaceae (Pink)	*Cerastium arvense* (field checkweed)	NP
	Stellaria media (common chickweed)	IA
	* *Silene noctiflora* (night flowering catchfly)	IA
Solanaceae (Potato)	* *Datura stramonium* (jimson weed)	IA
	Physalis longifolia (long leaf groundcherry)	NP
	†* *P. heterophylla* (clammy groundcherry)	NP
	P. subglabrata (smooth groundcherry)	NP
	* *Solanum elaeagnifolium* (white horsenettle)	NP
	†* *S. nigrum* (black nightshade)	IA
	* *S. rostratum* (buffalobur)	NA
	S. triflorum (cutleaf nightshade)	NA
Leguminosae (Pea)	*Medicago hispida* (California burclover)	IA
	M. lupulina (black medic)	IA
	M. sativa (common alfalfa)	IP
	Melilotus officinalis (yellow sweet clover)	IB
Portulacaceae (Purslane)	* *Portulaca oleracea* (purslane)	IA
	Calandrinia menziesii (redmaids)	IA
Cyperaceae (Sedge)	†* *Cyperus esculentus* (northern nutgrass)	NP
	C. rotundus (nutgrass)	IP

TABLE 4.1. *(cont.)*

Family Name	Binomial and Common Name	Origin and Life-span
Euphorbiaceae (Spurge)	* *Euphorbia esula* (leafy spurge)	IP
	E. serrata (toothed spurge)	NA
Verbenaceae (Vervain)	*Verbena hastata* (blue vervain)	NP

NOTE: I = introduced, N = native; A = annual, B = biennial; and P = perennial.
* Fifty most troublesome weeds.
† Twenty-five most unwanted weeds.

weeding in all crops eventually results in weed seed depletion and vegetative shifts within a field to clean culture.

WEED BIOLOGY

Weed seed numbers are usually more than several thousand per cubic foot of soil. These large populations result from short maturity cycles, repeated flowering, and heavy seeding tendencies of annual weeds.

In general, several factors affect weed seed germination, and among these, seed maturity, dormancy, depth of burial in soil, temperature, moisture, aeration, and light exposure are important (1). Usually after ripening, mature weed seed is viable and germinates readily under favorable conditions, although some species seeds become viable before maturity. Seed dormancy in temperate regions is naturally broken by cold, moist weather, but adverse germination conditions may induce seed dormancy, a natural survival mechanism analogous to hibernation. Plowing, deep tillage practices, and burrowing animals and earthworms bury weed seed deep within the soil profile. Many weed seeds remain dormant yet viable when buried for long periods and herbicides do not generally affect dormant weed seed, although seed coats are permeable to some herbicides. Deep burial promotes seed dormancy and inhibits germination by restricting oxygen, especially in compacted soil, and exposure to red and far red light waves (18).

Many weed seeds require light for germination; for example, lambsquarters, wild mustard, kochia, and purslane are light sensitive. Alternating temperatures within favorable levels and adequate moisture (particularly wetting and drying) are other main requirements for germination of most annual weeds (9). Chepil (11) showed that weeds have inherent periods of high germination unrelated to daily weather variations although related to seasonal conditions of spring, summer, and fall. It is well known that lambsquarters, kochia, and wild oats germinate early in the spring, while pigweed and nightshade germinate later, and cocklebur even later. Smartweed and nutgrass germinate and thrive after soil disturbance in wet places.

Chemical control measures must accompany seedbed tillage because soil disturbance places weed seed in a favorable position near the soil surface for germination. In a sugarbeet rotation study, chemical weeding significantly reduced weed seed density when compared to mechanical cultivation (17). A weed emergence test conducted with field soil before planting can reveal expected species infestation, thereby, herbicide selection beforehand. Reference to King (33) will provide a more comprehensive review of weed biology.

COMPETITIVE EMERGENCE

Three competitive weed conditions are recognized in sugarbeet fields, namely, weeds emerging before crop emergence, with or shortly after the crop but before thinning, and after thinning. Selective herbicides and cultural practices currently available provide relatively reliable crop protection against any of these three conditions. Early weed growth should be prevented because plants that occupy an area first are in a better competitive position than those that become established later. Suppression of germination, growth, and photosynthesis, resulting from soil-acting residual herbicides, is the basis for chemical weed control. Further suppression of tolerant and chemically weakened plants by a foliar-acting or postplanting herbicide in the presence of crop shading prevents regeneration and the establishment of new weeds.

HERBICIDE ACTIVITY AND SELECTIVITY

Herbicides may damage sugarbeet seedlings as well as the weed they are designed to control, since selectivity responses between sugarbeets and weeds are quite narrow. Knowledge about how herbicides act and how to use them is necessary to minimize poor herbicide activity or damage to the sugarbeet plant.

To be selective, a herbicide must completely or partially inhibit weed growth, while at the same time do little or no damage to the crop stand. In theory (7, 13, 24), differential susceptibility responses are complex, involving chemical structure, formulation, placement and timing, and plant morphology and physiology. In addition, herbicide dosage, soil composition, and weather variables are contributing factors.

ORGANIC CHEMICAL GROUPS

Organic chemicals may produce specific responses on plants. Biological screening of chemicals indicates their activity on the tolerant and intolerant species. Slight changes in molecular configuration of a chemical often produce marked changes in species responses. Progress in chemical weeding on sugarbeets is enhanced by the discovery of analogs and new chemical groups. The parent EPTC and the analogs, pebulate and cycloate, are examples of this on sugarbeets.

Organic chemicals vary in structural complexity, molecular size and weight, and analog configuration. A brief description of some representative herbicides selective on sugarbeets may help explain the basis for herbicide selection (Table 4.2).

Chlorinated aliphatic acids exemplified by trichloroacetic acid (TCA) have a relatively simple structural formula. TCA is made by chlorinating acetic acid as shown below.

$$\underset{\text{Acetic acid}}{H - \underset{\underset{H}{|}}{\overset{\overset{H}{|}}{C}} - \overset{\overset{O}{\|}}{C} - OH} + 3Cl \rightarrow \underset{\text{TCA}}{Cl - \underset{\underset{Cl}{|}}{\overset{\overset{Cl}{|}}{C}} - \overset{\overset{O}{\|}}{C} - OH} + 3H$$

Sodium TCA is soluble in water and is particularly effective for the control of grass seedlings when applied as a preemergence spray. On sugarbeets, TCA may be applied preemergence after a preplanting applica-

TABLE 4.2. Herbicides and suggested dosages useful on sugarbeets grown for roots

Herbicide Group	Herbicide Member		Application Method*	Dosage Range,lb/A
	Common name	Trade name		
Aliphatic acid	benzadox	Topcide	po	1.5–2
	dalapon	Dowpon	po	3–6
	TCA	TCA	pre	6–8
Amide	propachlor	Ramrod	pre,ppt	3–4
Aniline	trifluralin	Treflan	popt	0.5–0.75
	nitralin	Planavin	popt	0.5–0.75
Bipyridylium	paraquat	Paraquat	ccul	0.25–1
Carbamate	barban	Carbyne	po	0.75–1
	IPC	IPC	ppt,po	3–6
	phenmedipham	Betanal	po	1–1.5
Dicarboxylic acid	endothall (dipotassium)	Endothal–273	po	0.75–1.5
			pre	3–6
	endothall (amine)	Endothal–283	pre,ppt	3–4.5
Nitrophenylether	nitrofen	TOK–25	pre	2–4
Pyridazinone	pyrazon	Pyramin	pre,ppt,po	3.2–4
Thiocarbamate	diallate	Avadex	ppt	1.5–2
	EPTC	Eptam	ppt,popt	2–3
	cycloate	Ro–Neet	ppt	2.5–4
	pebulate	Tillam	ppt	3–5

*Symbols refer to: chemical cultivating (ccul); preplanting (ppt); preemeregence (pre); postemergence (po); and post planting (popt).

tion of EPTC, or TCA may be applied in a preemergence combination with pyrazon and endothall for broad-spectrum weed control in humid areas. Selective TCA dosages on sugarbeets range between 3 and 8 lb/A active ingredient alone or in combination. Dalapon is a close relative of TCA.

Thiocarbamates are molecularly more complex than the aliphatic acids. Typical thiocarbamates are insoluble in water but soluble in organic solvents such as kerosene and alcohol. EPTC may burn sugarbeet leaves when applied postplant over the row because of certain petroleum solvents. The structural formula for diallate is somewhat simpler than that for cycloate. Notice the similar presence of sulfur, carbon, and nitrogen in the molecular core.

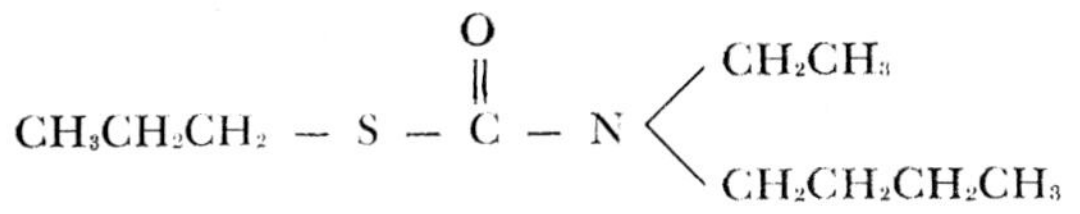

$$CH_3CH_2CH_2-S-\overset{\overset{O}{\|}}{C}-N\begin{pmatrix}CH_2CH_3\\CH_2CH_2CH_2CH_3\end{pmatrix}$$

diallate

$$C_2H_5-S-\overset{\overset{O}{\|}}{C}-N\begin{pmatrix}C_2H_5\\\text{(cyclohexyl)}\end{pmatrix}$$

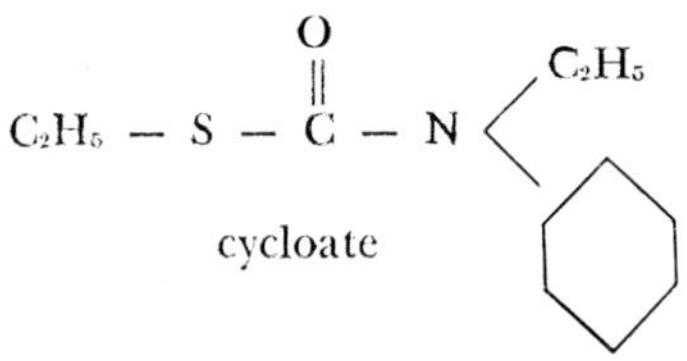

cycloate

Diallate is generally used for spring preplanting wild oat control at dosage ranges on sugarbeets from 1.5 to 2 lb/A active. Cycloate may be used alone, or in combination with diallate, preplanting at 2.25 to 4 lb/A, depending on soil type. Selective dosages of the thiocarbamate herbicides range from 1.5 to 5 lb/A on sugarbeets (4).

Pyrazon is a nitrogen heterocyclic compound derived from the pyridazinones. Over 500 substituted pyridazinones were evaluated before pyrazon was selected for use on sugarbeets. Pyrazon has low water solubility (28), but it is soluble in methanol.

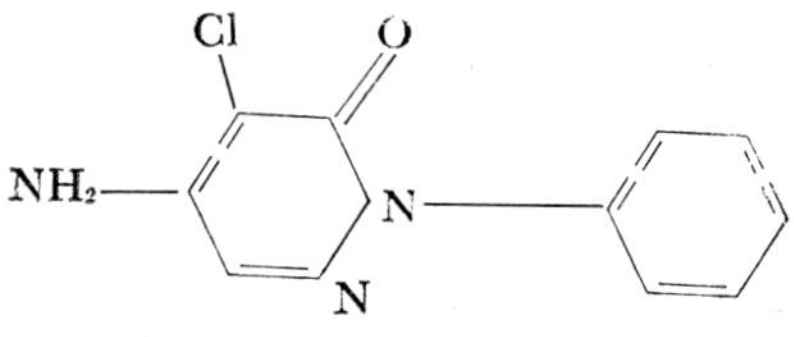

pyrazon

Pyrazon resists leaching and microbial breakdown, but requires adequate moisture for performance. Certain analogs of pyrazon containing bromine in place of chlorine have decreased the water solubility of the basic molecule.

Specific reactions can be expected within and between chemical groups and their members. Commonest are chemicals within groups for which sugarbeets have tolerance in relation to weeds, whereas others are species specific. For instance, the amide, carbamate, and aliphatic acid groups contain member herbicides that are differentially responsive on grasses. Among the thiocarbamates, EPTC appears most toxic to barnyardgrass, whereas diallate is least toxic (31). Pyrazon is active on certain broadleaf weeds, whereas the thiocarbamate and aniline groups contain members which affect both weed types. Some herbicides are quite species specific, and among these on sugarbeets, endothall controls smartweed and benzadox controls kochia (Table 4.2). Groups like the phenoxyaliphatic acids (2,4-D) and the triazines (atrazine), among others, are nonselective when applied even in very small dosages on sugarbeets.

FORMULATION

Physical-chemical characters of a herbicide decide formulation and influence selectivity, longevity, and utility of a herbicide. Some chemicals are more readily and economically formulated in solutions or emulsifiable concentrates than in solid form as granular, wettable powder, or flowable materials. Gaseous herbicides exist, although none is used directly for weed control on sugarbeets.

Machinery availability and management affects application and marketable formulations on sugarbeets. Frequently, granules or wettable powders extend residual life of a herbicide, although longevity is usually a function of chemical structure and dosage rather than of formulation. Grower preference usually decides what formulation he uses, although granular material is more costly per unit of active ingredient.

HERBICIDE PLACEMENT

Application methods can influence herbicide selectivity and activity. Preemergence (soil-surface applied before crop or weeds emerge), preplanting (soil incorporation before planting), and postplanting (after emergence of crop or weeds) are selective herbicide timings and placements. Postplant treatments are foliar applied (postemergence) or soil applied (layby at or after thinning). Postemergent nonselective herbicides such as paraquat may be made selective by protecting the crop with a hood during application.

PLANT MORPHOLOGY AND PHYSIOLOGY

Plant structure and physiology during development can affect selectivity by regulating herbicide entry into tissue and toxicant transfer to the site of action or kill. Growing point exposure of certain broadleaf weeds and growing point protection by the leaf sheath of grasses may regulate selectivity between plant types. In soil, depth of weed seed germination, weed seed dormancy and germination periodicity, and the rapidity and character of root growth in relation to herbicide placement are additional causes of variable toxicity and selectivity.

Water-soluble herbicides (polar) are absorbed more slowly through cell walls and membranes than fat or organic soluble herbicides (nonpolar) because of the nonpolar and electronegative nature of leaf tissue (40). As water is absorbed by roots, polar-oriented (water-diluted and water-soluble) herbicides are transported to sites of absorption. The soil-acting herbicides, diallate and pyrazon, gain entry into plant systems mainly via the emerging coleoptile in grasses and through the roots in broadleaves. Generally, most herbicides, including certain thiocarbamates, enter roots, although selective uptake of some nonpolar chemicals occurs. The structural order of ease of herbicide penetration is roots, stems, and leaves. Aging of leaves and other tissue sites slows penetration of herbicides regardless of type. Uptake mechanisms are further conditioned by environment, including management. For example, sugarbeet genetic composition may inhibit crop damage as shown by the work of Nelson et al. (42) with endothall.

Once inside the plant, systemic herbicides are usually transported at varying rates to the site of action through plant conductive tissues. Some intracellular transport may occur. Soil-acting herbicides which enter roots are translocated in the water-transpiration system (xylem) to the site of action, whereas foliar-acting herbicides may move through cells to the food conductive tissue (phloem) of leaves and then to the site of action (40). Toxicant leakage between the two vascular systems can occur, but final kill takes place at growing or food storage points within the roots or shoots. Some chemicals such as TCA translocate readily; conversely, IPC is not translocated.

Method of kill varies among herbicides, and it can be complex with some. Generally, growth is altered or cell division affected by poisoning vital metabolic or enzymatic processes, or coagulation of cell contents, including protein breakdown. Natural plant resistance to lethal responses is partial or complete, and tolerance is complex, involving absorption and translocation rates and degradation or other metabolic alteration of the toxicant (40). For example, diallate toxicity is more marked during cell division than during cell elongation (28). The experimental phenmedipham strongly inhibits photosynthesis or the Hill reaction in many species; sugarbeets and pigweed remain unaffected because the herbicide is inactivated by metabolism (5). Pyrazon combines with glucose to form the conjugate,

N-glucosylpyrazon, within beet leaf cells, which imparts partial selectivity (45). The close relative lambsquarters becomes susceptible because of pyrazon accumulation in roots, slow translocation, and slow breakdown in leaf tissue (22). Dalapon is not degraded within plants (2).

PRINCIPLES OF HEBICIDE USE

Effective chemical weeding results relate to the expertise of the technician and his knowledge of the weed-control fundamentals.

HERBICIDE CLASSIFICATION

Although classification is relative, herbicides are classified as selective or nonselective. They are classified further as contact or systemic, and residual or nonresidual.

Selective herbicides control susceptible weeds without serious injury to tolerant crops. Nonselective herbicides kill all susceptible vegetation. Soil sterilants and nonselective foliar sprays are used mainly on waste, utility, and recreational areas. Directed application of nonselective herbicides are used on certain economic crops. Massive dosage or misapplication may cause a selective herbicide to become nonselective and residual.

Contact herbicides kill plant tissue on contact by burning or caustic action and are usually foliar acting. Endothal 273, benzadox, and phenmedipham are quick-killing selective herbicides in this class, although partial systemic action may occur also.

Systemic herbicides penetrate cell tissue and translocate within the plant to the killing point. Site of action may be remote from site of application or plant entry. Systemic herbicides are also named translocated or hormone-type herbicides. Many herbicides used on sugarbeets are systemic. Pyrazon, cycloate, and the candidate CP-52223 are examples.

Residual herbicides remain soil active and kill weeds as they germinate and emerge. Some nonselective residual herbicides are active for several months. Residual properties of a herbicide may be beneficial to one crop yet harmful to another, particularly if the herbicide persists to the next planting season. Pyrazon has significantly more soil residual on sensitive species than cycloate which persists in killing activity for about six weeks. Nonresidual herbicides are best exemplified by paraquat and certain carbamate herbicides which persist in soil for less than three weeks. Half-life of a herbicide, the time required to dissipate 50 percent of its initial effectiveness, is a measure of persistence.

FORMULATION AND RATE EXPRESSIONS

Herbicides are usually formulated as emulsifiable concentrates (E or EC), wettable powders (W or WP), and granulars (G). The concentration of soluble concentrates may be given in percent of the parent acid, termed acid equivalent (a.e.). Concentrations of other formulations are given in percent of active ingredient (a.i.) by weight or pounds active ingredient per gallon.

Application rates are given on a material basis as total-coverage rate or band-coverage rate. Band-coverge rate is the total amount of material used per acre when applied in a restricted area over the row (Fig. 4.1). The band-coverage rate may be found by multiplying the total coverage rate by the fraction of the row width covered by the band. Dosage refers to concentration of active ingredient in pounds per acre. Injector applications are figured in the same manner as band applications. Surfactants are usually applied volume of delivery to volume of surfactant (V/V) or volume of delivery to volume of surfactant per acre (V/A).

RATE CALCULATIONS

Examples for computing rates of material to apply total coverage and band coverage are given below.

Apply 6E or EC formulation of cycloate at a 4 lb/A dosage, in a 7-inch band on 22-inch row spacing.

FIG. 4.1. Benzadox control of kochia *(Kochia scoparia)* when applied postemergence in a band at 2 lb/A.

Step 1. $\dfrac{128 \text{ fl oz/gal}}{6 \text{ lb/gal active}} = 21.3$ fl oz/lb active.

Step 2. 21.3 fl oz × 4 lb/A dosage = 85.3 fl oz cycloate 6E per acre, total coverage.

Step 3. 85.3 fl oz × 7/22 = 27.3 fl oz per band-acre.

Apply 80W or WP formulation of Pyramin at a 4 lb/A dosage, in a 7-inch band on 22-inch row spacing.

Step 1. $\dfrac{4 \text{ lb/A dosage} \times 100}{80\% \text{ active/lb}} = 5.0 \text{ lb} \times 16$ oz/lb
= 80 oz Pyramin 80W per acre, total coverage.

Step 2. 80 oz × 7/22 = 25.6 oz per band-acre.

Apply 10G formulation of Avadex at a 2 lb/A dosage, in a 7-inch band on 22-inch row spacing.

Step 1. $\dfrac{2 \text{ lb/A dosage} \times 100}{10\% \text{ active/lb}} = 20$ lb Avadex 10G per acre, total coverage.

Step 2. 20 lb × 7/22 = 6.4 lb per band-acre.

Apply 0.75% V/V of surfactant in 20 gal per acre of spray solution.

Step 1. 128 fl oz/gal × 20 gal per acre (gpa) delivery = 2,560 fl oz/A.

Step 2. 2,560 fl oz × 0.75% V/V concentration = 19.2 fl oz/A surfactant per 20 gpa delivery.

Apply 1 gal per acre (V/A) of a nonphytotoxic oil in a 7-inch band on 22-inch row spacing.

Step 1. 1 gal oil or 128 fl oz per acre, total coverage.

Step 2. 128 fl oz × 7/22 = 41 fl oz per band-acre.

APPLICATOR CALIBRATIONS

Uniform application of herbicides in the field requires a thorough check and calibration of equipment beforehand. Several terms are used in methods for determining rates of application: fom (fluid ounces per minute); gpa (gallons per acre); mph (miles per hour); and psi (pounds pressure per square inch). Gallons per acre delivery (gpa) of a sprayer is a function of ground speed, nozzle size, nozzle spacing, and nozzle pressure. Once in the field, changes in speed may conveniently regulate spray dosage for within-field variations in soil type. Less dosage may be obtained by acceleration and more dosage by deceleration with other factors remaining constant.

Several useful calibration methods follow:

Method I—Stationary nozzle delivery formula for boom or band spraying.
　Step 1. Fill tank 2/3 full (and spray system) with clean water.
　Step 2. Adjust line pressure to 30 to 40 psi.
　Step 3. Test each nozzle tip for 2-minute delivery and discard tips with 5% ± error.
　Step 4. Then:

$$\text{gpa} = \frac{\text{fom/nozzle} \times 46.4}{\text{mph} \times \text{nozzle spacing in inches}}.$$

Method II—Field delivery formula for boom spraying.
　Steps 1–3. Same as Method I.
　Step 4. Spray water over measured distance of 660 ft in a fitted field at normal planting speed.
　Step 5. Measure and record the amount of water needed to refill tank.
　Step 6. Then:

$$\text{gpa} = \frac{\text{gal used in 660 ft} \times 66}{\text{spray pattern width in ft}}.$$

Method III—Field delivery formula for band spraying.
　Steps 1–5. Same as Method II.
　Step 6. Then:

$$\text{gpa} = \frac{\text{gal used in 660 ft} \times 66}{\text{band width in ft} \times \text{number of bands}}.$$

　Step 7. Repeat procedure, using herbicide solution, and check for accuracy.

Method IV—Field delivery formula for band application of granules.
　Step 1. Adjust spout setting on each hopper and fill hoppers with granules.
　Step 2. Collect granules from each spout in a suitable container while operating tractor over a measured course of 660 ft in a fitted field.
　Step 3. Weigh the material discharged from each spout separately and adjust each spout to deliver an equal quantity of granules.
　Step 4. Repeat, but weigh all granules discharged together.
　Step 5. Then:

$$\text{band-acre rate of material} = \frac{\text{pounds of granules discharged in 660 ft} \times 66}{\text{band width in ft} \times \text{number of bands}}.$$

The total quantity of herbicide required, and tank fill and refill quantities of herbicide to add to water are computed as follows:

Step 1. Acres per tank = tank capacity in gallons ÷ gpa.
Step 2. Quantity of herbicide to add to tank = total or band-coverage herbicide rate × tank acres or fraction thereof.
Step 3. Herbicide required = total or band-coverage rate × number of acres.

Check the sprayer and granular applicator frequently for uniform coverage and rate accuracy over a given acreage. Thorough knowledge on parts, operating procedure, and care and cleanliness of herbicide applicators prevent time loss in the field. In addition, rechecking dosage calculations may prevent crop injury.

FACTORS AFFECTING SOIL-APPLIED HERBICIDES

TILTH AND MOISTURE

Well-tilled soils promote even distribution of water and herbicide in close proximity to germinating weed seeds. Clods prevent uniform distribution of herbicides, and when the clods "melt down," the protected weed seeds can germinate without herbicide contact. A firm seedbed that produces rapid germination on a given soil type is favorable for herbicide action, regardless of flat planting, bed planting, or humid or irrigated conditions.

Seldom are environmental conditions so adverse that acceptable herbicide performance cannot be obtained initially or from subsequent treatment, including favorable changes in the natural environment. In humid areas, spring moisture generally is favorable for soil-acting herbicides, while in semiarid regions, timely irrigation may be necessary to offset dry weather. Reception of sufficient moisture to completely wet the soil profile within a week of preplanting application promotes herbicide action. If the soil moisture level permits unimpeded herbicide incorporation, the moisture level may be too low for good seed germination and herbicide movement. Soil wetting completely across the rows or beds prevents dry places and loss of herbicide activity. Repeat irrigations are often required to allow chemical weeding processes to proceed to completion without interruption from surface soil drying.

Many sugarbeet herbicides move laterally and upward by capillary action under furrow irrigation; however, in most instances, this movement is not adverse. Lateral diffusion of thiocarbamates in soil is less than leaching downward (31), and incorporation depth seems to have little effect on their lateral or leaching character.

Magnitude of movement is related to water solubility of the herbicide, vapor pressure, soil organic matter content, and initial direction of the wetting front before soil saturation. After soil wetting, herbicide movement in solution is from high to low concentration (51). Wide bands or total coverage applications modify chemical migration; this expanded coverage may lower the dosage required, extend residual somewhat, and allow a cultivation delay, depending on soil type.

Soil wetting in excess of needs may cause highly soluble preemergence herbicides to penetrate soil too deeply. On light soil types, this leaching effect may cause substandard weed control, crop injury, or both. Wet soils impede incorporation and allow the escape of volatile chemicals from soil by codistillation with water vapor. For instance, Gray (26) showed that critical moisture percentage before loss for incorporated EPTC was 10 percent on light clay loam and 25 percent on heavy clay loam. EPTC effectiveness on giant foxtail increased when silt loam soil moisture was increased from 25 to about 37 percent, whereas response to trifluralin decreased with increasing moisture, and propachlor toxicity remained relatively unaffected within this narrow moisture range (52).

Herbicides applied to the soil are not weatherproof. Precipitation distribution and temperature, soil composition, penetration depth, solution dilution, leaching, and persistence rates of herbicides regulate performance. Usually about one-half to three-quarters of an inch of water is sufficient to activate preemergence herbicides like pyrazon and TCA.

PREEMERGENCE APPLICATION

Herbicide placement at the soil surface after planting is an accepted practice in humid areas. Precipitation or sprinkler irrigation must follow soon after application to move the herbicide into the soil. Effectiveness and selectivity are related to the degree of herbicide dilution in solution at the soil surface and uniformity and depth of penetration within the germination zone.

PREPLANT APPLICATION

In arid regions where furrow irrigation is practiced and the precipitation pattern is erratic, most herbicides, including pyrazon, require soil incorporation to provide weed control at crop emergence. Surface tillage places the herbicide in the weed germination zone for solution distribution by irrigation water, if necessary.

Fall plowing and bedding are favorable practices preceding preplanting incorporation. Often herbicide application, incorporation, and planting can be accomplished in one operation. Immediate incorporation of volatile herbicides such as cycloate, pebulate, and EPTC avoids vapor loss from the soil surface. Vapor loss for EPTC is greater than for cycloate. Power-driven rotary tillers with hoods are the most effective tools for incorporation. Rolling cultivators and disc harrows have been used successfully. Light-textured soils respond more favorably to nonpower incorporation. Cycloate appears more reactive than pebulate or diallate to disc incorporation.

Herbicides applied preplant respond favorably when incorporated in dry soil to depths between 1 and 3 inches. Pyrazon performs best at a shallow depth, while the thiocarbamates react effectively to depths between 1½ and 3 inches. Herbicides are unevenly distributed vertically under incorporation with more at the top of the incorporation zone than below. Subsurface layering of herbicide increases weed kill when compared to incorporation by a power-driven tiller or harrow. Nevertheless, sugarbeet seedling stands may be severely thinned by pyrazon and thiocarbamate herbicides placed in a thin layer beneath the soil surface (36).

LINE OR SIDE INJECTION

This technique of modified layering is based on vapor distribution in water of herbicides like cycloate in the weed-germination zone. Volatile herbicides are placed in narrow bands or lines 1 to 2½ inches deep and up to 2¼ inches each side of the seed row. Horizontal herbicide movement distributes the herbicide in a 5- to 7-inch band across the planted row. Recent evidence indicates that certain carbamates form a thin layer in the upper soil profile (29). Perhaps the absence of a continuous thin layer in cloddy soils and presence in fine tilth soils may partially account for differences in herbicidal performance between rough and fine seedbeds.

POSTPLANTING APPLICATION

Incorporation of EPTC, trifluralin, and nitralin directed postemergence to beets and preemergence to weeds provides seasonal control of susceptible weeds. These herbicides will not control established weeds; therefore, application must be made after early weed removal. Sectioned rolling cultivators and power tillers are effective tools for incorporating postplant herbicides. Two incorporations, made in opposite directions, are more effective than a single operation. Cultivators should be adjusted to throw some treated soil into the crop row. Persistence of trifluralin applied after thinning increases with incorporation depth (47).

In particular, EPTC can be applied in irrigation water for control of summer annual weeds following thinning and cultivation. Uniform distribution of water and thorough wetting across the row are essential for weed control. Because of its short residual, repeated applications of EPTC at 4- to 5-week intervals may be necessary to maintain control.

HERBICIDE ADSORPTION

Soils vary in their chemical and physical makeup (8, 24). These factors in turn influence the adsorption characteristics for herbicides. Once adsorbed, a balance is established between the herbicide fixed and herbicide remaining free. Herbicide is released (desorbed) from soil particles into soil water to replace herbicide removed by absorbing plant tissue and other dissipation processes, such as vapor loss, chemical breakdown, and leaching.

Herbicides vary as to how strongly fixed they are on soil colloids. Adsorption forces and microbial degradation govern herbicide availability and persistence. Several ideas on adsorption mechanisms have been advanced. Generally, these involve competition between herbicide and water molecules for adsorption sites on soil particles. Pebulate and EPTC have affinity for clay surfaces and these herbicides compete more effectively for binding sites when soil is relatively dry. The degree of thiocarbamate adsorption varies among chemicals with pebulate being held tightest and EPTC the least (31).

HERBICIDE RESPONSE TO SOIL TYPE

Correct dosages are closely related to soil texture (clay, silt, and sand), particularly when percent organic matter is low. Application of soil-acting herbicides to coarse-textured soils containing less than 1 percent organic matter is risky, depending on conditions and past management experiences. Soils such as these respond more effectively to postplant applications, particularly postemergence to the crop and weeds. Soils containing 1.5 to 4 percent organic matter and at least 20 percent clay are generally favorable for preplant or preemergence placement. In soils having low organic matter content, soil texture divides preplanting herbicide rate range into about five dosages; namely, low (sandy loam), medium-low (loam), medium (silt loam), medium-high (clay loam), and high (clay soil).

FACTORS AFFECTING FOLIAR-APPLIED HERBICIDES

It is difficult to predict field performance of postemergence herbicides beforehand, because of the large number of interacting factors that influence activity. Species presence is obvious, but ensuing environmental conditions and growth rates are not. Plant structure, spray retention and penetration, weed maturity, weather, and herbicide formulation regulate effectiveness to a large degree.

AGE AND GROWTH RATE

Young, actively growing weeds are more sensitive and generally require lower dosages for control than older weeds and weeds growing under drouthy conditions. Aging of leaves and stem tissue slows and sometimes prevents control response. Certain herbicides act specifically on certain weeds, often with ease, even at somewhat advanced stages of growth (Fig. 4.2), whereas other species may escape unless timely foliar applications are made before the central axis elongates or axial buds appear (20). Susceptible broadleaf weeds treated at the cotyledonous to 2–4 true leaf stage are usually controlled.

Rapidly growing seedling sugarbeets also become more susceptible to certain postemergence herbicides. To avoid undue retardation, the seedlings should have the first true leaves extended about ½ to 1 inch before herbicide treatment. Apparently, the presence of true leaf tissue and active photosynthesis prevents buildup of toxicant within the crop, particularly of pyrazon and phenmedipham.

STRUCTURAL FEATURES AND LEAF PERMEABILITY

Leaf angle, arrangement and area; thickness and chemical properties of surface cells; wettability of leaf surface and amounts of pubescence and waxy cuticle; and the quantity and position of stomatal openings are operative in herbicide uptake by leaves (40). The size and distribution of guard cells can influence rate of toxicant uptake and translocation. For efficient penetration, a uniform herbicide film in close contact with the leaf

FIG. 4.2. Pyrazon control of black mustard (*Brassica nigra*) at a somewhat advanced maturity stage.

surface is desirable. This means complete coverage with spray, and recommendations to accomplish this vary from 5 to 40 gpa.

Penetration of foliar-applied herbicides is related to high humidity and water-film life at the leaf-herbicide interface. Rapid spray drying time on leaves impedes penetration. Surfactants (wetting agents) and nonphytotoxic oils are commonly used to promote leaf permeability by reducing surface tension between polar (water) and nonpolar (leaf) surfaces. Surfactants keep spray droplets moist for a longer period; hence, they increase herbicide penetration. Stomatal penetration occurs mainly with water-soluble herbicides, although this pathway is frequently limited because of stomate location and opening time, particularly in drouthy or semiarid climates.

WEATHER

Temperature and moisture influence activity and selectivity of foliar-acting herbicides. High dosage may behave somewhat marginally on sugarbeet seedlings at air temperatures above 85° F, especially in direct sunlight, although effects on weed species become irreversible more quickly than under cooler temperatures. Temperatures below 55° to 60° F retard herbicidal activity. Generally, high relative humidity and high air temperature increase sensitivity, especially to dalapon, pyrazon, and phenmedipham. Normally, pyrazon and dalapon are relatively slow acting when foliar applied, although the combination, pyrazon + dalapon, increases rate of activity, residual, and control spectrum.

HERBICIDE COMBINATIONS

Tank mixtures of herbicides are mainly used to widen the scope of species control and increase performance from a single application. Compatible herbicides are combined that have the same or dissimilar modes of action. Mixtures often interact favorably with sites of action within the plant and external environmental conditions; namely, variations in moisture and temperature. A herbicide of low water solubility may be mixed with one having high water solubility. Sometimes a toxic chemical is added at a very low dosage to activate the selective constituent of the mixture. Mixtures make possible a reduction in residual carry-over by reducing the rate of the more persistent herbicide. Adjuvants are added to postemergence herbicides to aid plant penetration and response. A slow-acting residual herbicide may be mixed with a fast-acting herbicide which kills weeds while they germinate. Perhaps the most effective herbicide combinations are those applied in split application, because time delay and placement separation are involved. Most simultaneously applied combinations on sugarbeets are 2-way mixtures, although 3-way mixtures are common in Europe (Table 4.3).

TABLE 4.3. Herbicide combinations and suggested dosages useful on sugarbeets grown for roots

Combination	Application	Dosage Range, lb/A
endothall (Na) + TCA	preemergence	3 + 6-4 + 7
pyrazon + TCA	preemergence	3 + 4-4 + 6
pyrazon + propachlor	preemergence	2 + 2-3 + 3
cycloate + diallate	preplanting	2.25 + 0.75-3.75 + 1.25
pebulate + diallate	preplanting	1.65 + 0.85-3.25 + 1.75
pyrazon + endothall (283)	preplanting	2.4 + 1.6-3.75 + 2.5
pyrazon + diallate	preplanting	2.5 + 1.5-4 + 2
pyrazon + cycloate	preplanting	3 + 2-4 + 3
cycloate + endothall (283)	preplanting	2.25 + 1.5-3 + 2
cycloate + EPTC	preplanting	2 + 1-3 + 1
dalapon + endothall (273)	postemergence	2 + 1-3 + 1
pyrazon + dalapon	postemergence	3 + 2-4 + 2.25
pyrazon + endothall (273)	postemergence	3 + 1-4 + 1.5
dalapon + phenmedipham	postemergence	2 + 1-1.5 + 1.5
* pyrazon + phenmedipham	postemergence	2.5 + 1-3 + 1.5
dalapon + barban	postemergence	2 + 0.5-2 + 0.75
pyrazon + dalapon + endothall (273)	postemergence	3 + 2 + 1
EPTC + TCA	split-placement	1 + 4-2 + 6
cycloate + pyrazon	split-placement	2 + 3-3 + 4
cycloate + diallate + benzadox	split-application	2.25 + 0.75 + 1.5-3 + 1 + 1.5
pyrazon + TCA + phenmedipham	split-application	2 + 4 + 1-4 + 6 + 1.25
cycloate + phenmedipham	split-application	2.5 + 1-3.5 + 1.25
cycloate + pyrazon + dalapon	split-application	2.5 + 3 + 2-4 + 4 + 2.25
pyrazon + dalapon + phenmedipham	split-application	3 + 2 + 1-4 + 2.25 + 1.5
pebulate + diallate + EPTC	split-application	1.65 + 0.85 + 2-3.25 + 1.75 + 3
cycloate + trifluralin	split-application	2.5 + 0.5-4 + 0.75

* Somewhat incompatible depending on conditions.

The most successful combinations of broad-spectrum herbicides are pyrazon + TCA, pebulate + diallate (54), and pyrazon + dalapon. Herbicides in these mixtures compliment each other among species, especially when broadleaf and grassy weeds coexist. Certain mixtures may produce additive responses, although field results suggest that selective additivity above 85 percentage points total weed control is a matter of condition rather than design.

Additivity occurs when total weed control from the mixture is equal to the control sum of the herbicides in the mixture (34). Additivity under susceptible infestations has been proved for pyrazon + dalapon + wetting agent (49), pyrazon + TCA, and in particular, pebulate + diallate. The formula for derivation of expected toxicity by Gowing (25) was used to evaluate the results from a field experiment comparing pebulate and diallate as follows (53):

x = observed toxicity for the most toxic unit pebulate at 3 lb/A.
y = observed toxicity for the least toxic unit diallate at 3 lb/A.
then,

(1) $$\frac{x + (100 - x)\, y}{100}$$

(2) $$\frac{56 + (100 - 56)\, 44}{100} = 75 \text{ percent expected toxicity.}$$

The observed toxicity was 81 percent at the combined dosage of 6 lb/A. Therefore, it was concluded that the mixture had at least additive properties which was proved further by predictable wide-scale field results. An additive combination has a definite ratio of active ingredients for optimal response. The cycloate + diallate ratio is 3:1, whereas the pyrazon + dalapon ratio is 1.75:1. Generally, an increase in dosage of the least active component reduces the complementary effect until antagonism results. A mixture is antagonistic when its weed control is less than the sum of the component herbicides.

Combination dosages or ratios are defined by comparing dosage-response curves of the single chemicals and the mixture. The principle of practical herbicide usage infers that equal increments of dosage do not produce equal increments of response because response is nonlinear in relation to dosage. Minimum disproportionality of the dosage-response curve occurs within the middle range of the curve near the 50 percent level of effectiveness. This middle position or lethal dose that controls 50 percent of the weed species (LD_{50}) is where statistically valid comparisons of herbicide worth can be made. Direct comparisons made at the LD_{50} position are more accurate than comparisons made higher or lower on the curve (19, 25). Herbicides are combined at LD_{50} dosages and the effectiveness of the combination determined. This is usually accomplished by comparing graphically the relative slopes and characteristics of the dosage-response lines. Parallel dosage-response lines suggest similar mode of action of the individual herbicides, hence substitution possibilities because of dependency.

Real additivity or apparent synergism occurs when less combined dosage produces selective control response greater than the sum of the separate herbicides. Herbicide synergism of this type is rare. Logarithmic screening of herbicides and probit or regression analysis by arithmetic-graphic means (19) are useful tools for determining effective dosage range, and the additivity, antagonism, and synergism of combinations.

SPLIT APPLICATIONS

Complete broad-spectrum control from a single application of herbicide is not common. The most suitable treatment may lack control of a specific weed; for example, kochia and wild mustard tolerance of cycloate. Split application of herbicides is often more effective and is accomplished by two methods: delayed sequence application and split placement. Schweizer and Weatherspoon (48) showed that a preplanting application of cycloate preconditioned remaining weeds to a later application of pyrazon + dalapon applied postemergence. Cycloate or EPTC applied preplanting, followed by pyrazon or TCA applied preemergence before crop emergence, are effective split-placement systems without time lapse.

Sequence-application schemes are designed primarily for regions where weeds emerge with or after the crop. Preemergence treatments may control up to 80 percent of the weeds in complex infestations. Tolerant and susceptible species escape, many in close proximity to seedling beets, but these escapes are weakened by chemical exposure. Some preemergence chemicals may reduce cuticular wax of weed seedlings, thereby increasing penetration of a foliar-acting herbicide applied subsequently. Single doses of certain herbicides may be reduced up to one-third of normal when applied in sequence. Frequently, weather interacts favorably with chemical preconditioning, causing conditions for maximum kill. Some current examples of near complete chemical weeding are cycloate + phenmedipham or benzadox (Fig. 4.3) and pyrazon + dalapon + phenmedipham.

Systems employing two separate herbicide applications have no adverse effect on harvest yield or crop quality. These systems are based on soil-acting + foliar-acting, and soil-acting + soil-acting, and foliar-acting + foliar-acting responses, respectively. Split application of postplant herbicides such as trifluralin extend weed control activity. Postplant soil-acting herbicides should be applied to weed-free fields since they do not control emerged weeds.

Any split-application system should provide a relatively weedless row for free exposure of a uniform crop stand to selective thinner operation. Maximum selective weed control is seldom obtained from a single herbicide. Schemes designed on the principle of full-season weed control with minimal hand labor appear practical for widespread use during this decade. Dawson

Fig. 4.3. Split-application control of weeds from cycloate applied preplanting, followed by benzadox placed postemergence in a band.

suggests the following schedules for weeds that emerge with or after crop emergence (15):

Scheme 1. Apply a soil-acting herbicide with at least 12 weeks of residual effectiveness.

Scheme 2. Apply a soil-acting herbicide with 8 weeks persistence, followed by a foliar-acting herbicide about 4 weeks later or a second soil-acting herbicide with at least 4 weeks persistence during or after thinning.

INSECTICIDE OR FERTILIZER MIXTURES

Crop protection chemicals mixed with herbicides may alter herbicide performance. Organophosphate insecticides and herbicides applied together can increase phytotoxicity under adverse growing conditions or on light-textured soils. Knowledge is limited on these reactions on sugarbeets. Krionderis (35) showed that sugarbeet seedlings were unaffected by cycloate + Thimet, whereas pyrazon + Thimet was injurious. Enzyme hydrolysis of propanil in rice is inhibited by organophosphates applied in mixture, and crop injury occurs (39). However, mixtures of Sevin and carbamate herbicides, and phenmedipham + Diazinon or parathion are compatible.

Physical compatibility of tank mixtures should be known beforehand. Excessive salt content of a fertilizer solution may cause the herbicide to settle out or curdle in the tank, particularly emulsifiable and wettable powder formulations. Spray adjuvants may aid in keeping fertilizer and herbicide mixtures in solution and avoid emulsion breakup. When tank mixing a wettable powder with a liquid herbicide, always add the solid in a slurry to the tank first while keeping the contents under constant agitation. Herbicides like dalapon decompose by hydrolysis on prolonged storage in water, and salty water and scaly tanks may coagulate or chemically reduce herbicidal activity. Stability data indicate that phenmedipham herbicide is rather short-lived in basic solution with a half-life of 135 days at pH 4, 5 hours at pH 7, and 10 minutes at pH 9. Spray solutions and mixtures should be mixed fresh before use without storage overnight in the tank to avoid possible decomposition and activity loss.

TROUBLESHOOTING

MISAPPLICATION

Failure to read and heed label instructions, improper herbicide selection, incompatible mixtures, inaccurate equipment calibration and adjustment, and faulty application timing cause dosage mistakes. Soil incorporation too deep or too shallow changes applied dosage. Incorporation in soils too moist, causing chemical throw-out by the planter discs, and covering the treated band with incoming untreated soil decrease weed control in the row. Volatile chemicals may escape, reducing chemical rate, particularly in wet soil, unless incorporation ensues immediately. Total coverage applications are best made perpendicular to planting direction to avoid chemical overlap and skips from affecting an entire row. In addition, incomplete agitation of wettable material or liquid material with specific gravity greater than water causes settling out and severe overdosage (Fig. 4.4). Incompatible mixtures of herbicides and fertilizers can reduce chemical weeding results or can cause severe crop injury. Worn or incorrect nozzle tips and dirty line and tip screens contribute to misapplication failure. Incorrect nozzle tip distance above the ground can change the chemical dosage in the band. Line pressure above 40 psi, irregular pressure, and application in wind above 15 mph without wind shields may result in uneven herbicide distribution and drift.

MISJUDGMENT

Faulty observations of soil type, organic matter content, expected species infestation, and weed size are primary causes for unsatisfactory results. Cloddy seedbeds, particularly under adverse moisture and temperature, are detrimental to herbicide activity. Excessive rain or sprinkler irrigation after preemergence application can cause crop injury by concentrating herbicide near crop seed. Once germination begins, dry soil conditions can interrupt herbicide activity, allowing weeds to emerge with or before the crop. Rain shortly after emergence can dilute application rate and reduce performance of some herbicides, particularly if weeds are growing rapidly beyond the optimum stage for control. Phenmedipham usually performs normally when the rainless period exists for at least six hours after application. Many postemergence herbicides are temperature-sensitive, with less

FIG. 4.4. Pyrazon + endothall (283) damage on sugarbeets from overdosage caused by improper agitation and settling out of the mixture during application.

control below 55°–60° and more crop injury above 85° F. Most failures of postemergence herbicides are caused by improper application or timing. Abnormal temperatures may delay germination of some weed species yet promote early establishment of others. When soil temperatures stay too low for an extended period, herbicide dissipation below toxic level may take place or the crop may remain in the chemical zone too long and become damaged. Many weedy or chemical-damaged fields are destroyed and replanted each year before the herbicide has time to work or the crop has recovered from injury.

CONTROL SYMPTOMS

Familiarity with chemical control symptoms on weeds and seedling beets is helpful in avoiding hasty management decisions. Two types of control expressions are common. Weeds fail to emerge because of the lethal effect of herbicide on germinating seed, or weeds emerge and die from cumulative effect of the toxicant. Chemically sickened weeds are pliable and nonturgid. Pyrazon-affected weeds are chlorotic and leaf tissue becomes progressively necrotic, commencing at leaf margins. Carbamates swell and shorten the primary roots, and the root tissue becomes darkened until death. Sugarbeet seedlings treated with thiocarbamates often have thickened cotyledons and the true leaves may fuse (Fig. 4.5). Pyrazon toxicity symptoms on seedling beets are similar to those on broadleaf

FIG. 4.5. Leaf fusing and retardation of sugarbeet seedling (left) caused by cycloate compared to untreated (right).

weeds. Under normal dosages, formative effects and crop retardation symptoms are temporary and seldom is harvest yield or quality affected adversely, unless retardation persists for a lengthy period or vigor loss is caused by a persistent herbicide damaging to sugarbeets.

Daily examination of chemically retarded fields is wise management because an irrigation or rain, coupled with a nitrogen fertilization, often improves crop vigor, if placed when seedling beets show evidence of recovery. Often cultural delays are confused with chemical retardation on treated fields. Uneven planting depth, salt accumulation, and fertilizer movement on bed plantings may delay crop seedling emergence and development.

Troubleshooting herbicide failure should be accomplished in a systematic manner. First, establish if known causes were operative, then establish their order of occurrence or absence. Usually, a human error or environmental adversity is discovered in the process.

HERBICIDE DISSIPATION

Dissipation may consist of vapor loss, photochemical decomposition, chemical degradation, plant assimilation, adsorption on soil particles, cultivation and microbial breakdown, and leaching, or any combination of these (40). Microbial breakdown is the principal dissipation pathway. Most sugarbeet herbicides have no lasting effect on microbial populations in the soil (21). Chemical decomposition by hydrolysis at clay particle interfaces appears slight for most sugarbeet herbicides. Degradation by ultraviolet light is almost absent, perhaps with the exception of trifluralin (43).

Aerobic fungi, bacteria, and actinomycetes metabolize carbon and nitrogen in herbicide molecules for energy. Kearney and Kaufman (32) have shown that an enzyme of the bacterium *Pseudomonas* sp. isolated from soil culture detoxifies to aniline many biologically active phenylcarbamates, including IPC. Microbial degradation of dalapon is known to occur in many soils in less than three weeks, while TCA degrades microbially at a slower rate. Diallate has a half-life of about 30 days because of attacking microbes, but pyrazon and trifluralin resist rapid microbial breakdown and remain active for about 12 weeks. EPTC and pebulate, unlike diallate, dissipate from soil by physical means, including vapor loss and leaching (28). Rapid adsorption of paraquat on soil particles completely inactivates the herbicide.

Herbicides released into the ecosystem are naturally dissipated in the air and soil before harmful residues can occur. Proper application and handling prevent accidental spillage and excessive dosage. The use of wind shields and hoods, and ground application at low wind velocities, prevent air drift.

Seldom does the herbicide reach streams because of the sequence of water movement through soil. If complete sequence occurs, herbicides move in percolating water from the soil surface → zone of aeration (area between soil surface and water table) → zone of saturation (area of groundwater movement to stream) → stream course → sea (37). Herbicidal breakdown occurs in the aeration zone. Most herbicides leach to a depth of about one foot within the aeration zone or area of clay accumulation where herbicidal wastes are deactivated in a short time. Some herbicides may leach to depths where microbial activity is low, which prolongs persistence. In most instances, this depth is well above the water table, thus significant lateral herbicide movement in groundwater is avoided.

HERBICIDE RESIDUAL

Herbicide selection should be planned to minimize carry-over and the possibility of residual chemicals interacting with a current application. Particularly in semiarid regions, herbicide carry-over from preceding crops can damage sensitive crops. Atrazine and dicamba carry-over from treated cornfields and trifluralin persistence on bean and cotton land, and from misapplication during postplanting on beets, may become troublesome in the succeeding year on sugarbeets, alfalfa, tame barley, and sorghum. Conversely, pyrazon applied on sugarbeets may injure catch crops such as dry bean, and thiocarbamates can damage tame oats. Arp and Dotzenko (6) concluded that herbicides applied to dry beans, corn, and tame barley left no damaging residuals or produced no adverse effects on the following sugarbeet crop and vice versa.

Chemical-cultural spot treatments in other crops, employing 2,4-D, dalapon, amitrole, and atrazine, among others, are effective for control of perennial weeds. Until recent restriction on cropland, picloram and dicamba were effective for the control of certain perennial weeds. These treatments can persist and migrate in irrigation water, killing sugarbeets grown subsequently. Field records of weed infestations and herbicide usage and results enable the grower to choose herbicides wisely and to select effective cropping sequences and herbicides that minimize carry-over and costs.

RESIDUAL DETECTION

Sometimes persistent herbicides accumulate temporarily in the aeration or rooting zone. Biological and chemical tests are used to determine if soil residue levels remain high enough to be harmful to sugarbeet seedlings and the environment in general.

In bioassay tests, sensitive indicator species grown in pot tests can detect a very small amount of chemical in soil. A simulated control, prepared by mixing 0.5 g activated carbon with five pounds of soil, is also potted for comparison. Atrazine and many other persistent herbicides are inactivated by carbon. Sugarbeets and at least one indicator species are grown together in the pots. For example, tame oats and turnips, dry beans and tomatoes, safflower and dry beans, sorghum and foxtail millet, dry beans and turnips, and tame oats and ryegrass are indicator crops for atrazine, dicamba, picloram, trifluralin, pyrazon, and EPTC (thiocarbamates), respectively. Periodically, observations are made for formative effects on plants and non-emergence. If response symptoms appear acute between sugarbeets and the indicator species, a damaging residue is indicated and crop planting should be deferred. Slight symptoms on the indicator species without crop injury suggest a noninjurious residue level unless uptake is cumulative. Usually, plant uptake, growth dilution, and natural breakdown processes will dissipate the residual further during the growing season.

Chemical tests, including gas chromatography and colorimetric and infrared spectrum analysis (23), are used to verify results of bioassay tests and to monitor the presence of chemical residue in plant material and soil. Field soils are sampled at various depths for these tests, commonly at the 0–3- and 3–6-inch depths. Chemical tests measure minute amounts of chemical or derived metabolite in roots, tops, thick juice, and pulp of sugarbeets.

HERBICIDE REGISTRATION

Federal certification, registration, and labeling procedures for pesticides are established by law, and these regulations on performance and safety are administered by the U.S. Department of Agriculture cooperating with the U.S. Department of Health, Education and Welfare, Food and Drug Administration. State departments of agriculture have adopted regulations governing pesticide use patterned after the federal law. In addition, international residue tolerances and daily intake levels for pesticides, including some sugarbeet herbicides, are established by the worldwide Codex Alimentarious Commission, sponsored by the United Nations (FAO). These state, federal, and international tolerances or acceptable residue levels are stipulated for plant products to protect against repeated intake and misuse from endangering human and animal health or otherwise contaminating the environment (40).

TOXICOLOGICAL PROPERTIES

In addition to efficacy performance, toxicological properties of a herbicide are established before labeling and release. These properties include acute oral toxicity, toxicity to skin and eyes, and inhalation toxicity. The lethal oral dose is usually given as the LD_{50} on white rats. Doses are recorded in milligrams of chemical per kilogram of body weight (mg/kg) that will kill 50 percent of the rats. Mammalian toxicity ratings for sugarbeet herbicides range from extremely toxic for endothall acid (35 mg/kg) to slightly toxic for phenmedipham (> 5,000 mg/kg). The bulk of sugarbeet herbicides is moderately toxic with LD_{50} ranges from 500 to 5,000 mg/kg, and most of these have less ingested toxicity than aspirin, which is rated as 1,200 mg/kg. There is little danger to man from the use of herbicides, but the herbicide label should always be consulted before use and the usage directions and precautions should be abided by.

EFFECTIVE AGRONOMIC PRACTICES

SYSTEM MANAGEMENT

Chemical weeding, supplemented by tillage, forms the basic plan for residual control of weeds in sugarbeets. Selection of the most effective weed control system is predicated on:
1. An effective cropping sequence, placing sugarbeets after an intertilled crop.
2. Careful field selection, free of perennial weeds.
3. Proper land preparation to facilitate uniform herbicide application and operation of cultivation equipment.
4. Use of proper tools for herbicide application and cultural weed control.
5. Correct timing of application relative to planting.
6. A knowledge of potential weed infestation, growth cycle, and growth rate to enable selection of the most effective herbicide or combination of herbicides for local conditions.
7. A vigorously growing crop in a uniform stand that permits successful use of a selective thinner and provides competition to late-germinating weeds.
8. Use of selective herbicides and control measures for weed control in other crops.
9. Sanitation in land areas and water near sugarbeet fields.

OUTLOOK

A general reduction in sugarbeet production costs depends on the successful integration of chemical weeding, spaced planting, and selective thinning practices. Rate of progress in problem solving depends on increased manpower and funding for weed science at all levels, and the rapid extension of useful information to the grower. Patent position on new herbicides, marketing decision, plant construction cost, public concern for environmental pollution, and intensification of regulations on labeling may cause delays in releasing effective new herbicides.

DISCOVERY OPPORTUNITIES AHEAD

Scrutiny of current practice and knowledge in weed control technology on sugarbeets reveals four main areas for promising research and discovery. Categorically, these are new herbicide evaluation and residue detection, mechanisms of herbicide action, herbicide application methods, and the study of weed biology and competition (56).

Herbicides must be improved for broad-spectrum weed control and problem situations, particularly postemergence control of broadleaf weeds and perennials. Discovery and development of new herbicides, formulations, adjuvants, and management systems should remain a directed and continuing process. Sugarbeet acreage is small when compared to crops such as corn and soybeans, and beet income is less than the high-return horticultural crops. Hence, herbicide synthesis for use on sugarbeets may be handicapped by potential returns to the originating company in relation to development costs. Nevertheless, chemical companies are actively seeking new candidates for sugarbeets each year, often with success.

The fate of herbicides in farm and large-scale ecosystems, and the effect of individual herbicides, repeated herbicide application, and crop protection chemical mixtures on sugarbeets and subsequent crops, require further and more detailed study. These basic studies should include evaluations of herbicide persistence and degradation rates in sugarbeet soils and in plants. Investigations on interactions between herbicides and other crop production chemicals (insecticides and fertilizers) on performance and attenuation of pesticide residues in soil and water should be initiated.

Herbicide activity is difficult to understand because basic information on a particular herbicide is lacking in many instances. Knowledge regarding herbicide-plant interactions would improve synthesis of new herbicides and field performance. In particular, mechanisms regulating postemergence control of broadleaf weeds and adsorption-desorption equilibria affecting dosage response of soil-acting herbicides should receive study. These detailed studies should include uptake and penetration, site of absorption by plants, site of action within plants, and herbicide movement and dissipation in soil and foliage to obtain information on the basis of selectivity.

Application methods require further improvement to enhance chemical weeding performance from a single application. Placement on a particular soil type or on a particular plant part may influence performance more than type of toxicant or application rate. Incorporation affects performance differently, depending on chemical characteristics of the herbicide, incorporation device, and soil composition. Specific placement and application methods should be determined for each herbicide in different soil types. The development of improved or new machines and methods of application, such as high pressure, low-volume liquid, foam and injector applicators might enhance performance. Improvement of cultivating equipment depends on the magnitude of physical control of weeds required to supplement chemical weeding rather than to supersede it. Other cultural practices, including seedbed type, planting time, irrigation type, and drainage, should be investigated in relation to herbicide placement.

Life-cycle features, including weedy plant succession, should be examined to determine competitive and sensitivity differences between weeds and sugarbeets in their relation to herbicides and timed application. Physiological and morphological growth studies may reveal inherent weaknesses in weeds that permit selective weed control by chemical or cultural practices. Study of the weed-free period and its position in the growth cycle of sugarbeets necessary for maximum yield and additional control measures remains urgent. The introduction of soil-acting chemicals which govern dormancy, induce germination, or regulate fertilization of weed propagules could effect complete weed absence on cultivated land. Weed responses to habitat and habitat changes from tillage and sugarbeet monoculture should be examined.

In the final analysis, it appears feasible that treatment programs will be developed for essentially complete weed control in sugarbeets, allowing a replacement of hand labor for weeding and thinning.

REFERENCES

1. Andersen, Robert N. Germination and establishment of weeds for experimental purposes. Weed Soc. Am. Handbook. W. F. Humphrey Press, Inc., Geneva, N.Y. 1968.
2. Andersen, Robert N., A. J. Linck, and R. Behrens. Absorption, translocation, and fate of dalapon in sugar beets and yellow foxtail. Weeds 10:1–3. 1962.
3. Anderson, G. W. Weed control in sugar beets in 1963. In Natl. Weed Comm. Canada Res. Rept., Eastern Sec., p. 38. Can. Dept. Agr., Ottawa. 1963.
4. Appleby, A. P., W. R. Furtick, and S. C. Fang. Soil placement studies with EPTC and other carbamate herbicides on Avena sativa. Weed Res. 5:115–22. 1965.
5. Arndt, F. and C. Kotter. Selectivity of phenmedipham as a post-emergence herbicide in sugar beet. Weed Res. 8:259–71. 1968.
6. Arp, A. L. and A. D. Dotzenko. The effect of crop sequences, nitrogen fertilizer, and herbicides on the yield and quality of sugar beets. Colo. Agr. Exp. Sta. Prog. Rept. 69-19. 1969.
7. Audus, L. J. The Physiology and Biochemistry of Herbicides. Academic Press, Inc., London and New York. 1964.
8. Bailey, G. W., J. L. White, and T. Rothberg. Adsorption of organic herbicides by montmorillonite: role of pH and chemical character of the adsorbate. Soil Sci. Soc. Am. Proc. 32:222–34. 1968.
9. Barton, Lela V. The germination of weed seeds. Weeds 10:174–82. 1962.
10. Brimhall, Phil B., E. W. Chamberlain, and H. P. Alley. Competition of annual weeds and sugar beets. Weeds 13:33–35. 1965.
11. Chepil, W. S. Germination of weed seeds. I. Longevity, periodicity of germination, and vitality of seeds in cultivated soil. Sci. Agr. 26:307–46. 1946.
12. Cormany, C. E. Preemergence applications of a combination of TCA and endothall show promise for control of grassy and broadleaf weeds in sugar beets. Proc. Am. Soc. Sugar Beet Technologists 8:135. 1954.
13. Crafts, A. S. The Chemistry and Mode of Action of Herbicides. Interscience Publishers, New York-London. 1961.

14. Davis, J. F. The effect of cultivation on yield of sugar beets. Proc. Am. Soc. Sugar Beet Technologists, Eastern U.S. and Can., pp. 52–54. Detroit. 1941.
15. Dawson, J. H. Competition between irrigated sugar beets and annual weeds. Weeds 13:245–49. 1965.
16. Deming, G. W. Use of IPC for weed control in sugar beets. Proc. Am. Soc. Sugar Beet Technologists 6:453–55. 1950.
17. Dotzenko, A. D., M. Ozkan, and K. R. Storer. Influence of crop sequence, nitrogen fertilizer, and herbicides on weed populations in sugar beet fields. Agron. J. 61:34–37. 1969.
18. Downs, R. J., H. A. Borthwick, and A. A. Pirringer. Light and plants. USDA Misc. Publ. 879. 1966.
19. Finney, D. J. Probit Analysis. Cambridge Univ. Press, London. 1952.
20. Fischer, Bill B. and L. K. Beutler. Effect of postemergence applications of herbicides on sugar beet development and weed control in central California. Proc. Am. Soc. Sugar Beet Technologists 15:200–208. 1968.
21. Fletcher, W. W. Herbicides and the bio-activity of soil. Dutch working party for weed control, pp. 274–81. Wageningen, Neth., Mar. 22, 1966.
22. Frank, R. and C. M. Switzer. Adsorption and translocation of pyrazon by plants. Weed Sci. 17:365–70. 1969.
23. Freed, V. H. Determination of herbicides and plant growth regulators. In L. J. Audus, The Physiology and Biochemistry of Herbicides, pp. 39–74. Academic Press, Inc., London and New York. 1964.
24. Fryer, J. D. and S. A. Evans. Weed Control Handbook, vol. 1. Principles. 5th ed. Blackwell Scientific Publications, Oxford and Edinburgh. 1968.
25. Gowing, D. P. A method of comparing herbicides and assessing herbicide mixtures at the screening level. Weeds 7:66–76. 1959.
26. Gray, Reed A. and A. J. Weirerich. Factors affecting the vapor loss of EPTC from soils. Weeds 13:141–47. 1965.
27. Grigsby, B. H. Weed control in the sugar beet crop. Proc. Am. Soc. Sugar Beet Technologists, Eastern U.S. and Can., pp. 146–48. Detroit. 1951.
28. Herbicide Handbook. Weed Soc. Am. W. F. Humphrey Press, Inc., Geneva, N.Y. 1967.
29. Herrett, R. A. and J. A. Kramer. The thin-layer effect and herbicidal action of 3,4-dichlorobenzyl methylcarbamate. Weed Soc. Am. Abstr. pp. 51–52. 1966.
30. Jacobsohn, Ruben and R. N. Andersen. Differential response of wild oat lines to diallate, triallate, and barban. Weed Sci. 16:491–94. 1968.
31. Karen, Ephraim, C. L. Foy, and F. M. Ashton. Phytotoxicity and persistence of four thiocarbamates in five soil types. Weed Sci. 6:172–75. 1968.
32. Kearney, Philip C. and D. D. Kaufman. Enzyme from soil bacterium hydrolyzes phenylcarbamate herbicides. Science 147(3659):740–41. 1965.
33. King, L. J. Weeds of the World. Biology and Control. Plant Science Monographs. Interscience Publishers, Inc., New York. 1966.
34. Klingman, Glenn C. Weed Control: As a Science. John Wiley & Sons, Inc., New York-London. 1961.
35. Krionderis, Dennis J. Response of sugar beet seedlings to combinations of herbicides and insecticides. Crops Paper 651M. Plant Sci. Div., Univ. Wyo. 1968.
36. Lee, G. A. and H. P. Alley. Effects of four methods of mechanical incorporation on the phytotoxicity of pyrazon. J. Am. Soc. Sugar Beet Technologists 14:248–53. 1966.
37. LeGrand, H. E. Movement of pesticides in the soil. In M. E. Bloodworth, ed., chr., Pesticides and Their Effects on Soils and Water, pp. 71–77. Am. Soc. Agron. Spec. Publ. 8. Soil Sci. Soc. Am., Madison, Wis. 1966.

38. McBirney, S. W. Weed control studies on sugar beets using pre-emergence treatments. Proc. Am. Soc. Sugar Beet Technologists 5:453–63. 1948.
39. Matsumaka, Shooichi. Propanil hydrolysis: inhibition in rice plants by insecticides. Science 160(3834):1360–61. 1968.
40. National Academy of Sciences. Principles of Plant and Animal Pest Control, vol. 2. Weed Control. 1968.
41. Nelson, R. T. Trials with herbicides for weed control in sugar beets. Proc. Am. Soc. Sugar Beet Technologists 7:121–25. 1952.
42. Nelson, R. T., R. R. Wood, and R. K. Oldemeyer. Selection of sugar beets for tolerance to endothall herbicide. J. Am. Soc. Sugar Beet Technologists 11:155–59. 1960.
43. Parka, S. J. and T. B. Tepe. The disappearance of trifluralin from field soils. Weed Sci. 17:119–23. 1969.
44. Reeve, Perc A. Results from the use of TCA and endothall for weed control in sugar beets. Proc. Am. Soc. Sugar Beet Technologists 7:90. 1953.
45. Ries, S. K., M. J. Zabik, G. R. Stephenson, and T. M. Chen. N-glucosyl metabolite of pyrazon in red beets. Weed Sci. 16:40–41. 1968.
46. Robbins, W. W. Weed investigations by the California Agricultural Experiment Station. Proc. Am. Soc. Sugar Beet Technologists, pp. 19–20. 1938.
47. Savage, K. E. and W. L. Barrentine. Trifluralin persistence as affected by depth of soil incorporation. Weed Sci. 17:349–52. 1969.
48. Schweizer, E. E. and D. M. Weatherspoon. Herbicide control of weeds in sugar beets. J. Am. Soc. Sugar Beet Technologists 15:263–76. 1968.
49. Smith, D. T. and W. F. Meggitt. Pre and postemergence sugarbeet trials in Michigan in 1967. North Central Weed Control Conf. Res. Rept. 24:79. 1967.
50. Smith, P. B. A milestone in sugar beet production. Proc. Am. Soc. Sugar Beet Technologists, Eastern U.S. and Can., pp. 1–7. Detroit. 1951.
51. Splittstaesser, W. E. and L. A. Derscheid. Effects of environment upon herbicides applied preemergence. Weeds 10:304–7. 1962.
52. Strickler, R. L., E. L. Knake, and T. D. Hinesly. Soil moisture and effectiveness of preemergence herbicides. Weed Sci. 17:257–59. 1969.
53. Sullivan, E. F. and H. L. Bush. Probability analysis of herbicide response. J. Am. Soc. Sugar Beet Technologists 13:721–26. 1966.
54. Sullivan, E. F., R. L. Abrams, and R. R. Wood. Weed control in sugar beets by combinations of thiocarbamate herbicides. Weeds 11:258–60. 1963.
55. Sullivan, E. F., R. R. Wood, R. L. Abrams, and S. G. Walter. Preplant and postemergence chemical weeding in sugar beets. North Central Weed Control Conf. Res. Rept. 20:68–75. 1963.
56. USDA and the State Univ. and Land Grant Col. A national program of research for sugar. RPA209-control of weeds in sugar beets, pp. 24–28. Res. Program Develop. and Evaluation Staff, Washington, D.C. 1969.
57. Varma, S. C. On the nature of competition between plants in the early phases of their development. Ann. Botany 2:203–25. 1938.
58. Wicks, G. A. and F. N. Anderson. Weed control in sugar beets with herbicides and cultivation. Weed Sci. 17:456–59. 1969.

5

Nitrogen Nutrition

F. JACKSON HILLS
University of California
Davis

ALBERT ULRICH
University of California
Berkeley

NITROGEN IN THE SOIL	112
NITROGEN IN THE PLANT	113
Deficiency Symptoms	113
Absorption by Roots and Movement within the Plant	114
FERTILIZING FOR OPTIMUM PRODUCTION	114
Factors Affecting Nitrogen Requirement	117
Effects of Nitrogen on Root Diseases and Pests	118
Applying Nitrogen	119
Nitrogen Source	121
Fertilize Crop—Not Residue	124
Length of Nitrogen Deficiency Prior to Harvest	126
Amount Required for Optimum Production	128
GUIDES TO FERTILIZATION	130
Soil Tests for Nitrogen Availability	130
Plant Analysis	131
An Integrated Approach	131
FUTURE DEVELOPMENTS	133

OF the mineral elements essential for sugarbeet growth, nitrogen, by far, has the greatest influence on root quality and sucrose production. This implies that the fertilization of sugarbeets should be centered around the use of nitrogen after making certain that all the other nutrients have been adequately supplied to the crop either by natural fertility of the soil or by fertilization.

Large amounts of nitrogen are required to insure adequate top and root growth, but if storage roots are to be high in sucrose concentration, the plants must be deficient in nitrogen prior to harvest to retard the utilization of sucrose for growth and to allow sucrose to accumulate in the storage roots. Thus, nitrogen management becomes more exacting than with many crops in that overfertilization with nitrogen can reduce economic yield as well as increase the cost of production.

NITROGEN IN THE SOIL

The ultimate source of soil nitrogen used by plants is the earth's atmosphere, and although there are about 35,000 tons of nitrogen over each acre, the majority of cultivated soils contain only 0.5 to 5 tons of nitrogen in the top foot of soil. Most of this occurs in the soil organic matter complex, and at any one time only a few pounds will exist in forms readily usable by plants.

There are many excellent accounts of soil nitrogen and the processes by which it becomes available or unavailable for plant growth (1, 2, 5, 24). In this chapter we will briefly review only the processes affecting the flux of soil nitrogen between organic and inorganic compounds.

Mineralization or mobilization of nitrogen refers to the conversion of organic nitrogen to inorganic forms. There are three phases to this conversion: aminization, ammonification, and nitrification. The first involves the breakdown of organic matter into various amino compounds and the second to the further breakdown of these compounds to release ammonium ions. In well-aerated soils, both steps are carried out by soil microorganisms that decompose organic matter to furnish energy and carbon compounds for their own growth. The third step, nitrification, is the biological oxidation of ammonium to nitrite and nitrate. This step is carried out by microorganisms that require molecular oxygen and are able to build their carbon structure from the reduction of CO_2 of the soil atmosphere, using energy derived from the oxidation of ammonium.

Immobilization refers to the incorporation of ammonium and nitrate into organic matter by higher plants or soil microorganisms. Organic material added to a well-aerated soil serves as food for soil fungi and bacteria, thus stimulating their growth. These organisms also require nitrogen, and when the organic material is low in this constituent, they use soil nitrate and ammonium and thus compete with crop plants. Much of the nitrogen used by these soil microorganisms will again become available to crop plants through the mobilization processes. Broadbent and Nakashima (6) have shown, however, that even without the addition of straw to a soil, more than 20 percent of added fertilizer nitrogen can be immobilized and remain in the soil even after seven cuttings of sudan grass. With large additions of low nitrogen organic matter, as much as 70 percent of fertilizer nitrogen may be immobilized for a considerable period of time.

In some soils, especially those containing vermiculite or related clay minerals, ammonium nitrogen may be fixed within the lattice structure of the colloidal minerals. Ammonium so fixed is not available to soil microorganisms and cannot readily be exchanged for other cations. In a study of 21 Wisconsin soils, nitrogen fixed as ammonium varied from about 210 to 450 pounds per acre-foot (1).

The nitrate ion is completely mobile in soil and readily moves with rain or irrigation water. It can be moved to the surface of planting beds or leached below the effective root zone, and in both cases be lost for crop production.

Under anaerobic conditions, denitrification may occur and nitrate may be reduced to gaseous forms of nitrogen by certain bacteria and thus lost to crop production.

In summary, ammonium or nitrate nitrogen in the soil, whether from chemical fertilizers or the mineralization of organic matter, can be absorbed by higher plants, used by soil organisms, fixed in the soil, lost by leaching, or lost as a gas. When we fertilize a crop we want to maximize the use of applied nitrogen by the crop and minimize its loss by other processes.

NITROGEN IN THE PLANT

More atoms of nitrogen are required within the plant than for any of the other essential mineral nutrients. It is an essential constituent of proteins, amides, amino acids, coenzymes, nucleic acids, certain hormones, and clorophyll. Thus, in the sugarbeet, as in other green plants, nitrogen is important to the synthesis of sucrose as well as to the many reactions involving the utilization of sucrose as an energy source for plant growth and cell maintenance.

DEFICIENCY SYMPTOMS

A sugarbeet plant well supplied with nitrogen, water, and other essential nutrients grows rapidly in a favorable climate and forms large leaves of a good, green color. With the onset of nitrogen deficiency, however, leaves approaching full size become light green, turning to yellow. Yellowing continues as the plant ages, accompanied by wilting and an accelerated death rate of the older leaves. The newly formed leaves in the center of the plant are then smaller, often lanceolate in shape, with an intense green color, and may lie nearly parallel to the soil surface, with petioles curved slightly upward. The actual size of the center leaves depends upon the rate at which nitrate is formed in the soil and is absorbed by the plant. If this rate is slow, the leaves will be small. When the rate of nitrate supply is fairly rapid, yet still insufficient, the new leaves will be larger—perhaps almost large enough to cover the soil surface and to give a field a uniformly green appearance. Only a close inspection of the plant and a chemical test for nitrate in the petioles can reveal the deficiency (28).

ABSORPTION BY ROOTS AND MOVEMENT WITHIN THE PLANT

Nitrogen can be absorbed by plant roots as nitrate, ammonium, urea, water soluble amino acids, and as nucleic acid (24). However, by far the largest amounts are taken up as nitrate, and for all practical purposes, the other forms of nitrogen may be ignored for soils normally cultivated for sugarbeet production. After nitrate is absorbed, it moves to the petioles of the young leaves, and from there to the corresponding blades. Nitrate in leaf blades is reduced and metabolized as amino acid supplies are depleted. Usually, petioles contain three to six times as much nitrate as do the blades until the leaves become deficient in nitrogen. When this occurs, both petioles and blades of young leaves and leaves approaching full size contain very little nitrate, but nitrate may still be high in the petioles and blades of older leaves where it remains trapped. Contrary to the situation with other plants such as cotton, oats, and tobacco, there apparently is not a rapid movement of nitrate from old to young leaves in the sugarbeet plant (13). The failure of nitrate to move rapidly from the older to the younger leaves implies that a continuous supply of nitrate must be available to the plant if a nitrogen deficiency of young leaves is to be prevented at all times.

When nitrogen deficiency is severe, the rate of growth of both tops and roots declines quickly. When the deficiency is less severe, however, top growth slows faster than root growth. This indicates that when the nitrogen supply is limited, sugar utilization is located primarily in the root, where sugar from the top first meets the nitrate from the soil. Under these conditions, nitrate is apparently reduced in the roots, and root growth takes precedence over top growth for the use of nitrogen. Conversely, when nitrate is abundant, much moves directly to the young leaves, and top growth resumes at a rapid rate as the raw materials for sugar utilization, nitrogen, and sugar meet in the young leaves.

FERTILIZING FOR OPTIMUM PRODUCTION

Without nitrogen fertilization many soils could not produce profitable beet crops. Figure 5.1 (Field A) illustrates the extent to which production

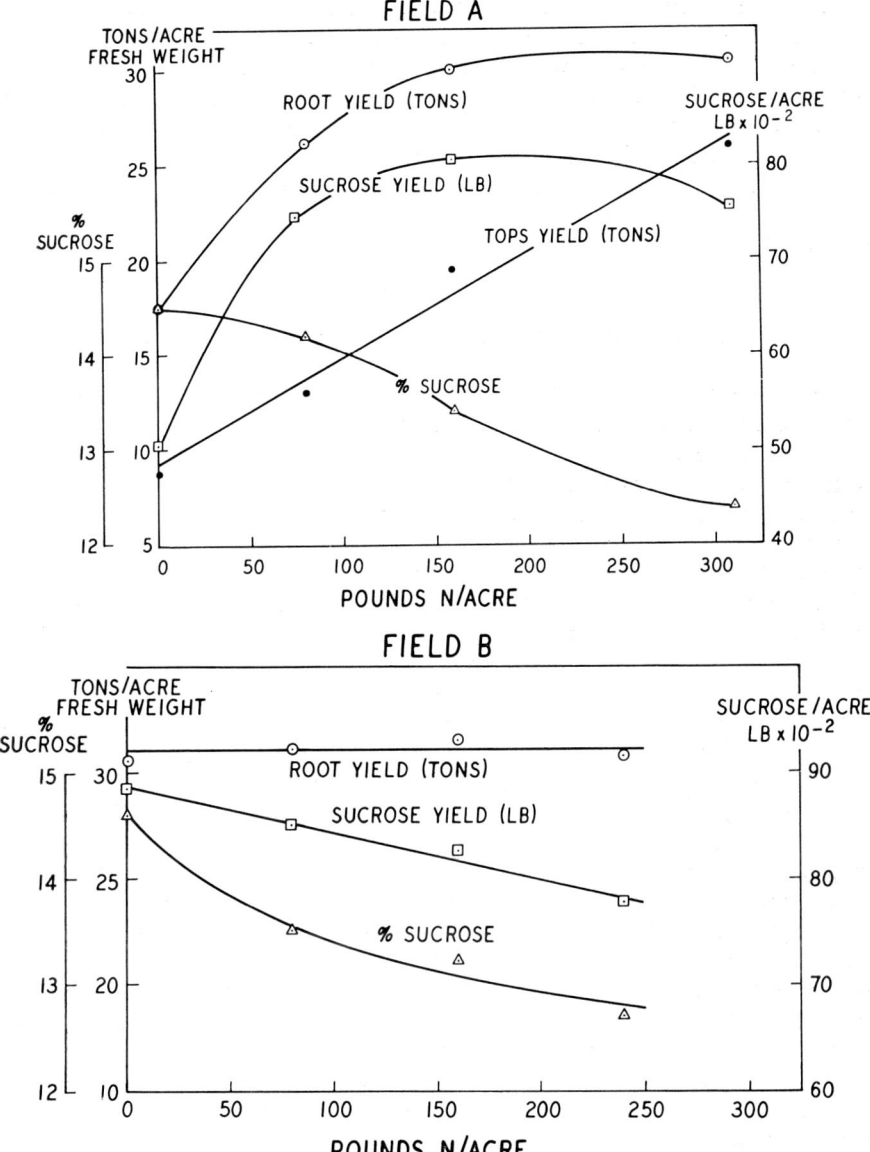

FIG. 5.1. Field responses to nitrogen fertilization. *Field A.* One hundred sixty pounds of nitrogen per acre resulted in optimum production. *Field B.* Nitrogen fertilization reduced sucrose concentration and sugar production.

can be lost by inadequate fertilization. Plants that were not fertilized produced 5,000 pounds of sugar per acre, a gross value of about $260. Plants that were properly fertilized produced 8,000 pounds of sugar per acre, a gross value in excess of $400. Note that root growth and sucrose production were greatest at 160 pounds of nitrogen per acre. Higher rates of fertilization did not increase root growth but decreased sucrose production, due to increased top growth at the expense of sugar storage.

There are many fields, however, on which sugarbeets do not respond profitably to nitrogen fertilization because the soil is already too high in nitrogen. Field B of Figure 5.1 illustrates such a case. Nitrogen fertilization was an unnecessary expense and reduced sugar production.

Numerous field experiments have established optimum rates of nitrogen fertilization for specific fields (7, 10, 11, 29). Responses vary widely, even within a geographically narrow production area, from no response to increases in sugar yield of 50 percent and more. For example, Walker et al. (29) conducted 11 fertilizer trials in Yolo and Solano counties of California during a two-year period and estimated nitrogen rates to maximize economic returns to growers that varied from 50 to 240 pounds of nitrogen per acre. James (10) and James et al. (11) completed 16 field trials over a two-year period in central Washington and determined that the optimum nitrogen rate varied from 0 to 200 pounds of nitrogen per acre. Such variability in optimum nitrogen needs makes it hazardous for a grower to rely on an average rate of nitrogen fertilization. To apply such a rate could result in considerable economic loss for a particular field, due to underfertilization or overfertilization.

Table 5.1 indicates the range and the most usual rate of nitrogen fertilization for the major U.S. sugarbeet production areas.

TABLE 5.1. Estimates of fertilizer nitrogen rates used in the United States

Area	Lb N/A	
	Range	Most usual
North Central		
Red River Valley: East N.Dak., Minn.	0–50	0*–50
Ohio, Michigan	40–180	100
Great Plains		
Montana, Wyoming, Nebraska, Colorado	40–180	100
Texas	120–200	150
Intermountain		
Washington, Oregon	80–300	160
Utah, Idaho	80–160	120
Pacific Southwest		
California, Arizona	50–325	150

SOURCE: Estimates by sugar company agriculturists.
* Zero nitrogen when a legume green manure plus summer fallow precede the beet crop.

FACTORS AFFECTING NITROGEN REQUIREMENT

Time of harvest can have a considerable effect on optimum nitrogen rate, particularly in areas where the harvest campaign extends over several weeks when climate is favorable for beet growth. This is well illustrated in Figure 5.2, the results of an experiment conducted in Kern County, California, where the harvest season begins in early July and usually extends well into September. The optimum nitrogen rate for the field on which this experiment was located was about 80 pounds of nitrogen per

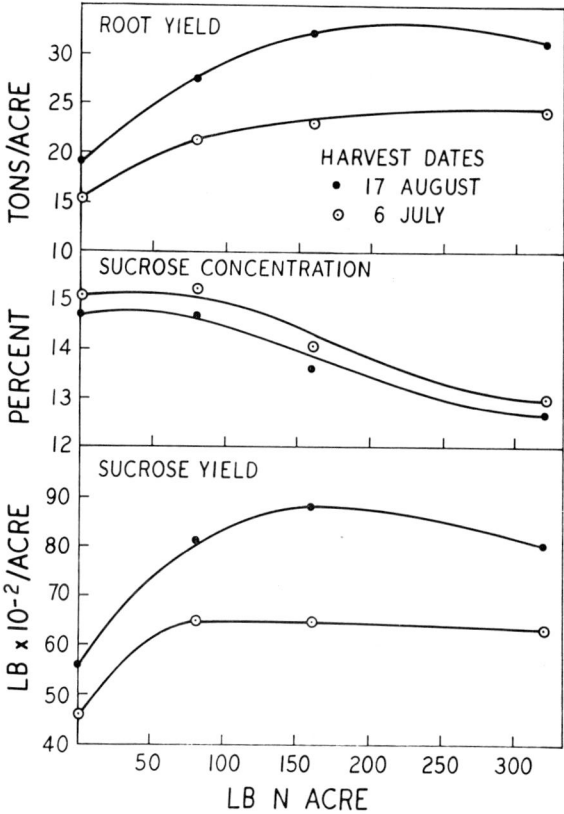

FIG. 5.2. When harvest extends over a period of favorable growing weather, the rate of nitrogen fertilization for maximum production must be adjusted accordingly. For this experiment in Kern County, California, 80 pounds of nitrogen per acre maximized production for an early harvest, but 160 pounds were needed for a harvest 6 weeks later.

acre for an early July harvest but increased to about 160 pounds of nitrogen if the field was harvested in mid-August. The shift in optimum nitrogen rate over the six-week period was due to a more rapid rate of root growth for plants receiving 160 pounds of nitrogen compared to plants fertilized with 80 pounds of nitrogen. The prolonged period of nitrogen deficiency did not increase the sucrose concentration of the plants receiving 80 pounds of nitrogen compared to the higher nitrogen rate. Sucrose concentration declined uniformly for both nitrogen rates over the six-week period of high summer temperatures (9).

Under other conditions, however, a prolonged harvest period may have little or no effect on the optimum rate of nitrogen fertilization. In Yolo and Solano counties in California, a considerable portion of the crop is overwintered. The optimum nitrogen rate for a fall harvest is usually the same as for a spring harvest four months later. As the experiment of Figure 5.3 illustrates, about 125 pounds of nitrogen per acre produced maximum sugar yield on October 5 as well as on February 13 of the following year. Even though there was considerable root growth during the period between the two harvests, the rate of growth, limited by cold temperatures, short days, and reduced light intensity, was slow enough to be sustained by nitrogen already present in the root zone, leached from the tops of planting beds into the active root zone, or made available by soil mobilization processes. Note that the nitrogen response curves are parallel for the harvest dates and there is considerable increase in petiole nitrate from the early to late harvest for all rates of fertilization.

Deficiences of other mineral nutrients, inadequate soil moisture, excessive weed growth, plant diseases, and pests can be factors limiting plant growth and therefore the response of a crop to nitrogen. As an example, sugarbeets at Davis, California, 53 percent infected by yellows viruses, grew at a rate of 0.9 ton per acre per week from August 21 to October 22, but plants protected from virus infection (10 percent yellows) grew at a rate of 1.5 tons per acre per week and produced 40.8 tons of roots per acre on October 22 compared to 32.2 for the diseased plants. Obviously, more nitrogen is required to produce the disease-free crop.

An important consideration in planning a fertilizer program, although a difficult one to evaluate, is to consider the limiting factors that usually exist and to fertilize according to the crop yield that reasonably can be expected.

EFFECTS OF NITROGEN ON ROOT DISEASES AND PESTS

Sugarbeet diseases are fully discussed in Chapter 9, but the effect of nitrogen fertilization on certain root diseases should be briefly mentioned here. Keeping plants well supplied with nitrogen is helpful in the alleviation of the effects of sclerotium root rot (15), dry rot canker (8, 29), and root-knot nematode (32). Fertilization with nitrogen cannot be relied on to control these diseases but is a factor in reducing damage.

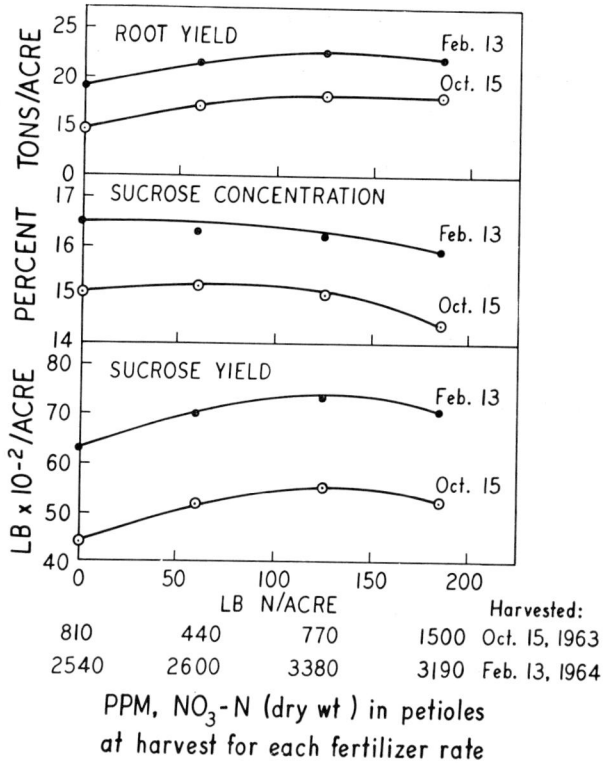

Fig. 5.3. A prolonged harvest period had no effect on the amount of fertilizer nitrogen required for optimum production during a winter period when growth was slow. (Solano County, California, 1963–64.)

APPLYING NITROGEN

In general, it is not good practice to apply nitrogen fertilizer in the fall or winter preceding the planting of a sugarbeet crop. It has been argued that ammonia applied in the fall will not nitrify until the following spring and thus will be protected from leaching by adsorption on soil colloids. It has been shown, however, that nitrification of ammoniacal fertilizers is a rapid process, even at soil temperatures of 45° F. Tyler et al. (25) showed that the nitrate produced from an application of 200 pounds of ammonium-nitrogen per acre in six weks at 45° F was equivalent to the quantity produced in two weeks at 75° F. At 37° F the amount of nitrate produced in four weeks was about one-half of that produced at 45° F. This indicates that fall-applied nitrogen may be lost through nitrification and leaching by winter and spring rains. In addition to losses by leaching, anaerobic conditions brought about by water saturation of soils can result in the reduction of nitrate to gaseous forms and a loss of nitrogen to the atmosphere.

If plants will need fertilizer nitrogen soon after germination, a portion of the total amount to be applied can be broadcast and plowed down in the early spring or broadcast and listed into planting beds. This method of application is particularly effective if phosphorus or potassium is to be applied also. It is not advisable to apply more than 40 pounds of nitrogen per acre for incorporation into planting beds unless irrigation will be by sprinklers. Fertilizer nitrogen will move with the water from furrow irrigation to the tops of beds where it cannot be absorbed by plants (23). Subsequently, this nitrogen may be leached into the active root zone by fall rains, may stimulate vegetative growth, and may reduce the sucrose concentration of storage roots.

A common practice is to inject anhydrous ammonia or aqua ammonia during seedbed formation. Injection shanks are usually as far apart as the rows to be planted and the ammonia is placed six to eight inches deep. High concentrations of anhydrous ammonia in contact with germinating seed are toxic and can be avoided by placing the ammonia at least four inches outside the seed row and six inches from the top of planting beds so that the material is an inch or two below the irrigation furrow. Placed in this manner, as much as 100 to 200 pounds of nitrogen per acre from anhydrous ammonia can be used with little probability of damage to germination or seedling growth. Placement slightly below the level of the irrigation furrow insures the movement of nitrate into beds and down rather than predominantly into beds and up as would be the case for placement in beds above the furrow level. Most damage from anhydrous ammonia is the result of careless placement, especially in sandy soils.

Many soils are sufficiently fertile to supply enough nitrogen to meet the early needs of a crop. In these soils, nitrogen can be side-dressed at thinning time from six inches outside the beet row to as far out as half-way between the rows or irrigation furrows. Two or three inches below the furrow level is sufficiently deep for nitrogen salts; ammonia should be placed about six inches deep to avoid volatilization losses. Ammonium or urea forms of nitrogen are quickly nitrified to nitrate and are carried to roots by rain or irrigation water. The rapid proliferation of sugarbeet feeder roots and the mobility of nitrate make close placement generally unnecessary.

Application Schedule. Most field experience indicates that a single application at planting or thinning time produces results comparable to splitting the total amount required into two or three applications. Apparently, with judicious irrigation, little nitrogen is lost throughout the growing season by leaching. There may be some conditions, however, such as a sandy soil, where it may be difficult to avoid leaching, and thus a split application may be desirable. When a second application is made, it should be applied at or before midseason.

A midseason nitrogen deficiency, followed by refertilization, can seriously reduce sugar yield. Loomis and Nevins in pot culture (17) and

Loomis and Worker in field plots (19) have shown that storage roots do not exhibit an accelerated rate of growth when nitrogen is resupplied to deficient plants compared to plants continuously well supplied with nitrogen. Root production is lost in proportion to the length of the deficiency period. The resupply of nitrogen results in a rapid drop in the sucrose concentration of storage roots, with the danger of a possible further loss in sugar production if nitrogen cannot be again depleted four to six weeks before harvest.

Foliar Feeding. Repeated "spoon feeding" of a beet crop by applications of nitrogen to the leaves has an appeal as a useful technique for the careful regulation of nitrogen levels for this nitrogen-sensitive crop. In California experiments, however, it was concluded that there was little to be gained from this practice. As shown in Table 5.2 the crop does respond to foliar applications but no better than to soil applications. Not much more than 20 pounds of nitrogen per acre can be applied to leaves without seriously burning them. When beets are deficient in nitrogen it usually takes 40 pounds of nitrogen per acre, and often 80, to produce maximum returns. Applying this amount by foliar spray would take several applications and therefore does not appear to be economically practical.

NITROGEN SOURCE

Experimental results and field experience indicate that there are no important differences in sugarbeet response to various nitrogen carriers applied to furnish equal amounts of nitrogen (3, 8, 16). In the Imperial Valley of California, sugarbeets were side-dressed with 120 pounds of nitrogen per acre from several nitrogen carriers at midseason soon after depletion of a preplant application of 80 pounds of nitrogen per acre from ammonium nitrate. The effect of the nitrogen carriers on the concentration of nitrate-nitrogen in petioles is shown in Figure 5.4. The estimated

TABLE 5.2. Effect of foliar and side-dressed nitrogen on sugarbeet root yield

Thinning side-dressed	Lb N/A		Roots, Tons/A California	
	Midseason			
	Side-dressed	Spray	Davis	Imperial Co.
0	0	0	21.1	15.0
80	0	0	23.6	23.6
80	40	0	25.9	28.6
80	0	10 + 10	...	26.1
80	0	20	...	27.1
80	0	20 + 20	26.6	26.6
80	80	0	28.5	30.9
LSD, 5%			2.4	2.5

SOURCE: D. Ririe, Univ. of Calif., Davis.

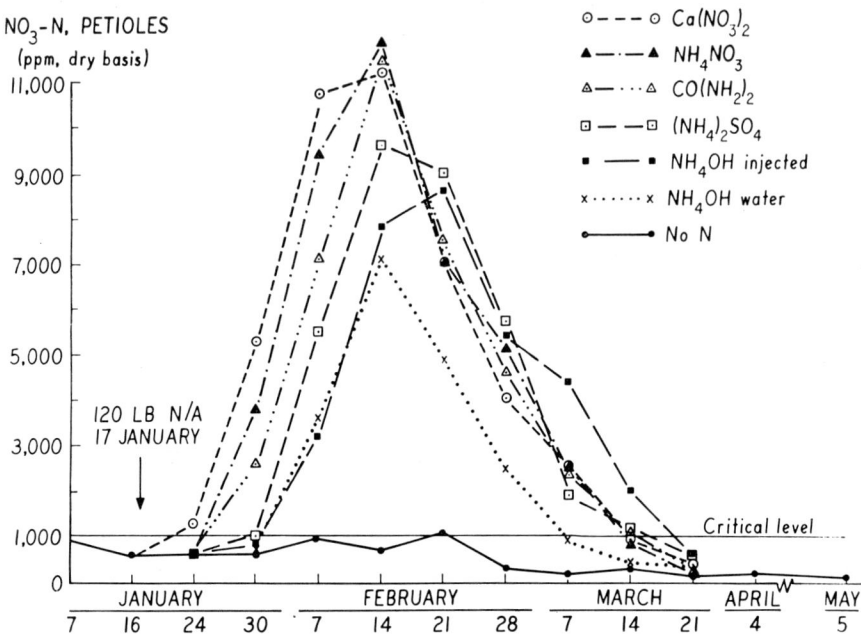

FIG. 5.4. Nitrate-nitrogen concentration in beet petioles from several nitrogen fertilizers. (After Loomis et al., 16.)

time lags to raise the nitrate-nitrogen concentration to 1,000 ppm by the various nitrogen carriers are given in Table 5.3, along with sugarbeet root yields and root sucrose concentrations determined at harvest. This experiment illustrates several points important to the use of fertilizer nitrogen sources at midseason: (1) Nitrification of urea and ammoniacal sources is rapid. Even at soil temperatures averaging 50° F, as was the case with these mid-January applications, there were only very short time lags in the appearance of nitrate-nitrogen in petioles from ammonium sulfate and urea compared to the appearance of like concentrations from calcium nitrate. Nitrification of anhydrous ammonia was also rapid but lagged slightly. (2) Considerable nitrogen loss can occur when ammonium hydroxide (aqueous ammonia) is applied in irrigation water. Losses are due to volatilization of ammonia from water and the soil surface as water evaporates. An additional disadvantage to the application of any soluble nitrogen material in irrigation water is that its distribution will be no better than that of the irrigation water, which can be quite erratic. (3) If midseason nitrogen applications are depleted prior to harvest, there is little danger of seriously depressing the sucrose content of storage roots. Note that petiole nitrate-nitrogen from all nitrogen carriers was low for about six weeks prior to harvest and that none of the midseason applications affected root sucrose concentrations.

TABLE 5.3. Effect of nitrogen fertilizers on time required to raise the nitrate-nitrogen concentration of sugarbeet petioles of mature leaves to 1,000 ppm (dry wt.) and on root yield and sucrose concentration

N Source	Days to Reach 1,000 ppm NO$_3$-N (dry basis) in Petioles of Mature Leaves	Roots	
		tons/A	% sucrose
Control	...	14.6	14.8
NH$_4$OH-irrigation	14.0	18.2	14.7
NH$_4$OH-injected	15.5	23.9	15.0
(NH$_4$)$_2$SO$_4$	11.5	22.9	15.4
CO(NH$_2$)$_2$	7.0	21.6	14.9
NH$_4$NO$_3$	5.0	22.2	15.0
Ca (NO$_3$)$_2$	3.5	22.0	15.9
LSD, 5%		2.1	n.s.

SOURCE: Loomis et al. (16).

Alfalfa and Green Manure as Nitrogen Sources. Alfalfa (*Medicago sativa* L.) has long been valued for its contribution to soil fertility. For example, in Washington, Boawn et al. (4) have shown that turning under a three-year stand of alfalfa with four inches of spring growth furnished 340 pounds of nitrogen per acre more for corn production over a subsequent five-year period than was furnished by previously uncropped virgin soil. During the first and second years of corn production following alfalfa, over 100 bushels of corn per acre were produced without nitrogen fertilization. The nitrogen uptake was equivalent to that from 200 pounds of nitrogen per acre applied to the virgin soil. Many experiments have shown that alfalfa included in a rotation will contribute to higher sugarbeet yields. In most cases, higher yields are attributed to nitrogen fixation by the symbiotic bacteria associated with alfalfa roots and its release to subsequent crops. There is a danger, however, that the sucrose concentration of beet roots may be rduced by late-season release of nitrogen and its uptake by the sugarbeet plants.

Other leguminous green manure crops turned under prior to a sugarbeet crop can furnish a substantial portion of the nitrogen required to produce a beet crop. In the Red River Valley of North Dakota, a legume, usually sweet clover (*Melilotus indica* All.), plowed down in early July and followed by summer fallow is the principal means of nitrogen fertilization for the beet crop planted the subsequent spring.

In areas where winters are mild, fall-planted legumes plowed down the following spring can furnish a considerable portion of the nitrogen required for sugarbeet production. At Santa Maria, California, Williams and Ririe (32) showed that purple vetch (*Vicia apropurpurea* Tesf.) produced 3,100 pounds of dry matter per acre of shoot growth at 4.66 percent nitrogen, and when turned under one week prior to planting, sugarbeets increased sugar production equivalent to 80 pounds of nitrogen per acre from ammonium nitrate side-dressed to sugarbeets that followed winter fallow. In this experiment it was concluded that the green manure influenced sugarbeet production primarily by its effect on nitrogen nutrition. However, when low-cost fertilizer nitrogen is available, it is doubtful that green manures can be justified solely on the basis of the nitrogen they furnish subsequent crops.

FERTILIZE CROP—NOT RESIDUE

The incorporation of plant materials into surface soil can improve soil structure; but Williams and Doneen (31, Fig. 5.5) have demonstrated a marked inverse association between improvement in water infiltration and the nitrogen concentration of plant material. Nitrogen fertilization of low-nitrogen plant materials to hasten their decomposition when incorporated into soil may also reduce their effectiveness in increasing water infiltration (30, Table 5.4). In another experiment, Williams (30, Fig. 5.6) found that nitrogen fertilizer applied to barley crop residue in the fall was of no benefit to a succeeding sugarbeet crop. It is likely that substantial amounts of the nitrogen applied to the residue were leached from the root zone by winter rains.

Based on the results of these experiments as well as others cited by Krantz et al. (14), there is little justification for the application of nitrogen to crop residues. It appears more desirable to apply nitrogen directly to

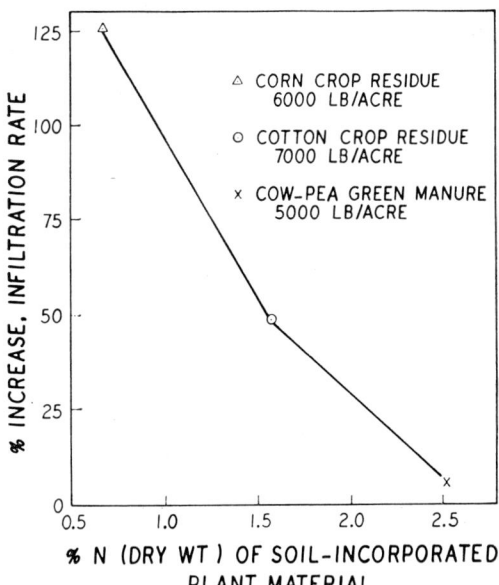

Fig. 5.5. Inverse relation between nitrogen concentration of soil-incorporated plant material and improvement in water infiltration. (After Williams and Doneen, 31.)

TABLE 5.4. The effect of barley green manure and crop residue, with and without fertilization, on infiltration rate over a 10-month period following incorporation into soil

Treatment	Dry Weight of Tops, Lb/A	Nitrogen in Tops, %	Infiltration Rate,* Inches/Hr
Control	...	...	0.54
Barley green manure	7,620	1.26	1.72
Barley green manure + 120 lb N/A	6,910	1.25	1.47
Barley crop residue	7,600	0.47	1.57
Barley crop residue + 120 lb N/A	7,210	0.54	1.31
LSD, 5%	n.s.	0.20	0.19

SOURCE: Williams and Doneen (31).
* Average of 6 tests from October through July.

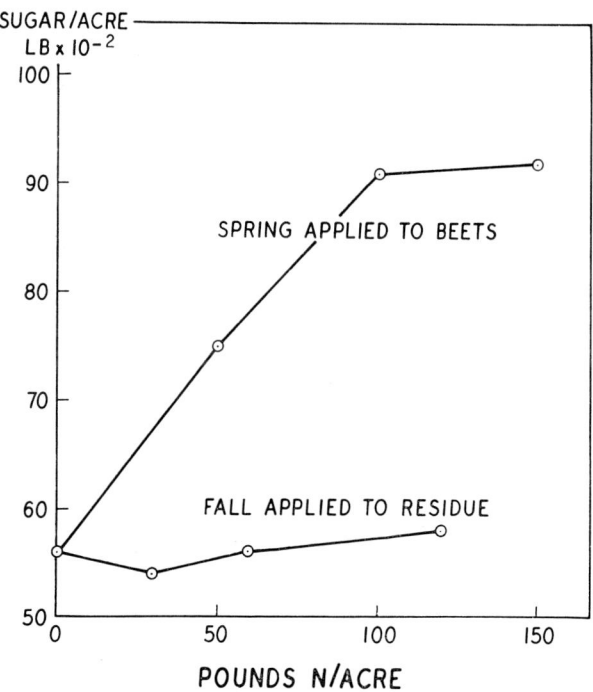

FIG. 5.6. The effect of nitrogen applied to barley crop residue in the fall versus side-dressing to the succeeding sugarbeet crop. (After Williams, 30.)

LENGTH OF NITROGEN DEFICIENCY PRIOR TO HARVEST

A prominent feature of the response of sugarbeets to nitrogen deficiency is an increase in the sucrose percentage in the storage roots. The change may be visualized as resulting from inhibition of vegetative growth, which permits a higher proportion of the sucrose produced in the leaves to accumulate in the roots rather than to be utilized in the growth processes. The degree of the shift in this growth-storage balance is dependent upon the length and degree of the deficiency, root size, the amount of photosynthesis, and the temperature regime under which the response is studied. Effects of photoperiod, light intensity, and temperature are discussed in Chapter 2.

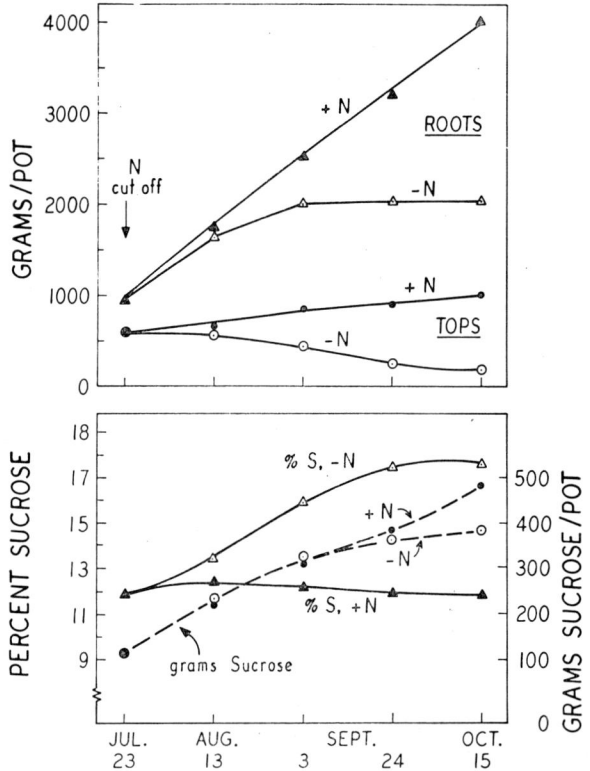

Fig. 5.7. Effect of nitrogen nutrition on sugarbeets in vermiculite-nutrient solution culture. (After Loomis and Nevins, 17.)

As Ulrich (26) and Loomis and Nivens (17) have shown, depriving sugarbeets of nitrogen in solution culture experiments results in a rapid rise in the percentage of sucrose in roots. Maximum increases are obtained in 6 to 8 weeks (Fig. 5.7). In such experiments, the degree of deficiency is severe, and growth rates of tops and roots quickly decline. For periods of about 8 weeks of nitrogen deficiency, nitrogen-deficient plants often contain more sucrose in storage roots than nitrogen-sufficient plants. Thereafter, more sucrose is contained in nitrogen-sufficient plants, due to their more rapid root growth.

In field soils the degree of nitrogen deficiency is seldom, if ever, as severe as in culture solutions, and root growth continues, often nearly as rapid as nondeficient plants (Fig. 5.8). However, even when nitrogen deficiency is not severe, the sucrose concentration of storage roots of deficient plants increases sharply, reaching a new plateau in about 4 weeks, and sucrose production will exceed that from nondeficient plants for periods of 12 weeks or more. Thus, an objective of nitrogen management should be to allow plants to become nitrogen deficient 4 to 6 weeks prior to harvest. If an error is to be made, it appears desirable to make it on the side of underfertilization, as under most field conditions, deficiency periods of 8 to 12 weeks can be tolerated without reducing total sugar yield.

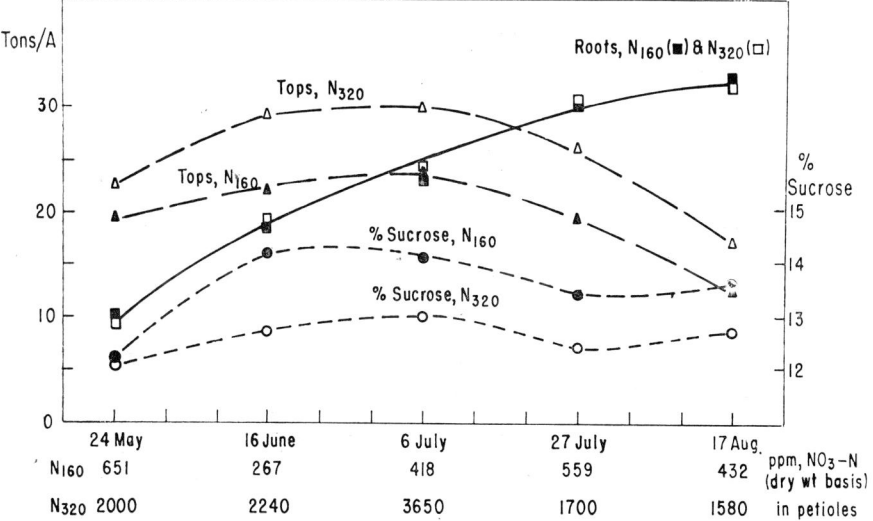

Fig. 5.8. Marginal nitrogen deficiency. In this field experiment, plants that received 160 pounds of nitrogen per acre were nitrogen deficient for 12 weeks prior to harvest. Top growth declined, sucrose concentration increased to a maximum within 3 weeks from the onset of deficiency, but root growth was not affected. Thus the lower nitrogen rate produced more sugar per acre over the 12-week period.

TABLE 5.5. Diphenylamine test ratings on root pulp related to percent sucrose of beet samples delivered from 508 grower contracts in the south San Joaquin Valley of California in 1968

Nitrate Rating, Diphenylamine	Contracts, %	Mean % Sucrose
Low	27	15.2
Medium	57	14.2
High	13	12.8
Very high	3	12.5

SOURCE: L. M. Burtch, Spreckels Sugar Co.

The diphenylamine test,[1] applied to pulp from roots being analyzed for sucrose, has been useful in revealing areas of low sucrose concentrations due to excessive nitrate uptake at harvest. This test is also useful in evaluating how well crops in an area have been fertilized with nitrogen to meet desirable quality standards. In 1968 Burtch[2] tested all sugar samples from more than 500 contracts in California's San Joaquin Valley. The diphenylamine test (Table 5.5) revealed that only 16 percent of the contracts were overfertilized with nitrogen. The year 1968 was outstanding for superior beet yields. The same levels of fertilization in a year of lower root yields could result in more extensive overfertilization.

Effect of Root Size. Dense sugarbeet populations, with larger and denser foliage canopies, usually produce roots with somewhat higher sucrose concentrations (18). Smaller roots produced by competition appear to respond more rapidly to changes in environment that affect sucrose concentration. Growing sugarbeets in pots of vermiculite, watered with culture solution, Loomis and Ulrich showed that starting at high nitrogen status, small beets increased in sucrose concentration to a greater extent than larger beets after 8 weeks of nitrogen depletion (Fig. 5.9). Thus, dense field populations with smaller roots might somewhat shorten the duration of nitrogen deficiency required for maximum increase in sucrose concentration.

AMOUNT REQUIRED FOR OPTIMUM PRODUCTION

One of the most critical questions a sugarbeet grower must answer is, How much nitrogen is needed for maximum net return for my beet field? Unfortunately, this question is also one of the most difficult to answer. It is doubtful if it ever will be possible to answer it precisely, but an approxi-

---

1. Diphenylamine reagent is prepared by dissolving 0.2 g diphenylamine in 100 ml concentrated sulfuric acid. The reagent can be used as a semiquantitative test for nitrate. The intensity of blue color developing upon the dropwise application of diphenylamine to thinly spread pulp is related to the relative amount of nitrate present (Alexander, Holly Sugar Corp., unpublished).
2. Spreckels Sugar Co., unpublished data.

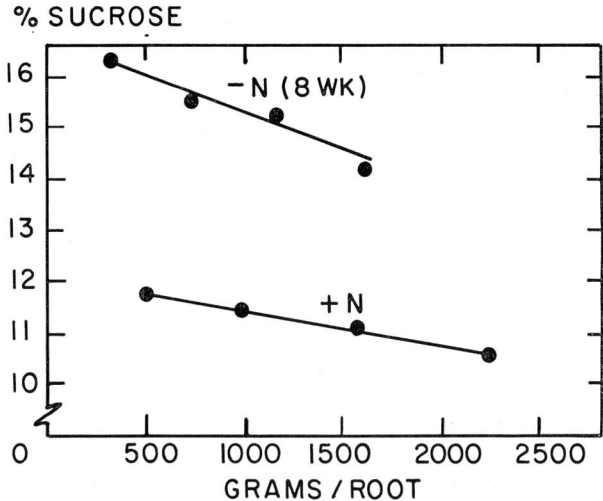

FIG. 5.9. Effect of root size and nitrogen deficiency on the sucrose concentration of beets grown in vermiculite and watered with nutrient solution. Root sizes were achieved by growing 1, 2, 4, and 8 plants per pot. (After Loomis and Ulrich, 18.)

mate answer can be given by careful consideration of the variables involved and by making full use of available technology.

The principle of fertilization is deceptively simple: the amount of fertilizer nitrogen to be applied is the difference between the total amount of nitrogen required to produce a given crop yield and the amount that will be supplied to the crop through the reserves already present in the soil, for example, fertilizer nitrogen equals total nitrogen demand minus available soil nitrogen.

Estimates of the amount of nitrogen required for an expected crop yield can be obtained from nitrogen rate field experiments as the amount of nitrogen taken up by the crop that produces the maximum sugar yield resulting from the addition of the smallest amount of fertilizer nitrogen. When such a determination is made, there must be a response to nitrogen to assure that the soil nitrogen supply is limiting yield and not already furnishing nitrogen in luxury amounts. Fertilizer rates must explore a sufficiently wide range of nitrogen additions to identify the nitrogen rate that produces the maximum amount of sugar per acre. Such approximations usually underestimate nitrogen requirements as they usually do not consider fibrous root yield or the old leaves that die throughout the season. The latter may be of minor importance as most of this material can be recovered by careful harvest. The fresh weight of fibrous roots, however, may be as much as 25 percent of the yield of storage roots (13), and it appears reasonable to increase the nitrogen demand to reflect the nitrogen required to produce this crop component.

It is convenient to express the nitrogen requirement as pounds of nitrogen per ton of fresh roots, a figure that readily can be used to approximate nitrogen demand for any expected root yield. An experiment by Boawn et al. in central Washington (3) indicates a nitrogen demand of 8.76 pounds of nitrogen per ton of roots for a crop that produced 29.4 tons of roots per acre. This is probably too small, since root yield was still increasing at the highest nitrogen rate. Haddock (7) reported a demand of 10 pounds of nitrogen per ton, based on a crop producing 20 tons of roots per acre in Utah. Data of MacKenzie et al. (20) in the Imperial Valley of California indicate a demand of about 14 pounds of nitrogen per ton of roots. This amount of nitrogen produced a root crop of 30.7 tons per acre and appeared to be the crop that produced maximum sugar per acre. About 40 percent of total nitrogen uptake was in storage roots and 60 percent in tops.

Based on MacKenzie's data, a 30-ton root crop requires about 420 pounds of nitrogen per acre. Estimating fibrous root production at 25 percent of storage roots increases the nitrogen required by about 40 pounds to 460 pounds of nitrogen per acre or about 15 pounds of nitrogen per ton of storage roots.

Estimating the amount of nitrogen that will be taken up by a crop from the soil is even more difficult, as this amount will depend on the amount of ammonium and nitrate ions present at planting, the distribution of these ions in the soil profile, the net amount of nitrogen resulting from mineralization and immobilization during the current season, and the root distribution of the crop.

GUIDES TO FERTILIZATION

Over the years many tests have been developed and used for estimating the nutrient requirements of crops. Various methods of evaluating soil fertility are discussed at length by Tisdale and Nelson (24). Of all methods, the field trial is the oldest for evaluating the amount of a nutrient to be added as fertilizer to maximize crop production. The most important objection to the field trial is that the results apply only to the soil area tested and for only the year or years the experiments are conducted. As discussed earlier, the variability encountered in the response to nitrogen among fields within a narrow geographical area in a single season makes the practice of adding an average amount of nitrogen to all fields a poor one in terms of meeting the needs of crops on specific fields.

To meet the need for broader methods of estimating fertilizer requirements, two general types of tests have been developed, namely, soil tests and plant tests.

SOIL TESTS FOR NITROGEN AVAILABILITY

Soils may be tested by chemical or biological means. A principal objective is to obtain an index of nutrient availability upon which a fertilizer recommendation can be based. Chemical tests are favored over biological tests in that they are more rapid, less expensive, and usually easier to carry out. With respect to nitrogen, soil tests have centered on determining mineralizable nitrogen or initial inorganic nitrogen. Since the advent of high fertilizer nitrogen consumption, mineralizable nitrogen alone has not correlated well with crop production. Current trends are to determine nitrate-nitrogen shortly before planting (12). Recently, however, Stanford and Legg (22) have shown improved correlation with nitrogen uptake by considering both initial nitrate nitrogen and nitrogen subsequently mineralized. To date, no soil test for nitrogen has been developed that correlates well with sugarbeet production. Research efforts are continuing, however, and some progress is being made in Washington by James and co-workers (11, 12). An easy-to-conduct soil test or tests that would allow a classification of a particular soil into three or four categories of nitrogen fertility would be valuable to use in estimating fertilizer nitrogen requirements, especially when used in conjunction with plant analysis on the current season's crop.

PLANT ANALYSIS

The analysis of petioles of recently matured leaves of the sugarbeet plant has been shown to be a good indicator of the nitrogen status of beet plants. When samples can be collected in a timely manner, properly analyzed, and promptly reviewed, this procedure can be used to determine the adequacy of an initial nitrogen application and whether or not a supplemental application should be made (21, Fig. 5.10).

Petiole analysis can be useful in deciding on the order of harvest of sugarbeet fields. If fields have been periodically sampled throughout the season, the approximate time that each crop became deficient can be determined. Other factors being equal, those fields that have been deficient the longest should be harvested first.

AN INTEGRATED APPROACH

Despite the difficulty in answering the question of how much fertilizer nitrogen is needed for this crop, it cannot be avoided and an answer must be given for each field. At present the answer is usually based on local experience, on what neighbors are doing, and on a consideration of crop history and soil characteristics. There is great need for a procedure that will give a more precise answer. We suggest the following:

1. *Estimate nitrogen demand.* This should be based on the root yield to be expected under the conditions the crop is to be grown.

2. *Evaluate the nitrogen fertility status of the soil.* This might be done by soil analysis to a depth of several feet to determine initial nitrate as well as mineralizable nitrogen. But until such tests are developed to the

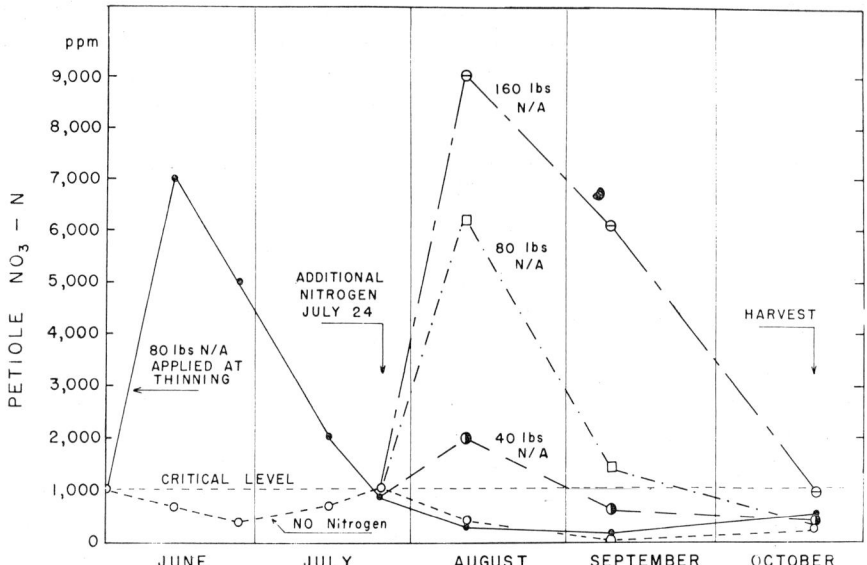

Fig. 5.10. Sugar yields associated with 0, 80, 80 + 40, 80 + 80, and 80 + 160 pounds of nitrogen per acre were 66.8, 72.8, 84.0, and 84.8 pounds $\times 10^{-2}$ sugar per acre, respectively. Eighty pounds of nitrogen per acre at thinning were not enough. In mid-July, when petiole samples indicated that plants would soon be deficient in nitrogen, a second application of 80 pounds resulted in maximum sugar production.

extent that they are reliable in approximating the amount of nitrogen that the soil will furnish, fertility evaluation must depend on more subjective procedures. An "improved guess" as to fertility status of a field can be made by considering the following characteristics of high vs low fertility fields. Indications of high fertility are one or more of the following: (1) high yield of previous crop, coupled with a lack of deficiency symptoms; (2) evidence from plant analyses of the previous crop that nitrogen status was high late in the season; (3) heavy applications of nitrogen fertilizer to previous crop; (4) alfalfa or another leguminous green manure crop turned under; (5) soil high in organic matter. Low fertility is indicated by one or more of the following: (1) low yield of previous crop, coupled with symptoms of nitrogen deficiency and a lack of other major limiting factors; (2) evidence from plant analyses of low nitrogen status of the previous crop; (3) little or no fertilizer nitrogen applied to the previous crop; (4) large amounts of low nitrogen residue turned under; (5) sandy soil low in organic matter.

From such considerations the field to be planted can be classified as low, medium, or high fertility. Experience suggests the following percentage ranges of the estimated nitrogen requirement to be added at planting or by thinning time: low fertility, 40 to 60 percent; medium fertility,

20 to 40 percent; high fertility, 0 to 20 percent. Thus, for an expected yield of 20 tons of roots per acre and a nitrogen demand of 300 pounds per acre (e.g., 15 pounds of nitrogen per ton of roots), a low fertility field would receive 120 to 180 pounds of nitrogen per acre, a medium fertility field 60 to 120, and a high fertility field 0 to 60.

In making the initial nitrogen application, it is preferable to be conservative, and if an error is made, it should be made on the side of underfertilization. Overfertilization with the initial nitrogen application cannot be undone but underfertilization can be corrected.

3. *Evaluate the adequacy of the initial fertilization.* This can be done best by petiole analysis, as indicated in Figure 5.10, by taking samples from recently matured leaves at two-week intervals, starting about two weeks after thinning (27). Indications of impending deficiencies 10 weeks or more prior to harvest can be corrected by additions of 40 to 80 pounds of nitrogen per acre.

When plant analysis facilities are not available, the adequacy of the initial fertilization can be evaluated by test strips. When the initial nitrogen is applied, the rate can be doubled on a strip through the field, and alongside nitrogen may be applied to another strip at only one-half the rate that is used for the field. Careful observation of these strips, particularly with the aid of the diphenylamine test for nitrate on fresh petioles, can reveal the adequacy of the initial application. If the field has been seriously underfertilized, the strip with the one-half nitrogen rate will appear lighter green than the balance of the field fairly early in the season. If this occurs, plants receiving the field nitrogen rate can be evaluated with the diphenylamine test and corrective measures can be taken if it appears that these plants will soon be deficient.

A blue color with a drop of diphenylamine reagent on the exposed surface of a freshly cut petiole of a recently matured leaf indicates that the plant is well supplied with nitrogen. An impending nitrogen deficiency is usually indicated by a negative test (no color) on 20 percent or more of the petioles from a random sample of at least 20 plants.

4. *Postseason evaluation.* A consideration of crop yield, sucrose concentration, and plant anaysis results will indicate how well the field has been fertilized and will serve as a log of fertility evaluation for future use.

FUTURE DEVELOPMENTS

Sugarbeet crops of the future will be fertilized with nitrogen according to their needs as determined from plant analysis records of preceding crops, soil analysis values, weather forecasts, and estimates of anticipated harvest yields. On the average, more nitrogen will be applied to sugarbeets at planting and during the early leaf and root development stages of the crop, and applications of nitrogen beyond midseason will be avoided. Other nutrient deficiencies will not be permitted at any time during the growing period, since these would reduce yield and prevent ripening, normally induced by a period of four to six weeks of nitrogen deficiency and low fall night temperatures prior to harvest.

Nitrogen fertilization of sugarbeets in the future will be carefully budgeted, not because of its cost, but because excesses will reduce the sucrose concentration and processing quality of the beet root. Consequently, the development of nitrogen-tolerant sugarbeet varieties and the use of growth regulators to induce ripening under high nitrogen conditions become nearly meaningless as a solution to the high nitrogen problem of sugarbeets at harvest. Only by carefully budgeting nitrogen to the needs of the crop or by considering sugarbeets as a scavenger crop for nitrogen can we meet our economic and ecological obligations in efficient beet sugar production.

The selection of varieties with small maximum top size and vigorous root growth adjusted to an appropriate plant spacing, in contrast to varieties with large maximum top size, should lead to earlier root size development, earlier ripening, somewhat smaller root tonnages, and much higher sucrose concentrations. Much of the sugar now used by the plant for large top size development would be diverted to root size development and to sucrose storage upon nitrogen depletion prior to harvest.

REFERENCES

1. Bartholomew, W. V. and F. E. Clark, eds. Soil Nitrogen. Am. Soc. Agron. Monograph 10., Madison, Wis. 1965.
2. Black, C. A. Soil-plant Relationships. John Wiley & Sons, Inc., New York. 1957.
3. Boawn, L. C., C. E. Nelson, F. G. Viets, Jr., and C. L. Crawford. Nitrogen carrier and nitrogen rate influence on soil properties and nutrient uptake by crops. Wash. State Agr. Exp. Sta. Bull. 614. 1960.
4. Boawn, L. C., J. L. Nelson, and C. L. Crawford. Residual nitrogen from NH_4NO_3 fertilizer and from alfalfa plowed under. Agron. J. 55:231–35. 1963.
5. Bould, C. Mineral nutrition of plants in soils. In F. C. Steward, ed., Plant Physiology (A Treatise), vol. 3, pp. 15–96. Academic Press, Inc., New York. 1963.
6. Broadbent, F. E. and T. Nakashima. Reversion of fertilizer nitrogen in soils. Soil Sci. Soc. Am. Proc. 31:648–52. 1967.
7. Haddock, Jay L. The nitrogen requirements for sugarbeets. Proc. Am. Soc. Sugar Beet Technologists 7:159–65. 1952.
8. Hills, F. J. and J. D. Axtell. The effect of several nitrogen sources on beet sugar yields in Kern County, California. Proc. Am. Soc. Sugar Beet Technologists 6:356–61. 1950.
9. Hills, F. J., G. V. Ferry, Albert Ulrich, and R. S. Loomis. Marginal nitrogen deficiency of sugarbeets and the problems of diagnosis. J. Am. Soc. Sugar Beet Technologists 12:476–84. 1963.
10. James, D. W. Growing sugarbeets for maximum sugar production. Wash. Agr. Exp. Sta. Circ. 464. 1966.
11. James, D. W., D. C. Kidman, W. H. Weaver, and R. L. Reeder. Predicting the nitrogen fertilizer requirements of sugarbeets grown in central Washington. Wash. Agr. Exp. Sta. Circ. 488. 1968.
12. James, D. W., C. E. Nelson, and A. R. Halvorson. Soil testing for residual

nitrates as a guide for nitrogen fertilization of sugarbeets. Wash. Agr. Exp. Sta. Circ. 480. 1967.
13. Kelley, J. D. and A. Ulrich. Distribution of nitrate nitrogen in the blades and petioles of sugarbeets grown at deficient and sufficient levels of nitrogen. J. Am. Soc. Sugar Beet Technologists 14:106–16. 1966.
14. Krantz, B. A., F. E. Broadbent, W. A. Williams, K. G. Baghott, K. H. Ingebretsen, and M. E. Stanley. Fertilize crop—not crop residue. Calif. Agr. 22(8):6–8. 1968.
15. Leach, L. D. and A. E. Davey. Reducing southern sclerotium rot of sugarbeets with nitrogenous fertilizers. J. Agr. Res. 64:1–18. 1942.
16. Loomis, R. S., J. H. Brickey, F. E. Broadbent, and G. F. Worker, Jr. Comparisons of nitrogen source materials for midseason fertilization of sugarbeets. Agron. J. 52:97–101. 1960.
17. Loomis, R. S. and D. J. Nevins. Interrupted nitrogen nutrition effects on growth, sucrose accumulation and foliar development of the sugarbeet plant. J. Am. Soc. Sugar Beet Technologists 12:309–22. 1963.
18. Loomis, R. S. and A. Ulrich. Responses of sugarbeets to nitrogen deficiency as influenced by plant competition. Crop Sci. 2:37–40. 1962.
19. Loomis, R. S. and G. F. Worker, Jr. Restitution of growth in nitrogen-deficient sugarbeet plants. J. Am. Soc. Sugar Beet Technologists 12:657–65. 1964.
20. MacKenzie, A. J., K. R. Stockinger, and B. A. Krantz. Growth and nutrient uptake of sugarbeets in the Imperial Valley, California. J. Am. Soc. Sugar Beet Technologists 9:400–407. 1957.
21. Ririe, D., A. Ulrich, and F. J. Hills. The application of petiole analyses to sugarbeet fertilization. Proc. Am. Soc. Sugar Beet Technologists 8:48–57. 1954.
22. Stanford, G. and J. O. Legg. Correlation of soil N availability indexes with nitrogen uptake by plants. Soil Sci. 105:320–26. 1968.
23. Stout, M. Redistribution of nitrate in soils and its effects on sugarbeet nutrition. J. Am. Soc. Sugar Beet Technologists 13:68–80. 1964.
24. Tisdale, S. L. and W. L. Nelson. Soil Fertility and Fertilizers, 2nd ed. The Macmillan Co., New York. 1966.
25. Tyler, K. B., F. E. Broadbent, and G. N. Hill. Nitrification of fertilizers. Calif. Agr. 13(7):10, 13. 1959.
26. Ulrich, A. Influence of night temperature and nitrogen nutrition on the growth, sucrose accumulation and leaf minerals of sugarbeet plants. Plant Physiol. 30:250–57. 1955.
27. ———. Plant analyses in sugarbeet nutrition. *In* W. Ruther, ed., Plant Analysis and Fertilizer Problems, Publ. 8, pp. 190–211. Am. Inst. Biol. Sci., Washington, D.C. 1960.
28. Ulrich, A. and F. J. Hills. Sugarbeet Nutrient Deficiency Symptoms—A Color Atlas and Chemical Guide. Univ. Calif., Div. Agr. Sci. 1969.
29. Walker, A. C., L. R. Hac, A. Ulrich, and F. J. Hills. Nitrogen fertilization of sugarbeets in the Woodland area of California. I. Effects upon glutamic acid content, sucrose concentration and yield. Proc. Am. Soc. Sugar Beet Technologists 6:362–71. 1950.
30. Williams, W. A. Management of nonleguminous green manures and crop residues to improve the infiltration rate of an irrigated soil. Soil Sci. Soc. Am. Proc. 30:631–34. 1966.
31. Williams, W. A. and L. D. Doneen. Field infiltration studies with green manures and crop residues on irrigated soils. Soil Sci. Soc. Am. Proc. 24:58–61. 1960.
32. Williams, W. A. and D. Ririe. Production of sugarbeets following winter green manure cropping in California. I. Nitrogen nutrition, yield, disease, and pest status of sugarbeets. Soil Sci. Soc. Am. Proc. 21:88–94. 1957.

6

Phosphorus and Potassium Nutrition

W. R. SCHMEHL
Colorado State University
Fort Collins

D. W. JAMES
Utah State University
Logan

Phosphorus Nutrition 138
    Physiological Role of Phosphorus . . 138
    Phosphorus Uptake 139
    Phosphate Fertilization 143
    Effect of Phosphate Fertilizer on
        Crop Response 149
    Residual Value of Fertilizer Phosphate 150
    Determining the Need for Phosphorus 151
    Future Considerations 153

Potassium Nutrition 153
    Physiological Role of Potassium . . . 153
    Potassium Uptake 156
    Root and Sucrose Yield Interactions
        between Potassium and Other
        Nutrients 159
    Predicting Potassium Needs 163
    Potassium Fertilization 164
    Outlook 165

PHOSPHORUS and potassium are major plant nutrients needed in sugarbeet production. Phosphate fertilizer has been used in most sugarbeet-producing areas of the United States for more than 40 years. Potassium has been used less extensively, but it is required frequently in the Midwest. A survey of sugarbeet production in the United States has shown that yields have increased slowly over the past 25 years (see Chapter 1 by Theis). Improved management as the result of research in phosphorus and potassium nutrition of the crop has contributed considerably to this increase. The objective of this chapter is to review the present status of phosphorus and potassium in sugarbeet nutrition and their relation to current fertilizer practices and to possible future needs.

PHOSPHORUS NUTRITION

Phosphorus is one of the essential plant nutrients and the need for this fertilizer element for sugarbeets grown in the United States has been recognized for many years. Maxson (43) reported that a physiological disorder observed in sugarbeets as early as 1916 was diagnosed subsequently as phosphate deficiency. During the past 30 years, many greenhouse and field experiments have been conducted to study rates and methods of application and phosphate materials for sugarbeets. This bank of knowledge will be reviewed in relation to current sugarbeet production practices.

PHYSIOLOGICAL ROLE OF PHOSPHORUS

Phosphate is present in the plant both in inorganic form and in organic combination. Inorganic phosphate is found in the orthophosphate radical, principally as the $H_2PO_4^-$ ion. The inorganic form in sugarbeets ranges from about 60 percent of the total phosphorus in the harvested root to about 40 percent of the total in the leaf blades. The inorganic form occurs principally in vacuolar sap and in conducting tissue. It is a component of the buffer system in the plant and has an important role in cellular metabolism of the plant (4).

The organic compounds that contain phosphorus include phytin, phospholipids, nucleoproteins, nucleic acids such as ribose nucleic acid (RNA) and deoxyribose nucleic acid (DNA), certain enzymes and coenzymes (ATP, NAD, NADP), and the phosphorylated sugars (4).

Phytin is a form of phosphorus stored in seed and is the principal source of phosphorus for the germinating seedling. The phospholipids, along with protein, are constituents of cell membranes and have a role in the transport of ions across membranes. The nucleoproteins and nucleic acids constitute a large part of the nucleus and also are present in cytoplasm. DNA is confined to the chromosome and primarily functions in the transmission of hereditary characteristics. RNA is distributed more widely and is associated with the synthesis of proteins. The coenzymes NAD (nicotinamide-adenine-dinucleotide) and NADP (nicotinamide-adenine-dinucleotide phosphate) are important in oxidation-reduction reactions in plant processes such as photosynthesis, glycolysis, respiration, and fatty acid synthesis. Adenosine triphosphate (ATP), another coenzyme, is one of the most important phosphorus-containing compounds in the plant. It functions as an intermediate energy transfer compound which receives energy from one reaction and by transfer of the energy causes another reaction to take place. Among the many plant processes in which ATP is involved are carbohydrate metabolism, respiration, photosynthesis, the synthesis of sucrose and many other compounds in the plant, the opening and closing of stomata, and the movement of sugars across cell membranes.

Phosphorus is found in highest concentrations in meristematic regions of actively growing tissue where it is involved in the synthesis of nucleoproteins. It is not permanently fixed in a cell or tissue but is a mobile nutrient which is redistributed in accordance with phosphorus demands within the plant. Thus, in deficient plants, phosphorus moves from the older to the younger tissues where the demands are greater, and deficiency symptoms appear first on the older leaves.

Phosphate deficiency in sugarbeets is very difficult to diagnose by foliar symptoms (54, 68). When the soil is deficient in available phosphorus, the plants grow very slowly in the seedling stage and the leaf area is reduced throughout the deficiency period. If the deficiency is not too severe, the plants have a normal green color and stunting is the only indication of the deficiency. Early in the season the stunted growth (Fig. 6.1) may appear similar to the reduced growth resulting from late planting or from a long, cool spring. As the soil temperature increases, root growth increases and the plant may recover from the early stunting if the deficiency does not persist.

Where phosphate deficiency is severe, the leaves may develop a metallic luster and the color may range from a dull, gray-green to a blue-green (68). In older, severely deficient plants, a brown netted veining may develop on the older leaves. Wallace (71) reports that the leaves may exhibit a slight bronzing or reddish coloration on the leaf margins.

Maxson (43) and Hull (29) report that an acute phosphate deficiency may cause a brown or black interveinal necrosis and the petioles and blades tend to curl upward. In advanced stages the plant dies, leaving a whorl of dead leaves around the crown. Acute deficiencies of this type, however, are seldom observed in the field.

PHOSPHORUS UPTAKE

Phosphorus is absorbed by plants principally in inorganic form as the orthophosphate ion. Research with other species has shown that the $H_2PO_4^-$

FIG. 6.1. Response of sugarbeets to the application of phosphate fertilizer. Control plots left foreground and right background. Phosphate fertilizer at the rate of 250 lb P_2O_5 per acre applied right foreground and left background.

ion is absorbed more readily than the $HPO_4^=$ ion (35), possibly as the consequence of more absorption sites on the plant roots for $H_2PO_4^-$ ions (24). Although phosphate absorption tends to maximize at about pH 6.5 to 7.0, it will depend somewhat upon other ions in the system (26, 35). The major influence of pH on phosphate absorption apparently is that of ionic species in the system. In a solution at pH 7.2, $H_2PO_4^-$ and $HPO_4^=$ are found in equivalent amounts, but with a further increase in pH, the proportion of $H_2PO_4^-$, the more readily available form, decreases rapidly. There have been no specific studies with sugarbeets, but the general behavior of phosphate absorption is similar for other plant species that have been tested.

The amount of phosphorus absorbed by a crop of sugarbeets will depend upon factors that influence the seasonal growth characteristics of the plant. At a given level of soil phosphorus, available soil nitrogen and climatic environment generally are the principal determinants of phosphorus uptake because of their influence on total dry matter production.

The accumulation of dry matter shown in Figure 6.2 is typical for a crop grown in a favorable climatic year in northeastern Colorado where the frost-free period is about 140 days (59). The curves show total dry mat-

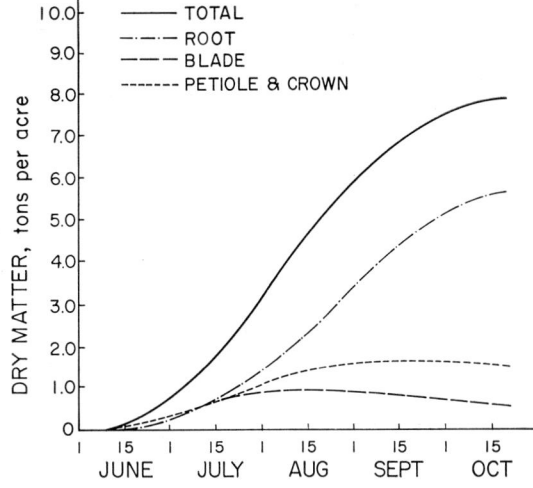

FIG. 6.2. Seasonal dry matter accumulation by the sugarbeet crop. (Storer, 59.)

ter production and the seasonal distribution of dry matter between the root and top parts of the plant for an optimum level of available nitrogen. The dry weight of the root maximizes at harvest but the top portion (blades, petioles, and crowns) maximizes during August. The dry weight distribution between top and root is favorable for maximum sugar accumulation per acre at harvest. When soil nitrogen is too high, the top will represent an excessive proportion of the total dry weight, and when nitrogen is deficient, the top-root ratio will be too low.

The total phosphorus concentration (percent P) in the sugarbeet plant during the season is shown in Figure 6.3 (59). The data are typical for the growing conditions represented by the dry matter production shown in

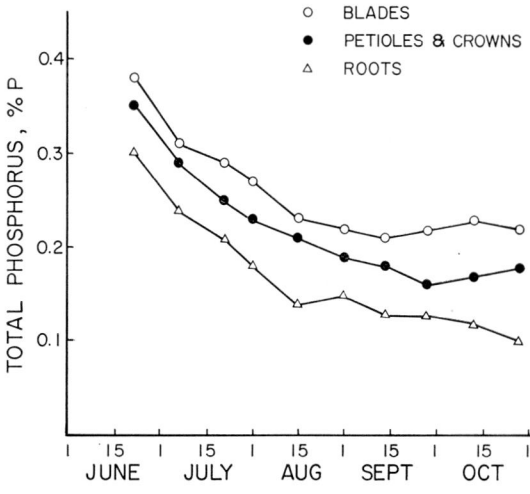

FIG. 6.3. Total phosphorus content of the sugarbeet crop during the season. (Storer, 59.)

Figure 6.2. Two significant relationships shown in Figure 6.3 generally hold for sugarbeets: (1) the total phosphorus contents of the plant parts are in the order, blades > petioles plus crowns > roots; and (2) the phosphorus content decreases as the season progresses. The second generalization may not occur if new supplies of phosphorus become available from a localized fertilizer placement as the season progresses or if the roots extend into a soil zone high in available phosphorus. The general relationships do hold, however, for variations in growth factors that do not influence the availability of soil phosphorus. By considering both dry matter production and phosphorus content, it is evident that the top-root ratio may have a marked influence on total phosphorus uptake per ton of roots.

A typical phosphorus accumulation curve for sugarbeets is shown in Figure 6.4 for a 25-ton crop. Phosphorus is absorbed for a longer period during the season than is nitrogen or potassium (59). The seasonal uptake curve for phosphorus has the same general shape as the dry matter accumulation by each of the plant parts, but uptake reaches a maximum sooner during the season than does dry weight. Total root phosphorus maximizes at harvest, but phosphorus in the top may maximize during August or September and decrease later in the season when leaves are lost as the result of nitrogen deficiency or freezing temperatures.

About 50 percent of the total plant phosphorus is in the root, about 30 percent in the petioles and crowns, and about 20 percent in the blades where the dry weight root-top ratio is approximately 2 at harvest. For the example in Figure 6.4, the total phosphorus content of the crop is about 1¼ pounds of phosphorus per ton of roots produced. The value is typical

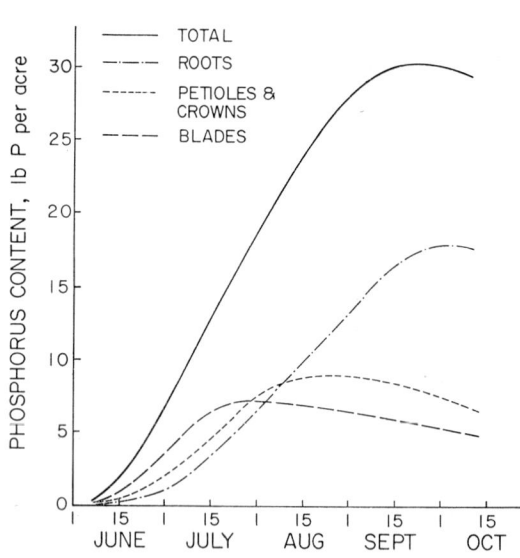

FIG. 6.4. Seasonal accumulation of phosphorus by the sugarbeet crop. (Storer, 59.)

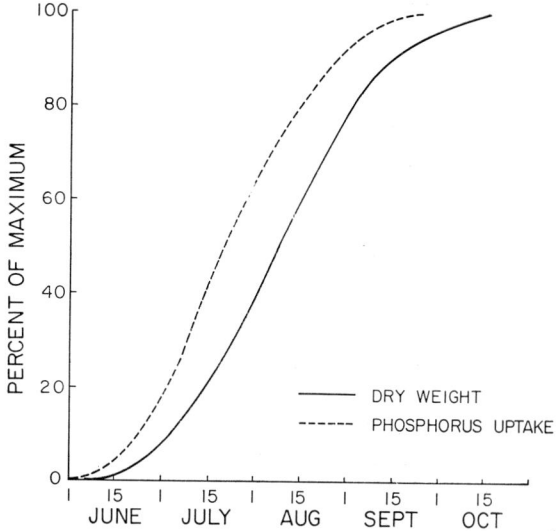

FIG. 6.5. Relative rates of phosphorus uptake and dry weight production for sugarbeets.

for sugarbeets under good growing conditions with optimum levels of nitrogen and phosphorus for yields of 20 to 30 tons per acre. With a large increase in top growth from excessive soil nitrogen and adequate available phosphorus, uptake of phosphorus may increase to 1½ pounds per ton of roots produced. With phosphorus deficiency, uptake may range as low as 0.75 pound of phosphorus per ton of roots.

Total dry weight production and total phosphorus uptake during the season (Fig. 6.2 and 6.4) were replotted in Figure 6.5 to compare the relative accumulation of dry weight and phosphorus during the season. The data show that the relative phosphorus uptake precedes dry matter production. For example, 50 percent of the total phosphorus was absorbed by July 20, whereas only about 25 percent of the total dry matter was produced at that time. This is consistent with the common observation that the influence of phosphate is greater when the fertilizer is applied early in the season.

PHOSPHATE FERTILIZATION

Soil Phosphorus. Plant roots absorb phosphorus from the soil solution. The amount of inorganic phosphorus in solution at any time is very small and ranges from about 0.01 ppm to 2 or 3 ppm. The soil-solution phosphorus is in equilibrium with solid-phase phosphorus. The equilibrium between the solid-phase, solution, and plant-absorbed phosphorus can be represented by the equation

$$\begin{pmatrix}\text{Inorganic phosphate}\\\text{and}\\\text{Organic phosphate}\end{pmatrix} \rightleftarrows \begin{pmatrix}H_2PO_4^-\\\text{and}\\HPO_4^=\end{pmatrix} \rightleftarrows \begin{pmatrix}\text{Plant-absorbed}\\\text{phosphate}\end{pmatrix}$$

Solid-phase phosphate (Capacity)   Soil-solution phosphate (Intensity)

The soil-solution phosphorus is frequently designated as the "intensity factor" and the solid-phase phosphorus as the "capacity factor" for soil phosphorus.

Solid-phase compounds in the soil may be placed in four broad groups: (1) organic phosphorus, (2) phosphates adsorbed on soil particles (surface phosphate), (3) sparingly soluble calcium, iron, and aluminum phosphates, and (4) occluded phosphate (22). Organic phosphorus, which becomes available as the result of microbial activity, and the adsorbed phosphate (surface phosphate) are the principal groups that determine the level of soluble phosphorus in the soil solution. The capacity factor is determined principally by the adsorbed phosphate and the organic phosphorus.

Since the concentration of phosphorus in the soil solution is so low, it must be resupplied many times a day to meet the needs of the sugarbeet plant. If the rate of absorption by the plant is 0.4 pounds per day (Fig. 6.4), and if the concentration of phosphorus in the soil solution of a loam soil is 0.2 ppm, phosphorus in solution in the top foot of soil has to be replaced about six times daily to supply the phosphate needed during the period of maximum uptake.

The equilibrium between the solid-phase and solution phosphorus is very important in determining the phosphate fertility level of the soil. Soils are deficient in available phosphorus when the solid phase cannot supply the soil solution with phosphorus required for the maximum rate of growth, for example, with 0.4 pound phosphorus per day as shown in Figure 6.4. This is the equilibrium that is altered by the application of phosphate fertilizer.

When phosphate fertilizer is applied, it rapidly reacts chemically with the soil. A small part of the phosphate fertilizer remains in solution and increases the intensity factor, but the major portion is adsorbed on the surfaces of soil particles or is precipitated as discrete calcium, aluminum, or iron phosphates. Both reactions result in "phosphate fixation," a term used to describe the chemical processes that remove fertilizer phosphorus from solution. Phosphorus fixation has considerable significance because it not only increases the capacity factor of the soil but also reduces the leaching of phosphate fertilizer. The capacity factor is increased more where a larger proportion of the phosphate is adsorbed on the soil surfaces than where it it is precipitated as calcium, aluminum, or iron phosphate. With an increase in the capacity factor, solid-phase phosphorus is released into solution more rapidly, thereby increasing the phosphate fertility level. The effect of application of phosphate fertilizer on this equilibrium will depend upon both soil and fertilizer characteristics as well as on the method and rate of application of the fertilizer.

Phosphate fertilizer moves only a few millimeters from the point of application because of phosphate fixation. Hence, there is very little loss of phosphate by leaching from the soil. Phosphate fixation also is responsible for the residual or carry-over effect of applied fertilizer.

Phosphate Materials. Phosphate fertilizer is sold on the basis of available phosphorus content, which usually is determined in the laboratory by an empirical chemical procedure. The neutral normal ammonium citrate method (5) is used in the United States to define available phosphorus for the administration of control laws. Many years of study have shown that this chemical analysis correlates highly with the effectiveness of the phosphate in a fertilizer when applied under a wide range of cropping conditions. This method does not necessarily give the best value for all situations, but it does give an index of available phosphorus which has been successful for most soil and crop conditions.

On the basis of the neutral normal ammonium citrate procedure, the total phosphorus in a fertilizer may be classed as (1) water soluble, (2) citrate soluble (citrate soluble but not water soluble), and (3) citrate insoluble. Available phosphorus content of a fertilizer is the sum of the water-soluble and citrate-soluble fractions. The method of application and possibly the rate of application of a phosphate fertilizer should be based on the relative amounts of the water-soluble and citrate-soluble fractions.

Phosphate fertilizers are produced for sale as straight goods and as mixed fertilizers. Straight goods that contain 20 percent available P_2O_5 (8.7 percent P) or more include the superphosphates, calcium metaphosphate, and phosphoric acid. The water solubility of the phosphate in calcium metaphosphate is less than 1 percent, but superphosphate is about 90 percent water soluble and phosphoric acid is 100 percent water soluble. Phosphate rock contains less than 5 percent available phosphorus with no water-soluble phosphorus. The latter material should not be used as a phosphate fertilizer for sugarbeets when applied at usual rates of fertilization.

The mixed fertilizers include nitrogen and/or potassium along with phosphorus. Among such materials are the ammoniated superphosphates, ammonium phosphates, nitric phosphates, ammonium polyphosphates, potassium metaphosphate, and the various dry and liquid blends. The water solubility of the phosphorus in mixed goods may range from a low of several percent for some of the nitric phosphates and ammoniated superphosphates to 100 percent water solubility for the phosphorus in the ammonium phosphates.

The superphosphates, concentrated or ordinary, are the commonest single nutrient phosphate fertilizers used for sugarbeets. Concentrated superphosphate contains 45 to 55 percent P_2O_5 (20 to 24 percent P) and ordinary superphosphate has about 20 percent P_2O_5 (8.7 percent P). Superphosphate is a suitable phosphate fertilizer for any of the methods of application commonly used for sugarbeets (37, 51, 55, 56). The ammonium phosphates and the polyphosphates of high water solubility are equally as satisfactory as superphosphate for most management practices (51, 55, 56).

When phosphate fertilizers of low water solubility such as calcium metaphosphate, the ammoniated superphosphates, and the nitric phosphates are applied to beets, the particle size of the fertilizer and method of application must be taken into consideration. Schmehl et al. (55, 56), Olsen et al. (51), and Peterson (52) note that these materials are unsatisfactory sources when applied in bands, but where the fertilizer is mixed into the plow layer, they are generally as suitable phosphate sources as concentrated superphosphate. The availability of low water-soluble-phosphate fertilizer also increases as the particle size decreases.

Calcium metaphosphate, as well as other polyphosphates, must first go into solution, then hydrolyze to the orthophosphate form before the phosphorus is absorbed by the root. Consequently, there may be a delay before an application of calcium metaphosphate becomes available to the plant. Hydrolysis of calcium metaphosphate is hastened by applying finely ground material and by mixing the fertilizer with the soil. To be as available a phosphorus source for sugarbeets as superphosphate, calcium metaphosphate must be minus-40 mesh or smaller particle size material for most soils (51, 55, 56).

Method of Application. The most effective way to apply phosphate fertilizer for sugarbeets depends upon the rooting characteristics of the plant, type of phosphate material, soil properties, and cropping system. Of these factors, the rooting pattern is one of the more important. The sugarbeet is a long-season crop that normally has an extensive root system, both vertically and laterally (34). Early in the season, however, the taproot grows more rapidly than do the lateral roots, and it may penetrate to depths of 12 to 18 inches before lateral roots extend more than an inch or so. Also, in cool soils typical of spring plantings, the root system develops slowly and phosphate absorption by the root is reduced. Although the total phosphorus requirement of the plant is not large at this time (Fig. 6.4), the intensity factor of the soil phosphorus should be high because of the small root system and cool soil. Later in the season during the grand period of growth, the plant has a more extensive root system. The daily phosphate requirement is highest during this time (Fig. 6.4), and the capacity factor of the soil phosphorus must be adequate to supply phosphorus at the maximum absorption rate for an 8- to 10-week period. Thus, because of the type of root system for the sugarbeet, some of the phosphate should be placed near the seed for early uptake by the rapidly developing taproot, but most of the phosphate should be made accessible to a larger root system for use during the peak uptake period. Since phosphorus moves no more than a few millimeters from where it is placed in most soils, distribution of phosphate fertilizer throughout the soil must be accomplished by mechanical methods such as plowing, discing, or banding. Fortunately, the results of many field experiments conducted under a wide variety of conditions have shown that several methods of application may be equally effective, but variations in soil and climatic conditions may dictate certain local preferences.

Phosphate broadcast before plowing is an effective method of application for beets grown in the alkaline and neutral soils under irrigation in the West (37, 53, 56). Similar results are reported for the slightly acid soils of Great Britain (11). Fall is about as effective as spring for plowdown applications of phosphate fertilizer (9, 56). The success of plowdown or deep incorporation of phosphate apparently is twofold. First, it gives a random placement of fertilizer particles throughout the plow layer so that some particles are near or below seed. This enables the rapidly extending taproot to contact the fertilizer sooner than lateral roots can contact a sideband placed fertilizer. Root contact with plowdown phosphate may not be as rapid as when the fertilizer is placed with or immediately below the seed, but the time interval does not seem to be sufficiently great to affect final crop yields (37, 53, 56). A second reason for the success of the plowdown application on these soils is the lack of rapid fixation of the phosphate into plant-unavailable form.

The incorporation of phosphate into the surface two to three inches of soil has been an effective method of application in Great Britain (1, 11) and in some parts of the United States (37, 53). To be effective, however, soil moisture in the fertilized zone must be favorable for root growth throughout the growing season. Surface-incorporated phosphate is unavailable to the plant when drouthy soil conditions prevent root growth into the fertilized zone.

Band-placed phosphate for sugarbeets has certain theoretical advantages because (1) the fertilizer is placed near the seed to stimulate early growth and (2) the band placement reduces phosphate fixation. Many experiments have shown that band placement of phosphate near or with the seed stimulates early growth and early phosphate uptake by the sugarbeet (13, 23, 37, 38). It also has been observed frequently that the early growth response to phosphate does not necessarily carry through the season to give higher final yields (11, 37). Thus, the effectiveness of band placement should be based on final harvest yields rather than on early growth effects. When based on final harvest yields, band-placed phosphate has been more effective than incorporation by discing or plowing down in some experiments (7, 14, 44). In other experiments either there has been little difference among application methods or the band placement has been less effective (7, 37, 56). To be most effective the band of phosphate should be placed as near the seed as possible (14, 37, 38). The National Joint Committee on Fertilizer Application (48) recommends that fertilizer should be placed no more than one and a half inches to the side and one to two inches below the seed. Also, highly water-soluble phosphate should be used when fertilizer is band applied. Apparently, differences in position of the fertilizer band, water solubility of the phosphate, rate of application, and possibly other factors have resulted in inconsistent results with band placement. Since the results of band placement have been inconsistent, incorporation methods are more widely used, primarily because of easier application. Band-placed phosphate generally is recommended, however, for minimal rates of phosphate or for application of "starter fertilizer."

A small amount of phosphate in a fertilizer mix placed near the seed, usually called "starter fertilizer," may give an additional increase in yield of beets (13, 14). The benefit from the starter will be greater when placed near and below the seed rather than several inches to the side, and is greater in soils of very low levels of available phosphate. The advantage of starter phosphate placed with or below the seed may not be realized if germination is reduced or if the seedbed is unduly disturbed. Surface incorporation or plowdown applications of a starter phosphate mixture may be as effective as band placement of starter fertilizer where the soil is medium low in available phosphorus and where good soil moisture is maintained (Table 6.1).

The method to apply a phosphate fertilizer will be determined also by the proportion of water-soluble phosphorus. Under most conditions, phosphate materials of low water solubility are less effective for sugarbeets when placed in band than when incorporated into the soil (51, 55, 56). Phosphate materials of low water solubility that should be incorporated into the soil are the calcium metaphosphates, the highly ammoniated superphosphates, and some nitric phosphates.

The literature on the method of application of phosphate fertilizer for sugarbeets may first appear to be contradictory. If, however, interpretations are based on final crop yields, management practices can be outlined by considering differences among soils, fertilizer materials, and cultural practices. Also, small differences in yield due to fertilizer placement for sugarbeets may be obscured by other factors when considered on a rotation basis.

The sugarbeet is a high-value crop, and for maximum efficiency in production the crop should not be grown at marginal levels of phosphate fertility. The phosphorus requirement is higher than for many field crops,

TABLE 6.1. Effect of method of application of starter fertilizer on yield of sugarbeets in tons per acre on Nunn clay loam, Colorado, 1962

Phosphorus Source	Method of Application of N-P Fertilizer*			
	Fall plow down	Disc after plow	Band†	P band N disc
	T/A	T/A	T/A	T/A
Conc. super	20.3	20.0	19.8	20.0
Am. poly	19.9	19.9	20.7	...
Mean	20.2	20.3	19.9	20.0

NOTE: 60 lb N and 30 lb P_2O_5 per acre in starter, 60 lb N uniform application preplant.

* No phosphorus check—18.1 tons per acre.
† Band 2 inches to the side and 1½ inches below the seed at planting.

and fertilizer applied to sugarbeets frequently has sufficient carry-over to supply the phosphate needs of the other crops in the rotation until returning to sugarbeets. Under most conditions, it will be more profitable to fertilize with phosphate on a rotation basis. Larger applications of phosphate are applied to supply phosphorus for maximum yields of sugarbeets. Carry-over phosphate from the beet crop is used to furnish the phosphate for the succeeding crops in the rotation. With larger amounts of phosphate, the method of application is not as important a consideration, and fertilizer usually can be incorporated into the root zone at the time of plowing. If a shallow incorporation is used, moisture relations must be optimum throughout the season. Where the fertilizer is incorporated, the water solubility of the phosphate is of small consequence and is applied on the basis of total available phosphorus content of the fertilizer. The cost of application is less for incorporation methods because of the savings in the cost of attachments and the cost of fertilizer bags, as well as a saving in labor in the spring when the schedule for cultural operations may be critical. Plowdown phosphate can be broadcast by bulk spread equipment in the off-season, when farm work is less pressing. Because of the various advantages, phosphate fertilizer is generally incorporated into the soil before planting sugarbeets by plowdown or deep discing.

If the phosphate rate is kept minimal, and particularly on acid soils having a high fixing capacity for phosphorus, the fertilizer possibly should be placed in a band below and to the side of the seed. Where band-placed fertilizer is applied, care should be exercised to prevent excessive disturbance of the seedbed. If a mixed fertilizer is placed near the seed, nitrogen and potash rates should be low to prevent injury to the germinating seed.

Nitrogen-Phosphorus Interactions. Incorporation of nitrogen into a phosphate fertilizer applied in band may increase the growth and early season uptake of the fertilizer phosphorus by sugarbeets (23, 70, 72). This may be one of the reasons for the starter effect previously noted. Under other situations, however, the incorporation of nitrogen has had no additional effect on yield or phosphate uptake (51, 55, 58). For example, the results in Table 6.1 show no advantage to a nitrogen-phosphorus combination in band for sugarbeets grown in the calcareous Nunn clay loam nor did the band application have an appreciable effect on the total phosphorus content of the leaves. Again, as noted with band-placed phosphate, early season uptake and growth are not reliable criteria for the evaluation of the effect of starter fertilizer on final yield. In general, if available soil nitrogen is limiting, the application of this nutrient element by any method of application will increase plant growth, total phosphate uptake, and crop yield.

EFFECT OF PHOSPHATE FERTILIZER ON CROP RESPONSE

When the soil is deficient in available phosphorus, the application of phosphate fertilizer produces a marked increase in early top growth in comparison with unfertilized soil (Fig. 6.1). If the deficiency is severe, top growth responses will persist throughout the season, but if the deficiency is only moderate, the visual differences may disappear by midseason or earlier, because with root extension there is greater exploitation of soil phosphorus. A response in root yield to application of phosphate usually will be obtained if the top response lasts to midseason and longer. Other growth factors should be near optimum to get maximum benefit from phosphate fertilization.

The effect of phosphate fertilizer on the sucrose content of the root is variable and difficult to define (12, 49). In phosphate-deficient soils, the application of phosphate fertilizer may increase the sucrose content of the root at intermediate levels of soil nitrogen but generally not at high levels of nitrogen. When applied to a nonresponding soil, additional phosphate has little effect on sucrose content of the root at low or high soil nitrogen. Thus, the application of high rates of phosphate fertilizer does not overcome the detrimental effects of excessive nitrogen on quality of the root.

The increase in top growth resulting from phosphate fertilization of deficient soils does not cause the decrease in sucrose and purity commonly associated with increased top growth when nitrogen fertilizer is applied. Nor is the root-top ratio changed much by phosphate fertilization, whereas it can be decreased markedly from the application of nitrogen.

RESIDUAL VALUE OF FERTILIZER PHOSPHATE

The uptake of fertilizer phosphorus by sugarbeets ranges from about 10 to 15 percent of that applied the year of fertilization (12, 51). Thus, if phosphate is applied to crops annually or even every two or three years, crop removal will be less than that applied. Because there is little movement or leaching of phosphate from most soils, residual phosphate accumulates in the soil. At least part of the residual phosphate increases the capacity factor of the soil phosphate and therefore the level of available soil phosphate increases. The carry-over effect of phosphate fertilizer may continue for several years after a single application, and continued frequent applications may result in a residual phosphate level so high that succeeding crops no longer respond to further application. Adams (3) reports that phosphate carry-over from continued use of fertilizer in Great Britain has resulted in very little or no response for sugarbeets to further application of phosphate fertilizer on most of their farms . The residual value of phosphate fertilizer should be recognized when developing a fertility management system on a rotation basis. All crops in the rotation should be adequately supplied with phosphate, but excessive rates of application should be avoided to prevent possible micronutrient imbalances in the plant. Such imbalances are commoner for the other crops in the rotation than for sugarbeets.

DETERMINING THE NEED FOR PHOSPHORUS

The sugarbeet crop has a high phosphorus requirement, and loss in yield because of an inadequate supply of phosphorus is observed with beets before most other crops grown in the rotation. There may be several reasons for poor yields, however, and phosphorus deficiency must be identified as the specific limiting factor. Visual diagnosis can be used to identify phosphorus deficiency for some crops, but it is not a suitable method for determining the phosphorus needs of sugarbeets. There are no characteristic symptoms unless the deficiency is severe; by then the crop will not fully recover even if fertilizer phosphorus is added. Thus, more definitive methods must be used, such as field testing, plant tissue analysis, and soil analysis.

Field testing is the classical method for assessing the fertilizer requirements of sugarbeet soils and for determining the effects of added nutrients in commercial fields. These tests consist of randomized replicated experiments that usually involve one or more nutrient elements. This kind of investigation has been conducted in the United States, in Western Europe, and in Great Britain over the years. A simpler and commonly recommended practice is for the individual grower to establish strips of treated and nontreated areas in beet fields and to compare the yield and quality effects of the fertilizer.

Continuous field plot work has the effect of eventually upgrading the yield and quality of a crop. Disadvantages of field-testing procedures are that they are time-consuming and costly, and the results are not known until the end of the season. Field testing is used principally as a research tool to determine rates of fertilization and interaction among nutrients and to correlate the results with alternate testing procedures, such as soil and plant analyses.

Tissue Testing. Plant analysis is a useful tool for diagnosing the nutrient status of the sugarbeet. In addition, analyses for nutrient concentrations within the plant have been used successfully to predict a crop's needs or to predict the ultimate performance of a crop in terms of yield and quality. Ulrich (66) has discussed the practical and theoretical aspects of the analysis of sugarbeet tissues.

Sugarbeets are notorious for their genetic variability in the characteristics that control nutrient uptake. Accordingly, before any credence can be given to results of plant analysis, careful attention must be devoted to the problem of sampling. Ulrich and co-workers (67, 69) studied field-sampling techniques and have outlined good sampling procedures. The size of field and pattern followed in the field in collecting a sample, the number of plants represented by a sample, the specific plant part collected, the relative age of the plant part, the time of season, the method of preparing the tissue and subsample for analysis, and the analysis itself are each important in establishing the integrity of a given result.

In the technique developed by Ulrich and co-workers (67, 68, 69), the first fully matured leaves are removed from 30 to 50 plants in a sampling unit. The petioles are separated, diced, dried at about 65° C, and analyzed for phosphorus soluble in 2 percent acetic acid. Soluble phosphorus below 750 ppm indicates a deficiency in the plant at any stage of growth between thinning and harvest. In the seedling stage, the critical phosphorus level is about 1,500 ppm. If tissue analysis indicates a phosphorus status below the critical level, the crop will respond to the application of phosphate fertilizer. Phosphate should be applied as soon as the deficiency is detected to prevent further reduction in growth. Because phosphorus moves very little in most soils, the fertilizer should be placed in the root zone.

An optimum supply of phosphorus to the plant is needed during the seedling stage and throughout the growth period. When plant analysis alone is used, a delay of 4 to 8 weeks may result before the deficiency is identified and phosphorus is applied. Also, applications of phosphate fertilizer after the crop is growing usually are not as effective as preplant applications. Such delays may cause the loss of several tons of beets at harvest. Consequently, the identification of the phosphorus status of the soil should be made early so that phosphate can be applied preplant or no later than at planting. Plant analysis of the crop preceding sugarbeets may be a useful fertilizer guide for preplant applications of phosphate.

Soil Analysis. Soil tests for nutrient availability have a distinct advantage over other methods of prediction because information can be obtained beforehand, and fertilizer can be applied before planting to preclude the possibility of nutritional stress. This assumes, of course, that background information is in hand for interpreting soil test results.

Soil testing involves two separate and important steps. First, a field must be sampled in such a way as to provide soil samples that truly represent the area. Because of natural variation, it is often desirable to subsample large fields in a manner that will allow pinpointing or mapping of major fluctuations in the soil test results. Second, the chemical procedure must extract from the soil an amount of the nutrient which is proportional to that available to the crop.

The chemical extractant for the phosphorus test is selected on the basis of the chemistry of the soil. The level of soluble phosphorus in the extract is correlated with the plant growth response to applications of phosphate obtained in carefully conducted field or greenhouse experiments. After the correlation has been derived, the soil test is used to define the phosphorus status of the soil for any given crop.

Two chemical soil test procedures for phosphorus are now used in most sugarbeet-growing areas of the United States. The sodium bicarbonate method (50) is used for calcareous and neutral soils and the Bray method (6) for neutral and acid soils. The principal limitations of both methods are operational rather than technical. First, the soil sample may not represent the field unit, and second, the sugarbeet crop is not grown under the optimum management conditions used for correlating the soil test. Although yield responses are not always consistent with the soil test results, an agronomist can obtain a high degree of success in his recommendations as he gains experience with the management practices of farmers in a given area.

FUTURE CONSIDERATIONS

Phosphorus nutrition of sugarbeets presents no major problems at current yield levels. General fertilizer recommendations concerning method and rate of application, phosphate material, and diagnostic methods are sufficiently well understood for most situations.

Improved production techniques are expected in the future, however, that will give significant increases in yield. New production levels will be attained because of improved hybrids, better management, and innovations. The new hybrids will be higher yielding and of higher quality. Management will include improved cultural practices and better fertility, water, and pest control. Innovations will be devised to exert some control of the microenvironment to fit the plant more effectively to local climatic conditions.

As the result of increased production, the phosphorus needs of the plant will be greater. The relation of phosphorus to other nutrients and to other production practices will become more important. Soil and plant testing methods will need to be reevaluated for the new hybrids and for the new production levels. More precise control of the soil phosphate level will be required to provide an optimum level of phosphorus for the sugarbeet crop and to prevent excessive rates that may cause micronutrient imbalances in succeeding crops. Refined soil and plant analysis techniques will permit the grower to follow the plant nutrition and to guide phosphate fertilization of all crops in a rotation on a continuing basis.

POTASSIUM NUTRITION

Sugarbeet is classified as a plant that has a high requirement for potassium, and more potassium is absorbed by sugarbeets than any other mineral nutrient element. Accordingly, the literature is voluminous on the subject of soil-plant relationships involving potassium and sugarbeets. This section will deal only with the salient features of the subject, but sufficient references are given to guide the reader to more detailed information on each facet.

PHYSIOLOGICAL ROLE OF POTASSIUM

The specific role or action mechanism of many nutrient elements is known in great detail. Despite the fact, however, that potassium is recognized as being absolutely indispensable and that it is present in high concentrations in plants, most physiology textbooks state simply that the precise function of potassium is unknown. The metabolic role of potassium has nevertheless been inferred from growth disorders or malfunctions that occur when potassium is present in limiting quantities.

Physical Appearance. The most obvious consequence of potassium deficiency in sugarbeets is stunted growth. Other external evidences are chlorosis and necrosis of leaf tissue. Ulrich and Hills (68) state that the first signs of potassium deficiency in sugarbeets appear as a leathery tan color at tip and leaf edges of recently matured leaves. As the deficiency increases, the discoloration extends inward between the veins, the edges turn brown, and a marginal necrosis develops. Brown et al. (8) attributed a "bronzing" appearance of sugarbeet leaves to potassium deficiency. They further indicated that the beet leaves appeared dark bluish green and appeared grainy when observed with transmitted light. Leaves later curled downward and became necrotic around the edges. Finally, the whole leaf became necrotic. Plants were stunted and the leaves were generally smaller than normal.

It is common in potassium-deficient sugarbeets for the young leaves to appear normal while the old leaves show the deficiency symptoms. Nason and McElroy (47) point out that potassium is highly mobile and is readily redistributed within plants during the growing cycle. Metabolically active regions—buds, new leaves, root tips—are particularly rich in potassium. It is evident that potassium accumulates in new tissue at the expense of the old where the overall potassium status is poor.

Plant Anatomy. Looking inside the plant, several kinds of abnormalities occur as a result of potassium deficiency. Potassium is not built into any discrete structure or compound. It usually is observed as a completely soluble component of the cell sap and cytoplasm. Therefore, such changes as do occur are regarded as secondary effects of potassium stress.

According to Hewitt (27), lack of potassium leads to changes in lignification and the thickness of cell walls. Collenchyma cells decrease in number and xylem cells tend to become parenchymatous. In leaves, necrosis of palisade cells is followed by collapse of the spongy parenchyma, with concomitant shrinkage and desiccation of the necrotic areas.

Plant Biochemistry. Biochemically, potassium deficiency causes many changes in the vital processes of plants. Some of these changes have been analyzed in great detail. They are summarized in Table 6.2 under broad groupings.

All of the processes listed in Table 6.2 are catalyzed by plant enzymes. Potassium is thus readily implicated in enzymatic reactions. In fact, potassium is frequently referred to as an enzyme activator. Research specifically directed toward the relationships between potassium and enzyme

TABLE 6.2. Alterations in plant biochemistry processes and compounds which result from potassium deficiency

Changes in carbohydrate metabolism
    Initial increase in soluble carbohydrates
    Ultimate decrease in soluble carbohydrates
    Decreased translocation of carbohydrates
    Decreased storage of carbohydrates
Changes in nitrogen metabolism
    Decreased protein synthesis
    Increased soluble nitrogen compounds—amino acids, amides
Decreased photosynthesis
Decreased chlorophyll formation
Increased respiration

SOURCE: Adapted from Evans and Sorger (20) and Nason and McElroy (47).

chemistry has been very fruitful. Evans and Sorger (20) tabulated some 40 different enzymes that either do not function at all or function at a reduced rate in the absence of potassium.

From the foregoing it becomes evident that potassium is very intimately associated with the substances that are responsible for plant metabolic processes. This is despite the fact that potassium ions exist in plants essentially in solution in ionic form.

Recent Developments. Consideration of the foregoing ideas led Evans and Sorger (20) and Wilson and Evans (73) to postulate a discrete function for potassium in plant physiology. Their theory states that enzymes (which are large protein molecules) are flexible bodies and that their shape or conformation is subject to environmental conditions. In addition, a particular conformation is most advantageous for a catalysis to proceed. Thus, peculiar surfaces, bonding sites, or active groups are "available," depending upon their exposure in folds or depressions of the enzyme body. According to this model, environmental requirements for optimum enzyme activity include the potassium ion in relatively high concentrations with its peculiar electrostatic-hydration radius characteristics.

The physicochemical properties of divalent cations eliminate them as enzyme activators. However, certain other univalent cations do elicit similar, though not necessarily identical, responses as does potassium in terms of enzyme activity. From a practical point of view sodium is the most important alternate univalent cation in the mineral nutrition of plants. This is especially true for sugarbeets, because this plant has been recognized for a long time as a sodium accumulator.

Sodium and Potassium Nutrition of Sugarbeets. Studies on the mineral nutrition of sugarbeets indicate that sodium may be considered from two points of view. First, sodium is considered to be able to substitute for part of the sugarbeet's potassium needs. This school of thought is exemplified by Cook and Davis (10) and Dorph-Petersen and Steenbjerg (16).

The second point of view holds that sodium does more than simply substitute for potassium; it also serves a unique role and thus should be classified as an essential nutrient element. El-Sheikh et al. (19) give strong evidence for the latter theory. There is abundant evidence of a circumstantial nature supporting this idea (25, 39, 63). Harmer et al. (25) classify 48 different plants as to their responses to sodium. Members of the sugarbeet species are most prominent in that category where large responses to sodium have been measured in the presence of ample potassium supplies.

For practical purposes it is clear, regardless of the exact function of sodium, that the potassium nutritional status of sugarbeets cannot be intelligently assessed without including sodium in the analysis. Of necessity, therefore, much of the discussion which follows will include sodium, while delineating the various factors of potassium in sugarbeet production management practices.

POTASSIUM UPTAKE

Critical Level and Seasonal Pattern. Ulrich (66) defines critical nutrient level as that concentration of a nutrient in plant tissue at which growth begins to be reduced in comparison to plants adequately supplied. In the case of potassium in sugarbeets, Ulrich (66) points out that the estimation of critical level is complicated by the potassium-sodium interaction. The same total amount of potassium may occur in two different sugarbeet plants, but one may be sufficient and the other severely deficient in this element. The existence of stress is related to the amount of sodium in the respective plants. According to Ulrich (66), when sugarbeet petioles contain more than 1.5 percent sodium, the critical potassium level is approximately 1 percent. When petioles contain less than 1.5 percent sodium, the critical potassium concentration is 2 percent or higher. He suggests that leaf blades be sampled in preference to petioles, because potassium concentration of blade tissue is much less sensitive to sodium. The critical potassium level in blades is given as 1 percent regardless of the sodium concentration.

The nutrient concentration versus expression of deficiency symptoms has even more contrast in a report by El-Sheikh et al. (19). They report that potassium was as high as 2.55 percent in some sugarbeet petioles where the corresponding blades were deficient in potassium. At the same time, other leaf blades appeared normal where the petioles contained only 0.73 percent potassium. In the first instance, sodium concentration was low, and in the second, sodium was high.

The foregoing reports were from solution-culture experiments. Brown et al. (8), also experimenting in the greenhouse but using low-potassium soil, reported sugarbeet yield responses and elimination of deficiency symptoms where the leaf blade tissue was initially below 1 percent potassium content.

It is evident from the data that sodium can cause a redistribution of potassium within the sugarbeet plant. When potassium is limiting (and sodium is low), introduction of sodium apparently causes movement of potassium from the vascular tissue to the leaf mesophyll tissue. This phenomenon has been referred to as a conserving or sparing action whereby sodium extends the supply of potassium in the plant.

Climatic factors can influence the potassium content of sugarbeets. Ulrich (65) showed that night temperatures of 4°, 10°, and 17° C in the greenhouse resulted in successively higher levels of potassium in petioles. In this same experiment, a constant supply of potassium and other nutrients was available to the plants in the nutrient solutions. The result was that the potassium content of blades and petioles was consistently high during the period June to October. Under field conditions (2, 31), it has been observed that potassium levels in beet petioles were high in June but decreased gradually for the remainder of the season.

Adams (2) showed that the total content of potassium in the whole plant increased steadily throughout the season. In his experiments, the largest increases in total potassium coincided with the time of most rapid growth.

Relationship to Nitrogen. The effect of nitrogen on potassium uptake should be considered on the basis of the nitrogen status of the plant. Husseini (30) demonstrated that when nitrogen availability was increased from limiting to moderate levels, the potassium concentration in the blades decreased. When nitrogen increased still further, the potassium concentration increased. These results could be explained as (1) a dilution effect on potassium where the plants are responding to increased nitrogen, and (2) an attempt by the plants to maintain balance between anions and cations where growth increase was not proportional to increased nitrate uptake. The subject of ionic balance is treated at length by de Wit et al. (74).

An effect of potassium on nitrogen analogous to the effect of nitrogen on potassium also was demonstrated by Husseini (30). A word of caution is necessary at this point on interpretation of the potassium-nitrogen mutual interaction. There is a pronounced negative interaction or antagonism between chloride and nitrate ions for plant uptake. Since chloride is very commonly introduced as potassium chloride fertilizer, it should not be surprising if there is a negative correlation between potassium treatments and nitrate uptake. The effect of potassium in this instance would obviously be an indirect one. Such a phenomenon is shown for sugarbeets in Figure 6.6 (32). The results in Figure 6.6 will be discussed in detail later.

Relationship to Sodium. It is generally recognized that there is a reciprocal relationship between potassium and sodium uptake by beets. As potassium availability and uptake increase, sodium uptake decreases and vice versa (19, 31, 41, 64). On the basis of chemical equivalents of the elements, there are wide variations in the sodium-potassium content of sugarbeets.

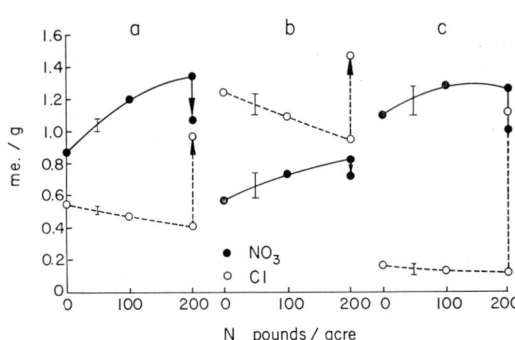

FIG. 6.6. Effect of NH_4NO_3 and KCl fertilizer on NO_3 and Cl concentrations of sugarbeet petiole at three locations. Petioles were sampled during June. Three treatments (N_0, N_{100}, N_{200}) had no KCl. A fourth treatment, as indicated by the arrows at N_{200}, had KCl at the rate of 200 pounds of K per acre. Vertical line segments are relative $LSD_{(.05)}$. Element contents at treatment (N_0, K_0) reflect differences in availability of soil-indigenous N and Cl. (From James et al., 32.)

Also, the ratio of these elements tends to shift during the growing season. Some sodium and potassium concentration values in sugarbeet petioles from various places are shown in Tables 6.3 and 6.4. It will be noted that sodium is frequently more concentrated than potassium.

Despite the strong tendency for sodium to decrease the concentration of potassium in the plant, applications of sodium have frequently resulted in greater total uptake of potassium (2, 64).

Relationship to Magnesium and Calcium. Tinker (61) and Draycott and Durrant (18) have shown that potassium concentration of sugarbeet tissues decreases with magnesium applications. There is a similar tendency with increases in calcium availability (8). The evidence on plant tissue concentration of nutrient elements in sugarbeets points to a close relationship between potassium, sodium, and magnesium + calcium (e.g., 19).

TABLE 6.3. Sodium and potassium contents of sugarbeet petioles and leaf blades at different times of the season and in different areas

Location	Time of Season	Plant Part	meq/g			Reference
			Na	K	Na:K	
Colorado*	late	petiole	1.12	0.47	2.38	Follett et al.
	late	blade	1.07	0.93	1.14	(21)
Washington	early	petiole	1.48	0.66	2.24	James et al.
	late	petiole	1.02	0.26	3.92	(31)
England	early	petiole	1.43	1.65	0.87	Adams (2)
	late	petiole	0.66	0.66	1.00	
	early	blade	1.81	1.19	1.52	
	late	blade	1.32	0.74	1.78	
England	middle	petiole	0.36	1.12	0.32	Draycott and
	middle	blade	0.64	1.39	0.46	Durrant (18)

* Average of six varieties.

TABLE 6.4. Ranges of sodium and potassium contents and ratios of the elements in sugarbeet petioles from commercial fields, Washington

	Concentration—meq/g*					
	1967 (9)†		1968 (14)		1969 (16)	
	Range	Average	Range	Average	Range	Average
Na	0.53–1.77	1.13	0.16–1.76	0.94	0.13–1.90	0.86
K	0.55–1.67	1.20	0.62–2.52	1.59	0.84–1.86	1.21
Na/K	0.31–3.15	1.26	0.06–2.84	0.79	0.07–1.94	0.79

* Sampled early to mid-June each season.
† Numbers in parentheses refer to number of locations.

Distribution of Potassium in the Plant and Total Uptake. On a concentration basis, potassium is least in sugarbeet roots and highest in the petioles, with leaf blades being intermediate (2, 30, 59). On the basis of total content, sugarbeet roots have been reported to contain 33 percent (2), 22 percent (30), and 35–40 percent (33) of the beet's total potassium. The results from various areas indicate that sugarbeet roots contain 3 to 5 pounds of potassium in each ton of beets fresh weight (33, 59). James et al. (33) point out that the yield of sugarbeet tops (and therefore the yield of potassium in the tops) bears little relationship to the yield of roots. In one experiment where the total dry weight yield of tops plus crowns was 5.5 tons per acre, root yields averaged 30 tons per acre. In another experiment, the dry weight yield of tops averaged 3.2 tons per acre but the root yield averaged 37.7 tons per acre. The reason for these differences was the nitrogen status of the crop at the end of the season. Excessive nitrogen results in much greater amounts of foliage. In summarizing the Washington results, it was indicated that 75 to 100 pounds of potassium per acre would be removed in the harvested roots. If the beet tops were removed, the total potassium removal could range from 220 to 400 pounds of potassium per acre. It is apparent, therefore, that the net effect of a crop of sugarbeets on the soil's potassium reserves depends heavily on management factors, especially the disposition of the above-ground parts and the nitrogen economy of the crop.

ROOT AND SUCROSE YIELD INTERACTIONS BETWEEN POTASSIUM AND OTHER NUTRIENTS

Munson (46) has discussed the broad aspects of potassium interactions with other elements in plant growth. The beneficial influence of potassium, in terms of crop performance, cannot reach full expression if one or more other nutrient elements are limiting. On the other hand, higher than normally optimum levels of potassium may be needed if another nutrient is present in excessive amounts. Special attention is given here to interactions between potassium and, respectively, nitrogen, sodium, and magnesium + calcium.

Nitrogen. McDonnell et al. (41), in a series of field experiments, and Husseini (30), in the greenhouse, have demonstrated that there is a pronounced interaction between nitrogen and potassium in terms of beet root and sucrose yield. When the nitrogen supply is low, increasing levels of potassium have little or no effect. When the nitrogen supply is high, root and sucrose yield increases steadily and maximum ouput occurs at high levels of both nitrogen and potassium. Muller et al. (45) stated that the nitrogen-potassium ratio in the plant is critical. At high nitrogen they proposed that widening the $N:K_2O$ ratio in the fertilizer would result in greater root and sucrose production. Tinker (60) reported a positive nitrogen-potassium yield interaction in sucrose production in a series of 42 field trials. He indicated, however, that the nitrogen-sodium positive interaction was more pronounced than was the nitrogen-potassium interaction.

It has frequently been observed that potassium fertilization results in decreased "noxious" nitrogen and also slight but consistent increases in sucrose content (17, 31). James et al. (32) interpreted this as an indirect effect of potassium, since the changes in sucrose percent were associated with potassium chloride fertilization. Results of this nature are summarized in Table 6.5 for experiments done in Washington over a period of three years. Of the 27 experiments showing positive changes in sucrose percentage, 22 were significant (difference of 0.2 percent or greater). Of the 9 experiments showing negative changes, 4 were significant.

One explanation of the potassium chloride fertilizer effect on sucrose accumulation relates to the chloride-nitrate antagonism in plant uptake. Figures 6.6a, b, and c show the results of fertilizer treatments on petiole nitrate and chloride concentrations in three different experiments (31). In each experiment the petiole nitrate increased with increasing levels of nitrogen fertilizer. At the same time, chloride levels decreased, although in Figure 6.6c the indigenous chloride was apparently too low to be much affected. With the combination ammonium nitrate-potassium chloride fertilizer treatment, chloride in the petiole increased sharply and nitrate decreased. In other experiments where potassium sulfate was included as a

TABLE 6.5. Changes in sucrose percent associated with KCl fertilizer applications: a frequency distribution over three years

Year	Number of Experiments Type of Response		
	Positive	Zero	Negative
1967	7	2	3
1968	9	1	4
1969	11	0	2
Total	27	3	9
Percent	69.2	7.7	23.1

SOURCE: Data for 1967 and 1968 adapted from James et al. (32). 1969 data are unpublished.

TABLE 6.6. Influence of sodium and potassium on yield of sugarbeet roots

Sodium Rate*	Potassium Rate* (lb/A)			
	100	200	300	600
	T/A	T/A	T/A	T/A
0	10.0	12.7	14.2	16.5
500	12.6	15.3	15.8	17.4

SOURCE: Adapted from Shepherd et al. (57).
NOTE: Average of four experiments in two years that tested both current and residual fertilizer applications.
* Sodium applied as agricultural salt and potassium applied as mixed grades of N-P_2O_5-K_2O fertilizers.

variable, there were no comparable effects of fertilizer treatments on nitrate uptake.

From the results shown in Figure 6.6 and Table 6.5, it is apparent that interactions between nitrogen and potassium chloride fertilizers must be carefully evaluated to avoid confusion on the true cause-and-effect relationships.

Sodium. In 1865 Lucanus (cited by Evans and Sorger, 20) first demonstrated that potassium was indispensable to plants. In that same year, Voelcher (cited by Lehr, 39) observed that the application of sodium increased the yield of mangels.[1] For many years the effect of sodium was assumed to be indirect. For example, sugarbeets were responding to potassium which became more available in the soil solution as a result of a cation exchange reaction with sodium. Plant tissue analysis and solution culture work have proved this assumption to be wrong, since increased growth followed increased uptake of sodium per se. The sparing action theory discussed previously is analogous to the soil potassium exchange theory except that it refers to exchange of potassium by sodium within the plant.

The positive sugarbeet yield interaction between sodium and potassium is exemplified in Tables 6.6 and 6.7. In both tables it appears that the yield was best with a combination of sodium and potassium fertilizers. In some cases it is reported that potassium fertilizer completely masks the effect of sodium in field experiments (16, 28, 63). Some results of a pot-culture study by Lehr (40) are given in Table 6.8. In this case there was only one beet plant per pot, but the main effects of sodium and potassium were both significant as also was the sodium-potassium interaction. It is apparent from Table 6.8 that the wide range of sodium and potassium treatments resulted in first an increase, then a decrease in yield at the combined higher levels.

There was a significant increase in sucrose percentage and total sucrose yield for the experiment reported in Table 6.6 when salt was added.

---

1. Mangels are a member of the same species as sugarbeet—*Beta vulgaris*.

TABLE 6.7. Influence of sodium and potassium on yield of sugarbeet roots

Sodium Rate*	Potassium Rate*			
	0	200	400	Mean
	T/A	T/A	T/A	T/A
0	12.19	15.82	16.87	14.96
200	13.95	16.61	17.42	15.99
400	14.42	17.15	17.76	16.44
Mean	13.52	16.53	17.35	

SOURCE: Adapted from Masterson (42).
NOTE: Average of two experiments on low fertility soils and two experiments on high fertility soils in each of two years.
* Given as pounds per acre, respectively, as muriate of potash and agricultural salt.

No analyses for sucrose were given for the experiments of Table 6.7. For Table 6.8, sucrose yield trends corresponded closely to the reported root yield trends but the sucrose results were not significant. In the solution-culture work reported by Husseini (30), maximum output of roots and sucrose occurred with a combination of high potassium and high sodium in the nutrient medium.

Magnesium and Calcium. Tinker (61) and Draycott and Durrant (18) reported that sugarbeets gave a higher root and sucrose yield with both potassium and magnesium treatments than were measured with either element alone. Both these series of experiments were done on soils that were apparently well supplied with calcium, since the soil pH averaged above 7.0. Sugarbeets are rarely grown on strongly acid soils, but it is to be expected that there would be a pronounced interaction between potassium and calcium regarding both the direct effects of calcium (availability as a nutrient) and indirect effects (changes in soil pH and related chemistries of aluminum and manganese).

TABLE 6.8. Yield of sugarbeet roots in grams per pot fresh weight as influenced by level of exchangeable potassium and sodium

Exchangeable Potassium Level*	Exchangeable Sodium Level*					Main effect of potassium
	0	4	8	16	24	
4	523	550	522	660	844	620
8	487	540	774	696	897	717
12	809	745	920	847	687	802
Main effect of sodium	670	611	739	734	809	

SOURCE: Lehr (40).
NOTE: Average of three replications.
* Percent of clay colloid saturated with Na or K in a clay-sand culture. Balance of colloid was Ca-Mg saturated.

PREDICTING POTASSIUM NEEDS

As with phosphorus, field testing has been used extensively to determine the need for potassium. The disadvantages for field testing have been noted previously in the phosphorus section, consequently there is need for greater reliance on plant and soil analyses to evaluate the need for potash fertilizer.

Tissue Testing. Reliable plant analysis techniques for potassium in the sugarbeet have been developed by Ulrich and co-workers (67, 68, 69). The field sampling and extraction procedures are described in the phosphorus section. Interpretation of plant potassium data is based on the concepts presented earlier in the section on potassium-sodium interactions for sugarbeets. Although the potassium requirement is determined, in part, by the sodium content of the plant, Ulrich and Hills (68) state that the blade should be used to determine the critical level of potassium in the sugarbeet because potassium in the blade is less affected by the sodium content of the plant. Their work shows that the critical level for potassium in the blade is 1.0 percent.

As noted previously, the highest rate of potassium uptake by sugarbeets coincides with the time of most rapid growth. It follows, therefore, that potassium must be abundantly available early and continuously in the season to preclude the possibility of growth slow-down from potassium stress. From this point of view, tissue testing has the disadvantage that results may come too late to signal the need for potassium fertilizer additions. This would be especially true in areas with short growing seasons.

Soil Testing. Because of the sparing action of sodium for potassium with sugarbeets, soil tests must give some information on both elements. Comparatively little work has been reported along this line. Kaudy et al. (36) published some results of soil-exchangeable sodium and potassium and the associated amounts of these elements found in sugarbeet tissue. Their results, shown in Table 6.9, give a fair correlation between plant and soil for sodium, potassium, and sodium + potassium. There is a rather weak correlation between plant and soil for the sodium-potassium ratio. Further examination of their data indicates that the potassium uptake tends to increase with increasing sodium in the soil and vice versa. This appears rather anomalous until the soil potassium-soil sodium correlation is determined. It is evident that there is a tendency for sodium and potassium to vary together in their soils. This feature should be taken into consideration when evaluating any simple or multiple correlation between plant and soil potassium or sodium or both.

Some comparative data from a preliminary investigation in Washington also are given in Table 6.9. These results are somewhat different from those of Kaudy et al. (36). There is a very weak association between plant and soil sodium and between plant and soil potassium. At the same time the ratio, sodium:potassium, and the sum, sodium + potassium, are fairly

TABLE 6.9. Simple correlation-coefficients (r) for potassium and sodium between plant and soil

Correlated Item	Data Source	
	Kaudy et al.* (36)	Washington† (unpublished)
Plant K vs soil K	0.84	0.21
Plant Na vs soil Na	0.74	0.27
Plant Na/K vs soil Na/K	0.52	0.78
Plant Na + K vs soil Na + K	0.82	0.72
Plant K vs soil Na	0.47	−0.14
Plant Na vs soil K	0.71	−0.16
Soil Na vs soil K	0.54	−0.01

* Data are means of seven soil series involving a total of 80 fields. Published data in terms of soil test pounds per acre and percent plant composition were converted to chemical equivalents of the elements in both plant and soil for use here.

† Data are means of four replications in each of 17 experiments.

well correlated between plant and soil. In this case, there is a weak negative relationship between plant potassium and soil sodium and vice versa. The correlation between soil potassium and soil sodium is zero, indicating that these elements varied independently of each other in this series of experiments.

Tinker (62) correlated sodium uptake in sugarbeets with sodium in the soil and found that the correlation was improved when the activity ratio $Na/(Ca + Mg)^{\frac{1}{2}}$ was used instead of exchangeable sodium.

It is apparent that testing for sodium and potassium in sugarbeet soils has some promise, but much work remains to be done, especially on extraction procedures. These kinds of investigations are worthy of pursuit because much can be learned about the factors that influence nutrient availability in the soil as well as the interaction between sodium and potassium in plant nutrient uptake.

POTASSIUM FERTILIZATION

Materials and Amounts. Potassium is most commonly applied in the chloride and sulfate fertilizer forms—frequently in a mix with nitrogen or phosphorus or both. Specialty potassium fertilizers are the nitrate and the double sulfate salt of magnesium.

Reports which indicate beet root and sucrose yield responses to potassium fertilization show also that potassium is needed in liberal amounts (3, 15, 31, 41). In various reports, optimum fertilizer rates varied from 130 pounds of potassium per acre (3) to 300 pounds of potassium per acre (41).

Methods of Application. For sugarbeets, potassium fertilizers are applied broadcast-plowdown in the fall or in the spring before seedbed preparation. Potassium is also shanked into the soil at planting time and following germination. Whatever the method, potassium should be placed into the root zone, because surface applications without incorporation have limited effectiveness.

From the information presented here on rate and time of potassium uptake, the need for timely potassium applications is clear. Fertilizer application efficiencies decrease rapidly with the advancing season.

OUTLOOK

Average sugarbeet yields will increase because of better management and improved varieties. It is expected, therefore, that potassium fertilizer rates will increase in those areas that now require regular applications. In those areas where potassium is required only occasionally, potassium fertilization will become more frequent.

"Fine-tuning" of all management factors will result in maximum production of beet sugar. Methods of predicting soil potassium availability, and therefore fertilizer requirements, should be investigated to facilitate these management decisions. This will require a thorough analysis of both sodium and potassium availability in the soil and their interactions in terms of competition for plant uptake and overlapping roles in plant metabolic processes. This kind of experimentation also will result in a better knowledge of optimum potassium nutrition under different environmental conditions, especially as it relates to nitrogen availability. Some excellent guidelines are available to aid in the design of future research to accomplish the desired goals.

REFERENCES

1. Adams, S. N. The effect of time of application of phosphate and potash on sugar beet. J. Agr. Sci. 56:127–30. 1961.
2. ———. The effect of sodium and potassium fertilizer on the mineral composition of sugar beet. J. Agr. Sci. 56:383–88. 1961.
3. ———. The response of sugarbeet to fertilizer and effect of farmyard manure. J. Agr. Sci. 58:219–26. 1962.
4. Arnon, D. I. The physiology and biochemistry of phosphorus in green plants. In W. H. Pierre and A. G. Norman, eds., Soil and Fertilizer Phosphorus in Crop Nutrition 4:1–42. Academic Press, Inc., New York. 1953.
5. Association of Official Agricultural Chemists. Official Methods of Analysis, 10th ed., pp. 11–15. Assoc. Off. Agr. Chem., Washington, D.C. 1965.
6. Bray, R. H. and L. T. Kurtz. Determination of total, organic, and available forms of phosphorus in soils. Soil Sci. 59:39–45. 1945.
7. Broadwell, C. E. Fertilizer placement tests in Ontario. Proc. 10th Regional Meeting, Am. Soc. Sugar Beet Technologists, East. U.S. and East. Can., pp. 90–92. 1959.
8. Brown, A. L., F. J. Hills, and B. A. Krantz. Lime, P, K, and Mn interactions in sugar beets and sweet corn. Agron. J. 60:427–29. 1968.
9. Caldwell, A. C. and D. B. Ogden. The response of sugar beets to phasphates applied at various times prior to planting. Seventh Phos. Conf. North Central Region, pp. 50–53. 1957.

10. Cook, R. L. and J. F. Davis. The residual effect of fertilizer. Advan. Agron. 9:205–16. 1957.
11. Cooke, G. W. Recent advances in fertilizer placement. J. Sci. Food Agr. 5:429–40. 1954.
12. ———. The Control of Soil Fertility, pp. 14–30. Crosby Lockwood & Son, Ltd., London. 1967.
13. Davis, J. F., G. Nichol, and D. Thurlow. The effect of phosphorus fertilization and time of application on chemical composition of foliage and on yield, sucrose content and percent purity of sugar beet roots. J. Am. Soc. Sugar Beet Technologists 11:406–12. 1961.
14. ———. The interaction of rates of phosphate application with fertilizer placement and fertilizer applied at planting on the chemical composition of sugar beet tissue, yield, percent sucrose, and apparent purity of sugar beet roots. J. Am. Soc. Sugar Beet Technologists 12:259–67. 1962.
15. Davis, J. F., W. B. Sundquist, and M. G. Frakes. The effect of fertilizers on sugar beets including an economic optima study of the response. J. Am. Soc. Sugar Beet Technologists 10:424–34. 1959.
16. Dorph-Petersen, K. and F. Steenbjerg. Investigations of the effect of fertilizers containing sodium. Plant Soil 2:283–300. 1950.
17. Draycott, A. P. and G. W. Cooke. The effects of potassium fertilizers on quality of sugar beet. Proc. 8th Congr. Intern. Potash Inst., Brussels, pp. 131–35. 1966.
18. Draycott, A. P. and M. J. Durrant. The effects of magnesium fertilizers on yield and chemical composition of sugar beet. J. Agr. Sci. 72:319–24. 1969.
19. El-Sheikh, A. M., A. Ulrich, and T. C. Broyer. Sodium and rubidium as possible nutrients for sugar beet plants. Plant Physiol. 42:1202–8. 1967.
20. Evans, H. J. and G. J. Sorger. Role of mineral elements with emphasis on the univalent cations. Ann. Rev. Plant Physiol. 17:47–76. 1966.
21. Follett, R. H., W. R. Schmehl, L. Powers, and M. Payne. Effect of genetic population and soil fertility level on the chemical composition of sugar beet tops. Colo. Agr. Exp. Sta. Tech. Bull. 79. 1964.
22. Fried, M. and R. E. Shapiro. Soil-plant relations in ion uptake. Ann. Rev. Plant Physiol. 12:91–112. 1961.
23. Grunes, D. F., H. R. Haise, and L. O. Fine. Proportionate uptake of soil and fertilizer phosphorus by plants as affected by nitrogen fertilization. II. Field experiments with sugar beets and potatoes. Soil Sci. Soc. Am. Proc. 22:43–48. 1958.
24. Hagen, C. E. and H. T. Hopkins. Ionic species in orthophosphate absorption by barley roots. Plant Physiol. 30:193–99. 1955.
25. Harmer, P. M., E. J. Benne, W. M. Laughlin, and C. Key. Factors affecting crop response to sodium applied as common salt on Michigan muck soil. Soil Sci. 76:1–17. 1953.
26. Hendrix, J. E. The effect of pH on the uptake and accumulation of phosphate and sulfate ions by bean plants. Am. J. Botany 54:560–64. 1967.
27. Hewitt, E. J. The essential nutrient elements: requirements and interactions in plants. In F. C. Stewart, ed., Plant Physiology. III. Inorganic nutrition of plants, pp. 137–360. Academic Press, Inc., New York. 1963.
28. Holt, M. E. and N. J. Volk. Sodium as a plant nutrient and substitute for potassium. J. Am. Soc. Agron. 37:821–27. 1945.
29. Hull, R. Sugar beet diseases. G. Brit. Min. Agr. Fisheries Food Tech. Bull. 142. 1960.
30. Husseini, K. K. Influence of nitrogen, potassium and sodium on sugar beet growth and quality. Ph.D. thesis, Colo. State Univ., Fort Collins. 1966.
31. James, D. W., D. C. Kidman, W. H. Weaver, and R. L. Reeder. Potassium

fertilization of sugar beets in central Washington. J. Am. Soc. Sugar Beet Technologists 14:682–94. 1968.
32. ———. Factors affecting chloride uptake and implications of the chloride-nitrate antagonism in sugarbeet mineral nutrition. J. Am. Soc. Sugar Beet Technologists 15:647–56. 1970.
33. James, D. W., W. H. Weaver, and R. L. Reeder. Soil test index of plant available potassium and the effects of cropping and fertilization in central Washington irrigated soils. Wash. State Univ. Agr. Exp. Sta. Bull. 697. 1968.
34. Jean, F. C. and J. E. Weaver. Root yield and crop yield under irrigation. Carnegie Inst. Wash. Publ. 357. 1924.
35. Jennings, D. H. The Absorption of Solutes by Plant Cells. Oliver and Boyd, London. 1963.
36. Kaudy, J. C., E. Truog, and K. C. Berger. Relation of sodium uptake to that of potassium by the sugar beet. Agron. J. 45:444–47. 1953.
37. Larson, W. E. Effect of method of application of double superphosphate on the yield and phosphorus uptake by sugar beets. Proc. Am. Soc. Sugar Beet Technologists 8:25–31. 1954.
38. Lawton, K., P. E. Erickson, and L. S. Robertson. Utilization of phosphorus by sugar beets as affected by fertilizer placement. Agron. J. 46:262–64. 1954.
39. Lehr, J. J. The importance of sodium for plant nutrition. I. Soil Sci. 52:237–44. 1941.
40. ———. The importance of sodium for plant nutrition. III. The equilibrium of cations in the beet. Soil Sci. 53:399–411. 1942.
41. McDonnell, P. M., P. A. Gallagher, P. Kearney, and P. Carroll. Fertilizer use and sugar beet quality in Ireland. Proc. 8th Congr. Intern. Potash Inst., Brussels, pp. 107–26. 1966.
42. Masterson, C. The effect of sodium and potassium on fodder beet. J. Dept. Agr. Publ. 54. 1958.
43. Maxson, A. C. Insects and Diseases of the Sugar Beet, p. 384. Beet Sugar Develop. Found., Fort Collins, Colo. 1948.
44. Miller, M. H. and C. E. Broadwell. Yield and sugar content of beets as influenced by fertilizer placement in Ontario. Proc. 11th Regional Meeting, Am. Soc. Sugar Beet Technologists, East. U.S. and East. Can., pp. 13–20. 1961.
45. Muller, K., A. Niemann, and W. Werner. The influence of the nitrogen-potassium ratio on yield and quality of sugar beet. Zucker 15:1–6. 1962.
46. Munson, R. D. Interaction of potassium with other ions. In V. J. Kilmer, S. E. Younts, and N. C. Brady, eds., The Role of Potassium in Agriculture, pp. 321–53. Am. Soc. Agron., Madison, Wis. 1968.
47. Nason, A. and W. D. McElroy. Modes of action of essential mineral elements. In F. C. Stewart, ed., Plant Physiology. III. Inorganic Nutrition of Plants, pp. 451–536. Academic Press, Inc., New York. 1963.
48. National Plant Food Institute. Methods of Applying Fertilizer. Natl. Fertilizer Assoc., Washington, D.C. 1958.
49. Nelson, R. T., R. Gardner, H. F. Rhoades, T. J. Dunnewald, J. C. Hide, R. R. Wood, J. A. Asleson, J. L. Mellor, and F. V. Pumphrey. Harvest results of inorganic fertilizer tests on sugar beets conducted in four states, 1947. Proc. Am. Soc. Sugar Beet Technologists, pp. 406–20. 1948.
50. Olsen, S. R., C. V. Cole, F. S. Watanabe, and L. A. Dean. Estimation of available phosphorus in soils by extraction with sodium bicarbonate. USDA Circ. 939. 1954.
51. Olsen, S. R., W. R. Schmehl, F. S. Watanabe, C. O. Scott, W. H. Fuller, J. V. Jordan, and R. Kunkel. Utilization of phosphorus by various crops as affected by source of material and placement. Colo. Agr. Exp. Sta. Tech. Bull. 42. 1950.

52. Peterson, H. B. A review of phosphate fertilizer investigations in 15 western states through 1949. USDA Circ. 927. 1953.
53. Romsdal, S. D. and W. R. Schmehl. The effect of method and rate of phosphate application on yield and quality of sugar beets. J. Am. Soc. Sugar Beet Technologists 12:603–7. 1963.
54. Schmehl, W. R. and R. P. Humbert. Nutrient deficiencies in sugar crops. In H. B. Sprague, ed., Hunger Signs in Crops, 3rd ed., pp. 415–50. David McKay Co., New York. 1964.
55. Schmehl, W. R., S. R. Olsen, and R. Gardner. Effect of type of phosphate material and method of application on phosphate uptake and yield of sugar beets. Proc. Am. Soc. Sugar Beet Technologists 7:153–58. 1952.
56. Schmehl, W. R., S. R. Olsen, R. Gardner, S. D. Romsdal, and R. Kunkel. Availability of phosphate fertilizer materials in calcareous soils in Colorado. Colo. Agr. Exp. Sta. Tech. Bull. 58. 1955.
57. Shepherd, L. N., J. C. Shickluna, and J. F. Davis. The sodium-potassium nutrition of sugar beets produced on organic soil. J. Am. Soc. Sugar Beet Technologists 10:603–8. 1959.
58. Soine, O. C. Uptake of phosphorus by sugarbeets. J. Am. Soc. Sugar Beet Technologists 15:159–66. 1968.
59. Storer, K. R. Growth studies with sugar beets. Ph.D. thesis. Colo. State Univ., Fort Collins. 1969.
60. Tinker, P. B. H. The effects of nitrogen, potassium and sodium fertilizers on sugar beet. J. Agr. Sci. 65:207–12. 1965.
61. ———. The effects of magnesium sulphate on sugarbeet yield and its interactions with other fertilizers. J. Agr. Sci. 68:205–12. 1967.
62. ———. The relationship of sodium in the soil to uptake of sodium by sugarbeet in the greenhouse and to yield responses in the field. Trans. Comm. II and IV, Intern. Soc. Soil Sci., 1966, pp. 223–31. 1967.
63. Truog, E., K. C. Berger, and O. J. Attoe. Response of nine economic plants to fertilization with sodium. Soil Sci. 76:41–50. 1953.
64. Tullin, V. Response of sugar beet to common salt. Plant Physiol. 7:810–34. 1954.
65. Ulrich, A. Influence of night temperature and nitrogen nutrition on the growth, sucrose accumulation and leaf minerals of sugar beet plants. Plant Physiol. 30:250–57. 1955.
66. ———. Plant analysis in sugar beet nutrition. In W. Reuther, ed., Plant Analysis and Fertilizer Problems, pp. 190–211. Am. Inst. Biolog. Sci. Publ. 8. 1961.
67. Ulrich, A. and F. J. Hills. Petiole sampling of sugar beet fields in relation to their N, P, K, and Na status. Proc. Am. Soc. Sugar Beet Technologists 7:32–45. 1952.
68. ———. Sugar Beet Nutrient Deficiency Symptoms—A Colored Atlas and Chemical Guide. Univ. Calif. Div. Agr. Sci., Berkeley. 1969.
69. Ulrich, A., D. Ririe, F. J. Hills, A. G. George, and M. D. Morse. I. Plant analysis: a guide for sugar beet fertilization. Calif. Agr. Exp. Sta. Bull. 766. 1959.
70. Vij, V. N. and M. H. Miller. The influence of nitrogen in a fertilizer band on the root growth and fertilizer phosphorus absorption by sugar beets. Proc. 11th Regional Meeting, Am. Soc. Sugar Beet Technologists, East. U.S. and East. Can., pp. 61–67. 1961.
71. Wallace, T. The Diagnosis of Mineral Deficiencies in Plants, 2nd ed. Chemical Pub. Co., New York. 1961.
72. Werkhoven, C. N. E., and M. H. Miller. Absorption of fertilizer phosphorus by sugar beets as influenced by placement of phosphorus and nitrogen. Proc. 10th Regional Meeting, Am. Soc. Sugar Beet Technologists, East. U.S. and East. Can., pp. 86–89. 1959.

73. Wilson, R. H. and H. J. Evans. The effect of potassium and other univalent cations on the conformation of enzymes. *In* V. J. Kilmer, S. E. Younts, and N. C. Brady, eds., The Role of Potassium in Agriculture, pp. 189–202. Am. Soc. Agron., Madison, Wis. 1968.
74. Wit, C. T. de, W. Dijkshoorn, and J. C. Noggle. Ionic balance and growth of plants. Versla. Landbouwk. Onderzoek. NR. 69.15. Wageningen. 1963.

7

Secondary Nutrients and Micronutrients

FRANK G. VIETS, JR.
Soil and Water Conservation Research Division, ARS, USDA
Fort Collins, Colorado

LYNN S. ROBERTSON
Michigan State University
East Lansing

DIAGNOSIS OF THE DEFICIENCY PROBLEM	173
Soil Acidity	174
Cations	177
Anions	182
Interrelations of Secondary and Micronutrients	185
Regional Problems	185
SUMMARY AND THE FUTURE	186

SUGARBEETS, like all higher plants, require at least 16 chemical elements for growth and reproduction. They get their carbon from carbon dioxide of the air, and their hydrogen and oxygen from air and water. The other 13 elements come primarily from the soil through the roots, but the leaves are capable of absorbing substantial amounts of nutrients from polluted air or from dusts or sprays applied to the leaves.

The mineral nutrients are often classed into three arbitrary groups: major, secondary, and micronutrients. The major or macronutrients—nitrogen, phosphorus, and potassium—are discussed in other chapters. The secondary nutrients are calcium, magnesium, and sulfur. The micronutrients are copper, iron, manganese, zinc, boron, chlorine, and molybdenum. The latter group are also called minor or trace elements. The arbitrary separation of the essential mineral elements into classes is a custom and a convenience. The plant must have a sufficient amount of each of them in relation to its requirements for high yields of roots and sugar. However, the amounts needed vary vastly, and that is the reason for the convenient classification. An acre yield of 25 tons of beets may contain in the roots and tops 250 pounds of nitrogen but only 0.005 pound of molybdenum.

Some other elements are known to have beneficial effects on growth of sugarbeets, but they are not regarded as essential by strict criteria, for beets can be grown without them. Sodium can perform some of the functions of potassium and can increase yields if a shortage of potassium exists. El-Sheikh et al. (4) showed that high concentrations of sodium or low concentrations of rubidium (Rb) increased both root and top growth of beets at either high or low potassium levels in solution cultures. High concentrations of rubidium were toxic, especially to fibrous roots. Common salt (NaCl) has been used for years in Michigan and Europe to increase beet growth. We now know that sodium was partially taking the place of potassium. In western United States, beets are one of the more tolerant crops to salinity. Perhaps part of this tolerance is related to the beneficial effects of sodium. Bromine as bromide can partially substitute for chlorine in meeting the plant's need for chlorine. However, since no areas deficient in chlorine are known, this substitution is of no practical value.

Perhaps other elements will be found to be essential, but there have been many attempts to show that ones such as silicon, vanadium, and cobalt are indispensable. To make the prediction that no others will be found would be a mistake. The essentiality of chlorine was not proved until 1954, even though it had been worked on sporadically for a century. Previous failures to show the need for chlorine were due to the universal contamination of salts, air, and water with chlorine.

Vanadium may prove to be important in making the sugarbeet "ripen." Singh and Wort (9) reported that a 0.01 M solution of vanadyl sulfate sprayed on the leaves of beets, 4.5 months old, decreased leaf growth, respiration rate, and nitrate reductase activity and increased photosynthetic rate and sucrose storage in the roots during the following three weeks. The chemical has not been field tested; its use may not be safe or economical; and other chemicals may be found that perform the same functions. Such a chemical is needed in the northern and intermountain areas for beet production. Although the micronutrient cations are so important in enzyme functions, vanadium may prove to be the only one that has the capacity to increase the activity of certain enzyme systems and to inhibit others.

Total uptake of secondary and micronutrients cannot, today, be used as a reliable guide to fertilizer practices or for diagnosing deficiencies because of lack of sufficient data.

Sugarbeets are among the crops least susceptible to secondary and micronutrient deficiencies, with the exception of beets' well-known susceptibility to boron and manganese deficiencies and their intolerance of acid soils. The latter may be related to their high calcium and magnesium requirements. For example, beans or corn may show severe symptoms of zinc deficiency on soils that would not produce deficiency symptoms in sugarbeets. Beets do not appear to have lower requirements than other crops; in fact, they may have higher ones because of the high tonnage produced. Beets may have greater capacity to absorb secondary and micronutrients from soils than many other crops because of their deep and extensive root system. Furthermore, there are many management factors that may favor better micronutrient supply for sugarbeets than for other crops: beets are usually grown on the most fertile and productive soils on the farm; they generally receive more fertilizer, particularly phosphate, which often contains micronutrients if it comes from western sources; and if manure is available, it is put on the beet land and plowed down. However, in recent years manure application is being shifted to corn land. Regardless of the causes, beets do not offer the problems of micronutrient nutrition that many other crops do, particularly tree fruits. In the field, deficiencies of only calcium, magnesium, sulfur, boron, manganese, and zinc have been reported in the United States. Deficiencies of other micronutrients may have been unreported or overlooked. Molybdenum deficiency occurs in Europe but not in the United States.

DIAGNOSIS OF THE DEFICIENCY PROBLEM

As in human health, timely and correct diagnosis of the problem is half of the cure. Unlike human health problems, all of the nutrient deficiencies affecting beets are curable and usually at very low cost. The grower that does not have a correct diagnosis, and does not use the right fertilizers to correct the deficiency, will suffer a severe financial loss that could easily have been avoided.

The diagnostic tools available include deficiency or toxicity symptoms, response of young plants to foliar or soil applications of the fertilizer nutrient suspected of being deficient, petiole or leaf analysis, and finally, soil analysis. Recognition of specific symptoms is by far the easiest of the methods of diagnosis. Such diagnosis must be done by an experienced observer. Care must be exercised to avoid confusion of nutrient deficiencies with disease or insect damage. A positive plant response to soil or foliage application of the nutrient suspected of being deficient is the ultimate proof that the diagnosis is right. Analysis of petioles or leaves and comparison with standard values may tell how deficient the plant is and tentatively confirm the visual diagnosis. Foliar analysis is particularly helpful when a deficiency is incipient—a deficiency not so bad that the plant shows distress symptoms, but bad enough to reduce its yield. Foliar analysis is also a valuable tool for detecting toxicity or serious nutrient imbalance. Soil tests and comparison with standard values derived from field experiments are most helpful in assessing how much fertilizer must be applied to correct a deficiency. Several of the methods mentioned should be used together for the best results. The time will come when we will use all of them for truly scientific management of the nutrient requirements for sugarbeet production. There is no *best* method. The devotee of plant analysis or soil analysis *alone* may make some serious mistakes.

A brief description of the foliage-deficiency symptoms for each of the elements is shown in Table 7.1. These descriptions are briefed from the excellent publication of Ulrich and Hills (10), which is also well illustrated in color. In our opinion, this publication is a "must" for every field man dealing with sugarbeets.

After a preliminary diagnosis of a nutritional problem has been made from plant-deficiency symptoms, the diagnosis should be confirmed by plant tissue analysis, and finally by a field test with the fertilizer nutrient suspected of being deficient. The plant tissue to be analyzed and the range of values for diagnostic purposes are shown in Table 7.2. This table is copied directly from Ulrich and Hills (10). California Agricultural Experiment Station Bulletin 766 (6, 11) is an excellent guide on methods of sampling and wet chemical analysis of leaves and petioles. Wet methods of analysis for the cations have been largely superseded in the last three or four years by physical methods. Potassium and sodium are determined by flame photometry or atomic absorption. Calcium, magnesium, copper, manganese, zinc, and iron are determined by atomic absorption. This instrumentation is expensive, but accurate, rapid, and essential when routine analysis involving a large number of samples is required.

No publication deals exclusively with soil testing for sugarbeets, but the interested reader can learn more about both soil testing and plant analysis from the references given at the end of this chapter (1, 3, 5, 8).

SOIL ACIDITY

Sugarbeets do not grow well on acidic soils. The harmful effects may be due to single or to multiple causes, depending on the soil. What went wrong in a particular soil is usually unknown, but the solution is simple. Keep the soil reaction right so that micronutrient nutrition will be right.

TABLE 7.1. A brief guide to deficiency symptoms on leaves and petioles

Element	Symptoms
	Uniform Yellowing
N Nitrogen	Overall yellowing of leaves. Young leaves may be small and dark green. Negative test with diphenylamine on petioles.
S Sulfur	Leaf symptoms similar to N deficiency, except young leaves tend to become light green or yellow. Brown blotches may develop on leaves and petioles. Positive test with diphenylamine on petioles.
Mo Molybdenum	Observed in the field only in Europe. Resembles sulfur deficiency closely, except pitting develops along the veins. Positive test with diphenylamine.
	Stunted Greening
P Phosphorus	Difficult to diagnose. Plants are small and have deep green leaves that may turn gray-green or blue-green. No purpling on most commercial varieties. Confirm with plant tissue analysis.
	Leaf Scorch
K Potassium	Tan color on margins of mature leaves that progresses between the veins to the midrib. Older leaves affected most. Young leaves may remain normal.
Mg Magnesium	Easily confused with potassium deficiency. Blades of recently mature leaves become chlorotic, then yellow, with scorching of interveinal tissue that dies. Triangular (Δ) shape of green tissue remains at base of leaf.
	Growing-Point Damage
Ca Calcium	First signs of deficiency are crinkling and downward cupping of young leaf blades. Progressive stages are: permanent damage to growing points, young petioles ending in blackened stubs. Dark rings of dead cambium in the root.
B Boron	First symptoms are white, netted chapping of upper blade surface or wilting of tops. Young leaves wilt the most, instead of least, as in dry soil. Progressive symptoms are: transverse cracking of petioles, death of growing point, and heart rot of root.
	Yellowing with Green Veining
Zn Zinc	Light greening or yellowing of larger leaves near center of plant followed by interveinal chlorosis and necrosis. Veins remain prominently outlined, turgid and green. Leaves are smaller than normal.
Mn Manganese	Chlorosis begins in younger leaves. A netted veining appears when leaf is held up to light. Early stages difficult to distinguish from chlorine, iron, or copper deficiency. Leaf blades gradually fade to a uniform yellow. In severe stages, gray to black freckling develops along veins, which coalesces into black, necrotic areas.
Fe Iron	Younger leaves turn light green, then yellow. Veins remain green unless deficiency is severe. Older leaves affected as deficiency worsens.
Cl Chlorine	Deficiency never observed in the field. Early stages resemble manganese deficiency. In advanced stages, interveinal areas appear as flat, yellow-green depressions which become dry between "raised" veins.
Cu Copper	Never observed in the field. Early symptoms resemble those for chlorine, manganese, and iron deficiencies.

SOURCE: Adapted from Ulrich and Hills (10).

TABLE 7.2. Chemical analysis to determine nutrient deficiency (and nondeficiency) in sugarbeet plants

Nutrient: Constituent Determined	Plant Part Tested*	Amount of Nutrition:		
		Critical concentration level†	Range showing‡ Deficiency symptoms§	No deficiency symptoms‖
Boron: B (ppm)	Blade	27	12–40	35–200
Calcium: Ca (percent)	Petiole	0.1	0.04–0.10	0.2–2.5
	Blade	0.5	0.1–0.4	0.4–1.5
Chlorine: Cl (percent)	Petiole	0.4	0.01–0.04	0.8–8.5
Copper: Cu	Blade	¶	¶	¶
Iron: Fe (ppm)	Blade	55	20–55	60–140
Magnesium: Mg (percent)	Petiole	¶	0.010–0.030	0.1–0.7
	Blade	¶	0.025–0.050	0.1–2.5
Manganese: Mn (ppm)	Blade	10	4–20	25–360
Molybdenum: Mo (ppm)	Blade	¶	0.01–0.15	0.2–20.0
Nitrogen: NO_2-N (ppm)	Petiole	1,000	70–200	350–35,000
Phosphorus: H_2PO_4-P (ppm)	Petiole	750	150–400	750–4,000
	Blade	¶	250–700	1,000–8,000
	Seedling: petiole (first leaf)	1,500	500–1,300	1,600–5,000
	blade (first leaf)	3,000	500–1,700	3,500–14,000
	cotyledon	1,500	200–700	1,600–13,000
Potassium: K (percent)				
> 1.5% Na	Petiole	1.0	0.2–0.6	1.0–11.0
	Blade	1.0	0.3–0.6	1.0–6.0
< 1.5% Na	Petiole**	**	0.5–2.0	2.5–9.0
	Blade	1.0	0.4–0.5	1.0–6.0
Sodium: Na (percent)	Petiole	¶	¶	0.02–9.00
	Blade	¶	¶	0.02–3.70
Sulfur: SO_4-S (ppm)	Blade	250	50–200	500–14,000
Zinc: Zn (ppm)	Blade	9	2–13	10–80

* Unless otherwise designated, blades and petioles are from a recently matured leaf.
† That concentration of a nutrient at which growth of a plant is retarded by 10%.
‡ All values are based on dry weight.
§ Leaf material for analysis must be collected shortly after symptoms appear. If this precaution is not taken, deficient plants may accumulate nutrients within the leaf without restoring "dead" tissues.
‖ The upper value reported is the highest observed to date for "normal" plants. Abnormally high values are often associated with other nutrient deficiencies; e.g., blades low in Fe may contain up to 4% Ca, 900 ppm Mn, and the like.
¶ Not yet determined.
** Because Na affects K content in petioles, blades must be used for K analysis when petioles contain less than 1.5% Na.
SOURCE: Ulrich and Hills (10).

Soil reaction is expressed in terms of "pH." A soil having a pH of 7.0 is considered to be neutral—neither acid nor alkaline. A soil with a pH of 6.0 is mildly acid; pH 5.0 is more strongly acid. In contrast, pH 8.0 is mildly alkaline. Sugarbeets seem to thrive best in a range between 6.0 and 8.0. For sugarbeets the upper limit of the pH range is not as specific as the lower. There is very little scientific literature to refer to on the effect of soil acidity on sugarbeets, but a review of successes and failures, over a period of time, of the sugar companies in the eastern region demonstrates the significance of soil acidity.

At one time, Indiana, Ohio, and Michigan had many more sugar processing plants than today. The plants and companies that are productive today are those located in areas where the soil is naturally not acid. Those plants located where the soils were naturally acid have gone out of business because sugarbeet yields were low. In the Netherlands, the senior author has seen excellent beets growing on field experimental plots maintained at pH 6.5. Poor growth occurred at pH 5.5, and at pH 4.5 there was almost no growth.

A high-yielding crop of sugarbeets requires as many or more nutrients than other field crops. Plant nutrients, particularly phosphorus, are most available in mineral soils having a pH between 6.0 and 7.0. Below pH 5.5 the solubility and availability of iron, aluminum, and manganese increase so rapidly that these elements may become toxic. Since the wild sugarbeet originally grew in the salt marshes of Europe, it is not likely that beets can tolerate high levels of these three elements. This concept should be recognized by people responsible for the location of new sugarbeet processing plants.

With time, the pH of sugarbeet soils tends to decrease. Sugarbeet soils become acid because farmers are now using high rates of ammonium-containing nitrogen and phosphorus fertilizers. Also, the fertilizers used today contain fewer bases. Thus, the acidity increases faster than in years gone by. In addition, the harvest of a crop removes bases such as calcium, magnesium, and potassium, thereby increasing the soil acidity.

In general, the amount of lime needed to neutralize soil acidity depends on three considerations: the soil pH, texture, and amount of organic matter. Soil testing is by far the most scientific and reliable way to determine whether the soil is acid, and if it is acid, the degree of acidity. Sugarbeet soils should be kept from becoming strongly acid, not only because beets do not grow well but because it is easier to make small than large adjustments in soil pH.

CATIONS

The cations with their positive electrical charge are grouped together because of some similarities in their reactions in soils and functions in beets. In soils, they are adsorbed on the negatively charged clays and some constituents of the soil humus. Adsorption means that they are less mobile in soils than the unadsorbed anions. Fertilizer placement may be more important than for soluble nutrients like nitrate and sulfate. They move more slowly in percolating water and are not readily leached into drainage water.

In plants, calcium, magnesium, potassium, and sodium appear to have a mutual and nonspecific function in neutralizing organic acids produced by plant metabolism. Expressed another way, organic acids balance the excess of cations over soluble mineral anions inside the plant. The excess of cations results because nitrate and sulfate absorbed as anions are reduced in the plant. Organic acids keep the plant sap at a reaction or pH normal for the species and preserve the electrical balance between positive ions (cations or C) adsorbed and the negative ions (anions or A) in the plant. This difference expressed on a chemical equivalent basis (usually milliequivalents per 100 g of dry tissue) is called C-A. In some species, C-A is known to be made up almost entirely of organic acids, either in solution or precipitated in the tissue as calcium oxalate. The experimental data and theory hold that yield is directly related to C-A or organic acid content, although some species may have an optimum. Sugarbeets are much higher in organic acids than most other species. Beets have 350 to 450 meq of C-A per 100 g of dry weight of foliage compared to 100 to 200 for many other species. This is perhaps not surprising because 80 percent of the organic acids of beet leaves are present as insoluble calcium oxalate. This theory is consistent with the high ash and cation contents of beets in relation to other plants and the substitution of one cation for another in performing nonspecific functions. It helps to explain why beets that readily take up sodium can use it beneficially in neutralizing negative charges and use it to stretch a limited supply of potassium. These functions of the cations appear to be nonspecific functions separable from their specific functions discussed later. Note in Table 7.2 that there are two sets of standards for potassium, depending on the concentration of sodium. The quantities of micronutrient cations absorbed are so small that they play an insignificant role in the cation-organic acid balance.

Calcium. Calcium occurs in the calcium pectate of the middle lamella which holds plant cell walls together. It occurs in calcium oxalate crystals in plant cells. Whether this is an important function of calcium is not known. Calcium is essential for the organizational integrity and function of membranes, organelles, and some enzyme systems in all cells. As in other plants, calcium is translocated only in the xylem and is not redistributed in the plant when calcium stress occurs. It does not move from old leaves to new ones. Hence, a continuous supply of calcium is essential. The seed contains too little calcium to supply the plant beyond emergence. Hence, adequate calcium is essential from the start for root and shoot growth. A 25-ton crop will contain about 35 to 200 pounds of calcium. This amount, as those mentioned later, includes that found in the tops, crowns, and storage roots.

When deficiency occurs, it shows first as a distortion and downward bending of the new leaves. As the deficiency progresses, the leaf blades blacken and die, leaving petioles with black tips. The symptoms at this stage may resemble acute boron deficiency and the older leaves may show scorched margins.

For most crops, calcium deficiency occurs on very acidic soils, particularly sandy ones. Such soils may also contain toxic concentrations of aluminum, manganese, and hydrogen (acid) ions, complicating the calcium deficiency with toxicities of other elements. In solution cultures, the very poor growth of beets at pH 4, and poor growth at pH 5 compared to pH 6, has been attributed to direct effects of acidity (12). The low tolerance of sugarbeets to soil acidity may be due partly to calcium deficiency, but toxicities of other elements may be involved.

Calcium deficiency is rarely important economically, except on acid soils that are very unfavorable for beets. It has been reported frequently as a temporary deficiency on neutral or even calcareous soils when an abundant supply of nitrogen induces luxuriant top growth. The symptoms are a "tip burn" of leaves. Western irrigated regions are adequately supplied with calcium as natural lime somewhere in the profile or by calcium adsorbed on the exchange complex. Irrigation water usually contains sufficient calcium to maintain the supply. In the acid soils of north-central and northeastern states, soil acidity should be corrected by liming (applying limestone) to a pH of 6.5 for beets. Application of neutral salts such as calcium chloride or gypsum to acid soils is not satisfactory.

Magnesium. Magnesium is an integral part of the chlorophyll molecule which makes up about 5 percent of the dry weight of sugarbeet leaves. The chlorophyll is contained in chloroplasts which make up about 35 percent of the dry weight. Although data are not available for beet leaves, corn leaves require about 200 ppm of magnesium on a fresh-leaf weight basis for maximum rates of photosynthesis. Magnesium is essential for the functioning of enzyme systems involved in energy transfer and respiration. A 25-ton crop will contain 27 to 85 pounds of magnesium.

Magnesium deficiency, like calcium deficiency, is rare in sugarbeets. When it occurs on acid soils, it is apt to be complicated, like calcium deficiency, with toxicities of aluminum and manganese. Western irrigated soils are generally well supplied with magnesium. Irrigation water may contain sufficient quantities. For example, in the Platte River basin, the irrigation water supplies have about 60 pounds of magnesium per acre-foot. A normally irrigated crop of beets would get 180 to 200 pounds per acre from this source. Since magnesium is more leachable than calcium, magnesium deficiencies can be expected to be more frequent than calcium deficiencies where irrigation water is low in magnesium. Under humid conditions, magnesium deficiency is uncommon but has been reported on acid peat and sandy soils. In general, magnesium deficiencies may occur on acid sandy soils containing less than 0.2 meq of exchangeable magnesium per 100 g of soil. Since magnesium deficiency, like calcium deficiency, occurs on acid soils, the best control is liming to pH 6.5 with dolomitic limestone. On limed soils, magnesium sulfate or potash containing magnesium can be used to maintain adequate magnesium supply.

Copper. In the United States, only Michigan makes recommendations for the use of copper on sugarbeets, and then only for those grown on organic soils. However, use of copper is negligible because the sugar companies do not like to contract for beets on muck and peat soils.

Since copper deficiency of sugarbeets has not been reported elsewhere in the field in the United States, there is little about copper that a grower or field man really needs to know. Copper is a constituent of the polyphenoloxidase in the chloroplast. Chloroplasts of the sugarbeet contain about 27 ppm of copper on a dry weight basis. About 64 percent of the total leaf copper is contained in the chloroplasts. Hence, copper has a vital role in photosynthesis. A 25-ton crop will contain about 0.07 to 0.11 pound of copper.

Iron. Iron is contained in and is essential to some of the enzymes involved in photosynthesis and respiration of sugarbeets. About 62 percent of the iron in the leaf is contained in the chloroplasts. The iron requirements of sugarbeets are higher than those for zinc, copper, manganese, boron, and molybdenum; yet reports conflict as to iron deficiency ever being observed in the field. If beets behaved like many other crops, you would expect deficiency on overlimed acid soils and on calcareous soils of the West prone to produce lime-induced or iron chlorosis. In the West many kinds of fruit trees, ornamentals, beans, corn, and sorghum fall victim to iron chlorosis on soils that let beets escape. The resistance of certain species to iron deficiency has been attributed to mechanisms such as excretion of complexing or reducing agents by the roots, or complexing organic acids in the plant sap that keep the iron in an available form. Nagarajah and Ulrich (7) found that iron-deficient beets increased the acidity of culture solutions and released riboflavin from the roots. They suggest that these effects may enable beets to get sufficient iron from calcareous soils that would produce chlorosis on many other crops. Little reliable data exist on the iron content of field-grown beets because of the labor involved in getting the soil washed off the roots and crowns, and dirt off the leaves.

Manganese. Like the other micronutrient cations, manganese is essential for the activity of enzyme systems. Unlike iron and copper, it does not concentrate in the chloroplast but is evenly distributed between the chloroplast and other leaf constituents on a dry weight basis. Nevertheless, about 35 percent of the leaf manganese is in chloroplasts. A 25-ton crop contains 0.27 to 0.89 pound of manganese.

Manganese deficiency in sugarbeets has been reported in Michigan, New York, and Montana. The deficiency is most apt to occur on sandy acid soils that have been overlimed to a pH of 6.8. Manganese availability is very dependent on soil pH. On soils below pH 5, manganese can become toxic to sugarbeets. The problem of manganese toxicity on such soils is often complicated with acidity, calcium deficiency, and aluminum toxicity. Too much manganese interferes with the normal iron metabolism so that arguments soon ensue on whether manganese toxicity or iron deficiency is involved. The continued use of acid-residual ammonium sulfate as a nitrogen source can lead to manganese toxicity on western irrigated soils that are sandy or medium textured and lime free. In Washington, use of ammonium sulfate dropped the soil pH from about 7.3 to 6.1. This drop in pH increased the leaf manganese content from 107 ppm to about 300 ppm on a dry weight basis, and manganese content of the total tops from 193 to 385 ppm. Continued use of ammonium sulfate probably would have produced manganese toxicity.

There are no generally accepted values for manganese used in soil testing. On soils of high pH, exchangeable manganese is generally regarded as the best test. On acid soils, water-soluble manganese is probably better. Since the oxides of manganese are regarded as the soil's reserve of manganese, they can be estimated by reducing the soil with hydroquinone or other reducing agent and extracting the soluble manganese with a 1 M salt solution.

Manganese deficiency can be corrected by the use of about 40 pounds per acre of manganese sulfate or its manganese equivalent in other manganese salts in the fertilizer.

Zinc. Zinc is essential for the activity of respiratory enzymes and for the production of auxin. It does not concentrate in the chloroplast as iron and copper do. The beet seedling must have a supply of available zinc from the time of emergence, since the seed contains no surplus. The normal range for zinc in a 25-ton crop is reported to be 0.16 to 0.53 pound. Evidence collected in Washington State indicates that 0.16 pound is sufficient for 25 tons. About 60 percent of this zinc is contained in the roots.

Zinc deficiency has been reported in Washington, California, and Nebraska. Often beets will not show zinc deficiency on soils that produce severe zinc deficiency in more sensitive crops such as corn, beans, and sorghum. The increasing frequency of zinc application to these sensitive crops grown in rotations with beets throughout all beet-growing areas will probably prevent spread of the deficiency and eliminate it entirely in sugarbeets. Sensitive crops are frequently zinc deficient following sugarbeets when they would not be deficient following other crops. Current evidence in Washington (2) indicates that this capacity of beets to produce zinc deficiency in succeeding crops is not related to high rates of phosphate application on beets or to higher zinc requirements of beets compared to other crops. However, in Michigan, beet-induced zinc deficiency is believed to be due to both high zinc requirements of beets and heavy phosphorus use on them. Zinc deficiency in a cropping system characteristic of western areas where beets are grown can be eliminated simply by recognizing the problem in time and applying about 10 pounds of zinc per acre as zinc sulfate, or some other zinc source, and thoroughly incorporating it in the soil before a zinc-sensitive crop is grown. One application will suffice for 3 to 5 years. Alternatives are use of zinc chelates at lower rates of zinc application or band placement of 3 to 5 pounds of zinc per acre along with other fertilization at planting. Banded applications are commoner in humid than in irrigated areas.

Zinc deficiency does not appear to be characteristic of any particular kind of soil. Sandy soils are probably more prone to show deficiency than heavier-textured ones when sensitive crops are grown. Zinc is generally more available on acid soils than on alkaline soils, unless the total supply becomes depleted on the acid soils. Overliming an acid soil can induce zinc deficiency.

No generally accepted soil test for zinc exists, but extractions with 0.1 N HCl, dithizone-ammonium acetate, or synthetic chelating agents have been used successfully.

ANIONS

The micronutrient anions are negatively charged and like nitrate are not adsorbed or are more weakly adsorbed by the soil clays than are the cations. At the soil pH at which beets can be grown, chloride and sulfate are not adsorbed and can be readily leached. Molybdate and borate are adsorbed on the soil clays and humus to some extent, but these ions also can be leached. Since molybdenum can be toxic to livestock and boron can be toxic to plants, the leachability and toxicity of these ions must be reckoned with in fertilizer practice.

Sulfur. Sulfur is contained in the amino acids, cystine and methionine, that are two of the building blocks of the cytoplasmic and enzymic proteins in the beet. Reduced sulfur compounds are also intermediates in metabolism. Leaf proteins of beets contain about 16.6 times as much nitrogen as sulfur. The sulfate is a storage form, like nitrate is for nitrogen, and is freely translocated. About 13 to 14 pounds of sulfur are the minimum needed in a 25-ton crop of beets, but the total uptake may range up to 50 pounds and higher.

Sulfur deficiency in beets grown for seed has been reported in western Washington, and in beets grown for sugar in central Washington and California. The normal sources of available sulfur are rainfall, irrigation water, manure, single superphosphate and ammonium sulfate, the mineralization of organic sulfur compounds in soil organic matter, and free gypsum in the soil. Many irrigation waters contain enough sulfur as sulfate for irrigated sugarbeets. The trend toward less sulfur in fertilizers, less dependence on organic matter and manure as sources of nutrients, and cleaner air will make sulfur deficiency commoner unless sulfur is naturally supplied in irrigation water or the soil contains gypsum.

Beets get practically all of their sulfur from the soluble sulfate in the soil, supplemented with some from the soil organic matter mineralized during the season. Sulfate is not adsorbed by the mineral fraction of the soil at pH values well suited to beet growth. Therefore, the available sulfur can be estimated from a water extract or from other mild extracting agents used in soil testing, such as sodium acetate, weak acids, or sodium bicarbonate.

Sulfur can be supplied as gypsum, ordinary superphosphate, ammonium sulfate, or even polysulfates or ground sulfur put into liquid fertilizers or aqua ammonia. Since sulfur in these materials is cheap and the fertilizer requirements are only about 20 pounds per acre, there is no excuse for having sulfur-deficient sugarbeets, except poor diagnosis of the problem.

Chlorine. Not until 1954 was chlorine proved to be essential for plants. Dr. Ulrich and Dr. Ohki (13) soon demonstrated that beets must have it too. The almost universal contamination of salts, air, and water with chlorine prevented the discovery of the essentiality of chlorine for almost a century, the span of time in which physiologists had been looking for the mineral elixirs of plant life. A 25-ton crop of beets must have about 1.7 pounds of chlorine in its roots and tops, but often has hundreds of times this much. Nevertheless, the plant's requirements for chlorine are large compared with its requirements for copper, zinc, manganese, molybdenum, and iron. Chlorine functions in one of the reactions of photosynthesis.

Chlorine deficiency is unknown in the field. Precipitation, irrigation water, air pollutants, fertilizers, and animal wastes provide adequate amounts. Unirrigated, sandy soils located far inland away from the influence of cyclic salt and fertilized with only nitrogen are sites to look for chlorine deficiency.

Molybdenum. The beet's requirement for molybdenum or "moly" is the least of any of the micronutrients known to be essential. It functions in the enzyme system that reduces nitrate to ammonium or amino compounds that enter into nitrogen metabolism. It performs additional functions, as plants given only ammonium nitrogen still require small amounts of molybdenum. Deficiency of molybdenum has not been identified in beets in the field in the United States, but it is known in Europe. The deficiency affects more sensitive crops such as cauliflower in the United States. Molybdenum availability is often the lowest on acidic soils and can usually be corrected by liming. Molybdenum absorption by plants is increased by high levels of phosphate and depressed with sulfate.

Excess molybdenum can be very toxic to livestock, particularly when grazed on legumes that accumulate molybdenum. So there is some hazard in incorporating molybdenum into fertilizers unless a need for it has been positively established.

Boron. Boron is a two-faced element—the ratio between the concentration needed in the plant to perform essential functions and the concentration that is toxic is narrower than for any of the secondary and microelements. Tolerances of various species grown in rotation with beets also differ. Hence, care must be used in correcting deficiencies.

The precise functions of boron in the plant are not known. Boron, along with calcium, is essential for the formation of new cells in meristems. Deficiencies show first in the growing points of both shoots and roots. Failure of root growth may be the reason new leaves of sugarbeets affected with boron deficiency wilt. Boron may have a vital role in sugar translocation. Since fungi and animals do not require boron, it may have an important and unrecognized function in photosynthesis.

Deficiency of boron is the commonest of the microelement deficiencies affecting sugarbeets. The deficiency has been reported in either garden beets or sugarbeets in Washington, Oregon, California, Montana, Texas, Wisconsin, Michigan, Pennsylvania, New York, Massachusetts, Alabama, and South Dakota. The deficiency is commonest on sandy, leached soils. The adsorption capacity of a soil for soluble borate becomes greater with rise in pH and increases of clay content, organic matter, and dryness of soil. The proportion of boron in soluble borate fertilizer adsorbed on the soil is low, and so it can be leached and can accumulate where drainage water evaporates. Some borates are less soluble than others and are used under leaching conditions. Boron is sometimes used in glass frits to release boron slowly and prevent leaching.

Since sugarbeets have a higher requirement for boron than many other crops, the usual recommendation is that the soil contain 0.5 ppm of hot-water-soluble boron, a value higher than for other crops. Beets are among the most tolerant crops to boron and can stand more boron in irrigation water than more sensitive crops such as beans. A class-1 irrigation water for beets should not have more than 1 ppm boron, but this is three times too much boron for water to be used on sensitive crops.

An amount of boron in fertilizer or a concentration of it in irrigation water that will not harm beets may harm most field and many truck crops that follow beets. Therefore, recommendations on the amount of boron or borated fertilizer to correct a deficiency in sugarbeets are best left to local fieldmen and advisors familiar with all aspects of the situation. The simplest and safest way to correct boron deficiency in beets is to use commercial fertilizer containing the right amount of boron.

INTERRELATIONS OF SECONDARY AND MICRONUTRIENTS

Not much is known about substitution reactions and the effects of a deficiency or an excess of one micronutrient on the availability or plant use of another for sugarbeets. These relations are somewhat clearer in other crops where micronutrient deficiencies are commoner and the opportunities to correct them in the field are greater. We have already recognized that sodium can partially substitute for potassium. This is shown in Table 7.2 where there are two sets of standards for evaluating potassium status in leaves and blades, the standards depending on the sodium present. We have noted the partial substitution of bromine for chlorine, but this substitution is now of no practical concern. We have noted that sodium, potassium, calcium, and magnesium are substitutable for each other, probably over a limited range, in the cation-anion balance. Other than these possible substitutions, antagonistic or synergistic relations among the elements are not recognized in sugarbeets. The standards for foliar evaluation in Table 7.2 show that other elements may have an effect on the needed level. The levels of the other elements are assumed to be nonlimiting. This is generally the situation. There is little practical point in finding out that you can get along with less zinc, for example, in phosphorus-deficient sugarbeets than you can in ones with sufficient phosphorus.

REGIONAL PROBLEMS

The beet-growing areas of the United States are naturally divided by precipitation into the subhumid and humid areas of the north central and northeastern states where beet culture depends on rainfall, sometimes supplemented with sprinkler irrigation, and the western states where irrigation is essential. In spite of the differences in source of water and the vast differences in soils associated with climate, the nutritional problems of the two distinct areas are not great.

Northern Humid Region. Secondary and micronutrient problems in the humid regions are rare, primarily because sugarbeets are a long-season and a deep-rooted crop. If deficiencies develop, the cause is frequently the low solubility and availability of the nutrient in a soil that has a relatively high pH.

None of the secondary nutrients are apt to be deficient in the humid region. Calcium and magnesium levels are adequate now in all soils used for sugarbeet production. Sulfur deficiencies never occur in those areas that are washed by rains containing sulfur from the fumes of industrial plants. Thus, sulfur is not considered to be in the deficient range in either Ohio or Michigan.

Manganese deficiency is likely to occur on organic soils with a pH of 5.8 or above. Such a deficiency can be corrected by applying manganese salts or adding enough elemental sulfur to acidify the soil. To obtain immediate results, the use of manganese salts is most frequently recommended. Use of sulfur is suggested where a more lasting effect is desired. Manganese deficiency also occurs on mineral soils with a pH above 6.5. It is most likely to occur on soils that are dark colored, high in organic matter, and have a grayish subsoil as occurs in some lake bed or glacial outwash areas.

Manganese deficiencies are easily corrected. Manganese salts, used at rates of 5 to 10 pounds of manganese per acre and applied in a band near the seed, have been effective. Broadcast applications usually are not recommended because most soils fix manganese so easily. Foliar spray treatments at the rate of 1 to 2 pounds of manganese per acre have been effective in the humid region. On soils that are very deficient, a combination band and spray treatment has been effective.

Boron is the only other micronutrient which might be deficient in the humid region. Boron deficiency is likely on organic soils if the pH is about 5.5 and on mineral soils if the pH is above 6.5. The use of 2 to 3 pounds of boron per acre is usually sufficient to correct any deficiencies, although up to 5 pounds per acre may be needed where sugarbeets are grown on organic soils with a pH of 6.5 or greater.

Boron and manganese deficiencies are not significant problems on organic soils, primarily because the sugar companies prefer not to write contracts for production on them.

Irrigated West. Problems peculiar to the irrigated area are nutrient deficiencies that may be induced by land leveling, and the beneficial effects of nutrients in irrigation water. In leveling or planing of land for surface irrigation, surface soil is frequently removed from the high places, leaving a subsoil that is deficient in many of the elements closely associated with organic matter. Removal of surface soil can leave low available supplies of nitrogen, phosphorus, iron, and zinc. This potential problem should be evaluated by soil testing the whole profile and conducting some relevant trials on what the situation may be and how it can be corrected before a large land-leveling program is initiated.

The beneficial nutrient content of irrigation water is often overlooked. Irrigation water may contain enough calcium, magnesium, boron, sulfur, chlorine, and sometimes potassium to supply all of these nutrients needed. It may contain too much boron and sodium. Irrigation water seldom contains enough manganese, iron, zinc, copper, and molybdenum to be of any significance.

SUMMARY AND THE FUTURE

Compared with the enormous problems of nitrogen nutrition, virus diseases, weeds, hail, early freezes, stand establishment, and labor that the grower has to contend with, secondary and micronutrient problems appear insignificant. Fortunately, sugarbeets are not plagued with many of the deficiencies, either in kind or extent, that affect many other kinds of crops grown in rotation with beets. Because a rotation is essential for sugarbeets, correction of these deficiencies in other crops helps to assure an adequate supply of secondary and micronutrients for beets, which are "good feeders." The symptoms of the deficiencies are known, and good plant tissue tests are available for diagnostic purposes. The outlook for better soil tests is encouraging. The nutrients needed in fertilizers to correct the known deficiencies are cheap, so there is no excuse for having secondary and micronutrient deficiencies in beets. We could not make such general statements if sugarbeets were severely afflicted with iron deficiency or lime-induced chlorosis, for which there is no really satisfactory control on field crops.

REFERENCES

1. Black, C. A., ed. Methods of Soil Analysis. II. Chemical and Microbiological Properties. ASA Monograph 9, pp. 771–1549. Am. Soc. Agron., Madison, Wis. 1965.
2. Boawn, Louis C. Sugar beet induced zinc deficiency. Agron. J. 57:509. 1965.
3. Chapman, A. D. Diagnostic criteria for plants and soils. Univ. Calif. Div. Agr. Sci., Berkeley. 1966.
4. El-Sheikh, A. M., A. Ulrich, and T. C. Broyer. Sodium and rubidium as possible nutrients for sugar beet plants. Plant Physiol. 42:1202–8. 1967.
5. Hardy, Glen, ed. Soil Testing and Plant Analysis. I. Soil Testing; II. Plant Anlaysis. Soil Sci. Soc. Am. Special Publ. 2. Madison, Wis. 1967.
6. Johnson, C. M. and A. Ulrich Analytical methods for use in plant analysis. Calif. Agr. Exp. Sta. Bull. 766, pp. 25–78. 1959.
7. Nagarajah, Sellappah and Albert Ulrich. Iron nutrition of the sugar beet plant in relation to growth, mineral balance, and riboflavin formation. Soil Sci. 102:399–407. 1966.
8. Schmehl, W. R. and R. P. Humbert. Nutrient deficiencies in sugar crops. *In* H. R. Sprague, ed., Hunger Signs in Crops, pp. 415–50. David McKay Co., New York. 1964.
9. Singh, B. and D. J. Wort. Effect of vanadium on growth, chemical composition, and metabolic processes of mature sugar beet (*Beta vulgaris* L.) plants. Plant Physiol. 44:1321–27. 1969.
10. Ulrich, Albert and F. J. Hills. Sugarbeet Nutrient Deficiency Symptoms—A Colored Atlas and Chemical Guide. Univ. Calif., Div. Agr. Sci. 1969.
11. Ulrich, Albert, F. J. Hills, D. Ririe, A. G. George, and M. D. Morse. Plant analysis, a guide for sugar beet fertilization. Calif. Agr. Exp. Sta. Bull. 766, pp. 3–24. 1959.
12. Ulrich, Albert and Kenneth Ohki. Hydrogen ion effects on the early growth of sugar beet plants in culture solution. J. Am. Soc. Sugar Beet Technologists 9:265–74. 1956.
13. ———. Chlorine, bromine, and sodium as nutrients for sugar beet plants. Plant Physiol. 31:171–81. 1956.

8

Irrigation and Water Management

MARVIN E. JENSEN
Snake River Conservation Research Center, ARS, USDA
Kimberly, Idaho

LEONARD J. ERIE
Water Conservation Laboratory, ARS, USDA
Phoenix, Arizona

DEFINITIONS	191
FACTORS INFLUENCING EVAPOTRANSPIRATION	192
DETERMINING EVAPOTRANSPIRATION	192
ESTIMATING EVAPOTRANSPIRATION	193
GROWTH STAGES AND CHARACTERISTICS	194
Spring-planted Beets	194
Fall-planted Beets	195
EVAPOTRANSPIRATION	196
Spring-planted Beets	196
Fall-planted Beets	202
Peak Rates	203
Seasonal Evapotranspiration	204
Estimating Evapotranspiration	204
MANAGING THE SOIL MOISTURE RESERVOIR	206
Soils and Their Water-holding Capacities	207

Irrigation Practices 208
Effects of Soil Moisture Levels on
    Yields 211
Allowable Depletion and Irrigation
    Intervals 214
Soil Moisture-Nitrogen Interactions . 215

IRRIGATION WATER REQUIREMENTS AND
    METHODS 215

PREPLANTING AND PREEMERGENCE
    IRRIGATIONS 219

EARLY-SEASON IRRIGATIONS 219

MIDSEASON IRRIGATIONS 220

LATE-SEASON IRRIGATIONS 220

THE SUGARBEET PLANT is adapted to a wide range in climate. It is a major crop in irrigated areas because of its tolerance to salinity, hardiness, and its productivity which makes it a good cash crop under various growing seasons and soils. The major areas of irrigated sugarbeets are in California, Colorado, Idaho, Nebraska, Washington, and Wyoming. The major non-irrigated sugarbeet areas are in the Red River Valley of Minnesota and North Dakota and in Michigan (37).

The objectives of this chapter are to summarize (1) measured evapotranspiration from the sugarbeet crop and to relate these values to meteorological conditions and degree of crop cover at various stages of plant growth, (2) the effects of various soil moisture levels on root yield and sugar production, and (3) the major characteristics of various irrigation methods and to recommend practical procedures for irrigation water management to assure near-optimum production of sugar.

Irrigation provides a favorable soil environment for the germination of seeds, emergence of seedlings, and development of the root system, and maintains favorable moisture levels and salt concentrations in the soil solution so as not to restrict significantly the photosynthetic process (15, 20). The amount of water retained within the tissue of the beet tops and the beet roots is only about 1 percent of the total evaporated from a field during the growing season. Therefore, irrigation replenishes water that has been transpired by the plants or evaporated from the soil surface. However, since water evaporating from the soil surface is salt free and plants selectively absorb dissolved salts, sustained crop production also requires the maintenance of a favorable salt concentration in the soil. The only practical means of controlling the salt concentration is to allow a fraction of the water applied to the soil to pass through the root zone (leaching). The fraction of the water applied that is essential (leaching requirement) to control salts is a beneficial and necessary use of water (14).

Managing the soil moisture reservoir can be a relatively complicated procedure if the basic principles involved are not understood. The irrigator cannot see when the reservoir is full or nearly empty. In addition, the reservoir is leaky. If the root zone is refilled at each irrigation, some leakage or waste occurs for several days following each irrigation. However, if leaching is required, some deep percolation is essential.

Management of the reservoir can be relatively simply if a few basic principles are understood. The quantity of water evaporated from the soil and transpired by the plants is largely determined by the meteorological conditions and the degree of plant cover during the cropping season. Most sugarbeet evapotranspiration data include the sum of evaporation and transpiration from planting to harvest. There may be additional evaporation losses before planting in the spring and after harvest in the fall which must be considered when accounting for the annual water requirement.

DEFINITIONS

The following definitions explain the terminology used in this chapter.

Transpiration is the loss of water in the form of vapor from plants. All aerial parts of plants may lose some water by transpiration, but most of the water is lost through the stomates.

Evapotranspiration is the sum of water lost by transpiration and evaporation from the soil or from exterior portions of the plants.

Consumptive use is essentially identical to evapotranspiration. It differs by the inclusion of water retained in plant tissues. For most agricultural plants, the amount of water retained by plants is insignificant when compared to the amount evaporated (15).

Deep percolation is the amount of water that passes beyond the root zone of the crop.

Leaching requirement is the relative portion of the irrigation water applied that must pass through the root zone in order to maintain a favorable salt concentration in the soil solution.

Effective rainfall is the amount of rainfall that is used in evapotranspiration and leaching.

Irrigation water requirement is the total amount of water required to

FACTORS INFLUENCING EVAPOTRANSPIRATION

Evapotranspiration is affected by many factors. Man can influence or control some of these; others he has no control over, such as climate. Climatic factors include precipitation, solar radiation, temperature, humidity, wind movement, and length of growing season. Man can influence or control such factors as water supply, water quality, date of planting, crop varieties, soil fertility, plant spacing, water management, cultivation, and chemical sprays.

All these factors may influence plant growth and, thereby, evapotranspiration. Thus, evapotranspiration may vary from farm to farm, season to season, and day to day. When plants are young, the amount of water used is small. The amount used increases with plant growth and reaches a peak during some part of the growth period. Nevertheless, for optimum production at a specific location, the sugarbeet crop requires a fairly definite amount of water during the growing season.

The magnitude of daily evapotranspiration is controlled primarily by meteorological conditions when a green crop has sufficient leaf area and soil moisture so as not to limit evapotranspiration. In contrast, deep percolation is influenced by precipitation, soil characteristics, the amount of water applied at each irrigation, the frequency of irrigations, and the available soil moisture at the time of irrigation. Evapotranspiration is predictable mainly because it is limited at any time by the heat energy available for evaporation. As a result, evapotranspiration for a given crop will be similar in areas of similar climate and growing seasons. In contrast, deep percolation and other losses may vary widely, depending on the irrigation practices and the management of the system.

DETERMINING EVAPOTRANSPIRATION

The commonest method of determining E_t rates under field conditions, which has been used over 70 years, is gravimetric soil sampling 2 to 3 days after an irrigation, followed by another sampling in 5 to 20 days, or just before the next irrigation. More recently, the neutron soil moisture probe has been used. From the change in soil moisture and the rainfall received during the interval, the average E_t rate is calculated as follows:

$$E_t = \frac{W_{et}}{\Delta t} = \frac{-\sum_{o}^{S_r} \Delta\theta \ \Delta S + R_e - W_d}{\Delta t} \tag{8.1}$$

in which W_{et} = the total water used in evapotranspiration, S = the distance from the soil surface, S_r = the depth of the effective root zone, $\Delta\theta$ = the volumetric change in soil moisture (a negative sign indicating a decrease), R_e = effective rainfall, Δt = the time interval between sampling dates (usually days), E_t = the average evapotranspiration rate, and W_d = the water drained from the 0 to S_r depth. Drainage from the effective root zone or upward movement into the effective root zone is the major unknown in Equation 8.1 and is generally assumed to be zero. Drainage may be as large as E_t in the spring when E_t rates are low and the entire profile is wet. During these periods, the first sample after an irrigation should be taken at least 5 to 7 days after an irrigation.

Lysimeters are also used for determining evapotranspiration rates. These consist of a large volume of soil contained within a tank that is weighed by a mechanical, a combination mechanical and electrical, or a hydraulic weighing system. When properly installed, operated, and instrumented, lysimeters provide the most accurate measurement of evapotranspiration. Most of the data reported in this chapter were determined by soil-sampling techniques.

ESTIMATING EVAPOTRANSPIRATION

Daily evapotranspiration may be estimated using one of many rational and empirical equations. Most methods attempt to separate meteorological factors from soil-crop factors. One such technique is:

$$E_t = K_c E_{tp} \tag{8.2}$$

in which K_c is a dimensionless crop coefficient relating the rate of evapotranspiration for a crop like sugarbeets to potential evapotranspiration (or the evapotranspiration for a reference crop such as well-watered grass or alfalfa, with at least 12 to 18 inches of growth). The crop coefficient is a parameter determined primarily by the wetness of the soil surface and the leaf area for a crop like sugarbeets. Potential evapotranspiration (E_{tp}) is primarily determined by the magnitude of solar energy absorbed by the crop surface, the vapor pressure deficit of the atmosphere, windspeed, and the resistance to the passage of water vapor through the stomata.

The total amount of water required for the growing season is generally determined by summing the daily, weekly, or monthly estimates:

$$W = \sum_{i=1}^{n} (K_c E_{tp})_i \tag{8.3}$$

where i is the number of the time increment involved.

GROWTH STAGES AND CHARACTERISTICS

Growth stages and characteristics of sugarbeets can best be illustrated by typical curves representing the rate of top and root growth when beets are either spring planted or fall planted. Full cover of sugarbeet leaves is achieved slowly because of the small size of the juvenile leaves and because only 3 to 4 new leaves are initiated each week (24, 27). A leaf area index of 3 to 5, which can be considered as full cover, may not be attained until 60 to 90 days after planting (leaf area index = area of leaves, one side only, per unit of land surface). The leaf area index may go as high as 6 to 12 (23). In northern climates 100 to 120 days may be required to attain full cover because of low air and soil temperatures in the spring.

SPRING-PLANTED BEETS

The relative cumulative top growth and the cumulative root growth for spring-planted sugarbeets at Twin Falls, Idaho, is illustrated in Figure 8.1. These data dramatically illustrate the slow top growth development when beets are first planted, especially in northern climates where soil and

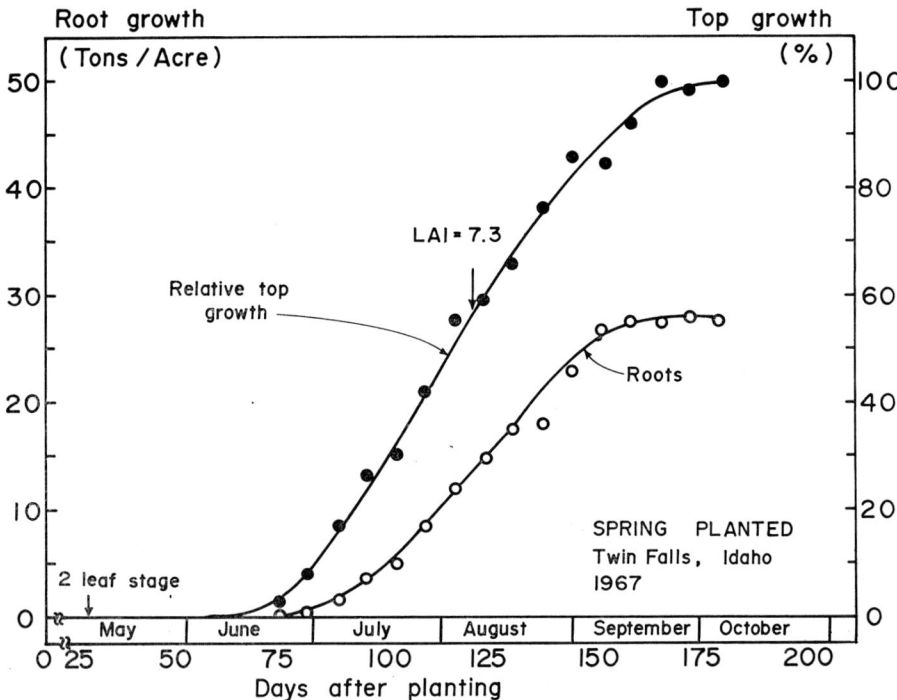

Fig. 8.1. Relative top growth and root growth for spring-planted beets at Twin Falls, Idaho. (After Carter et al., 2.)

air temperatures are low in the spring. In this example, the sugarbeet plants were at a two-leaf stage on May 8. Rapid top growth did not begin until the latter part of June. Full effective cover was attained near the latter part of July. A single leaf area index measurement of 7.3 was made on August 7. This value is generally greater than normally expected at this time of year (normally about 4).

Rapid root growth did not begin until July. The maximum growth rate occurred from the latter part of July to about September 1. During this period, root growth averaged about 3 tons per week. Most of the root growth occurred by mid-September.

Based on the leaf area attained between July 20 and August 1, evapotranspiration from sugarbeets would be essentially equal to the potential evapotranspiration rate from about July 25 until a severe frost. A frost may not kill the entire tops back, but research data indicate that a frost severely affects the rate of transpiration during the next few days.

FALL-PLANTED BEETS

Similar curves are presented for fall-planted beets at Phoenix, Arizona, in Figure 8.2. Data were not available on the top growth immediately

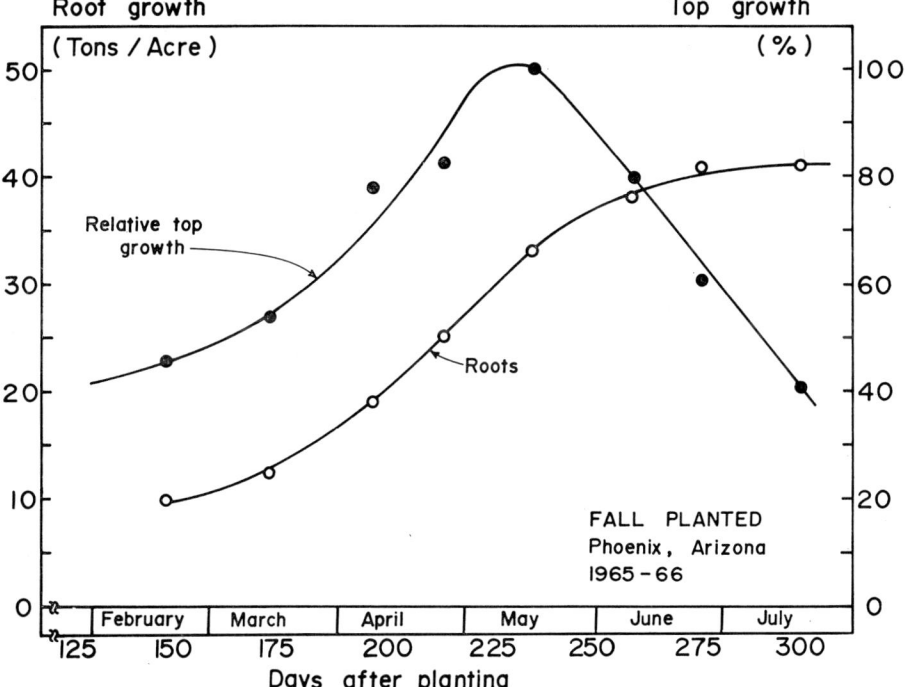

FIG. 8.2. Relative top growth and root growth for fall-planted beets at Phoenix, Arizona. (After Erie and French, 5.)

following planting on September 22. However, because of the warm soil, top growth increased rapidly until about January 20, when it began to decrease because of the low minimum temperatures (near 32° F). Top growth rapidly resumed early in March, reaching a maximum in mid-May. Mean maximum temperatures near mid-May ranged from 90° to 95° F. Following mid-May, maximum temperatures ranged from 98° to 110° F. Root growth increased in March and reached a maximum in April and May. The maximum rate during April and early May in this example was 2.6 tons per week. Root weight reached a maximum by the latter part of June. Similar decreases in top growth occur in the lower San Joaquin Valley in July and August when beets are planted early in January (8).

EVAPOTRANSPIRATION

Numerous studies on the rate of evapotranspiration or consumptive use have been conducted in western United States and in other countries. Sufficient data are now available so that if the effects of degree of crop cover, surface soil moisture, and available soil moisture are separated from meteorological effects (potential evapotranspiration), one can estimate the rate of evapotranspiration for practical purposes on sugarbeet fields in any climatic regime. Similarly, one can also estimate the total evapotranspiration for the season. Typical data are presented in this section to illustrate rates of evapotranspiration for spring- and fall-planted sugarbeets grown under medium soil moisture conditions in widely varying climates. The similarity of crop coefficients for spring-planted sugarbeets in two different climatic zones and fall-planted sugarbeets is presented. These data substantiate the previous statement that if meteorological and soils data are available, evapotranspiration rates can be estimated for practical purposes of management or estimating water requirements. Such estimates are sufficiently reliable to enable scheduling irrigations using soil-crop-meteorological data (17, 19, 21).

Evapotranspiration rates vary with meteorological conditions, which are influenced by latitude, elevation, cloud cover, humidity, and windspeed when sufficient leaf area has been attained and soil moisture is not limiting. The rate of evapotranspiration for sugarbeets increases as leaf area increases until the evapotranspiration rate is approximately equal to the potential rate existing in that area under given climatic conditions. The data in Figures 8.3 to 8.10 illustrate these characteristics.

SPRING-PLANTED BEETS

Evapotranspiration rates for spring-planted sugarbeets grown in an intermountain area (Twin Falls) at an elevation of about 4,000 feet above sea level are presented in Figure 8.3. In this area, beets are planted early in April and attain an effective full cover about August 1. These data were

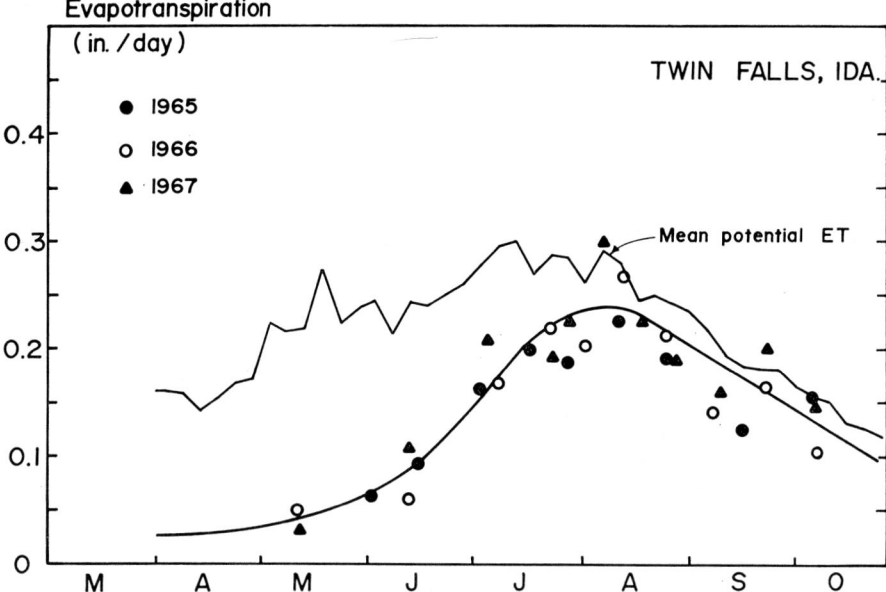

Fig. 8.3. Measured evapotranspiration and estimated mean potential ET during three growing seasons.

obtained in an irrigation experiment using a neutron soil moisture probe. Evapotranspiration rates early in May were only about 15 to 20 percent of potential evapotranspiration during this period, primarily because the soil was essentially bare (very limited leaf area). Although the evaporative demand increased slightly from the first part of June until mid-July, the rate of evapotranspiration increased more rapidly because the leaf area increased during this period. The rate of evapotranspiration was nearly equal to the potential evapotranspiration rate after the first of August. However, by this time the potential rate was decreasing as solar radiation and air temperatures decreased.

Similar data for the Northern Plains area are presented in Figure 8.4. In general, the same characteristics are prevalent in that area as at Twin Falls. Many of these data were obtained from old experiments conducted at Scottsbluff, Nebraska, in the early 1930s. These values may not represent the upper limit of evapotranspiration rates for sugarbeets under high soil moisture levels because the interval between irrigations may have been longer than desired for optimum soil moisture conditions. However, the curve as plotted in Figure 8.4 represents primarily the higher values and should be fairly representative of mean evapotranspiration rates expected in that area. Only a few points are presented for Newell, South Dakota. They are very similar but one would expect the rate for South Dakota to decrease more rapidly in the fall than at Scottsbluff.

These data were obtained by soil-sampling procedures. Scottsbluff is

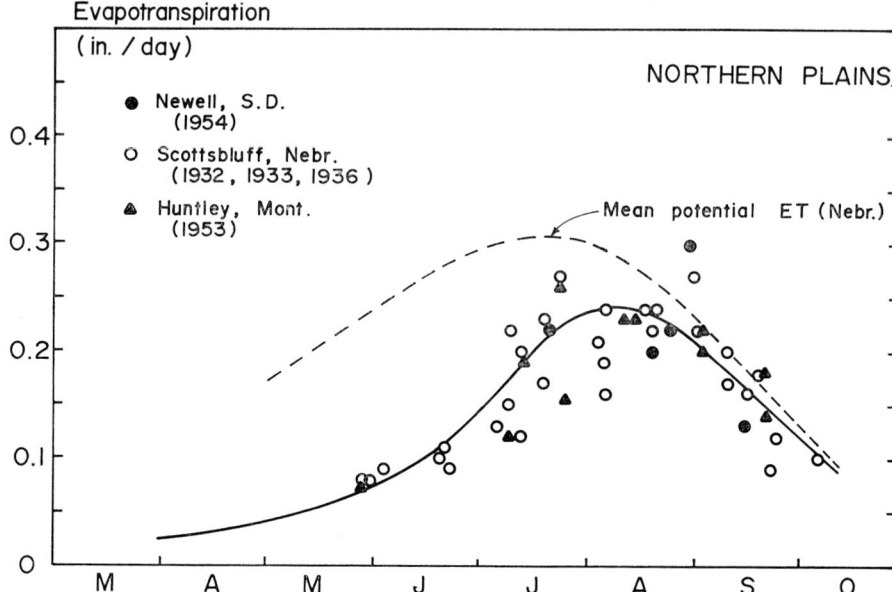

FIG. 8.4. Measured evapotranspiration at Newell, South Dakota (unpublished data, Baird and Bonneman), at Scottsbluff, Nebraska (unpublished data tabulated by O. W. Howe), and at Huntley, Montana (Larson and Johnston, 22).

located at a latitude of 41° and 52′ north and is about 4,000 feet above sea level. Newell is located 44° and 44′ north and is about 2,900 feet above sea level. Huntley, Montana, is located 45° and 55′ north and is about 3,300 feet above sea level.

Data for the Columbia Basin are presented in Figure 8.5. In this area, sugarbeets are planted early in March and attain an effective full cover about July 20. The elevation in the Prosser, Washington, area is approximately 840 feet. At these lower elevations, summer temperatures are higher even though the latitude is 46° and 15′ north. Consequently, with the clearer skies and similar summer clear day radiation values, peak evapotranspiration rates are higher than at Twin Falls and Scottsbluff. The earlier data (1940, 1941, and 1949) were collected by H. G. Nickle and summarized by S. J. Mech (25). Data collected in the mid-50s were obtained by Middleton et al. (26). In general, those values represent optimum soil moisture conditions and generally are somewhat higher than the values obtained by Mech. The data collected by Nickle and Mech were on medium soil moisture levels where the yields were essentially the same as those obtained on a higher soil moisture level. However, data from the high soil moisture level indicate that some deep percolation may have occurred during the sampling interval. Again, the trend is very similar to that presented for Twin Falls and Scottsbluff. Evapotranspiration rates are less than the

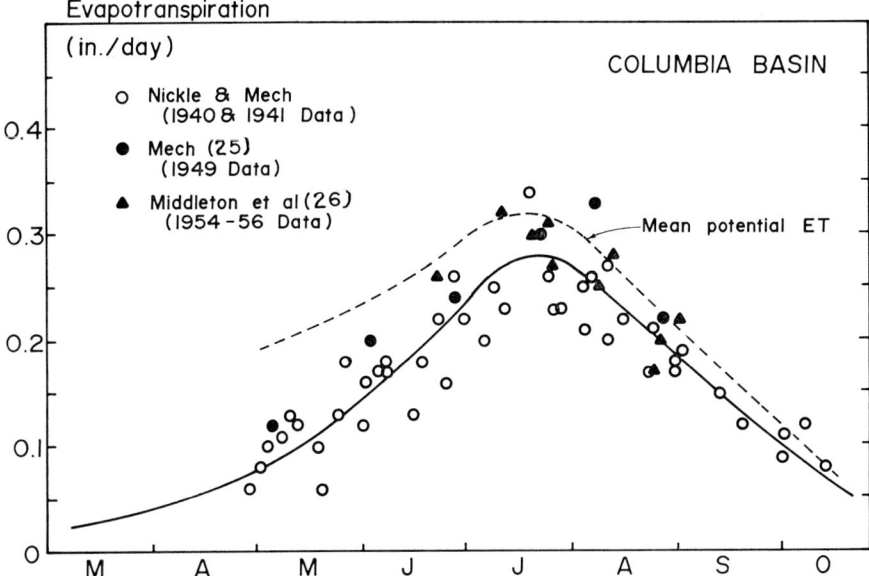

FIG. 8.5. Evapotranspiration and estimated mean potential ET for the Columbia Basin.

potential early in the year and approach the potential values, in this case, near mid-July. The decrease after July occurs primarily because of a decrease in potential evapotranspiration.

Similar data for the Southern Plains are presented in Figure 8.6 in which two years of evapotranspiration data determined at Bushland, Texas, using a neutron probe, are presented, along with the mean potential evapotranspiration for the two years involved. Evapotranspiration rates again show a similar trend in that they are significantly less than the potential in April and May, approach the potential early in June, and are about 90 percent of the potential value until approaching harvest in October. The values near the end of the season apparently represent the effect of decreasing soil moisture before harvest.

Recent studies of evapotranspiration from sugarbeets, using accurate weighing lysimeters at Davis, California, confirm these typical patterns. Evapotranspiration rates for beets planted March 25 were very similar to those obtained at Bushland (30).

If the ratio of evapotranspiration to potential evapotranspiration is calculated at various stages of growth for spring-planted sugarbeets in the Twin Falls area, a value referred to as a crop coefficient is obtained (15). Crop coefficients for Twin Falls are presented in Figure 8.7. When presenting the data in this manner, it is apparent that the rate of evapotranspiration ranges from 10 to 15 percent of the potential shortly after plant-

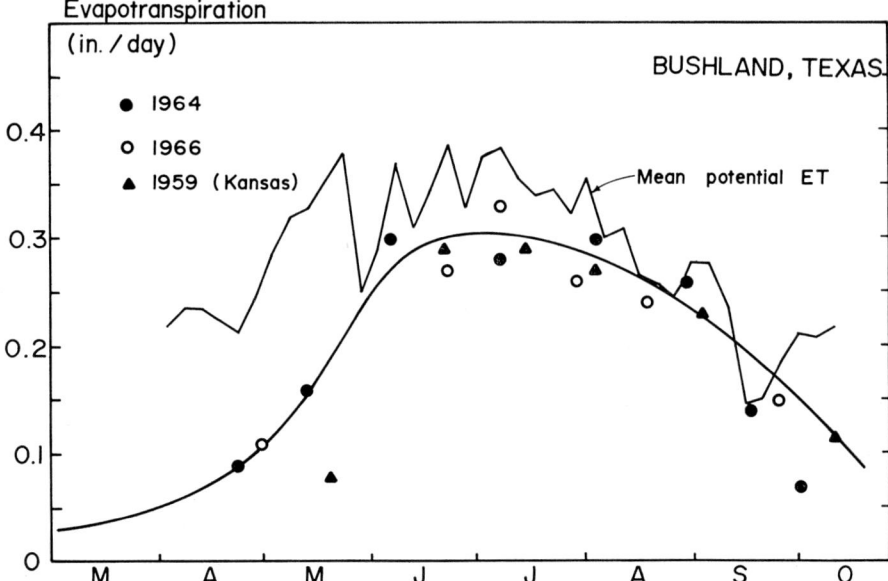

Fig. 8.6. Measured evapotranspiration and estimated mean potential ET for two years at Bushland, Texas (adapted from Schneider and Mathers, 35), and measured evapotranspiration for Garden City, Kansas (Herron et al., 13).

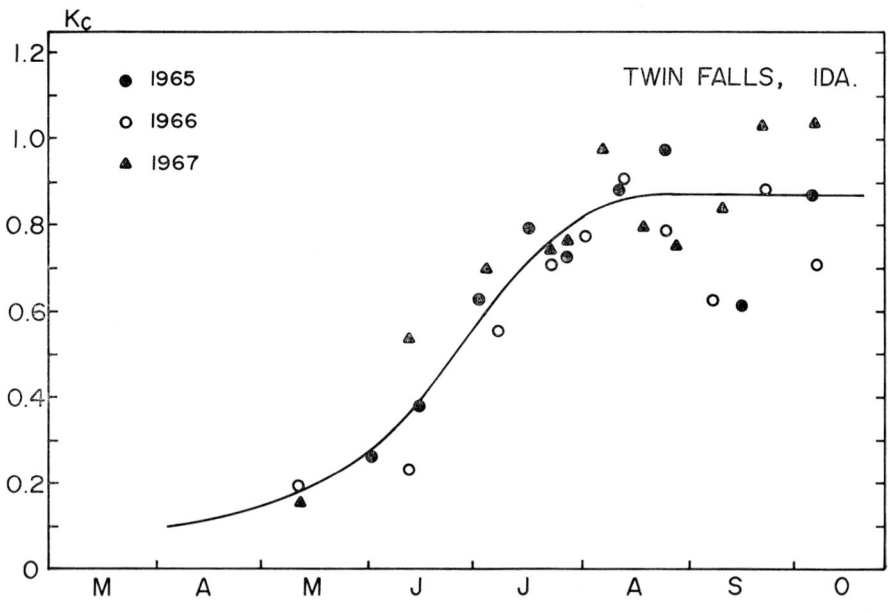

Fig. 8.7. Crop coefficient (E_t/E_{tp}) for spring-planted sugarbeets at Twin Falls, Idaho.

ing, which represents largely evaporation from the soil surface, and it increases to about 90 percent of the potential by August 10. Some of the scatter of data can be attributed to the measurement techniques involved. Similar but more accurate crop curves relating evapotranspiration from sugarbeets to evapotranspiration from grass are presented by Pruitt et al. (30).

Crop coefficients for Bushland are shown in Figure 8.8. In this case, the smooth line plotted is identical to the line plotted for Twin Falls, but with the horizontal axis shifted forward about 40 days. The two data points near the end of the season reflect the reduced evapotranspiration that takes place as soil moisture is allowed to decrease. In general, crop coefficients indicate that spring-planted sugarbeets in the western states require 3 to 4 months to attain full effective cover. Crop coefficients were not calculated for the Columbia Basin area, but Middleton et al. (26) presented ratios of evapotranspiration to pan evaporation, which would be very similar to the estimate of potential evaporation used in Figures 8.3 and 8.6. The average ratio of evapotranspiration to pan evaporation after full cover was attained was approximately 0.9, indicating that crop coefficients calculated for the Columbia Basin would probably be very similar to those presented in Figures 8.7 and 8.8 except the planting date and date of attaining full effective cover would need to be shifted accordingly.

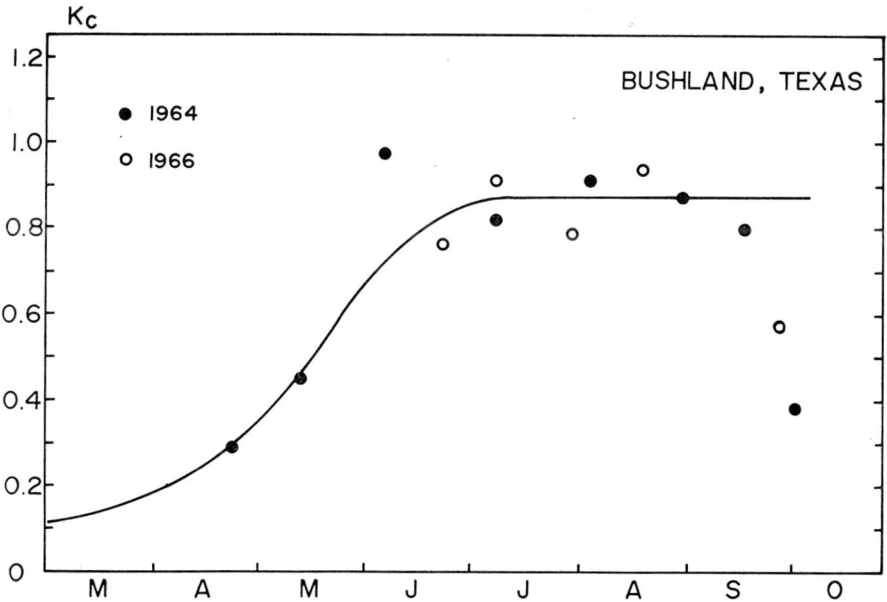

Fig. 8.8. Crop coefficients (E_t/E_p) for spring-planted sugarbeets at Bushland, Texas.

FALL-PLANTED BEETS

Evapotranspiration data for fall-planted sugarbeets near Phoenix are presented in Figure 8.9. These data were collected by L. J. Erie, using soil-sampling techniques. Phoenix is located at about 33° and 20′ north latitude at an elevation of about 1,100 feet above sea level. The beets in the Phoenix area are planted early in October. They attain effective full cover by January, but the leaves begin to disintegrate and the plants defoliate from mid-January through February, due to low temperatures. During the study period, the minimums were near 32° or lower. New growth begins early in March and the evapotranspiration rate rapidly increases as leaf area increases; the potential evapotranspiration increases with increasing solar radiation and air temperature in the spring. Observed evapotranspiration rates again are approximately 90 percent of the potential after the full cover is attained. Because of high temperatures (greater than 100° F) beginning in May, the leaves begin to defoliate and by the latter part of June and July the observed rate of evapotranspiration is greatly decreased relative to the evaporative demand.

Crop coefficients for the Phoenix area are shown in Figure 8.10. Although the horizontal scale is different, the rate of increase in the crop coefficient from near the first of October through mid-January is essentially identical to that for the spring-planted sugarbeets. The decrease in the crop coefficients beginning in mid-January through February clearly illus-

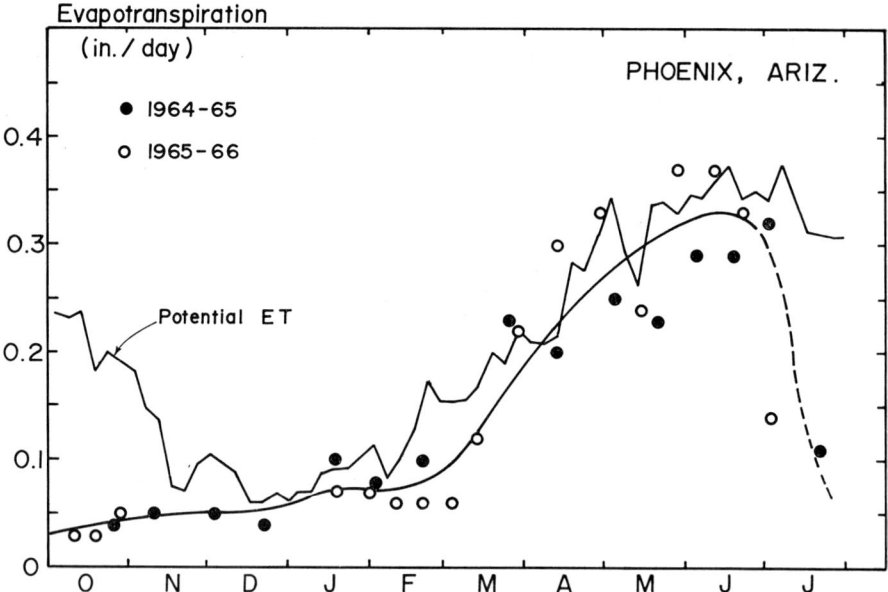

FIG. 8.9. Measured evapotranspiration and estimated mean potential E_t for the two-year period.

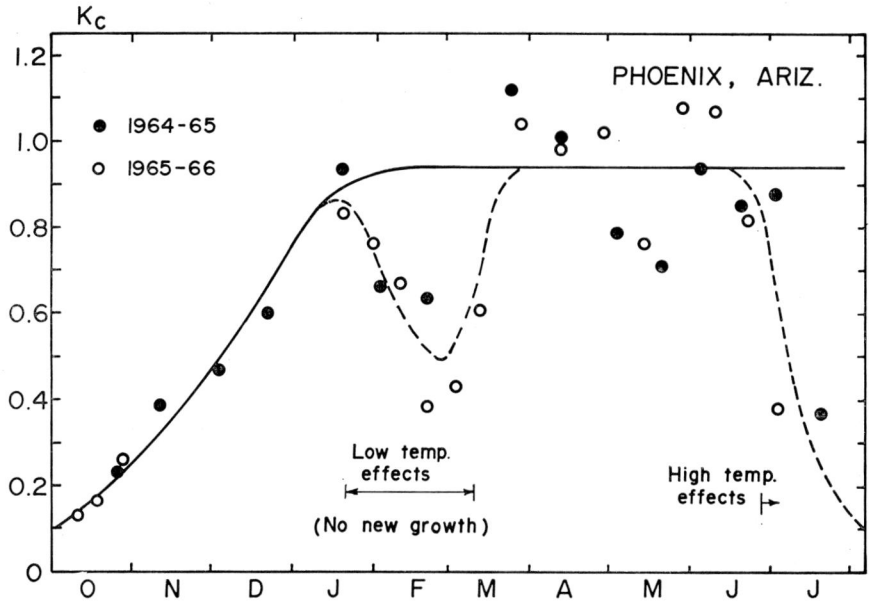

Fig. 8.10. Crop coefficients (E_t/E_{tp}) for fall-planted sugarbeets at Phoenix, Arizona.

trates the effect of inactive leaves caused by low temperatures. The crop coefficient rapidly increases in March to an average of about 90 percent of the potential until near the end of June, then rapidly decreases because of limited leaf area. The solid line is plotted to illustrate what might be expected in slightly warmer climates where the leaves do not drop off during the winter months because of low minimum temperatures or during June and July because of high temperatures. In the Phoenix area, harvesting of sugarbeets begins in April because beets cannot be stored for processing. Irrigation of sugarbeets in that area on a silt loam soil could be terminated about one month before the anticipated harvest date without a reduction in sucrose (5).

PEAK RATES

From the data presented in the previous section, the average rates of evapotranspiration for spring-planted sugarbeets in the intermountain and Northern Plains area range from 0.22 to 0.26 inch per day during the latter part of July and the early part of August. Average rates of evapotranspiration range from 0.26 to 0.30 inch per day from early July to mid-August in the Columbia Basin area. In the Southern Plains, average rates range from 0.28 to 0.32 inch per day from early June to mid-August. These values also would be applicable in the western Kansas area as indicated by the data presented by Herron et al. (13), shown in Figure 8.6. The values for Kansas were essentially the same as those for Bushland after the first of June. Average rates of evapotranspiration range from 0.30 to 0.35 inch per day in Arizona and southern California from mid-April to June. Pruitt et al. (30) measured a peak use rate of about 0.30 inch per day from mid-June to the latter part of July at Davis.

SEASONAL EVAPOTRANSPIRATION

Seasonal evapotranspiration for sugarbeets varies with the evapotranspiration rate for an area and the length of the growing season. Published values of seasonal evapotranspiration for sugarbeets from southern Alberta to Phoenix are summarized in Table 8.1. These data indicate that seasonal evapotranspiration ranges from about 22 inches in southern Alberta and the Northern Plains to about 40 inches in the Southern Plains for spring-planted sugarbeets, and about 42 inches for fall-planted sugarbeets in Arizona if allowed to grow until July. If fall-planted sugarbeets are harvested earlier, the seasonal evapotranspiration will be reduced accordingly.

ESTIMATING EVAPOTRANSPIRATION

Seasonal evapotranspiration can be estimated using a well-known procedure such as the Blaney-Criddle equation (1), or it can be estimated using Equation 8.3. The Blaney-Criddle formula for seasonal estimates is as follows:

$$U = KF = \sum kf \qquad (8.4)$$

where $U =$ estimated evapotranspiration (consumptive use) in inches for the growing period or season; $K =$ empirical consumptive use coefficient (irrigation season or growing period); $F =$ the sum of monthly consumptive use factors, f, for the season or growing period ($f = tp/100$ where $t =$ mean monthly air temperature, in degrees F, and $p =$ mean monthly percent of annual daytime hours); and $k =$ monthly consumptive use coefficient.

If Equation 8.3 is used, the average crop coefficient for the season will be approximately 0.75 or 75 percent of the potential evapotranspiration from planting to harvest. Monthly consumptive use coefficients for the Blaney-Criddle formula must be determined for each major area, or adjusted accordingly. For example, Erie et al. (6) summarized semimonthly coefficients for most crops grown in Arizona. These coefficients are based on average local planting to harvest periods. The Blaney-Criddle formula should not be used in climatic zones significantly different from those in western United States unless local coefficients are available.

TABLE 8.1. Summary of root and sugar yield and seasonal evapotranspiration

Source of Data	Location	Length of Season	Root Yield	Sugar Yield (gross)	ET
			(tons/acre)	(tons/acre)	(inches)
Ripley (31)	Tabor and Vauxhal, Alberta, Canada	1950–1961 Average	20.4	...	21.5
Larson and Johnston (22)	Huntley, Mont.	20 Apr.–27 Sept. 1953	23.4	3.5	23.0
		20 Apr.–27 Sept. 1953	22.0	3.1	22.5
Erie and Dimick (4)	Redfield, S. Dak.	1951	24.7	...	25.0
		1952	19.6	...	23.0
		1953	15.4	...	24.0
Jensen and Carter*	Twin Falls, Idaho	13 Apr.–21 Oct. 1965	22.9	4.0	24.0
		23 Apr.–17 Oct. 1966	25.5	4.4	23.0
		8 Apr.–12 Oct. 1967	24.7	3.6	26.0
Pruitt et al. (30)	Davis, Calif.	25 Mar.–21 Sept. 1966	29.0	3.3	33.5
Mech (25)	Prosser, Wash.	11 Mar.–26 Oct. 1949	36.4	5.9	37.0
Herron et al. (13)	Garden City, Kans.	4 Apr.–28 Oct. 1959	29.8	4.3	39.0
		15 Apr.–3 Nov. 1960	31.7	5.3	34.0
Schneider and Mathers (35)	Bushland, Tex.	9 Apr.–22 Oct. 1964	16.2	...	40.0
		16 Mar.–13 Oct. 1966	24.1	...	38.0
Erie and French (5)	Phoenix, Ariz.	12 Oct. 1964–27 July 1965	32.5	4.5	40.0
		22 Sept. 1965–7 July 1966	39.3	5.4	43.0

* Unpublished data.

If irrigations are to be scheduled using current meteorological conditions, estimates of evapotranspiration for periods as short as 5 to 10 days are required. These can be obtained using Equation 8.2 and a simple two-parameter approximate energy balance equation for estimating potential evapotranspiration:

$$E_{tp} = C_T(T - T_x)R_s \tag{8.5}$$

where C_T is an air temperature coefficient which is constant for a given area and is derived from the long-term mean maximum and minimum temperatures for the month of highest mean air temperature, T is mean daily air temperature, T_x is a constant for a given area and is merely the linear equation intercept on the temperature axis, and R_s is daily solar radiation expressed as the equivalent depth of evaporation. However, potential evapotranspiration can be estimated more reliably if additional climatological data such as windspeed and mean daily dew point temperature are also used. An equation of this type is known as a combination equation, since it involves a combination of energy balance and aerodynamic terms. There are several forms of this equation. The one listed below is a modified Penman equation:

$$E_{to} = \frac{\Delta}{\Delta + \gamma}(R_n - G) + \frac{\gamma}{\Delta + \gamma}(15.36)(1.0 + 0.01W)(e_s - e_d) \tag{8.6}$$

where E_{to} is potential evapotranspiration in langleys; Δ is the slope of the saturation vapor pressure-temperature curve, de/dT; γ is the psychrometric constant; e_s is the mean saturation vapor pressure in millibars (mean at maximum and minimum daily air temperature); and e_d is the saturation vapor pressure at mean dew point temperature in millibars. The parameters $\Delta/(\Delta + \gamma)$ and $\gamma/(\Delta + \gamma)$ are mean air temperature weighing factors whose sum is 1.0, W is total daily wind run in miles, R_n is daily net radiation in cal cm^{-2}, and G is daily soil heat flux in cal cm^{-2}. Details on the use of these equations were presented by Jensen, Robb, and Franzoy (19). Where meterological data are not available for an area, evaporation from a U.S. Weather Bureau pan can be substituted for the estimate of potential evapotranspiration if the pan is surrounded by an irrigated crop like grass.

MANAGING THE SOIL MOISTURE RESERVOIR

With irrigation, the grower can obtain excellent seed germination and stand establishment without depending on rain. A good stand is the first prerequisite for a good yield. Following stand establishment, irrigation enables the grower to regulate the soil moisture reservoir that is exploited by sugarbeet roots. Instruments such as soil moisture tensiometers and soil moisture blocks are available for directly or indirectly indicating the soil moisture status. An experienced irrigator can assess the probable level of the soil moisture reservoir by taking samples of soil down to a depth of at least two feet and evaluating the soil moisture content by feel and visual appearance. There are certain basic concepts that are generally accepted for practical purposes in describing the soil moisture status. These are described in the following section.

SOILS AND THEIR WATER-HOLDING CAPACITIES

Soil moisture cannot be maintained at a constant level under normal irrigation practices. Instead, the soil is wetted by an irrigation, dries as the crop extracts water, and is rewetted by the next irrigation. Following an irrigation, if an adequate amount of water is applied, the soil moisture content is very high, but generally the soil is not saturated except at the surface. In a well-drained, fine-textured soil that has been thoroughly irrigated, the water slowly drains by gravity for as much as 60 days if evaporation is prevented. However, the drainage rate is very slow after 2 to 6 days. The moisture content at this stage is called "field capacity." The suction required to extract water at this time is only about 0.2 to 0.33 atmosphere (3 to 5 pounds per square inch). Plants can extract water from soils until the soil moisture tension approaches 15 atmospheres (about 225 pounds per square inch). This value is referred to as the "wilting point" or the wilting percentage. The water held by the soil between these values is called "available water."

The amount of water extracted does not increase uniformly as the soil moisture tension increases from 0.2 atmosphere to 15 atmospheres. Most of the water is extracted at lower suctions, especially on sandy soils. The percentage of water removed as the soil moisture tension increases to various values is presented in Table 8.2 for various soil types. The amount of available water that can be held by these soils ranges from about 0.8 inch in the loamy sands to as much as 2.5 inches per foot in the clays. The available water in medium-textured soils in which most of the sugarbeets are grown ranges from about 1.5 to 2 inches per foot.

TABLE 8.2. Approximate percentage of available water that can be depleted from soils by the time the soil moisture tension reaches the values indicated

Soil Moisture Tension (Atmospheres)	Percentage of Available Water Depleted in:				
	Loamy sand	Fine sandy loam	Sandy loam	Loam	Clay
0.5	70	65	57	30	13
1.0	82	75	68	57	27
2.0	89	83	78	72	46
5.0	95	92	89	87	73
10.0	99	98	97	95	91

SOURCE: Adapted from Haise and Hagen (12).

IRRIGATION PRACTICES

During the past half century, there have been distinct changes in the recommendations of soil moisture levels to be maintained, the amounts of irrigation water to be applied, and the frequencies of irrigation. A summary of historical recommendations and more recent recommendations is presented in this section to illustrate these changes. To clarify some of the current conflicts in opinions, the recommendations are referenced to the time at which they were made.

Historical Guides to Irrigation. In 1899 Foster (9) made the following statement, which has apparently persisted with many sugarbeet growers even though later studies have shown this to be incorrect:

Never irrigate until beets show they require moisture, usually letting them suffer a few days, and by so doing it gives a nice long tapering root.

The following excerpt was taken from USDA Farmers' Bulletin No. 392, Irrigation of Sugar Beets (33):

It is essential that there be sufficient moisture in the soil at the time of seeding to bring the plants up, and it is better to irrigate before rather than after seeding. The next irrigation, or the first in the case it is not necessary to irrigate to bring the beets up, should be delayed as long as there is sufficient moisture in the soil to keep up a steady growth. Too early irrigation tends to make a turnip-shaped beet and produces an unusually heavy growth of leaves without a corresponding development of the root.

USDA Farmers' Bulletin No. 1645, Sugar-Beet Growing Under Irrigation in the Utah-Idaho Area, was released about 20 years later (28). The following excerpts illustrate some of the changes in the thinking during this period:

As a whole, growers in the irrigated areas put too much reliance on rainfall and fail to irrigate the beet crops sufficiently following planting and in the early stages of the crop. A forward step in sugar-beet growing in the irrigated area will be taken when the practice of prompt irrigation following planting is generally adopted.

. . . At one time there were popular opinions to the effect that irrigating the sugar beets early in the season prevented deep rooting, and that late irrigation reduced the sugar content of beets. It has been found that these opinions are not based upon facts. The beets should be irrigated whenever the leaves turn a dark-green color or begin to wilt in midday and do not quickly recover at night.

The beets should be irrigated late in the season, if necessary, to keep them in a growing condition. At digging time soil should be in a good friable tilth, since where the soil is very dry there is a loss of roots by breaking, and on the other hand, if the soil is wet, mud clings to the beets, thereby increasing the tare. Therefore, the last irrigation should be timed so as to bring the land into good condition for harvesting.

These excerpts are from USDA Farmers' Bulletin No. 1867, Sugar-Beet Culture Under Irrigation in the Northern Great Plains (29):

Irrigation should begin whenever the soil is deficient in moisture. The normal rainfall of this area may provide sufficient moisture for germination of the seed; however, in many years irrigation is needed. . . . It is advisable to apply water within 24 hours after planting. If application of water is delayed, a portion of the seed may be in soil moist enough to sprout it; the rest may remain ungerminated. Uneven stands, because of such unevenness in germination of the seed, are very common where irrigation is delayed. . . .

When growing of sugar beets first began in the irrigated districts, it was commonly believed that irrigating early in the season would prevent the beet roots from growing deeply and would cause branching. This has not been found to be correct. Branching of the main root of sugar beets arises from a number of causes, such as high water table, loose seedbed, and insect or disease injuries. . . .

In the Nebraska district, it is commonly recognized that after irrigation is begun, a series of light irrigations, about 10 days apart, give better yields than less frequent and heavier individual applications. One of the most serious faults in irrigation practice is the tendency to delay the first application of water.

. . . The sugar beet obtains approximately 65% of its water from the top foot of soil and 85% of its moisture from the top two feet. The natural storage of water in the soil and deeper penetration of excess waters usually provide for sufficient moisture in the area below two feet. Keeping the top foot moist is the most important factor in irrigation of sugar beets. This requires more frequent but less heavy irrigations. If too much water is used the soluble plant foods are carried out of the soil or to depths where they are not readily recovered.

The following excerpt is from USDA Farmers' Bulletin No. 1903, Sugar-Beet Culture in the Intermountain Area with Curly Top Resistant Varieties (36):

Beets should be irrigated whenever the leaves turn dark green or begin to wilt in midday and do not recover at night. This is true whether the beets have only a few leaves or are nearing maturity. There are still many farmers who believe that withholding water early in the season until the small beets suffer "drives the roots down to water," and thus produces a larger beet. This theory has been disproved many times in field tests. Young beets must not be allowed to suffer for water.

More Recent Recommendations. These excerpts are from the 1955 USDA Yearbook of Agriculture, The Irrigation of Sugar Beets (11):

Because the length of the growing season is probably the first limitation on yield, the crop should be irrigated before the soil gets so dry as to delay plant growth. Any delay in the rate of growth is equivalent to shortening the growing season.

The beet plant is especially sensitive to unfavorable moisture conditions for three or four weeks after it emerges. During that time, the soil should be kept moist in the upper 12 inches, so that growth will be continuous and fast. If irrigation is necessary, then it should be light, so as to avoid leaching of soluble plant nutrients.

The quantity of available soil nitrogen is closely related to irrigation practice. The yield of beets may be seriously depressed by too much irrigation water and by too little water.

The following excerpts are from Chapter 33, Sugar, Oil, and Fiber Crops, Part 1, Sugar Beets (23):

Under field conditions, considerable attention has been given to whether irrigation should be withheld so that the crop will wilt immediately prior to harvest. Such practices are based in part on a common misconception that the resulting increase in sugar concentration represents a real increase in quality. . . . While preharvest wilting may offer savings in irrigation and hauling costs it usually increases harvest costs on soils with a high clay content, reduces storage and slicing quality through lower turgidity of the root, and sometimes reduces the extraction of sucrose. Total sucrose yields are not increased and may be reduced significantly by single, brief cycles, and almost invariably are lowered by repeated wilting due to reduction in leaf area and rate of photosynthesis. . . .

Both sprinkler or furrow irrigations are satisfactory with sugar beets while flood irrigation may contribute to pathogen problems (fungal root rot) and is not suitable. Sprinklers provide the best control of water distribution, prevent surface isolation of nutrients, and, on coarse textuered soils, conserve water and reduce leaching of nutrients. The choice between sprinkler and furrow irrigation is an economic one dependent on the relative costs and advantages of the systems under conditions prevailing in a given situation.

The following excerpt is from Water Management of Fall-Planted Sugar Beets in Salt River Valley of Arizona (5):

Water management during the first six weeks is especially important from the standpoint of water conservation and early growth. It is during this period that leaching, germination, weed control, stand maintenance, fertilizer and thinning are accomplished. The plant's moisture needs are very low during this period, so any individual application of water greater than two inches, except for leaching, is probably wasted. . . . Sugar beets will use moisture from deep profiles. However, nearly 90% of the water used is from the top three feet of soil. . . . Sugar beets can extract water from a depth as great as five feet, but over 70% of use is from the top two feet of soil.

Summary of Changes in Irrigation Recommendations. The general concept of withholding irrigations until the plants suffer in order to obtain a long tapering root (9, 33), apparently initiated during the late 1800s, persisted for nearly a half century even though many experiments did not verify this theory. Later, many of the researchers stressed the need for early light irrigations to assure germination and establishment of good stands. This is still good advice today and has become a standard recommendation (23).

Many of the researchers stressed the need to irrigate when the beets do not recover from midday wilting at night. They also began to recommend the maintenance of soil moisture late in the season, but this seemed to be primarily for harvesting and processing purposes.

Irrigations at intervals of 10 to 14 days were recommended in the intermountain areas and the Northern Great Plains, with the last irrigation about September 15. This is a valid recommendation except that the interval may be increased after mid-August.

Most researchers stressed the fact that roots may penetrate to 5 or 6 feet. Doneen's data (3), for example, show more complete water extraction to greater depths than Haddock (11) found in Utah. This observation has been repeated extensively, but must be qualified because on some soils the rooting depth may be restricted by cemented material or dense silt and clay layers.

The statement by Nuckols that "very fertile soils require somewhat less water per pound of crop produced as compared with soils of low fertility" is valid, but should not be construed to mean less total water is needed.

EFFECTS OF SOIL MOISTURE LEVELS ON YIELDS

The results of irrigation experiments involving various levels of soil moisture depletion before irrigating are summarized in Tables 8.3 to 8.8. These experiments were conducted at Huntley, Montana; Newton and Garland, Utah; Prosser, Washington; Davis, California; Twin Falls, Idaho; and Phoenix, Arizona. The most significant aspect of these experiments is that the production of sugarbeet roots is generally not very sensitive to soil moisture stress. For example, in each of these studies, reducing the total number of irrigations about 35 percent by allowing greater depletion of soil moisture before irrigating reduced the root yield only about 5 to 8 percent. However, the amount of water applied per irrigation usually

TABLE 8.3. Summary of irrigation regimes and root and sugar yields in 1953 at Huntley, Montana

Irrigation Regime*	Number of Irrigations	Root Yield	Gross Sugar Yield
		(tons/acre)	(tons/acre)
High	8	23.4	3.46
Medium	5	22.0	3.08
Low	1	16.9	2.45

SOURCE: Larson and Johnston (22).

NOTE: Planted 20 Apr. and harvested 27 Sept. Soil type: Fly Creek clay loam; 12 tons of manure and 96 lb N and 54 lb P_2O_5 per acre.

* Approximate percentage of available water removed from the root zone was 36, 75, and 95 percent for the wet, medium, and dry treatments before irrigating, respectively.

TABLE 8.4. Summary of irrigation regimes and root yields at Newton, Utah, in 1946 and at Garland, Utah, in 1947

Irrigation Regime‡	1946*		1947†	
	Number of irrigations	Root yield	Number of irrigations	Root yield
		(tons/acre)		(tons/acre)
W_1	6	24.9	8	27.5
W_2	4	...	5	28.7
W_3	3	21.9	4	24.9
W_4	2	19.8	3	22.5

SOURCE: Haddock (10).
NOTE: Planted 22 Apr. 1946 and 12 Apr. 1947. Harvested 24 Oct. 1946 and 10 Oct. 1947. Soil type = Millville fine sandy loam at Newton and Millville silt loam at Garland.
* With 160 lb N and 15 tons manure per acre.
† With 80 lb N and 15 tons manure per acre.
‡ W_1 = continuously moist, below 0.75 atmosphere at the 8-inch depth in 1946 and the 6-inch depth in 1947.
W_2 = similar to W_1 to 12 Aug. 1946 and 28 July 1947.
W_3 = similar to W_1 to 15 July 1946 and 26 June 1947, then soil moisture was allowed to reach the wilting percentage at 18 inches before irrigating.
W_4 = similar to W_1 to 15 July 1946 and 26 June 1947, then soil moisture was allowed to reach the wilting percentage at 30 inches in 1946 and at 36 inches in 1947 before irrigating.

must be greater with less frequent irrigations since less frequent irrigation has only a small effect on total evapotranspiration.

Salter and Goode (34) in their review of crop responses to water at different stages of growth summarized studies in England and indicated that soil moisture deficit in midseason may be more important than those occurring in mid-September. They also reviewed recommendations made by individuals as early as 1892 that irrigations late in the season may raise the yield of roots in England but may not result in an increased yield of sugar. However, they also cited data from Canada, reported over a 9-year

TABLE 8.5. Summary of irrigation regimes and root and sugar yields at Prosser, Washington, in 1948

Irrigation Regime*	Number of Irrigations	Root Yield	Gross Sugar Yield
		(tons/acre)	(tons/acre)
Wet	12	36.9	6.1
Medium	8	36.4	5.9
Dry	6	33.6	5.5

SOURCE: Mech (25).
NOTE: Planted 11 Mar. and harvested 26 Oct. Soil type: Warden fine sandy loam.
* Irrigation water was applied when the average available water in the 4-foot profile was 60 percent, 35 percent, and 15 percent for the wet, medium, and dry treatments, respectively. Since the soil held about 2.3 inches of available water per foot, the amount of water extracted amounted to about 3.7 inches, 6.0 inches, and 7.8 inches for the wet, medium, and dry treatments, respectively.

TABLE 8.6. Summary of irrigation regimes and root and sugar yields at Davis, California, in 1961

Irrigation Regime*	Number of Irrigations	Root Yield	Gross Sugar Yield
		(tons/acre)	(tons/acre)
Wet	13	31.6	3.80
Medium	8	29.8	3.86
Dry	5	28.1	3.76
Dry + stress	3	26.4	3.60

SOURCE: Loomis and Haddock (23). Unpublished data of L. D. Doneen and R. S. Loomis. The crop was planted at Davis, Calif., 14 May 1961 (late) on a deep, well-drained, Yolo loam. Means in the table represent four replications of two harvest dates, 14 Sept. and 27 Oct., and two nitrogen levels.

* Wet = irrigated every 5 to 6 days after cover was nearly complete.
Medium = between the wet and dry.
Dry = irrigated at the first sign of wilting.
Dry + stress = allowed to wilt about three days before irrigation.

TABLE 8.7. Summary of irrigation regimes and root and sugar yields at Twin Falls, Idaho, from 1964 through 1967

Irrigation Regime*	Year	Number of Irrigations	Root Yield†	Gross Sugar Yield
			(tons/acre)	(tons/acre)
M–H–1	1964	11	20.7	3.57
	1965‡	13	20.8	3.49
M–H–2	1964	11	20.8	3.67
	1965‡	13	19.5	3.28
M–1	1964	7	19.6	3.35
	1965‡	9	22.6	4.01
	1966	10	25.5	4.43
	1967	10	24.7	3.64
M–2	1964	7	20.5	3.55
	1965‡	9	21.8	3.73
	1966	10	24.2	4.17
	1967	10	23.4	3.43

NOTE: Planting and harvest dates:

Year	Planted	Harvested
1964	19 Apr.	22 Oct.
1965	13 Apr.	22 Oct.
1966	26 Apr. (replanted)	17 Oct.
1967	8 Apr.	12 Oct.

Soil type = Portneuf silt loam, lime-silica cemented layer beginning at 16–18 inches restricts root development.

* M–H–1 = irrigated 12 hours when the soil moisture tension at 8 inches approached 0.6 atmosphere.
M–H–2 = same as M–H–1 except irrigation was applied 24 hours.
M–1 = irrigated 12 hours when the soil moisture tension at 8 inches approached 4 atmospheres.
M–2 = same as M–1 except irrigation was applied for 24 hours.
† For a nitrogen treatment of 100 lb/acre.
‡ Second year on the same plots.

TABLE 8.8. Summary of irrigation regimes and root and sugar yields at Phoenix, Arizona in 1964-65 and 1965-66

Irrigation Regime*	Year	Number of Irrigations	Root Yield† (tons/acre)	Gross Sugar Yield (tons/acre)
Wet	1965-65	14	31.6	3.89
	1965-66	12	37.8	4.99
Medium	1964-65	11	32.5	4.14
	1965-66	10	39.3	5.38
Dry	1964-65	8	32.9	4.47
	1965-66	7	38.8	5.08

SOURCE: Erie and French (5).
NOTE: Planted 12 Oct. 1964 and 22 Sept. 1965. Harvested 27 July 1965 and 7 July 1966.
Soil type = Cajon silt loam, 130 lb nitrogen applied.
* Irrigation regimes initiated 23 Apr. 1965 and Mar. 1966. Irrigated when 40 percent, 60 percent, and 80 percent of the water was depleted on the wet, medium, and dry treatments, respectively. These correspond to depletions of about 2, 3, and 4 inches, respectively, for the three regimes.
† Yields will be less with earlier harvest dates.

period, that no decrease in sugar content of roots resulted from continuing to irrigate until late in the season. Salter and Goode concluded:

From the experimental work cited there are no obvious indications that this crop, when grown for its roots, is especially sensitive to soil moisture conditions at any particular stage of growth, although few workers have critically studied this particular aspect of its water relations. The conflicting evidence for and against early and late irrigation can probably be explained by the overriding effect of variable soil and weather conditions, including temperature, which prevail during the different experiments.

ALLOWABLE DEPLETION AND IRRIGATION INTERVALS

Based on the results presented in the two previous sections, it is apparent that sugarbeets can be irrigated at a high soil moisture level with frequent, light irrigations, or they can be allowed to deplete 60 to 70 percent of the available water in the root zone between irrigations if each irrigation refills the amount depleted. The interval between irrigations cannot be estimated on a basis of evapotranspiration values alone (Fig. 8.3 to 8.9). Instead, the available water-holding capacity per foot of depth and the expected depth of rooting for the soil in question must be known. Rainfall, of course, can also influence irrigation frequency. The general recommendation for an irrigation frequency from July 14 to the latter part of August of 10 to 14 days now appears to be very reliable for near-maximum yields. As a general rule, if one assumes that 60 percent of the available water in the root zone can be depleted and uses the average evapotranspiration rates presented in Figures 8.3 to 8.8, then the desired irrigation interval can be obtained during periods when rainfall is not significant. However, with some sprinkler systems, the frequency of irrigation will be determined by the amount of soil moisture replenished at each irrigation and the evapotranspiration rate essentially independent of available water-holding capacities.

Greater depletion can be allowed near the end of the season if moist soil is not needed for harvesting the beet roots. Ferry, Hills, and Loomis (7) conducted a study in 1963 in the San Joaquin Valley where beets are harvested in July and August. Their results indicated that moisture stress before harvest decreased root yields but increased the sucrose percentage, and therefore, only slightly decreased the gross sugar production. Erie and French (5) found that late season soil moisture stress decreased root weight but increased the sugar percentage to compensate for the loss in root weight. They concluded that since sucrose is not reduced by soil moisture stress, irrigations can be discontinued 3 to 4 weeks before harvest for maximum water economy.

SOIL MOISTURE-NITROGEN INTERACTIONS

In general, if adequate soil moisture is provided so as not to restrict growth, uptake of available soil nitrogen does not appear to be restricted. Excessive water application will reduce yields where the nitrogen level is just adequate, that is, for comparable yields, more nitrogen fertilizer is required if excessive water is applied (32). At Twin Falls, Idaho, for example, the 4-year average yields were 22.7 and 23.2 tons per acre with about 50 and 100 lb of nitrogen, respectively, with 12-hour irrigation sets. With 24-hour sets the average yields were 21.4 and 22.5 tons per acre for 50 and 100 lb of nitrogen per acre, and 23.1 tons per acre for 200 lb of nitrogen per acre.

IRRIGATION WATER REQUIREMENTS AND METHODS

Irrigation water requirements will be greater than the evapotranspiration values shown in the previous section because there are unavoidable losses. With the present methods of irrigating, for example, the desired quantity of water cannot be applied uniformly over the entire field. Also, where leaching is required to control salts, additional water must be added if this is not met by the excess water applied under normal irrigation practices.

A schematic diagram is presented in Figure 8.11 to illustrate the major reason for irrigation water requirements being greater than evapotranspiration for surface irrigation systems and in Figure 8.12 for sprinkler irrigation systems. In Figure 8.11, if water is applied until an adequate amount is absorbed at the lower end of the field and the water reaches the end of the field in about one-third of the total time, the water application efficiency will be about 65 percent. Although this will vary with soils of different intake characteristics, this example illustrates that approximately 50 per-

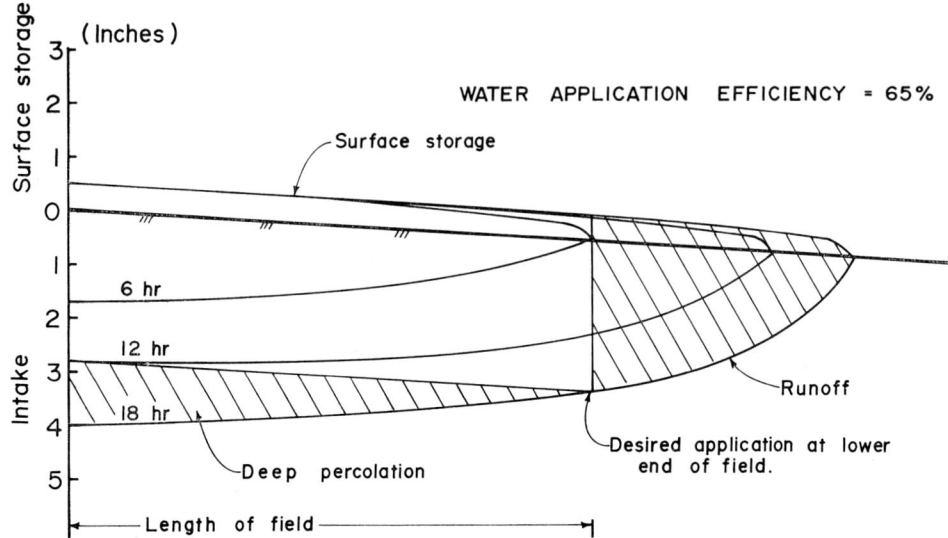

FIG. 8.11. If water is applied to furrows until an adequate amount is absorbed at the lower end of the field, the water application efficiency may be only 65%.

cent more water may be applied to a sloping field with furrow irrigation than that amount merely needed to replenish the depleted soil moisture.

Estimates of the advance of water in furrows for other soils and stream sizes can be made using the following equation:

$$x = \frac{qt}{0.8D_o + 0.67I_o} \qquad (8.7)$$

where $x =$ the distance to the advancing front (ft), $q =$ the average flow rate per foot of width (ft^3/ft-hr), $D_o =$ the average depth of water on the surface at the upper end of the field per foot of width (ft), $I_o =$ the depth of water infiltrated at the upper end at time t (ft), and $t =$ time in hours.

A typical example of the distribution of water from a sprinkler system for a single irrigation and the cumulative distribution after four irrigations is presented in Figure 8.12. To assure adequate water applied throughout the area for a single irrigation, a water application efficiency of about 77 percent would be needed. If one assumes that the evaporative loss during sprinkling will be about 7 percent, than a water application efficiency of 70 percent would be required to apply the desired amount of water to all areas. The improvement in water application efficiency after consecutive irrigations is the result of changes in wind direction and velocity from one irrigation to the next.

Because the evapotranspiration process concentrates the salts in the

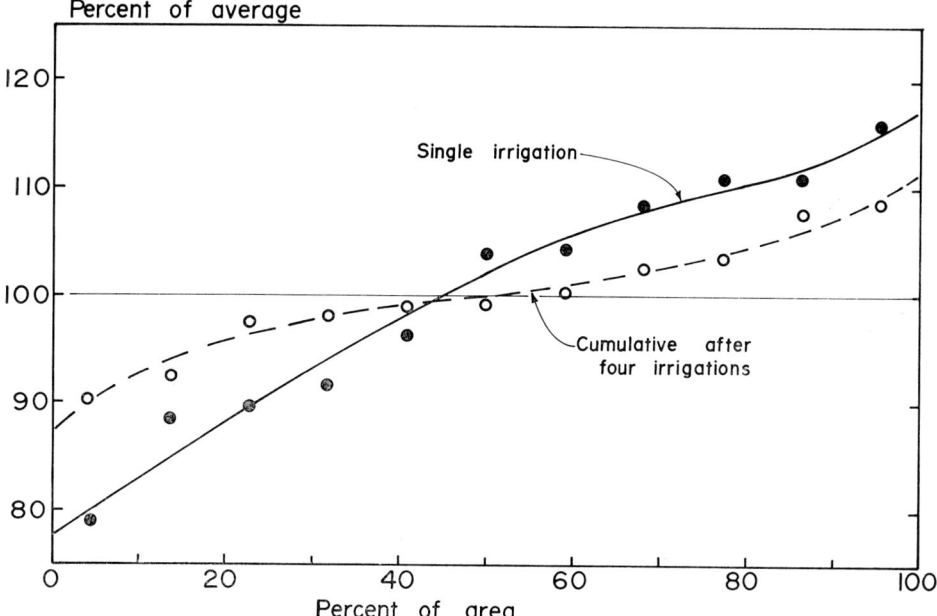

Fig. 8.12. Distribution of water from a typical sprinkler system for a single irrigation and after four irrigations. (Unpublished data from C. H. Pair.)

water applied to the soil, these salts must be removed by leaching. The quantity of water required for leaching is proportional to the weighted average concentration of soluble salts in the applied water, C_{aw}, the volume of water used by evapotranspiration, W_{et}, and inversely proportional to the concentration of salts that can be tolerated in the root zone, C_r. The salt concentration in the soil solution between irrigations increases as the plants extract water, but for convenience consider $C_r =$ the average salt concentration of water in the root zone when the soil is near field capacity. Where rainfall is negligible, the salt concentration in the applied water (rainfall + irrigation water) is essentially the same as in the irrigation water, C_{iw}.

The volume of water required for leaching, W_l, under steady-state conditions, or long periods of time, can be estimated as follows:

$$W_l = \frac{C_{aw}}{a(C_r - C_{aw})} W_{et} \qquad (8.8)$$

In this equation, $a =$ the leaching efficiency expressed as a fraction. The leaching efficiency is a coefficient expressing the ratio of the average salt concentration in the drain water to the average concentration of the soil water in the root zone at a soil moisture content near field capacity as stated in the following equation:

$$a = \frac{C_{dw}}{C_r} \qquad (8.9)$$

C_{dw} is the concentration of soluble salts in the drainage water. The ratio, C_{aw}/C_{dw}, is the same as the "leaching requirement" when the salt concentration in the irrigation water is taken as the weighted average of rainwater and irrigation water. The concentration given in Equation 8.8 and 8.9 can be replaced by the electrical conductivity, EC, of the water.

A more complete description of the leaching requirement, its limitations and the significance of the assumptions involved, can be found in USDA Handbook 60, published by the U.S. Salinity Laboratory Staff (38), and in a bulletin by Wilcox (39).

In the intermountain and Northern Plains areas, sugarbeets are either flat planted or planted on shallow beds with a furrow spacing of 22 inches, as illustrated in the upper portion of Figure 8.13. Water is usually applied in alternate furrows, at least for the first few irrigations. Typical wetting patterns from the furrows are shown. The second general procedure for irrigating, and this is probably used more frequently in areas where saline conditions exist, is to have wider beds and deeper furrows with two sugarbeet rows approximately 12 inches apart on each bed. When water is applied in the furrow, typical wetting patterns as illustrated occur and the

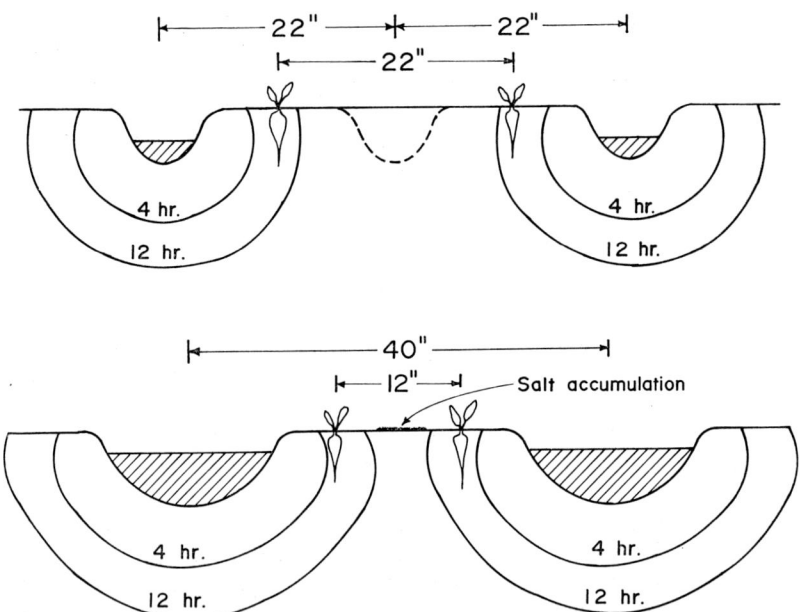

Fig. 8.13. Typical flat-planted, shallow furrows and bed-planted, deep furrows and expected wetting patterns.

salts tend to accumulate between the two rows and thereby result in better soil conditions for the germination and emergence of the seedlings. One other advantage of the large beds and the deep furrows in areas where land slopes are very small is that these systems can be installed perfectly level and no runoff occurs. Water must be applied to the deep furrows rapidly to move the water to the end of the field, and then must be reduced to maintain the depth of water close to the surface of the ridge to wet the soil for germination. When using this technique and nearly filling the furrow, one can apply a light irrigation of about $2\frac{1}{2}$ inches. If water is kept in the furrow, additional amounts can be applied.

Although border strips have been used for sugarbeet irrigation, they are not common. One of the problems encountered with the irrigation of sugarbeets in border strips is the large change in the retardance of the flow as the sugarbeet leaves develop. A summary of the increased depth of flow in the borders and the retardance in the rate of advance of water in a border is presented by Jensen and Howe (18). These data indicate that the time required for water to reach the end of the border strip may increase as much as $1\frac{1}{2}$ to 2 times as a result of additional retardance created by the sugarbeet leaves.

PREPLANTING AND PREEMERGENCE IRRIGATIONS

Preplanting and preemergence irrigations are usually applied to wet the subsoil and to wet the surface soil for germination and emergence. Water is usually kept in the furrow until it subs to the row. Irrigation for emergence may not be required in areas where frequent light precipitation occurs during the time of planting. Generally, very light irrigations are required for germination purposes, but these are difficult to apply with surface systems.

EARLY-SEASON IRRIGATIONS

An irrigation is usually needed after thinning and/or after side-dressing of nitrogen fertilizer. Thinning often delays a normal irrigation, resulting in low soil moisture in the seedling zone. Also, disturbed seedlings recover more rapidly if irrigated. An irrigation will move soluble fertilizers into the root zone of the seedlings. The movement of soluble materials is perpendicular to the wetting front illustrated in Figure 8.13. Therefore, nitrogen fertilizer should be applied to the side and just below the shallow furrows and to the side and about even with the bottom of deep furrows. If side-dressing is applied immediately below the furrow, a significant portion of the soluble material may be leached out of the root zone. Frequent, light irrigations may be required in high wind areas like the Texas High Plains to protect the seedling against wind damage.

MIDSEASON IRRIGATIONS

Midseason irrigations are largely for replenishing the soil moisture depleted by evapotranspiration plus leaching if necessary. Since most of the depletion takes place in the upper two feet of soil, these irrigations should apply sufficient water to refill the upper two feet to field capacity. If no leaching is needed, the frequency of the surface irrigations will depend on the rate of evapotranspiration during this period, rainfall, and the available water-holding characteristics of the soil. The net amount of water applied during midseason irrigations will probably vary between three and four inches on most soils and between four and five inches on deep soils. The number of irrigations needed can be approximated by dividing total evapotranspiration by the amount to be applied per irrigation where rainfall is negligible. In the Northern Plains where rainfall may partially replenish the depleted soil moisture, the irrigation interval may be more variable from one year to the next. With sprinkler systems, the frequency may depend on the amount applied per set.

LATE-SEASON IRRIGATIONS

In the northern areas and in the intermountain areas, the evapotranspiration rate is generally only about 0.15 inch per day late in the season. The time of the last irrigation before harvest depends on the soil moisture condition desired at harvest. If moist soil is desired, then an irrigation 10 days to two weeks before harvest is preferred. In Arizona and on deep soils where water may be expensive, the last irrigation can be applied three to four weeks before harvest if moist soil is not needed for harvesting. If an irrigation is applied close to harvest, then only a light amount should be applied.

REFERENCES

1. Blaney, H. F. and W. D. Criddle. Determining consumptive use and irrigation water requirements. USDA Tech. Bull. 1275. 1962.
2. Carter, J. N., M. E. Jensen, and S. M. Bosma. Interpreting the rate of change in nitrate nitrogen in sugarbeet petioles. (Scheduled for Sept. issue, Agron. J., vol. 63, no. 5, 1971.)
3. Doneen, L. D. Some soil-moisture conditions in relation to growth and nutrition of the sugar beet plant. Proc. Am. Soc. Sugar Beet Technologists 3:54–62. 1942.
4. Erie, L. J. and N. A. Dimick. Soil moisture depletion by irrigated crops grown in South Dakota. S. Dak. Agr. Exp. Sta. Circ. 104. 1954.
5. Erie, L. J. and O. F. French. Water management of fall-planted sugar beets in Salt River Valley of Arizona. Trans. Am. Soc. Agr. Engrs. 11:792–95. 1968.
6. Erie, L. J., O. F. French, and K. Harris. Consumptive use of water by crops in Arizona. Ariz. Agr. Exp. Sta. Tech. Bull. 169. 1965.
7. Ferry, G. V., F. J. Hills, and R. S. Loomis. Preharvest water stress for Valley sugar beets. Calif. Agr. 19(6):13–14. 1965.

8. Ferry, G. V., F. J. Hills, R. S. Loomis, and A. Ulrich. Sugar in beet roots. Limited by high temperatures and high levels of soil nitrogen in Kern County tests. Calif. Agr. 17(5):8–9. 1963.
9. Foster, L. Sugar beets in Sanpete and Sevier counties. Utah Agr. Exp. Sta. Bull. 63. 1899.
10. Haddock, J. L. Sugar beet yield and quality as affected by plant population, soil moisture condition and fertilization. Utah Agr. Exp. Sta. Bull. 362. 1953.
11. ———. The irrigation of sugar beets. In 1955 Yearbook of Agriculture, p. 400–405. 1955.
12. Haise, H. R. and R. M. Hagan. Soil, plant, and evaporative measurements as criteria for scheduling irrigation. In Irrigation of Agricultural Lands, pp. 577–604. Am. Soc. Agron. Monograph 11. 1967.
13. Herron, G. M., D. W. Grimes, and R. E. Finkner. Effect of plant spacing and fertilizer on yield, purity, chemical constituents and evapotranspiration of sugar beets in Kansas. I. Yield of roots, purity, percent sucrose and evapotranspiration. J. Am. Soc. Sugar Beet Technologists 12:686–98. 1963.
14. Jensen, M. E. Evaluating irrigation efficiency. Proc. Am. Soc. Civil Engrs. 93(IR1):83–98. 1967.
15. ———. Water consumption by agricultural plants. In Water Deficits and Plant Growth, vol. 2, pp. 1–22. 1968.
16. ———. Water requirements of crops. Intern. Comm. Irrigation and Drainage, 7th Congr. on Irrigation and Drainage, Question 23, Report E. 1969.
17. ———. Scheduling irrigations using computers. J. Soil Water Conserv. 24:193–95. 1969.
18. Jensen, M. E. and O. W. Howe. Performance and design of border checks on a sandy soil. Trans. Am. Soc. Agr. Engrs. 8:141–45. 1965.
19. Jensen, M. E., D. C. N. Robb, and C. E. Franzoy. Scheduling irrigations using climate-crop-soil data. Proc. Am. Soc. Civil Engrs. 96(IR1):25–38. 1970.
20. Jensen, M. E., L. Swarner, and J. T. Phelan. Improving irrigation efficiencies. In Irrigation of Agricultural Lands, pp. 1120–42. Am. Soc. Agron. Monograph 11. 1967.
21. Jensen, M. E., J. L. Wright, and B. J. Pratt. Estimating soil moisture from climatic data. Presented at Am. Soc. Agr. Engrs. Symposium, Modeling of Plant Growth, Chicago, Dec. 9–12, 1969.
22. Larson, W. E. and W. B. Johnston. The effect of soil moisture level on the yield, consumptive use of water, and root development by sugar beets. Soil Sci. Soc. Am. Proc. 19:275–79. 1955.
23. Loomis, R. S. and J. L. Haddock. Sugar, oil, and fiber crops. I. Sugar beets. In Irrigation of Agricultural Lands, pp. 640–48. Am. Soc. Agron. Monograph 11. 1967.
24. Loomis, R. S. and D. J. Nevins. Interrupted nitrogen effects on growth, sucrose accumulation and foliar development of the sugar beet plant. J. Am. Soc. Sugar Beet Technologists 12:309–22. 1963.
25. Mech, S. J. Minimizing irrigation water through management. Intern. Comm. Irrigation and Drainage, 7th Congr. on Irrigation and Drainage, Question 23, Report E, p. G23.36. 1969.
26. Middleton, J. E., W. O. Pruitt, P. C. Crandall, and M. C. Jensen. Central and western Washington consumptive use and evaporation data, 1954–62. Wash. Agr. Exp. Sta. Bull. 681. 1967.
27. Morton, A. G. and D. J. Watson. A physiological study of leaf growth. Ann. Botany, New Ser. 12:281–310. 1948.
28. Nuckols, S. B. Sugar-beet growing under irrigation in the Utah-Idaho area. USDA Farmers' Bull. 1645. 1931.
29. ———. Sugar-beet culture under irrigation in the Northern Great Plains. USDA Farmers' Bull. 1867. 1941.

30. Pruitt, W. O., F. T. Lourence, and S. von Oettingen. Water use by crops as affected by climate and plant factors. Calif. Agr. (In press.) 1971.
31. Ripley, P. O. The use of water by crops. Intern. Comm. on Irrigation and Drainage, Special Session, January 1966, SSR. 2. 1966.
32. Robins, J. S., C. E. Nelson, and C. E. Domingo. Some effects of excess water application on utilization of applied nitrogen by sugar beets. J. Am. Soc. Sugar Beet Technologists 9(2):180–88. 1956.
33. Roeding, F. W. Irrigation of sugar beets. USDA Farmers' Bull. 392. 1910.
34. Salter, P. J. and J. E. Goode. Crop responses to water at different stages of growth. In Research Review no. 2, Commonwealth Bureau of Horticulture and Plantation Crops, Commonwealth Agr. Bureaux, Farnham Royal, Bucks, England. 1967.
35. Schneider, A. D. and A. C. Mathers. Water use by irrigated sugarbeets in the Texas High Plains. Texas Agr. Exp. Sta. MP–935. 1969.
36. Tolman, B. and A. Murphy. USDA Farmers' Bull. 1903. 1942.
37. USDA Agricultural Statistics. Sugarbeets. Table 115. 1968.
38. U.S. Salinity Laboratory Staff. Diagnosis and improvement of saline and alkali soils. USDA, Agriculture Handbook 60. 1954.
39. Wilcox, L. V. Salt balance and leaching requirement in irrigated lands. USDA Tech. Bull. 1290. 1963.

9

Diseases and Their Control

C. W. BENNETT
Tobacco and Sugar Crops Branch, ARS, USDA
Salinas, California

L. D. LEACH
University of California
Davis

VIRUS DISEASES	225
CURLY TOP	226
OTHER CURLY TOP DISEASES	232
BEET YELLOWS	233
BEET WESTERN YELLOWS	239
BEET MOSAIC	242
CUCUMBER MOSAIC	246
YELLOW WILT	249
YELLOW NET	253
YELLOW VEIN	254
SAVOY	255
KRAUSELKRANKHEIT	256

OTHER VIRUS OR VIRUSLIKE DISEASES
THAT MAY OCCUR 257

FUNGUS AND BACTERIAL DISEASES
. . . . 260

SEEDLING DISEASES 260

ROOT DISEASES 265

VASCULAR DISEASES 271

OTHER ROOT DISEASES 273

FOLIAGE DISEASES 275

OTHER DISORDERS 281

THE SUGARBEET had its origin as a sugar-producing plant in northern Europe. From this place of origin it has spread around the world, mainly in the temperate zones. In each area of introduction it has encountered diseases, some of only local import and others involving large agricultural areas. In several of the areas into which the sugarbeet has been introduced, its value as an agricultural crop plant has depended in large measure on the success achieved in control of its diseases.

In Ireland, for example, the use of reduced seeding rates, processed seed, and precision seeding were delayed until a really efficient seed treatment for control of *Phoma* seedling infection was available.

Attempts to grow sugarbeets in Louisiana to extend the cane-sugar factory campaign have been a failure because of two serious fungus rots: *Rhizoctonia* crown rot during the cool weather and *Sclerotium* root rot during the warm season.

Cercospora leaf spot and *Aphanomyces* seedling disease are still limiting factors in the north-central area of the United States.

The success of the new sugarbeet industry in Chile was threatened by a serious seedling disease, called "caida," caused by a soil-borne fungus (*Aphanomyces cochlioides* Drechs.) which appeared in many fields during the first replanting of sugarbeets. The discovery that the disease could be controlled by a furrow application of a specific fungicide permitted profitable production of sugarbeets in infested areas.

Other diseases have been, and some continue to be, threats to the beet sugar industry over wide areas. Yellow wilt, for example, destroyed a budding beet sugar industry in the Rio Negro Valley of Argentina and has limited a young beet sugar industry in Chile to the southern third of the country, thus denying some of the best agricultural land of the central area to the industry. It is probable that until control measures are developed for this disease, profitable sugarbeet production will be impossible in many of the irrigated areas of both Chile and Argentina.

The beet sugar industry of western United States presented a history of alternate success and failure, accompanied by extensive losses, over a period of nearly 40 years as a result of the inroads of the curly top disease. The industry was placed on a secure footing only with the advent of curly top-resistant varieties in the early 1930s. In the absence of such resistant varieties, the sugarbeet industry in western United States would still be limited to areas where the disease is capable of causing only periodic damage of major magnitude on susceptible varieties.

Struggles against the incursions of plant disease have inevitably had marked influences in shaping the nature and evolution of the beet sugar industry as a whole—influences that have extended far beyond disease control alone. Some of these influences have been revolutionary as they have affected certain areas. In the early history of the beet sugar industry in the United States, seed supplies were imported from abroad and were composed of varieties highly susceptible to curly top damage. With the development and production of curly top-resistant varieties for western United States, it became necessary to establish a seed industry for increase of these varieties. This led to the introduction of the overwintering-in-the-field method of seed increase, which greatly reduced costs of seed production, and to the establishment of two major seed-producing areas—one in Arizona-Utah and the other in Oregon—which now produce an average of about 10 million pounds of seed annually.

Moreover, out of the breeding program in the United States, initiated principally for the development of varieties resistant to the curly top disease, have come other beneficial results such as the production of bolting-resistant varieties that have expanded planting dates, and monogerm seed and hybrid varieties that have reduced costs of production and significantly increased yields.

It is the purpose of this chapter to describe major and minor diseases of sugarbeets and to present suggestions for their control. Other treatises on sugarbeet diseases may be found in the list of references at the end of this chapter (18, 32, 42, 43, 50, 53, 54, 55, and 73).

VIRUS DISEASES

The sugarbeet is susceptible to infection by an unusually large number of viruses, many of which have not even been found occurring naturally on this host. Several viruses occur sporadically but cause little or no damage, and others cause injury locally under special conditions. However, several virus diseases, such as curly top, two types of virus yellows, mosaic, and yellow wilt, are highly destructive to sugarbeet and are highly important factors in sugarbeet production over the ranges of their occurrence.

CURLY TOP

HISTORY AND DISTRIBUTION

The early history of curly top disease is somewhat obscure because it was confused at first with other diseases and injuries. It was apparently observed on table beets in Nebraska as early as 1888. Extensive losses were reported in California in 1899, and the following year it was reported from all the western states where sugarbeets were grown, indicating it had a wide distribution before it was recognized as a distinct disease (22).

Within a few years, it was determined that curly top is caused by a virus that is transmitted by a leafhopper, *Circulifer tenellus* (Baker), that occurs widely distributed over western United States and parts of Canada and Mexico. There is evidence that this leafhopper even moves into eastern parts of the United States and occasionally produces curly top infection in such states as Wisconsin, Illinois, and Maryland, but it has not become permanently established in the eastern part of the United States.

For many years, curly top was known only in North America and it was generally assumed that the disease was native to western United States. However, it was determined in 1946 (59) that the beet leafhopper is an introduced species that probably came from the Mediterranean area where several other species of *Circulifer* occur. In 1958 (15) it was reported that curly top was widely distributed in Turkey, and later reports indicate that it may be present in other semiarid areas of Europe, Africa, and Asia. It seems probable, therefore, that both the curly top virus and its vector were introduced into western United States from abroad.

ECONOMIC IMPORTANCE

Curly top has been one of the most destructive diseases known to sugarbeet in western United States. The damage has varied in different areas in different years, but in many instances attacks have been disastrous. Average yields in southern Idaho from 1912 through 1934, for example, ranged from 4.88 tons per acre in 1934 to 16.15 tons per acre in 1932 (57). In many fields, the crops were a complete loss and were not harvested. Severe attacks occurred in other areas in western United States. Factories were closed and areas were abandoned for sugarbeet production. The introduction of resistant varieties, beginning about 1934, has reduced losses of sugarbeet to relatively low levels. At the present time, there are no areas in North America where curly top prevents the profitable production of sugarbeets, although marked damage may be produced locally in some years.

Other crops that have been severely damaged by curly top include bean, tomato, flax, and cucurbit. Resistant bean varieties are now in use, but the tomato crop may be damaged severely in some areas when leafhoppers are abundant. In parts of the Pacific Northwest, commercial tomato production is not attempted on an extensive scale because of the threat of damage by this disease.

HOST RANGE

Curly top virus has a very extensive host range, which includes more than 200 species of plants distributed through some 42 plant families. It has received most attention because of its attacks on sugarbeet, but it also causes a serious disease of tomato, bean, flax, cucurbit, and a number of ornamental plants. It infects many common weeds and desert plants, among which are Russian thistle, filaree, various species of mustard and plantain, and many plants of the Chenepodiaceae to which sugarbeet belongs.

SYMPTOMS

The symptoms of curly top on most susceptible species of plants are rather characteristic of the disease; however, there are variations in symptom expression on different host plants.

Sugarbeet. Usually, on plants of susceptible varieties, the first symptoms of curly top begin to appear 3 to 15 days after infection. Very young leaves show translucent veins and become twisted, rolled, or crinkled. All leaves produced later are characterized by rolling, crinkling, and dwarfing, as shown in Figure 9.1. The veins are irregularly swollen and may develop protuberances, especially on the lower side of the leaf. Plants are severely dwarfed and often die if infected in the seedling stage. Necrosis develops in the phloem areas of the vascular bundles which may appear as dark rings in transverse sections of beet roots or as dark lines in longitudinal sections as illustrated in Figure 9.2.

Resistant varieties may have a high percentage of infection but show only mild symptoms consisting of an inconspicuous vein translucency in young leaves and slight swelling of veins with little dwarfing or other evidence of disease.

Tomato. First symptoms of curly top on tomato, particularly under greenhouse conditions, are a slight twisting of one or more of the young leaves and inward rolling of leaflets, sometimes accompanied by vein translucency.

FIG. 9.1. Sugarbeet plant showing stunting and upward and inward rolling of leaves and swelling of veins caused by curly top disease.

Usually, the first symptoms observed under field conditions are a general yellowing of the affected plant, slight drooping of the leaves, upward rolling of leaflets, and the production of purplish coloration on veins and other parts of the leaf. As the disease progresses, yellowing increases and growth is markedly retarded. Plants infected in the early stages of development usually die before blossoming. Those infected in later stages may die or may continue to live as stunted yellow plants until the end of the season.

After midseason, tomato fields attacked by curly top often show plants in all stages of disease, ranging from dead plants through those showing different stages of yellowing and dwarfing.

Occasionally, diseased tomato plants may produce almost normal

FIG. 9.2. Cross section of a healthy and a diseased sugarbeet root showing dark rings in the diseased root caused by curly top.

shoots from near the base of the plant. Tobacco plants normally recover from the earlier more severe symptoms. The "recovery" condition in both tomato and tobacco can be perpetuated through vegetative propagation. Also, tomato plants do not show severe symptoms following inoculation by means of grafts from either recovered tomato or recovered tobacco. Plants from cuttings of recovered tomato plants have produced crops of fruit in the field under curly top exposures that killed most of the nonprotected plants in the same area.

TRANSMISSION AND SPREAD

No transmission of curly top virus has been obtained by the common methods of juice inoculation, but a low percentage of transmission has been obtained in some instances by inserting very fine needles into the vascular system of beet leaves through drops of inoculum. Transmission has been obtained by means of three species of dodder, the most efficient being *Cuscuta californica* Choisy which has given more than 50 percent infection in some tests with tobacco.

Under field conditions, curly top virus is spread by the beet leafhopper, *Circulifer tenellus*. This is the only known vector in North America, but in Turkey a related leafhopper, identified as *C. opacipennis* Lethierry, is a vector (15). There are several other species of *Circulifer* in the Mediterranean region that have not been tested for ability to transmit curly top virus. It is probable that some of these may prove to be vectors.

The beet leafhopper is primarily a desert insect. Breeding areas are distributed over much of western United States. In these areas, there are successions of host plants that provide for reproduction of the insect and for its persistence through the winter and other unfavorable periods. Leafhoppers overwinter as adults and move to spring host plants on which they reproduce. As the spring hosts mature and die, the leafhoppers migrate, sometimes for distances of many miles, to other areas where host plants are available. These include cultivated areas where curly-top-susceptible crops are grown. The leafhoppers usually arrive in beet fields in May or June. Some of the incoming leafhoppers carry curly top virus from desert host plants which they introduce into young beet plants. These first infected plants serve as sources of spread of virus to other plants both by leafhoppers that arrive from outside sources and by progeny of the leafhoppers that breed on sugarbeet.

One of the most important plants in the survival of leafhoppers in large numbers over much of the breeding area of western United States is Russian thistle, *Salsola kali* L. var. *tenuifolia* G. F. W. Mey. This plant is drought resistant and capable of carrying large populations of leafhoppers through periods in late summer and fall when other host plants are scarce or not available.

CAUSAL AGENT

Curly top is caused by an agent that is presumed to be a virus, although it has not been purified and photographed.

Characteristics. Properties of the curly top virus have been given considerable attention (3, 65). The virus is readily carried out of suspension of plant juices by precipitation of proteins by heat or chemical agents, such as ethyl alcohol. It is highly resistant to common types of disinfectants; remains active in plant juice at room temperatures for several days or weeks, depending upon conditions; withstands dilutions of 1–1,000 to 1–22,000, depending on the medium in which it is contained; has a thermal inactivation point between 70° and 80° C; and has retained activity for two to five months in dried materials, including dried leafhoppers, and for eight years in dried beet leaves preserved over calcium chloride.

Strains of Virus. The virus exists in nature as a complex of races or strains that differ in host range, virulence on sugarbeet and other host plants, type of induced symptoms, and perhaps other characteristics (40). As yet there is little evidence of cross-protection between strains. Plants infected with one strain remain susceptible to infection with a second strain of the same virus.

Under some conditions the virus has been very stable and has remained in sugarbeet plants under greenhouse conditions for more than 20 years without detectable change. However, the presence of innumerable variants under natural conditions indicates a degree of instability. Moreover, mutants, some of which produce marked vein yellowing that would not ordinarily be recognized as a symptom of curly top, have been found on both tobacco and sugarbeet. Also, highly virulent strains of virus, capable of producing marked symptoms on the most resistant varieties of sugarbeet, have appeared in recent years (7).

Relation of Virus to the Host Plant. Curly top virus is closely associated with the food-conducting tissue and probably does not extensively invade other types of cells. It can be prevented from moving through a stem by a ring that destroys all phloem connections between two parts. It may move as fast as one inch per minute through beet leaves in the direction of major food transport. It accumulates in relatively high concentrations in seeds of diseased beet plants, but it is unable to enter the embryo, hence is not seed transmitted. Its presence in the phloem results in considerable phloem necrosis and stimulation of proliferation of cells in the immediate vicinity of the vascular region, resulting in vein swelling and other types of overgrowths.

Relation of Virus to the Vector. The beet leafhopper is able to acquire virus from diseased plants in a feeding time of one minute and transmit it in the fourth hour after feeding. It is a very efficient agent of virus transmission and can introduce virus into healthy plants in a feeding time of one minute. However, a feeding time of two or more days is required for the insect to acquire a maximum charge of virus. Infection increases with size of virus charge and with time of feeding on healthy plants (16). The amount of virus a leafhopper is able to acquire appears to be related to the concentration of virus in the host plant. The presence of one strain of virus in the leafhopper apparently does not interfere with the acquisition of a second strain, so individuals may, and frequently do, carry and transmit more than one virus strain (41).

After the virus is acquired by the beet leafhopper, it is retained for prolonged periods; more than three months in some instances. However, when a leafhopper feeds on virus-free plants, its virus content slowly diminishes to a low level, or even to a point at which it is no longer able to transmit. There is no evidence, therefore, that the virus increases in its vector. The virus does not pass through the egg stage of the leafhopper, so all young leafhoppers are at first free of virus (16, 36).

CONTROL

The chief measures for control of sugarbeet curly top consist of (1) use of varieties resistant to the disease, (2) certain cultural practices such as early planting to avoid infection of plants in the seedling stages, (3) use of insecticides to control leafhoppers in beet fields, and (4) use of insecticides to reduce leafhopper populations in their natural breeding grounds.

All varieties of sugarbeets now used in western United States are highly resistant to curly top, although some are more resistant than others. Over the great majority of the area where curly top occurs, this resistance is sufficient to provide adequate protection against curly top, and other measures of control are not required, particularly if plantings are made early enough for the plants to have attained appreciable size before the spring influx of leafhoppers from breeding areas.

In a few areas, however, notably on the west side of the San Joaquin Valley of California, resistant varieties have been damaged in years (such as 1966) when leafhoppers, carrying highly virulent strains of curly top virus, migrated to beet fields and produced heavy infection when the plants were small. In such years, tests have shown that timely uses of insecticides, especially systemic insecticides, placed at a depth of about five inches below the seed at planting time, may give marked reduction in infection (21). Seed treatment with systemic insecticides in the seed-producing areas of Arizona has given reduction in curly top infection in fields planted in August and September, especially in varieties that do not have a high degree of resistance to curly top that are reproduced there for use in eastern United States where curly top is not a problem.

For a number of years the California State Department of Agriculture has conducted a program for control of the beet leafhopper, designed to reduce leafhopper population levels in noncultivated areas at strategic periods in the annual cycle of the insect and thus minimize movement of leafhoppers into cultivated areas in the spring. The program has been concentrated largely in the San Joaquin Valley where leafhoppers collect in large numbers in the fall on Russian thistle plants. Thousands of acres of this host are treated, usually in October, and large numbers of leafhoppers are destroyed. This may be followed with treatment of desert shrubs in canyons and on foothills where leafhopper concentrations are found, and in foothills where spring broods are produced. The program is designed to give protection to tomato, potato, flax, and other susceptible crop plants, as well as to sugarbeet.

OTHER CURLY TOP DISEASES

Curly top diseases that appear to be related to North American curly top occur in Argentina and Brazil (10, 11). Two of these diseases have been described under the names Argentine curly top and Brazilian curly top.

ARGENTINE CURLY TOP

In 1925 a curly top disease of sugarbeet was described in Argentina in the Province of Tucumán (35). This disease has since been found widely distributed in Argentina and Uruguay and is known to occur in Brazil. It appears, however, not to have invaded Chile where there is a thriving beet sugar industry.

The disease is capable of producing heavy losses on susceptible varieties of sugarbeet, and extensive losses have been reported in Uruguay in certain years. The introduction of this disease into sugarbeet-producing areas of Chile might result in a considerable reduction in yield of varieties now grown in that country. It seems probable, also, that it might be capable of causing damage to beets in other semiarid areas if it were to be introduced and become widely established.

The known host range of the Argentine curly top virus is similar to that of the North American curly top virus, except that the Argentine virus is not known to attack certain species of solanaceous plants such as tomato, tobacco, pepper, and potato. Varieties of sugarbeet tested were resistant in the same order to the two diseases.

Symptoms of Argentine curly top on sugarbeet, table beet, and Swiss chard, consisting of vein clearing, leaf curling, vein swelling, and vascular necrosis, are indistinguishable from symptoms produced on these plants by North American curly top, although sugarbeet plants with Argentine curly top appear to show a greater degree of recovery.

Argentine curly top virus is transmitted by a leafhopper, *Agalliana ensigera* Oman, which occurs in Argentina, Brazil, Uruguay, and perhaps in certain other Latin American countries. It has not yet been found in Chile. Argentine curly top virus apparently is not transmissible by the beet leafhopper, *Circulifer tenellus* (10).

Control measures have not been employed in areas where this disease occurs, but it seems probable that losses could be reduced to a very low level by use of varieties already known to be resistant to North American curly top.

BRAZILIAN CURLY TOP

In 1939 a curly top disease was described on tomato and tobacco in the state of São Paulo, Brazil, that produces symptoms very similar to those produced on these hosts by North American curly top. Infected tomato plants yellow and die and tobacco plants show a marked crinkling and curling of leaves, often with a distinct curving of the top portion of the stem. It has proved to be a serious disease locally on both tobacco and tomato.

The disease also infects sugarbeet, table beet, and Swiss chard, but symptoms are relatively mild. Sugarbeet varieties resistant to North American curly top are resistant in the same order to Brazilian curly top.

The causal agent is transmitted by a leafhopper, *Agallia albidula* Uhler, which is prevalent in central Brazil, but little is known about its distribution elsewhere.

There is little information as to whether Brazilian curly top could become a serious disease if introduced into areas where sugarbeet is a commercial crop, but it seems probable that in any event the disease could be satisfactorily controlled by use of varieties already known to be resistant to North American curly top.

BEET YELLOWS

HISTORY AND DISTRIBUTION

The disease of sugarbeet now called beet yellows probably was one of the diseases of a yellowing complex first described in Europe, under the name "jaunisse," a number of years ago. About 1935, beet yellows was shown to be caused by a virus transmitted by aphids. It was soon found to occur throughout western Europe and was reported prevalent in England in 1948 and later years.

It was first identified in the United States in 1951 at which time it probably had already attained a wide distribution in certain western states such as California and Arizona. More recently the disease has been found in Japan and Chile. It probably occurs in most countries where sugarbeets are produced commercially.

ECONOMIC IMPORTANCE

Beet yellows is a disease of major importance in areas where conditions are favorable for infection of a high percentage of plants in early stages of development (5). In experimental tests, beet yellows has caused reductions in root yields up to 40 percent or more, and reductions in seed yields of more than 30 percent. It may also result in reduction of sucrose content up to 2 percentage points.

The destructiveness of this disease varies greatly under different conditions. Earlier estimates of crop losses have been confused in the United States, and probably in Europe, with losses caused by beet western yellows (also called radish yellows, probably mild yellows, and mild yellowing). The relative importance of the two diseases in areas where they both occur is still difficult to assess. Beet yellows causes greater reduction in yield on infected plants, but western yellows is likely to have a greater incidence of occurrence.

Greatest losses in the United States occur in the Sacramento, San Joaquin, Salinas, and Imperial valleys of California and in the Salt River Valley of Arizona. These are the areas most likely to provide year-round sources of infection from beets carried through the winter for spring harvest, from beets that have escaped from cultivation, or from plants of the previous season's crop. In such areas, losses may be severe. There is good evidence that in certain years beet yellows was the principal factor in the reduction of yields to unprofitable levels in areas such as Kern County, California.

Although beet yellows is present in other parts of western United States, such as Washington, Oregon, and the intermountain states, losses from the disease in these areas have not been highly significant. In the eastern part of the United States, beet yellows is of little or no importance.

The disease undoubtedly causes considerable loss to the beet crop in western Europe and in the British Isles. It has not been shown to cause appreciable losses in Japan or Chile.

HOST RANGE

The principal crop plants attacked by beet yellows are sugarbeet, table beet, Swiss chard, and spinach. Experimentally, beet yellows virus has been shown capable of infecting species of plants in more than a dozen plant families. The largest numbers of susceptible species are found in the families Chenopodiaceae and Amaranthaceae.

Several of the weed plants common to cultivated areas are hosts of the yellows virus. The commonest of these are lambsquarters (*Chenopodium album* L.), nettleleaf goosefoot (*C. murale* L.), redroot amaranth (*Amaranthus retroflexus* L.), and common groundsel (*Senecio vulgaris* L.). High percentages of infection occasionally occur in each of these species in some areas of California when they are growing in or near infected beet fields. However, infection of weeds in beet fields is usually less than that in sugarbeet. It seems unlikely, therefore, that weed hosts often serve as important sources for extensive spread of virus to crop plants.

Australian saltbush (*Atriplex semibaccata* R. & Br.) is a symptomless perennial host that is widely distributed throughout much of western United States. It may serve as a reservoir for carrying virus through unfavorable periods and throughout the year, but it probably does not serve as a source for extensive spread since it is an unfavorable host for known vectors.

SYMPTOMS

Symptoms of beet yellows vary, depending on the virulence of the strain of virus involved, variety attacked, light conditions, and possibly other factors. First, clearly obvious symptoms produced by the more virulent strains of virus usually appear on young leaves and consist of clearing or yellowing of the veins. Vein clearing may appear on two to five leaves. These leaves are usually dwarfed at maturity and show shrunken veins on the lower side which impart a roughened or "alligator skin" effect.

About the time of appearance of vein clearing in young leaves, two or three leaves just older may begin to yellow over the entire surface or in certain areas sharply delimited by larger veins.

Leaves produced later that those showing vein cleaning are more or less normal until they approach maturity, at which time they begin to show various types of yellowing that may be somewhat uniform or may consist of various types of ill-defined splotches. Usually, leaves tend to yellow first at the tips and in parts having greatest exposure to sunlight. As the leaves age, areas between the larger veins tend to fade and become necrotic. Often the larger veins remain green, as indicated in Figure 9.3. Leaves in the later stages of disease are thickened, leathery, and brittle, and die prematurely.

As already indicated, the severity of symptoms varies with the virulence of strain of virus involved. Some strains rarely or never produce vein clearing and strains capable of producing vein clearing do not do so under all conditions. Intensity of yellowing of affected plants also varies, depending on the strain of virus, as illustrated in Figure 9.4 from results of inoculation with two strains having different degrees of virulence.

Symptoms of beet yellows on the common commercial varieties of sugarbeet do not vary greatly, but certain selections from breeding stocks have shown a wide range of reactions. Some show very little yellowing,

Fig. 9.3. Sugarbeet leaf showing yellowing caused by beet yellows. In this stage, veins may still be green as indicated by their dark color. Similar symptoms are produced also by beet western yellows.

whereas others are very yellow. A few selections have shown marked necrosis in mature leaves, and plants with red pigment usually show an intensification of red color.

Other susceptible crop plants, such as spinach, garden beet, and Swiss chard, show symptoms similar to those found on sugarbeet.

TRANSMISSION AND SPREAD

The virus causing beet yellows can be transmitted by rubbing juice from diseased plants over the surface of leaves of healthy plants. Small necrotic lesions, about 1 mm in diameter, are produced on sugarbeet and *Chenopodium capitatum* (L.) Asch., beginning about four days after inoculation, some of which lead to systemic infection. However, mechanical transmission is of no importance in natural spread, and there is no evidence that the disease is seed transmitted.

In the United States, spread of the disease under field conditions is by aphids, at least six species of which are capable of transmitting the causal virus. Additional species have been reported as vectors in other parts of the world. Of these species, the green peach aphid (*Myzus persicae* [Sulz.]) is by far the most important agent of spread. The bean aphid (*Aphis fabae* [Scop.]) has been reported as an important vector in some parts of Europe, but it is of minor importance in the United States.

The green peach aphid is a highly effective agent of spread of beet yellows virus, due to its extensive geographical distribution, wide host range, abundant reproduction, and efficiency in transmission. Individuals can acquire virus from a diseased plant in a feeding time of 5 minutes and reach maximum ability to transmit in a feeding time of 4 to 6 hours. Aphids carrying virus can transmit it to healthy plants in a feeding time of 5 minutes, but percentage infection increases in feeding times up to 4 to 6 hours. Most of the virus carried by an aphid is lost during the first 24 hours if the aphids feed on virus-free plants, but they may occasionally carry virus for as long as 72 hours.

Fig. 9.4. Field plots showing yellowing resulting from inoculation with beet yellows virus. Two rows inoculated with a virulent strain of the virus (left). Two check rows not inoculated (middle). Two rows inoculated with a less virulent strain of beet yellows virus (right).

In much of western United States, particularly in those areas where beet yellows is important, the green peach aphid lives throughout the year on various host plants. In the spring it multiplies rapidly and may reach enormous numbers. Winged individuals move from diseased host plants to beet fields where they establish centers of infection, from which further spread occurs as aphids move outward from infected plants.

The infection centers usually begin to appear as yellow spots in late April or early May, 15 to 25 days after plants are infected. If aphid populations are drastically reduced, the yellow spots may spread slowly and persist for several weeks or even during the remainder of the season. Often, how-however, the yellow spots continue to increase in size until the entire field is yellow. In some instances, particularly in central California, spread is so extensive and rapid that centers of infection are hardly evident, or evident for only a few days, and fields may appear to yellow uniformly within a period of 2 or 3 weeks.

CASUAL AGENT

Beet yellows virus is a complex of strains that vary considerably in severity of their effects on sugarbeet. The virus has been partially purified and photographed both inside and outside infected cells. Particles are relatively long, crooked rods which occur abundantly in the food-conducting elements and in other living cells into which they move through protoplasmic connections between adjacent cells (33).

The virus is inactivated when subjected to 10-minute exposures in plant juice at temperatures between 50° and 55° C. It loses infectivity in about 48 hours in plant juice held at room temperature, and no infection has been obtained from juice of diseased plants diluted more than 1–10,000. It soon becomes inactive in dried beet tissue (5).

CONTROL

In general, the possibilities for the control of beet yellows are limited to four methods. These consist of (1) destruction of aphid vectors by application of insecticides, (2) elimination of sources of infection for initial spread of the virus to beet fields, (3) selection of planting dates that enable the beets to escape infection, and (4) development and use of varieties resistant to the disease.

Under certain conditions in Europe, application of insecticides was effective in giving a high degree of control of yellows. Best results were obtained by timing one or more applications to control aphids before they could produce extensive secondary spread from plants infected initially by winged form that carried the virus into the field from outside sources. Where spring aphid populations were high, however, control was not always highly effective.

In most areas of California where sugarbeets are grown extensively, populations of green peach aphids are usually very high in March, April, and May. During this period, large numbers of winged individuals may enter beet fields from outside sources. It is probable that where abundant sources of infection are available, the winged form alone may be capable of producing high percentages of infection by June in areas such as the Salinas Valley. Nevertheless, some tests have shown increased yields from use of insecticides.

In those areas where aphid populations are greatly reduced with the advent of high temperatures and low humidities, as in the central valleys of California, it would appear that use of insecticides to protect the plants through the early period of growth should give a high degree of control. In actual practice, however, control has been less than could be expected.

Beet yellows is likely to be a serious disease only in those areas where sugarbeet or other important host plants are present to serve as sources of infection throughout the year. In western United States, it appears that sugarbeet itself is the chief source of infection. Where a beet-free period is feasible at any time during the year, beet yellows can be controlled effectively by the elimination of beet as an important source of infection. A beet-free period, however, should include destruction of wild and escaped beets and beets left in fields after harvest to serve as hosts for both aphids and virus. This method of control has given excellent results where it has been used over a wide area, notably in Kern County, California.

Populations of the green peach aphid usually drop to very low levels in the central valleys of California by the first of June. Beets planted the latter part of April or in May tend to escape infection by beet yellows virus as well as other aphid-transmitted viruses, in most years. Under conditions of heavy infection of early-planted beets, late planting has given significantly higher yields. However, such gains must be balanced against reduced yields that result from a shorter growing season and the possibility of greater losses from curly top, nematodes, and other adverse factors in late-planted crops.

The most economical method for control of most plant diseases is by use of resistant varieties, where such varieties are available. There is much hope that losses from beet yellows will be considerably reduced by the introduction of yellows-resistant lines that have been produced through some 13 years of selection and breeding for resistance to this disease. New yellows-resistant varieties, now becoming available, may be expected to reduce losses as much as 50 percent, and varieties even more resistant are a possibility.

BEET WESTERN YELLOWS

HISTORY AND DISTRIBUTION

In the United States this disease was first separated from the yellows complex by Duffus (28) in 1960 and called "radish yellows," but owing to its prevalence on sugarbeet, the name was changed to beet western yellows (29). It probably is not basically different from the disease called "mild yellows" or "mild yellowing," which is widespread in Europe. In fact, the type of yellowing here designated "beet western yellows" seems to be caused by a virus or virus complex having a wide geographic distribution in weed hosts, even in some areas where sugarbeets are not grown commercially. It appeared in such areas as Argentina, Chile, and Turkey, apparently very soon after beet sugar industries were established, and remoteness from other sugarbeet-producing areas did not appear to prevent introduction of the virus into new plantings. The disease, therefore, probably occurs in all major sugarbeet-producing areas of the world, and the causal virus may have an even wider distribution.

ECONOMIC IMPORTANCE

Reduction in yields of sugarbeet by western yellows usually is not exceptionally high, but the widespread occurrence of the disease and high incidence of infection in some areas increase its overall importance as one of the major virus diseases of sugarbeet.

In California, reductions in yield under experimental conditions have ranged from 5 to 18 percent where plants were inoculated in the early stages of development (14). Somewhat higher yield reductions have been reported from tests in Utah. Variations in the amount of reported damage probably result, in part, from the use of virus strains of different degrees of virulence in different tests. Reductions in percentage sucrose have been less than those reported as produced by beet yellows. It is evident, however, both from experimental tests and from field observations, that losses from this disease may be substantial if infection occurs in the early stages of plant development.

Greatest losses from this disease occur in the Sacramento, San Joaquin, and Salinas valleys of California where often many fields show almost 100 percent infection by the middle of June. In the Imperial Valley of California, symptoms are usually delayed until about a month before the beginning of harvest. Under these conditions, it is not believed that yields are greatly reduced. Considerable infection occurs also in some parts of Washington and Oregon, particularly in the seed crop of the latter state. High percentages of infection may occur also in Arizona and Colorado. In general, the disease is less prevalent in the northern tier of states, probably because climatic conditions in this area limit host plants that carry the virus through the winter. The disease is of little or no importance in eastern United States.

Yellowing of this general type is a major disease in western Europe, the British Isles, and Ireland. It is considered to be more destructive than beet yellows in certain parts of England. It is reported to be able to reduce yields as much as 22 percent in infected areas in fields in Chile.

HOST RANGE

Beet western yellows virus infects a wide range of plants (17, 28, 29). In addition to sugarbeet, it attacks wild species of *Beta,* Swiss chard, garden beet, spinach, radish, lettuce, and perhaps several other crop plants on which it causes little damage. It is widely distributed in nature on many weed plants such as various species of mustards (*Brassica* spp.) and shepherdspurse (*Capsella bursa-pastoris* L.), various composits such as groundsel (*Senecio vulgaris*) and sowthistle (*Sonchus* spp.), and a variety of other weed plants. Several of these weeds are excellent hosts of the green peach aphid. Some are biennials or winter annuals that carry both the vector and the virus through the winter in mild climates and serve as sources of spring infection for sugarbeet. New Zealand spinach (*Tetragonia expansa* Murr.) and strawberry blite (*Chenopodium capitatum*), both susceptible to beet yellows, are highly resistant or immune to western yellows.

SYMPTOMS

Western yellows differs from beet yellows in that no signs of the disease appear on young leaves. Infected plants begin to show first symptoms usually 20 to 30 days after infection, but the delay in the appearance of symptoms may be even longer under some conditions, especially if the plants are growing rapidly. On leaves approaching maturity, a slight chlorosis begins to appear in the areas between the main veins. This discoloration becomes more marked as the leaf ages and may reach a stage in which the leaf has a distinct yellow color, often with green veins. Interveinal areas of the older leaves may be attacked by fungi of various kinds that produce necrotic spots. Leaves tend to thicken and become brittle and die prematurely.

Symptoms can be largely masked by the application of nitrogenous fertilizers. Plants often are greener around the borders of fields and in areas of poor stands than in other parts of the field where there is greater competition for nutrients. Yellowing is greatly influenced by light intensity; plants in areas shaded by trees often are noticeably greener than those in full sunlight. Garden beet and spinach show symptoms similar to those produced on sugarbeet.

TRANSMISSION AND SPREAD

Western yellows virus has not been transmitted to any of its host plants by juice inoculation nor is there evidence of seed transmission. The only known method of spread is through the agency of several species of aphids, the most important of which is the green peach aphid *(Myzus persicae)*. The bean aphid *(Aphis fabae)*, which is a vector of beet yellows virus, although an inefficient one, apparently does not transmit western yellows virus.

The green peach aphid is a highly efficient vector. It can acquire the virus by feeding only a short time on a diseased plant and can transmit it to a healthy plant also in a short feeding period. Apparently, once the aphid has acquired the virus it can transmit it throughout the remainder of its life.

The spread of virus into beet fields from outside sources and the spread of virus within beet fields are similar to that already described for beet yellows virus. Indeed, both viruses can be, and frequently are, transmitted by the same aphid. The chief differences in the epidemiology of the two viruses are associated with the greater importance of weed hosts as sources of western yellows virus and the longer period of retention of the western yellows virus by aphid vectors.

CAUSAL AGENT

Beet western yellows is caused by a virus that appears to be a complex of strains that vary in host range, virulence, and perhaps other characteristics. Some of the strains of the complex apparently do not infect sugarbeet, others produce mild symptoms and little damage, and still others produce significant reductions in yield.

As already indicated, the virus remains active in the green peach aphid, its principal vector, probably during the life of the insect. Recently, by use of a membrane-feeding technique, Duffus (30) determined some of the properties of the virus. Extracts of diseased plants were infective after 16 days but not after 32 days. The virus is highly resistant to alternate freezing and thawing, withstands desiccation in plant tissue for at least three years, and is inactivated by a 10-minute treatment at 60° C.

CONTROL

In general, the methods for the control of beet yellows, already outlined, consisting of destruction of aphid vectors, elimination of sources of infection, selection of planting dates that enable beets to escape infection, and development and use of varieties resistant to the disease, apply also to control of western yellows. The main difference in the effectiveness of these disease control methods is associated with a reduction in sources of infection by use of beet-free periods. This is less effective with western yellows than with beet yellows, due to the greater prevalence of weed hosts that serve as virus sources for spread of western yellows. However, elimination of beets as a source of infection may be expected to delay spread of the disease.

BEET MOSAIC

HISTORY AND DISTRIBUTION

Mosaic disease of beets was first reported in 1898 on garden beets in France. It was reported in the United States in Colorado in 1915 (63). It is now one of the most widely distributed of the sugarbeet virus diseases and probably is present in all major sugarbeet-producing parts of the world. In newly developed areas, it has appeared within a short time after the beginning of sugarbeet production. It was observed in the Rio Negro Valley of Argentina in 1939 after the beginning of sugarbeet production in that area in 1929, and it has appeared in other areas far removed from known sources of virus on sugarbeet. Its extensive distribution suggests the presence of other host plants that have a widespread occurrence and the possibility that such host plants may serve as sources for initial introduction of mosaic into beet fields of new areas.

ECONOMIC IMPORTANCE

In most of the sugarbeet areas of the United States, percentages of infection are low. High percentages of infection occur only in those areas where sources of infection, usually sugarbeet, are present in abundance throughout the year. The disease is of most importance in those areas where crops of two different years overlap or where climate and cultural practices permit the overwintering of infected plants. It is usually abundant where seed and root crops are grown in the same area.

Even when there are high percentages of infection, the disease is not highly destructive. Under the most favorable conditions, losses rarely exceed 5 to 10 percent in beets grown for sugar. It may be that the seed crop can be damaged somewhat more than this, and it has been found that if infected stecklings are transplanted for seed production the loss in seed yield may be severe.

HOST RANGE

Sugarbeet is the chief host of economic importance for beet mosaic, but attacks may occur also on garden beet, Swiss chard, and spinach (64). The virus has a fairly extensive host range among plants of the Amarathaceae, Chenopodiaceae, Compositae, and Leguminosae. The disease is not uncommon on redroot amaranth *(Amaranthus retroflexus)*, lambsquarters *(Chenopodium album)*, and several other weeds, when these are growing in or near infected sugarbeet fields. However, there is usually less infection on the weed hosts than on sugarbeet under these conditions. Occasionally, one may encounter appreciable infection in yellow sweet clover *(Melilotus indica* [L.] All.) and crimson clover *(Trifolium incarnatum* L.). It is known that a number of other legumes, such as the common garden pea, are also susceptible to infection, as are certain species of composits, such as zinnia, but these plants suffer little damage from the disease.

SYMPTOMS

Beet mosaic is a typical mottling disease similar to mosaics that are found on many other kinds of plants. On sugarbeet, symptoms begin to appear about seven days after plants are infected. They often consist of chlorotic spots on young leaves, but if the plants are growing rapidly, first symptoms may consist of yellowing or clearing of the veins of young leaves. Subsequent growth, however, will show only mottling. The chlorotic spots may be more or less circular, often with sharply defined margins as shown in Figure 9.5 (left). They may even take the form of definite chloroting rings with green centers. However, there is much variation in the type of mottling and under most conditions the chlorotic areas are more irregular and diffuse as illustrated in Figure 9.5 (right). Symptoms tend to be less evident as plants increase in size and they may be difficult to detect later in the season, particularly on mature leaves.

Marked chlorosis or necrosis of veins may occur on the inoculated leaves, starting at the point of virus introduction and extending downward, probably marking the path of movement of virus from the point of introduction to the growing point of the plant.

Occasionally, mosaic may produce marked dwarfing and distortion of leaves which may show conspicuous green blisters surrounded by chlorotic areas. Symptoms on such leaves may closely resemble those produced by cucumber mosaic virus. Symptoms of this type are associated with the more virulent strains of the virus that do not appear to be widely prevalent (8).

TRANSMISSION AND SPREAD

Beet mosaic virus is readily transmissible by rubbing juice from diseased plants over the surface of leaves of healthy plants. Rubbing in the

FIG. 9.5. Sugarbeet leaves showing types of mottling produced by beet mosaic virus.

presence of an abrasive such as carborundum increases the percentage of infection. However, it is unlikely that spread of virus through mechanical inoculation by ordinary cultural operations is of significance. There is no evidence that beet mosaic virus is transmitted through the seeds of any of its host plants. It appears to be inactivated in mature beet seeds, even in the portion of the seed ball that carried virus when green.

Spread of mosaic virus is brought about chiefly by the green peach aphid, although there are several other species of aphids capable of transmitting the virus. The green peach aphid is a highly efficient vector. It can acquire virus from an infected plant in a feeding time of 6 to 10 seconds and can transmit it immediately to a healthy plant in a feeding time as short as 10 seconds. The aphid retains virus for only a relatively short period, usually not more than 30 minutes. Winged aphids are very efficient in transmission of virus over short distances. Winged aphids may occasionally carry virus for appreciable distances under very favorable conditions, but as a rule, spread of mosaic into beet fields is from nearby sources.

CAUSAL AGENT

Beet mosaic is caused by a virus that is inactivated in sap extracts at about 60° C in 10-minute exposures. It is inactivated in plant juice held at room temperature in about 6 days and it does not persist in dried beet tissue. The virus particles are reported to be slender rods about 700 millimicrons in length.

CONTROL

The rapidity with which vectors pick up and transmit beet mosaic virus makes control measures by means of aphicides difficult and mostly ineffective.

Where control of beet mosaic is of economic importance, the best choice is reduction or elimination of sources of infection. In most cases this involves the destruction of wild and escaped beets in the immediate vicinity of young beet fields and the destruction of "ground keepers" from previous crops. Beet mosaic rarely affects high percentages of plants in areas where there is a beet-free period at any time during the year. Special care should be taken to avoid infection in steckling beets if they are to be transplanted for seed production.

Varieties highly resistant to beet mosaic are not now available, but a rather wide range of susceptibility to injury observed in certain varieties and breeding lines suggests that more resistant varieties could be developed through the usual types of breeding and selection.

CUCUMBER MOSAIC

HISTORY AND DISTRIBUTION

As the name suggests, this disease primarily affects cucurbits. It has been known for many years on cucumber, squash, and melon, and in certain areas, it has been a limiting factor in the production of these crops. It also attacks many other types of plants, having been reported on more than 100 plant species distributed through 32 plant families. It is practically worldwide in its distribution, and as would be expected, it has been reported on sugarbeet in many parts of the world where this crop is grown commercially.

ECONOMIC IMPORTANCE

Despite the fact that cucumber mosaic is widely distributed and is an important disease on a number of other crop plants, its occurrence on sugarbeet is usually limited to traces. Under conditions exceptionally favorable for spread, however, it can be highly destructive to sugarbeet. It destroyed several beet fields, and badly damaged others, in the vicinity of Firebaugh and Mendota in California in 1940 and was present in considerable quantity in the same areas in 1941 (65). Also, local damage from this disease was found in parts of seed fields in the Salt River Valley of Arizona in 1956. Infection was as high as 30 percent in some areas of certain fields. Infected plants were markedly dwarfed and yield of seed from diseased plants was reduced 30 to 40 percent.

SYMPTOMS

First symptoms of cucumber mosaic on sugarbeet may begin to appear about 10 days after infection. Young leaves show mottling, characterized by chlorotic spots of various shapes and sizes. Chlorotic spots may range from small to very large and they may vary greatly in shape (Fig. 9.6). As the disease progresses, mottling continues to be produced and leaves may be dwarfed and distorted, sometimes showing green "blisters," formed as a result of the more rapid growth of green tissue between chlorotic areas. Plants are markedly dwarfed and if infection takes place in the early stages of development, they may be a complete loss.

Some strains of cucumber mosaic virus produce chlorotic spots on mature leaves at the point of virus introduction by either the rubbing method of inoculation or by aphids. These spots continue to increase in size as long as the leaf is in good condition. Such lesions may or may not result in systemic infection.

In seed fields, plants infected in the early stages of bolting showed

Fig. 9.6. Sugarbeet leaves showing two types of chlorosis produced by cucumber mosaic virus.

mottled and distorted leaves on the seed stalks and reduction in size of the inflorescence.

TRANSMISSION AND SPREAD

Cucumber mosaic virus is readily transmissible by rubbing juice from diseased plants over the surface of leaves of healthy plants. The virus is transmissible through a low percentage of seeds of certain plants such as wild cucumber, but there is no evidence that it is transmitted through the seed of sugarbeet.

It is readily transmissible to high percentages of plants of susceptible species by several species of dodder (*Cuscuta* spp.), which are themselves hosts of the virus.

Field transmission occurs almost exclusively through the agency of aphids. The following species have been shown to be capable of transmitting cucumber mosaic virus: *Aphis fabae* (bean aphid), *A. gossypii* Glov. (cotton aphid), *Hysteroneura setariae* Thomas (rusty plum aphid), *Macrosiphum pisi* (Kelt.) (pea aphid), *M. solanifolii* (Ashm.) (potato aphid), *Myzus ascalonicus* Dorchester, *M. circumflexus* (Buck.) (crescent-marked lily aphid), *M. persicae* (green peach aphid), *M. solani* (Ktlb.) (foxglove aphid), and *Rhopalosiphum maidis* (Fitch) (corn leaf aphid) (66).

Of these species, only *Myzus persicae* and *Aphis fabae* breed extensively on sugarbeet, but it seems probable that any one of the vector species would be capable of transmitting cucumber mosaic to sugarbeet since infection comes from host plants usually outside the beet fields and only very short feeding periods are required for aphids to transmit the virus to sugarbeet. There is very little spread of virus from beet to beet after the virus enters the field. Owing to the low efficiency of vectors in transmitting cucumber mosaic virus to sugarbeet, enormous numbers of insects are required to produce high percentages of infection.

The destructive attacks on sugarbeets by cucumber mosaic virus in the San Joaquin Valley of California in 1940 and 1941 probably resulted from the development of high populations of aphids on large acreages of desert weeds surrounding beet fields and the movement of enormous numbers of winged form into beet fields as the desert vegetation dried. Damage to the sugarbeet seed crop in Arizona in 1956 resulted from a similiar concentration of aphids. Extremely high populations of the rusty plum aphid (*Hysterioneura setariae*) were produced on grain sorghum, and winged individuals migrated to adjacent beet fields, picked up virus from infected weed plants on their way, and transmitted it to sugarbeet plants when they reached the seed fields.

CAUSAL AGENT

Cucumber mosaic is caused by a virus that invades practically all the living cells of affected plants, causing distinct metabolic disturbances, especially in cells containing chlorophyll. The virus occurs in nature as a complex of strains that vary in virulence and in other characteristics. It is rather sensitive to a number of environmental conditions. It does not remain active in dried tissues. It retains activity for only 72 to 96 hours in plant juice at room temperature and is inactivated by a 10-minute exposure to temperatures of 60° to 70° C. Owing to its unstable nature, cucumber mosaic virus has been difficult to purify and photograph, but evidence indicates that particles are roughly spherical and perhaps range from 28 to 30 millimicrons in diameter.

CONTROL

Only in rare cases would it be expected that control measures for cucumber mosaic on sugarbeet would be of value. Since high percentages of infection occur only under unusual conditions, infection has occurred in most cases before the danger was recognized. If areas should develop where cucumber mosaic becomes a serious problem, the disease probably could be controlled by reduction of numbers of plants that serve as sources of virus and aphids.

YELLOW WILT

HISTORY AND DISTRIBUTION

Yellow wilt was first observed in 1926 at the Experiment Station of Rio Negro, near General Roca, Argentina, where it is reported to have destroyed all the experimental plantings of sugarbeets in the 1926–27 season. It was observed on a wider scale in the Rio Negro Valley of Argentina when an attempt was made to develop a beet sugar industry in that area, beginning in 1929. Yields of sugarbeets were low, both in this area and in the Rio Colorado Valley 90 miles to the north, apparently caused by wilting and death of plants.

In 1937–38 and 1938–39, yellow wilt reached a peak of severity in the plantings of the Rio Negro Valley and was responsible for the almost complete loss of the crop in both years. In 1939–40 and 1940–41, sugarbeets were grown in the Rio Negro Valley only on an experimental scale, and in 1941, sugarbeet culture in the area was discontinued.

With the collapse of the beet sugar industry in Argentina in 1941, nothing further was heard of yellow wilt until 1945 when it appeared in experimental plantings in Chile. It has been present in beet fields in that country each year since the establishment of a beet sugar industry (13).

ECONOMIC IMPORTANCE

In semiarid areas, yellow wilt probably is potentially the most destructive disease known to sugarbeet. The causal agent is so well distributed in Argentina that it is probable that any attempt to produce sugarbeets commercially in most of the irrigated areas of that country would be hazardous in the absence of control measures for yellow wilt.

The disease has limited sugarbeet production in Chile to the southern half of the country and has prevented use of the great middle agricultural area for production of this crop. Even in the present zones of production, the disease may cause severe losses in individual fields in some seasons. Losses from the disease are all the more severe because plants showing disease in midseason, or later in some cases, usually are a total loss, due to desiccation and rot of roots before harvest.

The capabilities of this disease for destruction are illustrated in Figure 9.7, which shows a field of sugarbeets near Conesa, Argentina, where more than 90 percent of the plants were destroyed during January 1941.

Perhaps at present the most important economic implication of the disease lies in the possibility of its introduction into other semiarid areas of the world where it might be even more destructive than it has been in South America. The introduction and widespread dissemination of this disease in North America, for example, would present a major challenge to the beet sugar industry of western United States.

Fig. 9.7. Field of sugarbeets near Conesa, Argentina, showing destruction caused by yellow wilt.

HOST RANGE

The full host range of yellow wilt has not been determined, but from the information available it is evident that the disease is capable of attacking a wide range of plants. It has been found on sugarbeet, garden beet, Swiss chard, and spinach. It has been found also on red-stem filaree (*Erodium cicutarium* L. L'Her.), white-stem filaree (*E. moschatum* L. L'Her.), and species of dock (*Rumex* sp.), and has been induced on tomato (*Lycopersicon esculentum* Mill.), Indian tobacco (*Nicotiana bigelovii* S. Wats.), New Zealand spinach (*Tetragonia expansa* Murr.), redroot amaranth (*Amaranthus retroflexus*), miner's lettuce (*Claytonia perfoliata* Donn.), chickweed (*Stellaria media* L. Cyr.), and several other weed plants. Species of the families Chenopodiaceae and Amaranthaceae that have been tested have shown a high degree of susceptibility to the disease. Several plants, such as the two species of filaree, are excellent hosts of both the causal agent and the leafhopper vector.

SYMPTOMS

Yellow wilt usually is observed in beet fields only after the plants have attained considerable size. There is a wide range of symptoms from yellowing and stunting through wilting and rapid collapse of the plant. Two more or less distinct sets of symptoms have been described as the "yellowing phase" and the "wilting phase."

Yellowing of infected plants, without marked wilting, may occur in areas where temperatures and transpiration rates are relatively low. First symptoms consist of dwarfing, yellowing, and downward turning of the tips of young leaves which may also show vein yellowing. Leaves half-grown at the time the young leaves begin to show symptoms may have yellow sectors, or they may show general yellowing, often accompanied by necrosis. As the plants develop, new leaves are dwarfed, yellow, and stunted, and the leaf blades tend to be straplike. There is a marked tendency for growth and development of axillary buds and the plants may become markedly rosetted. Root growth is retarded and tips of many rootlets die. Successive regeneration and death of rootlets result in production of tufts of rootlets with restricted range of soil penetration.

Under conditions of high temperatures and low humidities, infected plants may wilt and die within a few days without producing specific top symptoms other than wilting. When the tops begin to wilt, however, the tips of the main roots have already become flaccid and shrunken. Shrinking and softening of the root may progress upward until the entire root is involved. Roots often shrink to such a degree that they may be readily removed from the soil. The root may persist as a "mummy" or it may collapse into a rotted mass.

Where diseased roots have been held in place through the winter for thermal induction, some have produced numerous weak seed stalks, but few have produced seed.

TRANSMISSION AND SPREAD

Yellow wilt can be transmitted readily by grafts and by dodder (*Cuscuta californica* and *C. campestris* Yuncker). Under natural conditions, it is transmitted by a leafhopper, *Paratanus exitiosus* Beamer, which has been found only in Argentina and Chile.

The vector apparently persists on desert host plants and on weeds in waste areas, from which it moves into beet fields in the spring. In some areas, especially around Santiago, Chile, it remains on certain weeds in cultivated areas and breeds throughout the year, although its rate of reproduction is markedly retarded during the winter months. It increases rapidly during the summer on sugarbeet.

After initial centers of infection are established by incoming leafhoppers in beet fields in the spring, secondary spread takes place, often through the agency of nymphs that acquire virus from infected plants and move to adjacent healthy plants. The nymphs are sluggish and their movements are slow and over short distances. For this reason, diseased plants often occur in groups. It is not unusual to find a high percentage of infection in one row of beets and no infection in an adjacent row. This type of spread and the long incubation periods of the causal agent in the vector and in the beet plant account for the delay in the appearance of diseased plants in large numbers in beet fields until midseason or later.

CAUSAL AGENT

The causal agent of the yellow wilt disease has been considered to be a virus, but treatment of diseased plants with an antibiotic (chlortetracycline) by Roberto Ehrenfeld K. in Chile has given temporary recovery, suggesting that the causal agent may be a mycoplasmalike agent. Little is known of its physical properties. It is transmitted by the leafhopper, *Paratanus exitiosus,* in which it has an incubation period of more than 16 days. The causal agent apparently increases in its vector, but there is no evidence that it is transmitted through the egg stage. Increase of the causal agent in the beet plant is relatively slow, and usually symptoms do not appear until 30 days or more after infection.

CONTROL

The control of yellow wilt in Chile has been obtained by abandoning those areas where severe damage is expected and moving the industry south where conditions are unfavorable for reproduction of the vector, *P. exitiosus.*

Numerous field tests have been made to control the disease by means of various insecticides applied to the tops and to the soil around the beet plants. Results thus far have been inconclusive. Some of the tests have reduced the incidence of infection appreciably, but this has not been reflected in increased yields that would be expected. It would appear, however, that a high degree of control should be possible with an effective insecticide, since there is evidence that usually, in areas where the disease now occurs, there are not enough leafhoppers in beet fields to produce 100 percent infection. Under such conditions, it would seem that destruction of a high percentage of leafhoppers by an insecticide should result in a corresponding increase in yield.

Field tests of more than 300 American and European varieties and selections of sugarbeet in Argentina and Chile have revealed no varieties or selections that have appreciable resistance to this disease. Also, the wild species of *Beta* that have been tested have shown no greater resistance than sugarbeet.

Selections for resistance have been made in badly diseased fields, but thus far no diseased plants selected from fields and transplanted for seed production have produced seeds. Some diseased plants, overwintered in place in the field for seed production, have bolted and produced numerous weak seed stalks, but very few have produced viable seeds. Of 500 plants overwintered in Chile in the 1967–68 season, only 53 survived the winter. Of these, only 6 produced viable seeds, the total yield of the 6 plants being 66 grams. It would appear from results obtained thus far that a high degree of resistance to yellow wilt may be difficult to obtain.

YELLOW NET

Yellow net was observed in beet fields of California as early as 1941 and attracted attention in the delta area near Rio Vista, California, in 1945 (70). It has since appeared sporadically throughout California and it probably occurs in other western states. It has been reported also from England.

In the delta area of California, infections up to 25 percent have been noted on a few occasions. However, infected plants have shown little reduction in yield, and the disease, under present conditions, is of minor economic importance.

SYMPTOMS

The yellow net disease is characterized by the production of conspicuous chlorosis or yellowing of veins and veinlets of leaves of affected plants. The yellowing and chlorosis may begin to appear in young leaves about 10 days after infection and continue to be produced for a time in successive leaves as the plant grows. Yellowing may be very extensive, largely suppressing chlorophyll induction in the leaf, or it may be more closely restricted to veins and veinlets.

Usually in the spring of the year, infected plants are conspicuous in beet fields because of their marked light yellowish color. In some cases, diseased plants occur in clusters, forming yellowish areas that can be recognized from considerable distances. Later in the season, diseased plants usually discontinue the production of leaves with yellow veins and new leaves appear normal. Diseased plants are not usually easily recognizable after midseason. Under greenhouse conditions, yellow and green cycles of growth may alternate at irregular intervals. The factors involved in this type of recovery and relapse are not known.

TRANSMISSION AND SPREAD

Yellow net virus is transmitted principally by the green peach aphid, *Myzus persicae*. Other species of aphids have been shown to be vectors under experimental conditions but none is believed to be of significance in the transmission of the virus under natural conditions.

The green peach aphid can acquire yellow net virus by feeding on diseased plants for 5 minutes and can transmit it to healthy plants in a feeding time of 15 minutes. It increases in effectiveness, both in acquisition and transmission, as the feeding periods are increased. Once acquired, the virus is retained by the vector for long periods, perhaps for life (70).

Sources for spring infection of beet fields are not clearly defined. Undoubtedly, overwintered infected beets are important in spring spread, but the patterns of field spread sometimes indicate that weed hosts may be involved as virus sources. However, weed hosts that may serve as virus sources have not been clearly identified.

CONTROL

Under present conditions, losses from yellow net are so small that no control measures are required.

YELLOW VEIN

A disease of sugarbeet that causes conspicuous yellowing of veins and dwarfing of plants was observed as early as 1913 in New Mexico (55) and has since been found in California, Arizona, Utah, Colorado, Kansas, Nebraska, and Oklahoma (4). A similar or identical disease has been found in Turkey.

Individual plants are severely affected and root weights may be reduced as much as 50 percent. In most areas, however, where the disease occurs, the incidence of infection has been so low that no appreciable damage has resulted. However, in 1964 some fields in Kansas showed percentages of infection up to 31 percent (38). This level of infection could cause appreciable reduction in root yield.

SYMPTOMS

First symptoms of yellow vein appear on young leaves of infected plants as dwarfing and vein yellowing. As the disease develops, the main veins of affected plants are distinctly yellow and the yellowing often extends into the adjacent tissue a distance of a millimeter or more, producing relatively wide yellow bands. The yellow areas may be more or less uniformly continuous along the veins or they may have irregular borders. Frequently, on the smaller affected veins, there may be yellow spots isolated from other yellow areas. Often, in the early stages of disease development, dwarfing is more conspicuous on one side of the plant; later the entire top is dwarfed.

TRANSMISSION AND SPREAD

The disease is easily transmitted by graftage but has not been transmitted by juice inoculation. The common insects that feed on sugarbeet appear to be unable to transmit the causal virus. However, the disease is probably insect transmitted. The rather wide range of geographical distribution of the disease indicates an extensive distribution of agents of transmission. Low incidence of infection, however, indicates that such agents must be either scarce or inefficient in spread in almost all areas where the disease is known to occur.

CONTROL

No methods of control of yellow vein have been employed and it does not seem likely that such measures will be required, except possibly in very rare cases where exceptionally favorable conditions for spread may be encountered.

SAVOY

Sugarbeet savoy has been known in the United States since 1890, when it was reported as a disease "not rare" in Indiana. It was reported from Michigan in 1923 and has since been found in practically all of the sugarbeet-growing states east of the continental divide and in Ontario, Canada (25).

In commercial sugarbeet fields, infection has ranged from a trace to 1 or 2 percent. Occasional fields have shown an incidence as high as 10 to 15 percent in limited areas. The incidence of disease usually is higher in portions of fields adjacent to weed areas, especially areas that contain plants of various species of *Amaranthus* that are hosts of the vector of the causal agent (26).

Because of the low incidence of infection, savoy has been a serious disease of sugarbeet only in very limited areas. However, the effect of the disease on individual plants is severe. Tests have shown that roots of diseased plants weigh about 25 percent less than roots of comparable healthy plants. Sucrose content may be reduced more than 4 percentage points (38). Sucrose reductions of this magnitude are unusual with virus diseases. These reductions in sucrose and in root weight indicate that it is fortunate for the beet sugar industry that this disease does not have a greater incidence of occurrence.

Sugarbeet and garden beet are the two plants chiefly affected by savoy. Its potential host range has not been extensively tested, but it probably could be found on other close relatives of the sugarbeet. The causal agent may occur also in weed plants, particularly in certain species of *Amaranthus,* since it is carried by lacebugs that move from such plants into beet fields.

SYMPTOMS

Primary symptoms on sugarbeet plants are veinlet clearing, followed by vein thickening and growth retardation which gives the lower leaf surface a netted appearance. Leaves are dwarfed and curled downward at the edges. Roots of affected plants show phloem necrosis and discoloration similar to that produced by curly top. Both tops and roots are markedly stunted.

TRANSMISSION AND SPREAD

The causal agent of sugarbeet savoy is transmitted under natural conditions by the lacebug, *Piesma cinerea* Say, which is the only known vector. No transmission has been obtained with common species of aphids or with the beet leafhopper, *Circulifer tenellus*. The virus has not been transmitted by juice inoculation and apparently is not transmitted through the seed.

The vector, *P. cinerea*, overwinters in grassy, shrubby, or woody areas. It breeds on weed hosts, such as various species of *Amaranthus*, and moves into beet fields in the spring, carrying virus from its overwintering areas. The sources of virus for field infection are not definitely known, but it appears that the virus has a widespread distribution in eastern United States and southeastern Canada, since nearly all collections of the lacebug, *P. cinerea*, from weed hosts of these areas have had infective individuals. Most evidence indicated that flights of the vector from breeding and overwintering areas are short, since incidence of infection is likely to be greater at the edge of the infected fields and less as distance from breeding areas increases (25).

CONTROL

Where control measures appear to be desirable, the disease probably could be kept in check by sanitary measures directed against the vector or its host plants or by locating beet fields at a distance from the sources of infection. Also, there appears to be a range of susceptibility among beet selections that have been tested, and it seems probable that varieties resistant to this disease could be developed if the disease should become sufficiently destructive to warrant such measures for control.

KRAUSELKRANKHEIT

A disease of sugarbeet, called "kräuselkrankheit" in Germany, occurs in northern Europe. Other names of the disease are "rübenkräusel" and "beet crinkle." It is highly destructive locally in Germany and Poland. Yields of infected plants may be reduced as much as 50 percent, and occasionally plantings are destroyed by the disease.

Symptoms consist of curling of young leaves, beginning about 21 days after infection, and dwarfing of both tops and roots of affected plants. The leaf curling and crinkling, and other symptoms, are very similar to those produced by savoy in eastern United States.

The causal agent is spread by a lacebug, *Piesma quadrata* Fieb., which is closely related to *P. cinerea,* the vector of the causal agent of savoy in the United States. *Piesma quadrata* hibernates in the edges of groves, along ditch banks, and in other protected areas. In the spring, it moves into beet fields and transmits virus to young plants. Movements are by crawling and short flights, so plantings at a distance from overwintering areas of the vector tend to escape infection.

Virus-free lacebugs can be made viruliferous by injecting them with liquids which contain virus. The causal agent multiplies in the vector and is present in the salivary glands, walls of the intestines, and in the haemolymph. It occurs in relatively high concentrations in both the vector and the plant (61).

Various methods of control, consisting chiefly of destroying or avoiding overwintered vectors, have been employed. In some parts of Europe a trap crop, consisting of a few rows of beets, is planted as early as possible around prospective beet fields. Lacebugs that collect on these beets are destroyed by plowing under the beets or by use of chemicals before the main part of the field is planted.

OTHER VIRUS OR VIRUSLIKE DISEASES THAT MAY OCCUR

In addition to the diseases already described, sugarbeet is susceptible to infection with a relatively large number of viruses that primarily attack other crop plants. Also, viruslike disturbances, with various causal relationships, are not uncommon on sugarbeet. Some of the better known of these virus and viruslike disturbances are described briefly below.

ROSETTE

A disorder of sugarbeet, characterized by dwarfing and rosetting of tops of affected plants, has been observed annually over a period of years in sugarbeet fields in California. The disease produces severe damage on affected plants, but it is of little economic importance, due to its low incidence of occurrence. Evidence indicates that the disease is caused by a virus that is readily transmissible by graft, but not by juice inoculation. The causal agent has been transmitted to a low percentage of inoculated plants by dodder, *Cuscuta campestris.* Apparently, it is not transmissible by any of the insects that commonly feed on sugarbeet. The disease has been recognized only on sugarbeet (12).

TOBACCO RATTLE

A disease of sugarbeet, causing stunting of affected plants, has been attributed to infection with the tobacco rattle virus. This virus has been consistently isolated from stunted beet plants showing bright yellow "laurel-leaf" symptoms. Roots of affected plants were smaller than normal and much branched. Tobacco rattle virus infects tobacco, potato, and a number of weed species. It is transmitted to sugarbeet by the nematode, *Trichodorus pachydermus* Sinhorst, in the Netherlands and England and to tobacco by another nematode, *T. christiei* Allen, in the United States (39). The economic importance and geographical distribution of this disease on sugarbeet have not been clearly defined. It apparently has not been found on sugarbeet in the United States.

TOMATO BLACK RING

A disease which causes marked stunting of sugarbeets has been described in England where it has been prevalent in the East Anglia area. Symptoms are described as being more evident on some plants than on others and consisting of a chlorotic blotchy mottle on one or more leaves. Often primary roots are missing and affected plants have a "sprangly" branched root system.

The disease has been associated with the tomato black ring virus which is transmitted by at least two species of nematodes, *Longidorus elongatus* (de Man.) and *L. attenuatus* Hooper. This virus attacks several other crop plants as well as a wide range of weeds. Its full economic importance and distribution on sugarbeet have not been determined. It has not been recognized on sugarbeet in the United States.

The causal virus has been transmitted through seeds to 3 to 27 percent of the progenies of some infected beet plants and has been pollen transmitted to seeds of healthy plants in numbers up to 14 percent. Plants infected through seeds may show no leaf symptoms and grow normally under greenhouse conditions (39).

MARBLE LEAF

A juice-transmissible virus which causes vein yellowing and mottling of immature leaves and distinct yellowing of mature leaves of sugarbeet plants was found on sugarbeet in Oregon in 1959 (9). The virus induced local lesions in inoculated leaves of *Beta vulgaris* (L.), *B. macrocarpa* Guss., *Chenopodium amaranticolor* Coste & Reyn., and *C. murale*. Systemic infection was obtained in 6 species of *Beta*, 2 species of *Chenopodium* and 2 species of *Atriplex*. The virus appears to have a limited host range. *Myzus persicae, Aphis fabae,* and *Macrosiphum euphorbiae* Thomas are inefficient vectors of the virus. No evidence of seed transmission was found. The virus has a thermal inactivation point between 60° and 65° C, a dilution endpoint of about 1–1,000, and it remained active in extracted beet juice 24 hours or less. Under greenhouse conditions, the disease caused about 10 percent reduction in plant growth. Yield and size of seed of *Beta macrocarpa* plants were greatly reduced. The disease apparently has a very limited distribution and is not known to cause measurable loss to the sugarbeet crop.

FAMILY 41 YELLOWS

In 1948 a seed-transmitted yellowing condition was reported in Ireland in a strain of sugarbeet known as "Family 41" (23). Transmission occurred through up to 30 percent of seeds of affected plants. The malady has not been reported as occurring naturally in any country other than Ireland.

Segments of leaves of affected plants may show yellowing or more often the yellowing may affect the entire leaf, sometimes with obvious chlorosis of veins. Leaves of an affected plant are shown in Figure 9.8. Plants showing such symptoms are somewhat dwarfed under greenhouse conditions.

Because of the close resemblance of symptoms of this seed-transmitted malady to certain yellowing diseases, it was at first strongly suspected that "family 41 yellows" was caused by a virus. This concept was strengthened by later reports indicating transmission by the green peach aphid *(Myzus persicae)*. More recent tests, however, have failed to show that the green peach aphid is able to transmit a causal agent from affected to nonaffected sugarbeet plants or to any other plant of several species known to be susceptible to infection by viruses of beet yellows and beet western yellows.

FIG. 9.8. Sugarbeet leaves from greenhouse plants showing effects of a seed-transmitted disorder described in Ireland as "family 41 yellows." This type of yellowing, which closely resembles certain symptoms of beet yellows, appears not to be caused by a virus.

No causal agent has been transmitted by juice inoculation and all attempts to transmit a causal agent by graft have failed. In tests with grafts, affected plants have grown on nonaffected plants for more than two years, with development of no yellowing symptoms on the nonaffected plants.

In Ireland this type of yellowing apparently disappeared with the discontinuance of production of family 41 stock. It seems probable, therefore, that this abnormal condition was not produced by a seed-transmitted virus but resulted from genetic factors peculiar to a special breeding line of sugarbeets.

YELLOW SPLOTCH

In several seasons over the past 10 years, sugarbeet plants in various parts of California have shown a type of yellow splotching early in the season, which has alarmed some of the growers whose fields were affected. The same type of yellowing has been seen also in Washington and Oregon (6).

In affected fields, yellowing may begin to appear on the first pair of true leaves when they are less than half-grown and it may continue to appear on the next two to six leaves. First evidence of abnormality is a light yellow blotching that becomes more conspicuous as the spots take on a light golden hue. The yellow areas are of various sizes and shapes and the margins are not sharply defined. Yellow splotches may appear on any part of the leaf blade and may range from a single spot to many spots that coalesce and produce a yellow leaf. Affected leaves do not recover, but symptoms become less evident on successive leaves as the plants grow. Although in some instances fields have shown considerable yellowing just after thinning, there is no evidence of permanent injury.

The cause of this type of yellowing is unknown. It has been associated with low temperatures in the early stages of plant development and appears to be a nutritional disorder.

FUNGUS AND BACTERIAL DISEASES

SEEDLING DISEASES

Diseases affecting sugarbeet seedlings are often called damping-off, black root, or black leg. These terms designate infection by an individual organism in some cases, but in others describe the effects of two or more pathogens attacking simultaneously or successively.

PATHOGENS INVOLVED

During germination, emergence, or juvenile growth, sugarbeet seedlings are susceptible to infection by soil-borne fungi such as *Pythium ultimum* Trow., *P. debaryanum* Hesse, *P. aphanidermatum* (Edson) Fitzp., *Rhizoctonia solani* Kuhn [perfect stage, *Thanatephorus cucumeris* (Frank) Donk] and *Aphanomyces cochlioides* Drechs., and if infected or contaminated seed is used, by seed-borne *Phoma beta* Frank (perfect stage, *Pleospora bjorlingii* Byford).

Pythium ultimum and *P. debaryanum* are present to some extent in nearly all arable soils and attack unprotected beet seedlings at all temperatures that favor germination of beet seed. Both are favored by high soil moisture and attack many other crop plants. They produce, primarily, seed decay and preemergence damping-off, although postemergence damping-off may follow under moist conditions. *Pythium aphanidermatum,* a high-temperature fungus, attacks seedlings only in warm soils with abundant soil moisture. *Rhizoctonia solani* in soils is found most abundantly following production of a susceptible crop such as legumes (beans, alfalfa, or clover) or cotton. Most strains of this species do not infect beet seedlings at low temperatures (below 12° C) but do not require high soil moisture levels for infection. It causes both preemergence and postemergence damping-off. The same fungus, later in the season, may cause crown rot or dry rot canker on fleshy roots.

The beet water mold fungus, *Aphanomyces cochlioides,* causes an important seedling disease in suitable soils in northern Europe and in the United States from Michigan and Ohio to the Pacific Coast. It is the primary cause of the very serious "caida" disease of sugarbeets in Chile. It seldom causes preemergence seedling death but invades the cortex of young seedlings, causing a toppling over, and in extreme cases, death.

Aphanomyces infection of beet seedlings is severe only in warm, wet soil and occurs primarily in soils of high water-holding capacity and of neutral to slightly acid reaction. Surviving seedlings are often chronically stunted, because of continuing invasion of feeder roots. Tip rot caused by the same fungus may occur later in the season if soils are extremely wet or poorly drained.

The only important seed-borne disease of sugarbeet seedlings is caused by a fungus, *Phoma betae* (31). Although this fungus does not persist in the soil from one year to the next, it may develop on a high percentage of beet seed produced in regions with summer rainfall. Infected or contaminated seed, germinating in cool damp soil, may produce infected seedlings that fail to emerge or die following emergence. Infected seedlings that survive are stunted and retarded in growth until warm weather permits recovery.

Species of *Fusarium* are frequently isolated from infected beet seedlings, especially those previously invaded by *Aphanomyces*. Most such isolates are nonpathogenic or only weakly parasitic, and in the writer's opinion can usually be considered secondary invaders.

IDENTIFICATION OF THE PATHOGENS

Each of the specific organisms mentioned above produces characteristic symptoms on sugarbeet seedlings that can often be recognized by experienced observers. These symptoms are not only very similar but also quite variable, however. Hence, the only sure identification involves isolating the organism in pure culture for microscopic observation or examining (under low magnification) seedlings incubated in water culture. The most convenient procedure is to wash fresh infected seedlings in running water to remove dirt particles and then place them in a shallow film of water in a Petri dish at about 20° C. After 24 or 48 hours, the characteristic vegetative mycelium, and spores or fruiting bodies of the causal fungus, can be recognized under low magnification ($\times$ 100) even though more than one pathogen may be involved. The diagnostic features of six seedling pathogens are illustrated in Figure 9.9.

Pythium ultimum (Fig. 9.9A) and *P. debaryanum* (Fig. 9.9B) grow from the seedling tissue, primarily below the surface of the water, as branched, nonseptate mycelium. After 12 to 24 hours, spherical spores (sporangia) are produced, either terminally on the hyphae or intercalary. Sporangia of *P. ultimum* germinate directly by a germ tube that grows into mycelium or an infection tube, but *P. debaryanum* may produce zoospores in a vesicle. Both species produce sexual spores (oospores). In water culture, these two species of *Pythium* are practically indistinguishable. Both produce similar symptoms on beet seedlings and respond to the same control measures. Species identification, if desired, can be made on the basis of differences in sporangial germination and in the number and arrangement of antheridia around the oogonium (56).

Pythium aphanidermatum (Fig. 9.9C) in water culture also produces nonseptate mycelium, and later, comparatively large lobate sporangia, from which are formed vesicles containing an abundance of swarm spores (zoospores). The liberated zoospores swim for a period, then encyst, and germinate by the production of germ tubes. The characteristic lobate sporangia clearly distinguish this species from *P. ultimum* or *P. debaryanum*.

Aphanomyces cochlioides (Fig. 9.9D) produces abundant nonseptate mycelium within the cortex of young seedlings, whereas in water culture exterior growth consists only of unbranched cylindrical tubes which serve as sporangia. The separated plasmal units, arranged in a single row, emerge from the tip of the sporangium, followed by large numbers originating from mycelium within the seedling. All plasma units collect as spherical encysted spores at the tip of the emptied sporangium and later emerge as zoospores and swim in the surrounding water. They are chemically attracted to nearby beet seedlings, and upon germination serve as the in-

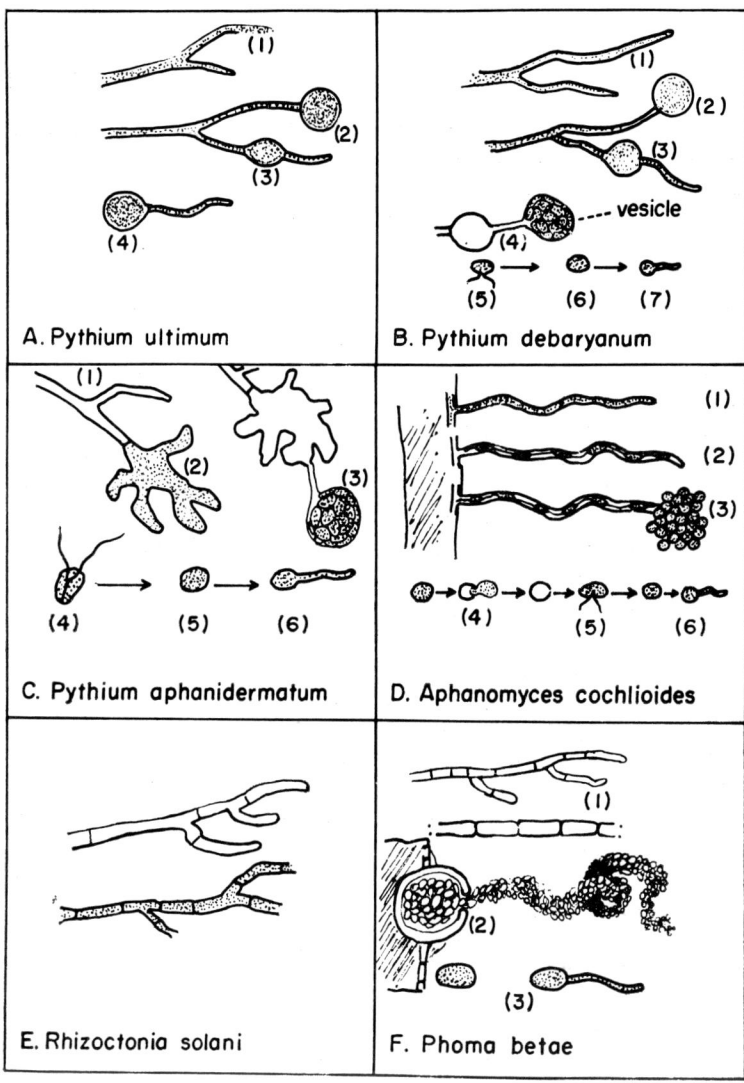

FIG. 9.9. Characteristics of 6 fungi commonly associated with sugar-beet seedling diseases. *A. Pythium ultimum*—*(1)* mycelium nonseptate, *(2)* sporangia terminal, or *(3)* occasionally intercalary, *(4)* direct germination of sporangium to form germ tube. *B. P. debaryanum*—*(1)* mycelium nonseptate, *(2)* sporangia terminal, or *(3)* intercalary, *(4)* germination of sporangium direct or indirect to form zoospores in a vesicle, *(5)* motile zoospore, *(6)* encysted spore, *(7)* germ tube. *C. P. aphanidermatum*—*(1)* mycelium nonseptate, *(2)* sporangium lobate, *(3)* forming zoospores in vesicle, *(4)* zoospore, *(5)* encysted spore, *(6)* germ tube. *D. Aphanomyces cochlioides*—*(1)* sporangium, *(2)* protoplasts divided, *(3)* evacuation tube with encysted spores, *(4)* emergence of zoospore from encysted spore case, *(5)* zoospore, *(6)* germ tube. *E. Rhizoctonia solani*—mycelium-branched, septate, sterile, no spores produced in water culture. *F. Phoma betae*—*(1)* mycelium, septate, constricted at septum of older hyphae, *(2)* pycnidium exuding pycnidiospores (conidia), in a cirrus (spore tendril), *(3)* conidia germinate to produce germ tube.

oculum source for secondary infections. After several days, oospores are formed within the cortex of infected seedlings, serving to perpetuate the fungus from one season to the next.

Mycelium of *Rhizoctonia solani* (Fig. 9.9E), in contrast to the *Pythium* species, grows largely on the surface of water and spreads widely, forming a floating colony. This characteristic often permits identification without magnification. No spores are formed, but the mycelium is clearly septate, with cross walls characteristically located just above the hyphal branches.

Phoma betae (Fig. 9.9F) is recognized most easily by culturing infected seedlings on agar, where the characteristic fruiting bodies (pycnidia) are produced after a few days. In water culture, the dark septate mycelium can be readily observed, and on many seedlings the pycnidia exuding conidia can also be found. A sexual stage produced on overwintering beet seed stalks was found first in Sweden and later in several other countries. This stage, consisting of large dark perithecia containing asci and multicellular ascospores, is named *Pleospora bjorlingii*.

CONTROL MEASURES

Rotation. Rotation with insusceptible crops or at least avoiding the culture of sugarbeets after the most susceptible crops will reduce the probability of severe infection. The inoculum potential of *Rhizoctonia* is strongly influenced by the previous crop. In irrigated areas the production of a dry-farmed crop preceding sugarbeets will minimize the activity of *Aphanomyces*. Research workers in Michigan found that the incorporation of residue from a corn crop tended to reduce the incidence of some seedling diseases (black root), whereas residues from legume crops increased severity unless the legume sod was plowed down the previous summer or early fall (24).

Temperature Limitations. *Pythium aphanidermatum, Aphanomyces cochlioides,* and *Rhizoctonia solani* do not grow well at low temperatures, and since sugarbeets do germinate well at temperatures below 15° C, it is possible in some areas to avoid severe infection by these three organisms by winter or early spring plantings (46). *Phoma betae,* in contrast, produces its most severe disease under low temperatures and therefore can be partially avoided with higher-temperature plantings. *Pythium ultimum* and *P. debaryanum* are active throughout the favorable temperature range for beet germination as long as the soil is sufficiently moist. Thus, modifications of seeding dates to avoid seedling infection must be worked out for each growing area, keeping in mind the pathogens involved and the local environmental conditions.

Chemical Control. Seed treatment is the most convenient method of controlling damping-off caused by *Pythium* sp. or *Rhizoctonia*. The usual practice in the United States is to apply a fungicide effective against *Pythium*, such as Dexon [p-(dimethylamino) benzenediazo sodium sulfonate], in combination with one effective against *Rhizoctonia*, such as PCNB (pentachloronitrobenzene). Infection by *Aphanomyces* is only partially controlled by seed treatment, but in Chile, where this seedling disease was a limiting factor in sugarbeet production, satisfactory control has been secured by adding Dexon to finely powdered phosphate fertilizer applied in the seed furrow. Several experiments have shown that fertilizer applications may have a profound effect on the severity of seedling disease and on the ability of infected seedlings to recover. This is particularly true with *Aphanomyces* infection, which is, at first, limited to the cortex. In Ohio it was shown that heavy applications of superphosphate strikingly increased the surviving stand. In Montana, where *Aphanomyces* appears to be the principal pathogen in heavy soils, complete fertilization with nitrogen, phosphorus, and manure reduced seedling disease significantly, whereas nitrogen alone gave little or no disease control (1). With seed lots moderately contaminated or infected by *Phoma*, dust or liquid seed treatment with organic mercury compounds or thiram (tetramethylthiuramidisulfide) provides partial protection, but the most effective control with heavily infected seed lots has been from steeping the seed for 20 minutes in a solution of ethyl mercury phosphate, as practiced in Ireland, and later, in England. Organic mercury treatments have the further advantage that they control seed-borne *Cercospora* on seed lots carrying that pathogen. Most seed treatments are applied as liquid or wettable fungicides in suspension through continuous spray treaters, mistomatic seed treaters, slurry, or liquid treaters.

Use of Disease-Free Seed. Most of the sugarbeet seed produced in arid regions of the western United States is free or nearly free of seed-borne *Phoma*, making unnecessary any special treatments for this pathogen. It is significant that most of the northern European countries are now producing their seed in southern Europe, especially Italy, Austria, or in Turkey, partly to avoid *Phoma* infection of seed.

Disease Resistance. Plant breeders have selected lines that show considerable resistance to *Aphanomyces* and some other seedling pathogens. These favorable characteristics have been incorporated in several commercial varieties in recent years.

The possibility of serious losses from seedling diseases in most sugarbeet areas is minimized by the use of disease-free seed with a high level of tolerance to some of the pathogens and protected by appropriate fungicides.

ROOT DISEASES

Some of the same fungus pathogens that cause seedling diseases also cause infection of older plants.

Aphanomyces TIP ROT

A tip rot of sugarbeets, caused by *Aphanomyces cochlioides*, occurs frequently in areas of heavy summer rainfall or in soils that are poorly drained or are waterlogged (20). Such losses are reduced by improved drainage, or in irrigated areas, bed planting or sprinkler irrigation.

Rhizoctonia CROWN ROT AND DRY ROT CANKER

The same strains of *Rhizoctonia solani* that cause damping-off of beet seedlings may, later in the season, cause crown rot or dry rot canker on maturing roots. The early symptoms of crown rot are the blackening of petioles of the outer leaves on half-grown plants, followed by extensive rotting of the entire crown and adjacent root tissue. When the rotted area is lower on the fleshy root, as sometimes happens, it is referred to as "side rot." The rotted tissue is brown, and deep horizontal cracks often develop, with a mass of brown feltlike mycelium, characteristic of *Rhizoctonia*, overgrowing the exposed surface. Crown rot is most frequent on beets growing in poorly drained heavy soil.

Dry rot canker, caused by a strain of the same fungus, *Rhizoctonia solani* (49, 62), is characterized by dry, sunken root lesions covering deep pockets of brown, spongy material sharply delimited from healthy beet tissue, as shown in Fig. 9.10. The surface tissues of cankers often show a series of concentric rings. This type of rot is most frequent in irrigated areas of western United States, including Utah and California. It is commonest in well-drained light soils and does not require high soil moisture. Lesions may occur on the taproot and even encircle and sever the root of young plants. Then the aboveground parts suddenly wilt. If, during cultivation or furrowing out for irrigation, infested soil is placed against or over leaf petioles, the *Rhizoctonia* fungus may attack the petioles and grow into the crown area, sometimes destroying the entire crown. Bed planting avoids the necessity for cultivating soil against susceptible tissue. *Rhizoctonia* infection is usually severe on sugarbeets following susceptible crops such as legumes or cotton, and is much less likely to occur following cereal, corn, or sorghum crops. Most reports indicate that the activity of *Rhizoctonia* mycelium in soil is favored by nitrogen applications and suppressed by incorporating crop residues which provide a high C:N ratio. In several field experiments, however, dry rot canker on sugarbeets was significantly reduced by heavy doses of nitrogenous fertilizers. This result could reflect increased resistance of the host rather than suppression of the pathogen.

Phoma ROOT ROT

The seed-borne fungus *Phoma betae* may also cause a crown or shoulder rot on maturing sugarbeets, though little damage to the crop results

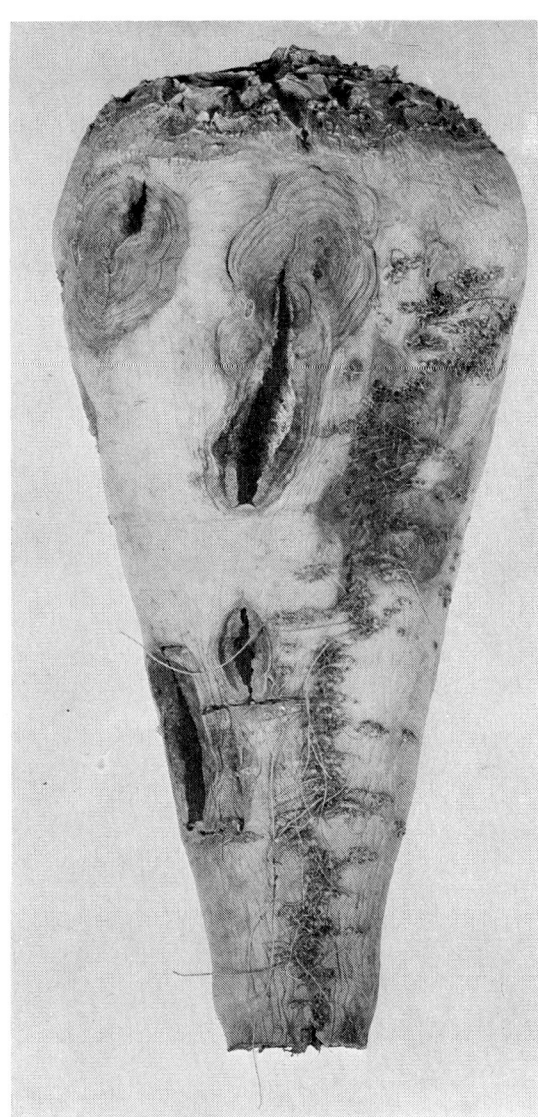

FIG. 9.10. Dry rot canker on sugarbeet caused by *Rhizoctonia solani*. Strains of the same fungus may cause seedling infection, crown rot, or leaf blight under some conditions.

from the shallow infected areas, usually dark brown or green. Necrotic areas are often invaded by a secondary organism, *Fusarium* sp., which contributes a pink or reddish-to-gold color to the surface. The important heart rot and dry rot of sugarbeets were at one time believed to be caused by *Phoma*, but research by Brandenburg (19), a German plant pathologist, clearly showed that the primary problem was not fungus infection but a deficiency of boron, correctable by adding boron in the fertilizer mixture. This subject is covered in greater detail in Chapter 7.

SOUTHERN SCLEROTIUM ROOT ROT

The most destructive root rot disease affecting sugarbeets in semi-tropical areas is caused by the fungus *Sclerotium rolfsii* Sacc. This disease is almost unknown in the northern parts of the United States and Europe but causes serious losses in the southwestern United States from Texas to California, in southern Europe, especially Spain and Italy, and in Israel, Korea, Japan, and Uruguay. Infected roots are covered by white strandlike mycelium upon which are produced numerous resting bodies or sclerotia. These are spherical, at first white and later tan or dark brown, about the size and appearance of mustard seeds (Fig. 9.11). From 5,000 to 10,000 sclerotia are produced on or near an infected root during its decay. They survive in soil from one year to the next, serving as an inoculum source on subsequent crops. The causal fungus grows actively only in moist soils at moderate to high temperatures. Winter-grown crops, therefore, escape infection.

FIG. 9.11. Sugarbeet infected by *Sclerotium rolfsii*. The characteristic signs are the white strandlike mycelium and the spherical white to dark brown sclerotia, about the size of mustard seed.

CONTROL MEASURES

A method of estimating the viable population of sclerotia in field soil and predicting losses in future sugarbeet crops was developed in California and is in use by some sugar companies (47). Soil samples are washed through a series of screens to remove sclerotia, which are then tested for germinability. It was found that the population of viable sclerotia following a diseased beet crop declined rapidly in succeeding years. Losses were usually severe with populations of over 100 viable sclerotia per square foot, and only low to moderate with lower numbers.

In most areas, rotation for two to four years with winter-grown, unirrigated or insusceptible crops reduced the fungus population to a level safe for another season of sugarbeet production.

To minimize losses from southern sclerotium rot, sugarbeets should receive sufficient nitrogen for vigorous root and top growth until late in the season, when temperatures in most areas are low enough to limit fungus growth (48).

A high water table or overirrigation without adequate drainage increases the incidence of infection, often in association with wet root rot. The fungus, as sclerotia or mycelium, may be spread from field to field in contaminated soil carried on farm equipment, in drainage water, or on infected transplants. Screenings from beet-loading stations should not be returned to cultivated fields.

Between 1930 and 1940, annual losses from sclerotium root rot in the Sacramento Valley of California were as high as 5,000 to 10,000 tons of sugarbeets, and the disease was considered a serious threat to the future of the sugarbeet industry of that area. Modification of cultural practices has drastically reduced losses, however, so that the severe form now occurs in only a few fields each year.

VIOLET ROOT ROT

Violet root rot is very descriptive of a disease which attacks sugarbeets as well as potatoes, alfalfa, and many other crops and weeds. It has been reported from all western European countries and has been carefully described from England by Hull (43) and from Ireland by McKay (54). In the United States it is found most frequently in the semiarid regions of the West, including Nebraska, Oregon, and Washington (34). The causal fungus, *Helicobasidium purpureum* Pat. (imperfect form, *Rhizoctonia crocorum* Pers. DC. ex Fr.), lives in the soil and usually attacks the taproot of the sugarbeet, spreading upward over the fleshy root. It produces a mycelial felt over the surface of the root, which is at first violet or red and later dark purple, quite unlike any other root disease affecting sugarbeets.

The disease in sugarbeet plantings is most likely following alfalfa or clover, but even so may be limited to localized areas. Growers should not replant infested areas to sugarbeets or other susceptible crops. Special control practices are usually not required in the United States, where violet root rot is considered a very minor disease.

Phymatotrichum ROOT ROT

Cotton root rot, caused by *Phymatotrichum omnivorum* (Shear) Dug., is a serious disease of cotton, alfalfa, and many other crop plants in southwestern United States from Texas to southern California. Sugarbeets are also attacked occasionally in Arizona, Texas, and New Mexico. Losses have not been severe, because in most infested areas sugarbeets are grown as a winter crop and harvested before the causal fungus has reached maximum activity. At higher elevations in Arizona, some loss occurs before harvest is completed. Infected roots are rotted completely and the fungus forms a thin feltlike covering of yellowish mycelium on the surface of the root. Strands of mycelium and occasionally brown sclerotia are formed in the soil. Sugarbeets should not be planted in infested fields, since no definite control measures have been developed.

CHARCOAL ROT

Charcoal rot of sugarbeets is observed occasionally in areas of extremely high temperatures. The infection usually occurs on the shoulder of the maturing root, following injury or some growth-retarding condition. Necrotic tissue is brown or black and converted to a dry mass resembling charcoal but remaining covered by a papery dry layer. Internally, the advancing zone of infection often shows a mustard-yellow color (71). This disease is caused by a soil fungus, *Macrophomina phaseoli* (Maub.) Ashby (imperfect stage, *Sclerotium batiticola* Taub.), which also attacks corn, beans, sweet potatoes, and other crops. No control measures are considered necessary.

WET ROOT ROT

Wet root rot describes a moist rotting of the sugarbeet root, usually from the tip upward (72). Infected tissue is first light brown in color and sharply delimited from healthy tissue by a zone of blackish brown-colored cells (Fig. 9.12). The rotting is incited by certain water mold fungi, such as *Phytophthora drechsleri* Tucker., *P. megasperma* Drech., or *Pythium aphanidermatum* (Edson) Fitzp. Later, bacteria convert the rotted tissue to a slimy mass, or secondary fungi, such as *Fusarium* species, grow abundantly. The primary pathogens mentioned above rarely attack uninjured beet roots unless they are exposed to excessive soil moisture. This disease

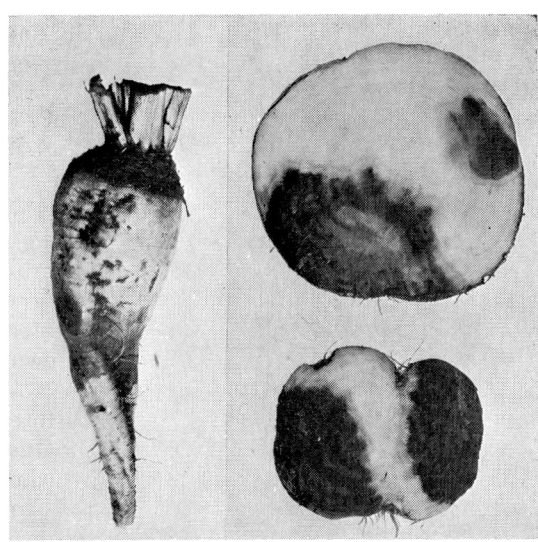

Fig. 9.12. Wet root rot incited by water mold fungi such as *Phytophthora drechsleri* or *Pythium aphanidermatum*. Usually in heavy, poorly drained soils or following heavy irrigation of wilted beets during a high temperature period.

usually develops in poorly drained low areas or where water penetration is impeded by a compact layer of subsoil. Even in medium-textured and well-drained soils, however, wet root rot may result from irrigation of wilted beets during very high temperatures.

Most of the losses from this type of rot can be avoided by good cultural practices such as proper land leveling, improved moisture penetration, bed planting, sprinkler irrigation, or proper timing and duration of furrow irrigation.

Rhizopus ROOT ROT

Rhizopus root rot occurs only after injury to the crown or after attack by chewing insects (Fig. 9.13). Spores of *Rhizopus* species, often called black bread mold, may initiate infection on injured tissue (if it is sufficiently moist), leading to rapid and complete decay. The same fungus may spread to adjacent roots in storage. The only precaution necessary is to avoid excessive injury or insect damage.

VASCULAR DISEASES

Fusarium YELLOWS

This disease, first described by Stewart (69), is known to occur only in limited areas of the United States, including parts of Colorado, Montana, Nebraska, New Mexico, South Dakota, and Wyoming.

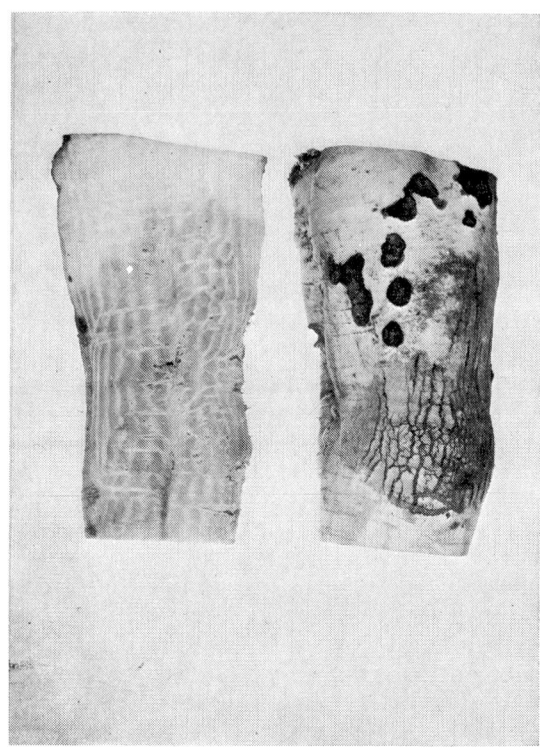

FIG. 9.13. Rhizopus root rot caused by the invasion of a species of *Rhizopus* following feeding injury by cutworms.

The older leaves of an affected plant show a yellowing of tissue between the large veins. Later these leaves become yellowish to gray and brittle in texture. The entire leaf then dies and drops to the ground while still attached to the crown. Younger heart leaves show inward rolling of the edges and twisting of the apex to one side. Large plants rarely wilt under field conditions. Roots cut in cross section show grayish to brown discoloration of the vascular system, often confined to only a few of the vascular rings or limited to one side of the root. This type of vascular discoloration differs from the internal root symptoms of curly top (virus), where discoloration is limited to the phloem portion of vascular bundles. The causal fungus, *Fusarium oxysporum* Schlecht. f. sp. *betae* (Stewart) Snyd. and Hans., originally described as *F. conglutinans* Wr. var. *betae* D. Stewart, survives in the soil as mycelium or spores and invades the host through roots. It grows in the water-conducting vessels and induces the production of toxic products responsible for the typical symptoms. Losses from this disease are serious only where sugarbeets are grown continuously for several years, thus promoting a buildup of the inoculum in the soil. Because the causal fungus attacks only sugarbeets or closely related crops, rotation for a few years with other crop plants is quite effective in reducing losses in future sugarbeet crops. No resistant varieties are available, and no control measures have been developed other than rotation.

Verticillium WILT

Another vascular disease, *Verticillium* wilt, has been reported on sugarbeets in Colorado, Nebraska, Idaho, and Washington (37). Caused by *Verticillium albo-atrum* Reinke and Berth., it also occurs on beets grown in Holland, but has not been reported from other areas. Foliage symptoms consist of wilting, and later, desiccation of the outer leaves and stunting and deformity of the inner leaves. The vascular discoloration consists of occasional black or brown vessels scattered through the fleshy root, sometimes traceable to a secondary root in contrast with the general vascular darkening characteristic of *Fusarium* yellows. The same species, *V. albo-atrum*, causes a serious wilt disease on other crop plants, including tomatoes, potatoes, strawberries, mint, cotton, stone fruit trees, and many ornamental plants. In some areas, sugarbeets have remained free of infection even when planted following serious *Verticillium* infection on the above-named crops. Hence, the race of *Verticillium* reported on sugarbeets is probably specialized as to host. Control measures have not been necessary for a disease of such limited occurrence and minor importance. Rotation with other crops for a few years would probably be sufficient.

BLACK WOOD VESSEL DISEASE

In Europe sugarbeets are occasionally observed with another type of vascular ailment, black wood vessel disease. Older leaves of affected plants show a distinct type of interveinal chlorosis, followed by necrosis of chlorotic areas. The most distinctive symptom is a blackening of vessels in the root, especially toward the tip. This results in concentric rings of black areas as viewed in cross section. This disease is caused by infection of feeder roots by a soil fungus, *Pythium irregulare* Buis. Toxins are produced and translocated upward, causing the blackened vessels and foliar symptoms. The disease is of little economic importance, with no control measures suggested.

OTHER ROOT DISEASES

COMMON SCAB

In some areas, sugarbeet roots exhibit scab lesions similar to those of common scab on potatoes. The infection is caused by a soil-borne organism, *Streptomyces scabies* (Thaxt.) Waks. and Henrici, which enters lenticels on the root, inducing excess multiplication of cells and the production of successive layers of cork cells. This produces circular or oval raised lesions, often in a band around the root. The disease, most apt to occur in alkaline soils, is increased by liming and suppressed by green manure crops or use of acid fertilizers. Soils known to produce scabby potatoes are most likely to yield sugarbeets with scab infection.

Severely affected roots may be slightly stunted, but only surface tissues are affected and scab on sugarbeets is not considered of economic importance. Other types of scab have been described in Europe and in the United States, also caused by species of *Streptomyces*.

CROWN GALL

Growers often notice an occasional beet with a large tumerous overgrowth attached to the side of the root by a small neck of tissue (Fig. 9.14). This striking overgrowth, sometimes nearly as large as the root itself, is caused by a soil-borne bacterium, *Agrobacterium tumefaciens* (E. F. Sm. and Town.) Conn, which enters through a wound. The infection results in abnormal cellular multiplication in susceptible tissue, leading to a large tumerous overgrowth. The weight of overgrowth and affected root approximates that of a healthy root, although sugar content may be somewhat less in the gall tissue than in a normal root. Because crown gall is comparatively

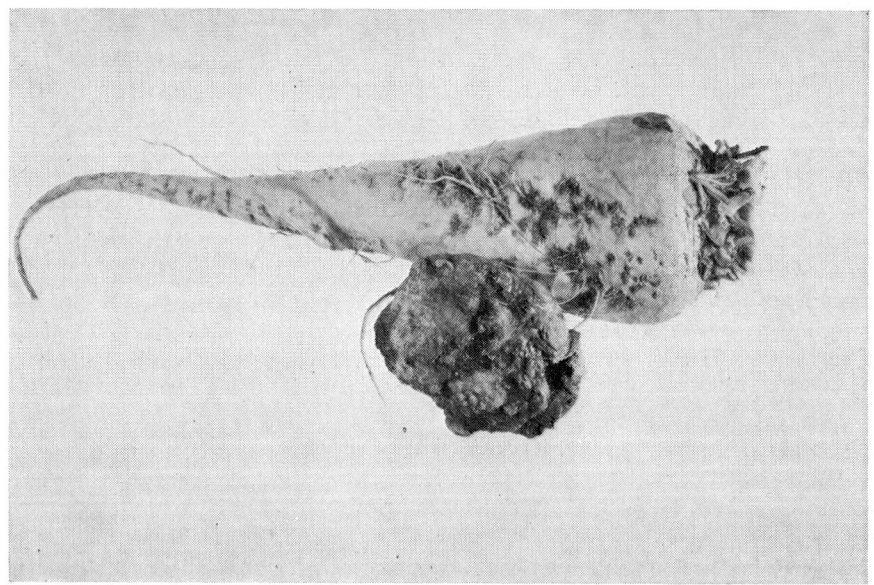

FIG. 9.14. Crown gall, an overgrowth disease incited by the bacterium *Agrobacterium tumefaciens*.

rare on sugarbeets, it is more of a curiosity than an economically important disease.

FOLIAGE DISEASES

Cercospora LEAF SPOT

One of the most widespread and destructive fungus diseases affecting sugarbeets is the leaf spot disease caused by *Cercospora beticola* Sacc. (58). In the United States it is most destructive in the central areas, from Ohio and Michigan to Colorado and in the high plains area of northwest Texas. Less frequently, it has appeared in localized areas in California and Arizona. In Europe, *Cercospora* leaf spot is most destructive in regions with warm, humid, summer weather, including Spain, Italy, Austria, and the southern portions of Germany and France. It is observed occasionally in Ireland, England, or the Scandanavian countries, but it is of little economic importance in northern Europe.

The individual leaf spots are nearly circular and measure about 3 to 5 mm in diameter when mature. Infected areas are light brown, later developing brownish or reddish purple borders. Under humid conditions, the center of the spot becomes ash gray from the production of conidiophores and spores of the fungus. The spots are isolated at first but coalesce as they increase in number; during the process, the entire leaf may turn first yellow and then brown, and finally collapse (Fig. 9.15A).

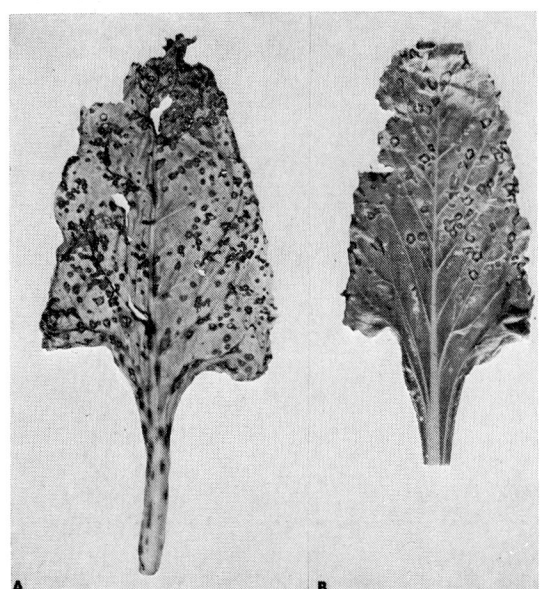

FIG. 9.15. Leaf spot diseases on sugarbeet. *A. Cercospora* leaf spot. *B. Ramularia* leaf spot. (From McKay, 54.)

The older dead leaves fall to the ground but remain attached to the crown while new leaves develop in the center of the crown. These, in turn, succumb to infection as they mature, leading to the production of an elongated crown. The continued replacement of the leaves is at the expense of stored food materials, reducing yield and lowering sugar content.

Rapid spread and development of the disease requires warm temperatures in combination with high humidity or free moisture on the leaves. The optimum conditions would be 75° to 85° F (25°–30° C) with above 90 or 95 percent relative humidity (60, 74). Growth of the fungus varies with temperature, with little or no infection occurring below 60° F (15° C). This explains why *Cercospora* leaf spot is serious in the warm, humid summers of central Europe and central United States.

In the lower valleys of Arizona and southern California, winter-grown sugarbeets are harvested before *Cercospora* becomes active, but in localized valleys at higher elevations in Arizona, where harvest of root crops is not completed before September, protective sprays are required against attack by *Cercospora* leaf spot. In the normally dry central valleys of California, this disease occurs about one year in five, and only in local areas when warm night temperatures (above 70° F) coincide with condensed moisture from irrigation and with the near-absence of drying winds.

Under favorable conditions, spores produced on infected leaves are wind-borne to susceptible foliage, where they germinate by means of a germ tube and enter the leaf through stomata and thus initiate additional leaf spots. Foliage appears predisposed to infection by *Cercospora* or other leaf spot fungi by chlorosis of leaf tissue, whether due to nutritional deficiency (such as sulfur deficiency) or to virus infection (virus yellows or western yellows).

The commonest source of infection is the sporulating fungus surviving on the remains of beet leaves and petioles left in the field. The spores are also carried on seed produced in areas where *Cercospora* is severe. The causal fungus has been shown to attack a number of weed hosts, and these may be a source of inoculum in some areas.

Control Measures. Rotation with other crops for at least one year will ensure that the spores and mycelium on infested plant debris will die out, for the causal fungus cannot survive in soil after plant tissue has decayed.

If seed fields are attacked by *Cercospora* leaf spot, as often occurs in southern Europe, it is considered desirable to treat the seed with an organic mercury fungicide to minimize the possibility of seed transmission. Most commercial sugarbeet seed fields in the United States, however, are located in the arid Southwest or in the Pacific Northwest, where *Cercospora* leaf spot is rare.

Where *Cercospora* leaf spot is prevalent, protective chemical sprays and dusts have been used extensively and have given beneficial results. Repeated applications of fixed copper or dithiocarbamate fungicides have reduced incidence of the disease and increased the yield and sugar percentage. Whether fungicide applications are economically feasible depends upon the severity of the disease and the regularity with which it occurs. Information on fungicide application and timing should be secured from advisory services in individual areas. Recently reported to be unusually effective were two new systemic fungicides, Thiabendazole [2-(4-thiazolyl benzimidazole)] (67, 68) and Benlate [methyl-1-(butylcarbamoyl)-2-benzimidazole carbamate] (27).

The most satisfactory control of *Cercospora* leaf spot comes from planting adapted resistant varieties, where available. As early as 1938 the United States Department of Agriculture released a leaf-spot-resistant variety. Improved varieties have since been released both by the USDA and by plant breeders employed by sugar companies. So far, varieties have not been as resistant as desired, but they have provided a large measure of protection, and varieties that are more resistant and have other desirable characteristics will very probably be available for most areas in the future. Varieties to be used in the leaf spot areas of the western states will also need resistance to curly top and virus yellows.

Ramularia LEAF SPOT

Ramularia leaf spot, found primarily in cool, damp climates, occurs in the United States most frequently in Oregon and Washington, and occasionally in Colorado and northern California at higher elevations. In Europe it is most common in Ireland, Scandanavia, and Russia, and less frequent in England and central Europe. Typical leaf spots are somewhat larger than those formed by *Cercospora*, being about 4 to 7 mm in diameter. The larger spots lose their circular shape and become angular (Fig. 9.15B). Mature spots are light brown and may or may not have a dark brown margin. The surface of the leaf spot is usually covered by grayish brown conidiophores, upon which the spores are borne.

Ramularia, favored by temperatures below 70° F (20° C), rarely causes enough crop damage to require control measures.

Phoma LEAF SPOT

Phoma leaf spot is caused by the same fungus, *Phoma betae,* which causes "black leg" of seedlings and one form of crown rot. This type of leaf spot is of little economic importance except that in seed fields it provides inoculum for seed-stalk and seed-cluster infection, the source of seed-borne inoculum. *Phoma* leaf spots are usually dark brown in color, round to oval, up to 1 or even 2 cm in diameter. Within the necrotic area they often show dark concentric rings bearing black dots, the pycnidia or spore cases. In the presence of moisture, the pycnidia exude a gelatinous mass of spores which are scattered by raindrops or by wind. Because most seed lots produced in the United States are free or nearly free of *Phoma* infection, leaf spots caused by this fungus are now comparatively rare in commercial fields.

Alternaria LEAF SPOT

In moist climates, mature leaves that are infected with virus yellows or have chlorosis induced by a nutritional deficiency are frequently invaded by *Alternaria tenuis* Nees ex Cda. This fungus is a saprophyte that attacks only dying or dead tissue. This type of leaf spotting, quite common along the Pacific Coast, is much less frequent in drier interior areas, but is found in most areas where sugarbeets are grown. The large, dark, irregular areas of fungus growth are separated by the veins of the leaf. Under moist conditions, the fungus sporulates abundantly on the necrotic tissue.

Another type of *Alternaria* leaf spot was reported by McFarlane et al. (52) from the Salinas Valley of California. They found certain inbred lines of sugarbeets to be quite susceptible to *Alternaria brassicae* (Berk.) Sacc. The inoculum apparently spreads from crucifers, especially wild radish, growing in the area. Although commercial varieties of sugarbeet proved to be resistant, those workers suggested that any genes for susceptibility should be eliminated from breeding materials.

BEET RUST

A typical leaf rust, *Uromyces betae,* frequently attacks winter-grown sugarbeets in the coastal areas of California, Oregon, Washington, and British Columbia, and less commonly in New Mexico and Arizona. Beet rust is frequent also in northern European sugarbeet areas. Leaves, petioles, and even seed stalks may be covered with small pustules filled with reddish brown spores (Fig. 9.16). Varieties and even individual plants show striking differences in susceptibility. Since plant breeders have selected for resistance to rust, along with other characteristics, varieties now in commercial use in coastal areas show very little damage from rust infection.

DOWNY MILDEW

Downy mildew, caused by a fungus, *Peronospora farinosa* (Fr.) Fr., occurs frequently in northern Europe but in the United States is found on sugarbeets only on the Pacific Coast (California, Oregon, and Washington). It has been reported on garden beets from New York and New Jersey. The fungus may attack beets in all stages of growth but is limited to cool, moist climates (44). The cotyledons or first true leaves of seedlings may be attacked, but more frequently the fungus invades the youngest leaves on the

FIG. 9.16. Beet leaf rust, *Uromyces betae*.

crown, grows down into the growing point, and systemically invades each new leaf as it develops. The result is a rosette of small, distorted, and mildewed leaves (Fig. 9.17A). Infected leaf tissue becomes light green, thickened, and puckered, with the outer edges of the leaf curling downward. Asexual spores, called conidia or sporangia, are formed in great numbers on treelike branched sporangiophores protruding from the stomata on the lower side of the leaf, and under moist conditions also from the upper side (Fig. 9.17B). With favorable weather conditions, these wind-blown spores may initiate localized infections 1 to 3 cm in diameter on fully formed leaves of very susceptible sugarbeets or garden beets.

Later in the season, when conditions become unfavorable for mildew development, infected leaves may die and a type of heart rot may occur. Older leaves on infected plants become chlorotic, a symptom that can be confused with virus yellows infection. As downy mildew activity declines, healthy leaves appear in the crown area, and some plants will show extensive recovery.

Effects on yield are related to the earliness of infection, the duration of mildew activity, and the extent of the recovery period before harvest (45). In field trials it appeared that root size was reduced to about half in plants that showed infection of the rosette leaves within 100 days of seeding, but was reduced only slightly if the disease did not appear until 150 days. Infected beets harvested less than two months after the termination of a severe mildew outbreak had considerably lower sucrose content and purity than did healthy beets or infected beets harvested later.

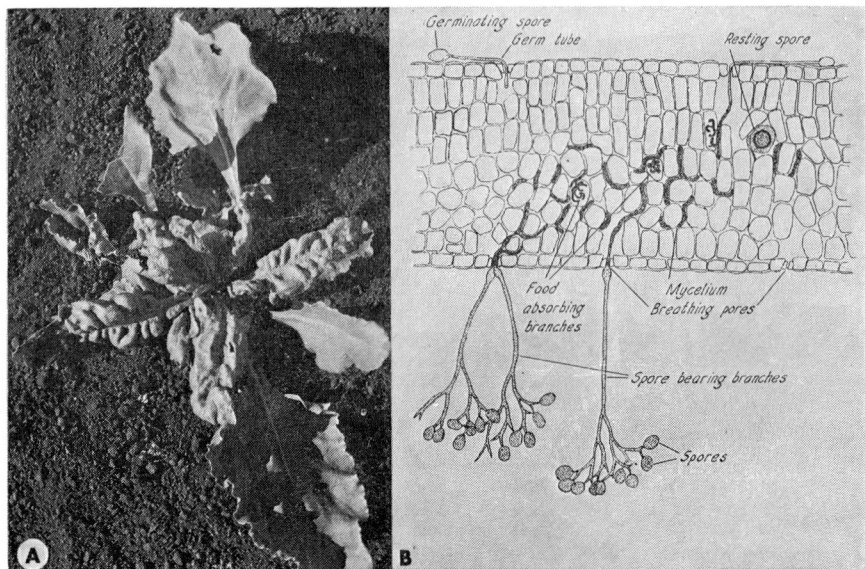

Fig. 9.17. Downy mildew caused by *Peronospora farinosa*. *A*. Sugarbeet showing systemic downy mildew infection of the heart leaves. *B*. Diagram showing entry of germ tube into the leaf, intercellular mycelium with food-absorbing haustoria, sporangiospores bearing asexual spores, and an intercellular sexual resting spore.

On seed beets, downy mildew infects young lateral branches, causing stunting and distortion, and may invade flower parts and produce mycelium and sexual resting spores (oospores) within seed clusters. Seed transmission may introduce the fungus into areas not previously infected, though it is not considered an important source of inoculum in areas already subject to mildew.

Control Measures. Protection of sugarbeets by fungicidal sprays or dusts has seldom proved satisfactory, because the disease appears sporadically, is internal and partially systemic, and requires protection over a prolonged period.

Outbreaks in European countries are most serious when root and seed crops are grown in the same area. Each crop acts as a source of inoculum for the other, and the seed crop carries the pathogen from one season to the next. Separation of seed and root crops reduces the inoculum available to root crops during spring and early summer.

Varieties differ strikingly in susceptibility to downy mildew, and selection under conditions of artificial epidemics has been shown to be effective in producing varieties with relatively high resistance (45, 51).

In the coastal valleys of California, losses from downy mildew were serious and frequent between 1925 and 1950, but only minor since 1954 (51). The most important factor leading to the decline of the disease has been the introduction of downy-mildew-resistant varieties by the United States Department of Agriculture and by the sugar companies. Of secondary importance is the elimination of inoculum sources by eradicating wild and volunteer beets (sources of virus yellows) and discontinuing farm gardens containing Swiss chard and garden beets (frequently infected by downy mildew and providing wind-blown spores to adjacent commercial fields of sugarbeets).

In summary, the best control measures are to use the most resistant adapted variety and eliminate the chief sources of inoculum.

POWDERY MILDEW

In central Europe and occasionally in western Europe, a thin white film of fungus mycelium may appear on leaves of sugarbeets. Spores are formed successively from unbranched conidiophores arising from the surface mycelium and are dispersed by wind. This fungus, in contrast to downy mildew, is favored by warm, dry weather. The mycelium draws nourishment from the surface cells of the leaf, which may turn pale yellow and eventually dry and collapse. The causal fungus, *Erysiphe polygoni* DC., has been reported in the United States only from California (75), Oregon, and Washington. An effective fungicide is sulfur as a spray or dust, but the infestation is usually not severe enough to warrant control measures.

BACTERIAL LEAF SPOT OR BLIGHT

Pseudomonas syringae van Hall, formerly referred to as *Bacterium aptata*, causes bacterial leaf spot or blight, characterized by black leaf spots on sugarbeets growing during winter or cool spring periods. It is often observed in connection with rust or downy mildew lesions. First reported from Utah and California, it is now commoner in the Pacific Northwest. Serious infections, consisting of leaf and stalk blighting, formerly occurred on seed crops in northern California and Oregon, especially at higher elevations and low temperatures. Seed clusters were attacked and the bacteria were found to be transmitted in the seed ball, producing infected cotyledons after germination (2). This phase was effectively controlled by seed treatment with organic mercury or thiram fungicide. This bacterial disease is not considered responsible for economic losses in sugarbeet root crops.

OTHER DISORDERS

DODDER

Dodder is a parasitic seed plant (*Cuscuta campestris* Yumker) that grows as a slender vine, entwining its yellowish stem, almost leafless, around the aboveground parts of its host plant. Dodder seed may survive several years in the soil and then germinate to form a yellowish shoot which makes contact with the nearest green plant. Except for food reserves in the seed, the dodder plant depends upon the host plant for food because, lacking chlorophyll, it cannot manufacture its own. The dodder vine extends from plant to plant, building up a tangled mass readily recognized by the bright yellow color of the parasitic vine. Various species of dodder attack a wide range of crop and weed hosts. After midseason, flowers and seeds are produced. Because the seeds are about the size of alfalfa or clover seeds, it is difficult to remove them from commercial seed lots of those crops. To prevent seed formation, dodder colonies and their supporting hosts can be burned with a flame torch or killed by chemicals.

In a few instances, dodder has been severe enough to affect yield and sucrose content, but usually dodder colonies are few in number and near field borders, thus causing little economic damage.

ALBINISM

A few plants in a sugarbeet field may show a variegated coloring of the foliage, part of a leaf being normally green and the remainder white, with sharply defined demarcation. In other cases only the outermost layers of cells are devoid of chlorophyll, while deeper layers of the leaf are normally green. These abnormalities are caused not by an infectious disease but by genetic aberration. The term "chimaera" is used to refer to plants exhibiting this condition.

LIGHTNING INJURY

Lightning striking the ground can cause a condition often attributed to leaf or root infection. Leaves on plants in a circular area may be killed, or roots may die over a period of days, progressing outward from a center. Secondary organisms invade the necrotic tissue, which also suggests a root pathogen. Beets beyond the border of the affected area remain healthy.

REFERENCES

1. Afanasiev, M. M. Effect of fertilizers on diseases and yield of sugar beets planted in depleted soil. Proc. Am. Soc. Sugar Beet Technologists 5:294–99. 1948.
2. Ark, P. A. and L. D. Leach. Seed transmission of bacterial blight of sugar beet. Phytopathology 36:549–53. 1946.
3. Bennett, C. W. Studies on properties of the curly top virus. J. Agr. Res. 50:211–41. 1935.
4. ———. Sugar beet yellow vein disease. Plant Disease Reptr. 40:611–14. 1956.
5. ———. Sugar beet yellows disease in the United States. USDA Tech. Bull. 1218. 1960.

6. ———. Noninfectious yellow splotching of leaves of young sugarbeet plants in western United States. Plant Disease Reptr. 47:63–64. 1963.
7. ———. Highly virulent strains of curly top virus in sugar beet in western United States. J. Am. Soc. Sugar Beet Technologists 12:515–20. 1963.
8. ———. Isolates of beet mosaic virus with different degrees of virulence. J. Am. Soc. Sugar Beet Technologists 13:27–32. 1964.
9. ———. Marble leaf of sugar beet, caused by a juice transmissible virus. J. Am. Soc. Sugar Beet Technologists 13:33–41. 1964.
10. Bennett, C. W., C. Carsner, G. H. Coons, and E. W. Brandes. The Argentine curly top of sugar beet. J. Agr. Res. 72:19–48. 1946.
11. Bennett, C. W. and A. S. Costa. The Brazilian curly top of tomato and tobacco resembling North American and Argentine curly top of sugar beet. J. Agr. Res. 78:675–93. 1949.
12. Bennett, C. W. and J. E. Duffus. Rosette disease of sugar beet. Plant Disease Reptr. 41:1001–4. 1957.
13. Bennett, C. W., F. J. Hills, R. Ehrenfeld K., J. Valenzuela B., and C. Klein K. Yellow wilt of sugar beet. J. Am. Soc. Sugar Beet Technologists 14:480–510. 1967.
14. Bennett, C. W. and J. S. McFarlane. Damage produced by beet yellows and beet western yellows under greenhouse and field conditions. J. Am. Soc. Sugar Beet Technologists 14:619–35. 1967.
15. Bennett, C. W. and A. Tanrisever. Curly top disease in Turkey and its relationship to curly top in North America. J. Am. Soc. Sugar Beet Technologists 10:189–211. 1958.
16. Bennett, C. W. and H. E. Wallace. Relation of the curly top virus to the vector, *Eutettix tenellus*. J. Agr. Res. 56:31–52. 1938.
17. Björling, K. and B. Nilsson. Observations on host range and vector relations of beet mild yellowing virus. Socker 21:1–14. 1966.
18. Björling, K. and F. Oasiannilsson. Pests and diseases of sugar beet. Hilleshog Brochure 10., N. V. Hollandsch. Zweedsche. Zaad Mallschappis, Amsterdam. An English edition of Gram E. and P. Bovien (1944). Rodfrugternes Syndomme og Skadedyr (in Danish). 1954.
19. Brandenburg, E. Die Herz-und Trockenfaule der Ruben als Bormangel-Erscheinung. Phytopathol. Z. 3:499–517. (The heart and dry rot of beets as a symptom of boron deficiency). Rev. Appl. Mycol. 11:147–48. (Abstr.) 1932.
20. Buckholtz, W. F. and C. H. Meredith. Pathogenesis of *Aphanomyces cochlioides* on tap roots of sugar beet. Phytopathology 34:485–96. 1944.
21. Burtch, L. The use of insecticides for curly top control. Spreckles Sugar Beet Bull. 32:15–17, 24. 1968.
22. Carsner, E. and C. F. Stahl. Studies of the curly top disease of sugar beet. J. Agr. Res. 28:297–320. 1924.
23. Clinch, P. E. M. and J. B. Laughnane. Seed transmission of virus yellows of sugar beet (*Beta vulgaris* L.) and the existence of strains of the virus in Eire. Scientific Proc. Royal Dublin Soc. 24:307–18. 1948.
24. Coons, G. H. Some problems in growing sugar beets. USDA Yearbook of Agriculture, 1953, pp. 509–25. 1953.
25. Coons, G. H., J. E. Kotila, and D. Stewart. Savoy, a virus disease of beet transmitted by *Piesma cinerea*. Proc. Am. Soc. Sugar Beet Technologists 6:500–501. 1950.
26. Coons, G. H., D. Stewart, H. W. Bockstahler, and C. L. Schneider. Incidence of savoy in relation to the variety of sugar beets and to the proximity of wintering habitat of the vector, *Piesma cinerea*. Plant Disease Reptr. 42:502–11. 1958.
27. Delp, C. J. and H. L. Klopping. Performance attributes of a new fungicide and mite ovicide candidate. Plant Disease Reptr. 52:95–99. 1968.

28. Duffus, J. E. Radish yellows, a disease of radish, sugar beet, and other crops. Phytopathology 50:389–94. 1960.
29. ———. Host relationships of beet western yellows virus strains. Phytopathology 54:736–38. 1964.
30. ———. Membrane feeding used in determining the properties of beet western yellows virus. Phytopathology 59:1668–69. 1969.
31. Edson, H. A. Seedling diseases of sugar beets and their relation to root-rot and crown-rot. J. Agr. Res. 4:135–68. 1915.
32. Ernould, L. and L. Van Steyvoort. Atlas des ennemis et maladies de la betterave. Institut Belge pour l'Amerlioration de la Betterave, Tirlemont (Belgium). 1958.
33. Esau, K., J. Cronshaw, and L. L. Hoefert. Relation of the yellow wilt virus to the phloem and to movement in the sieve tube. J. Cell Biol. 32:71–87. 1967.
34. Farris, J. A. Violet root-rot (*Rhizoctonia crocurum* DC.) in the United States. Phytopathology 11:412–22. 1921.
35. Fawcett, G. L. Encrespamiento de las jojas de la remolacha. Rev. Ind. Agr. Tucumán 16:39–46. 1925.
36. Freitag, J. H. Negative evidence on the multiplication of curly-top virus in the beet leafhopper, *Eutettix tenellus*. Hilgardia 10:305–42. 1936.
37. Gaskill, J. O. and W. A. Kreutzer. Verticillium wilt of the sugar beet. Phytopathology 30:769–74. 1940.
38. Gaskill, J. O. and C. L. Schneider. Savoy and yellow vein diseases of sugarbeet in the great plains in 1963–64–65. Plant Disease Reptr. 50:457–59. 1966.
39. Gibbs, A. J. and B. D. Harrison. Nematode-transmitted viruses of sugar beet in East Anglia. Plant Pathology 13:144–50. 1964.
40. Giddings, N. J. Studies of selected strains of curly top virus. J. Agr. Res. 56:883–94. 1938.
41. ———. Combination and separation of curly-top virus strains. Proc. Am. Soc. Sugar Beet Technologists 6:502–7. 1950.
42. Grente, J. Les maladies de la betterave. Cahiers de L' Institut Technique Francais de la Betterave Industrielle. No. 3. (Undated.)
43. Hull, R. Sugar beet diseases; their recognition and control. Gt. Brit. Min. Agr. Fisheries Food Tech. Bull. 142. 1950.
44. Leach, L. D. Downy mildew of the beet, caused by *Peronospora schachtii* Fuckel. Hilgardia 6:203–51. 1931.
45. ———. Effects of downy mildew on productivity of sugar beets and selection for resistance. Hilgardia 16:317–34. 1945.
46. ———. Growth rates of host and pathogen as factors determining the severity of preemergence damping-off. J. Agr. Res. 75:161–79. 1947.
47. Leach, L. D. and A. E. Davey. Determining the sclerotial population of *Sclerotium rolfsii* by soil analysis and predicting losses of sugar beets on the basis of these analyses. J. Agr. Res. 56:619–31. 1938.
48. ———. Reducing southern *Sclerotium* rot of sugar beets with nitrogenous fertilizers. J. Agr. Res. 64:1–18. 1942.
49. LeClerg, E. L. Studies on dry-rot canker of sugar beets. Phytopathology 29:793–800. 1939.
50. Lüdecke, H. and Chr. Winner. Farbtafelatlas der Krankheiten und Schädigungen der Zuckerübe. DLG-Verlag, Frankfurt am Main, Germany. 1959.
51. McFarlane, J. S. Elimination of downy mildew as a major sugarbeet disease in the coastal valleys of California. Plant Disease Reptr. 52:297–99. 1968.
52. McFarlane, J. S., R. Bardin, and W. C. Snyder. An *Alternaria* leaf spot of the sugar beet. Proc. Am. Soc. Sugar Beet Technologists 8:241–46. 1954.
53. McGinnis, R. A. Sugar Beet Technology. Reinhold Publishing Corp., New York. 1951.

54. McKay, R. Sugar Beet Diseases in Ireland. Irish Sugar Co. Ltd., Dublin, Ireland. 1952.
55. Maxon, A. C. Insects and diseases of the sugar beet. The Beet Sugar Develop. Found., Fort Collins, Colo. 1948.
56. Middleton, J. T. The taxonomy, host range and geographic distribution of the genus *Pythium*. Memoirs Torry Bot. Club 20:1–171. 1943.
57. Murphy, A. M. Sugar beet and curly top history in southern Idaho 1912–1945. Proc. Am. Soc. Sugar Beet Technologists 4:408–12. 1946.
58. Nagel, C. M. Epiphytology and control of sugar beet leaf spot caused by *Cercospora beticola* Sacc. Iowa Agr. Exp. Sta. Res. Bull. 338:680–706. 1945.
59. Oman, P. W. Notes on the beet leafhopper, *Circulifer tenellus* (Baker), and its relatives (Homoptera:Cicadellidae). J. Kansas Entomol. Soc. 21:10–14. 1948.
60. Pool, V. W. and M. B. McKay. Climatic conditions as related to *Cercospora beticola*. J. Agr. Res. 6:21–60. 1916.
61. Proesler, von G. Beziehungen zwischen der Rubenkrauselvirus II. Injectionversuche. Phytopathol. Z. 56:213–37. 1966.
62. Richards, B. L. A dryrot canker of sugar beets. J. Agr. Res. 22:47–52. 1921.
63. Robbins, W. W. Mosaic disease of sugar beets. Phytopathology 11:349–65. 1921.
64. Severin, H. H. P. and R. M. Drake. Sugar-beet mosaic. Hilgardia 18:483–521. 1948.
65. Severin, H. H. P. and J. H. Freitag. Some properties of the curly-top virus. Hilgardia 8:1–48. 1933.
66. ———. Outbreak of western cucumber mosaic on sugar beets. Hilgardia 18:523–30. 1948.
67. Staron, T. and C. Allard. Propiertes antifongiques du 2-(4thiazolyl) benzimidazole ou thiabendazole. Phytiat. Phytopharm. 13:163–68. 1964.
68. Staron, T., H. Darpoux, A. Lebrun, and B. de la Tullaye. Action remarquable du thiabendazole sur la cercosporiose de la betterave. Phytiat. Phytopharm. 15:113–19. 1966.
69. Stewart, D. Sugar-beet yellows caused by *Fusarium conglutinans* var. *betae*. Phytopathology 21:59–70. 1931.
70. Sylvester, E. S. The yellow-net disease of sugar beet. Phytopathology 38:429–39. 1948.
71. Tompkins, C. M. Charcoal rot of sugar beet. Hilgardia 12:73–81. 1938.
72. Tompkins, C. M., B. L. Richards, C. M. Tucker, and M. W. Gardner. Phytophthora rot of sugar beets. J. Agr. Res. 52:205–16. 1936.
73. Vallejo, V. M. Enfermedades de la remolacha en Chile. Industria Azucarera Nacional S.A., Santiago, Chile. 1966.
74. Vestal, E. F. Pathogenicity, host response and control of *Cercospora* leafspot of sugar beets. Iowa Agr. Exp. Sta. Res. Bull. 168. 1933.
75. Yarwood, C. E. Unreported powdery mildews—powdery mildew on sugar beet. Plant Disease Reptr. 21:179–82. 1937.

10

Insects and Mites and Their Control

W. H. LANGE
University of California
Davis

History	288
Crop Ecology	293
Descriptions of Pests and Injury	295
Pests Present at Time of Planting	295
Pests of Seedlings	301
Pests of Older Plants	305
Pests of Seed Production	311
Occasional Pests	312
Natural Enemies and Pathogens	313
Vector-Virus Relationships	315
Control Procedures	321
Outlook	325

THE estimated average annual losses caused by insects and mites to field crops in the United States are $1,482,325,000 of which $20,441,000 are due to pests affecting sugarbeets (Table 11, Agricultural Handbook 291). Sugarbeets are subject to attack from the time the seed is planted until the crop (roots, tops, seeds) is harvested. Damage to the plants is direct, such as defoliation, or indirect through the introduction of toxins or viruses. The history of sugarbeet production in the United States, particularly the West, is closely interwoven with the history of curly top, a disease caused by a virus transmitted by the sugarbeet leafhopper, and several yellowing viruses carried by aphids. Many pests such as wireworms, armyworms, cutworms, lygus bugs, spider mites, and aphids are general feeders on many field crops in addition to sugarbeets. Others such as sugarbeet root aphid, sugarbeet petiole borer, sugarbeet root maggot, and sugarbeet crown borer have narrower host ranges and are more specific to the sugarbeet.

Although current grower practices involve chemical control as the principal weapon in combating insects and other arthropod pests, there have been many problems associated with their use in recent years. Insect resistance to chemicals that formerly gave effective control, the possible presence of potentially hazardous residues in either roots or tops, the possibility of environmental pollution, adverse effects on predators and parasites, and the often transient nature of chemical control procedures are just a few of the problems which are confronting us. It is apparent that our objective should be the use of an integrated control program, a program which would incorporate all available methods and utilize them in a coordinated pest control management program. To effectively carry out such a program, we need much more information on the population dynamics of sugarbeet arthropod pests and more information on the parasites, predators, and pathogens associated with these pests. In addition, beets are grown under a wide range of cultural conditions over a large part of the United States, and for this reason control measures have to be modified for each individual area and for its associated pests. Usually grower practices differ in each area, depending upon a number of variable factors, and it is necessary for the sugarbeet grower to avail himself of local research results from the USDA, various state experiment stations, research departments of the several sugarbeet companies, and the agricultural extension service through its county agents or farm advisors.

HISTORY

Many insects and mites that attack sugarbeets are native to the United States, and some of these are known to have attacked table beets, spinach, and related cultivated plants long before the development of the sugarbeet

---

The assistance of B. J. Landis in the preparation of this chapter is gratefully acknowledged.

industry. Many subsist normally on wild plants in the same family—the Chenopodiaceae. With an increase from 3 sugar factories in 1891 to 42 in 1902, Chittenden by 1903 (23) had already accumulated records of approximately 150 insect pests of the crop, of which 40 to 50 were considered severe pests. Those he listed were strictly primary feeders and did not include some insects that were later found to be of much greater importance as vectors of certain virus diseases. One of these was curly top. Shortly before the turn of the century, Gillette and Baker (48) collected some strange leafhoppers from sugarbeets at Grand Junction, Colorado, but associated no damage with them. The next year Baker (6) found the same kind of leafhoppers in New Mexico and named them *Thamnotettix tenellus,* later changed to *Circulifer tenellus* (Baker). This insect is commonly known as the beet leafhopper or whitefly. A few years later, Ball (7) discovered that this leafhopper was responsible for the spread of the destructive "California beet seed disease," "western blight," or "curly-leaf" that appeared near Watsonville, California, in 1899. It is now known as curly top. Since then, the history of the beet leafhopper has been closely associated with the history of curly top.

The devastations caused by the beet leafhopper and curly top on sugarbeets within the area served by the Utah-Idaho Sugar Company have been chronicled by Arrington (4). At Lehi, Utah, the first known instance of beet leafhopper infestations in sugarbeet fields was in 1897, a year after the sugar factory was built. Plant officials noted that the crop was down 24,700 tons of beets or 7.3 tons per acre from the 1896 production. The next short crop was in 1900 when the yield dropped to only 6.7 tons per acre. In 1905 many farmers lost all or very heavy portions of their crops. A factory was built at Nampa, Idaho, in 1906. "Blight" affected the crop badly, and in 1910 less than 5,000 tons of beets were harvested from 3,600 acres.

In California, Idaho, Washington, and Oregon, sugarbeet factories were either abandoned or moved solely because of the leafhopper-curly top problem. Hawley (62) estimated that the loss to the sugar industry in this area from 1899 to 1916 was $10,000,000 and that in 1905 alone the loss was $500,000. The devastation caused by the beet leafhopper and curly top in Utah, Idaho, and Washington and the movement of the Utah-Idaho Sugar Company's factories are illustrated in Table 10.1.

An attempt has been made to compile a list of "bad" leafhopper years in the various western states in Table 10.2. In some of these years of curly top severity, there are little data on actual leafhopper abundance and it has been assumed, therefore, that leafhoppers were abundant at critical times, or perhaps more virulent strains occurred. In certain cases, some of the records may refer to more restricted parts of the state rather than a statewide outbreak.

The beet leafhopper thrives best in the arid regions of the West where it has been responsible for the greatest losses from curly top. Extensive breeding grounds exist in all the states bordering Mexico, and during favorable winters large numbers of leafhoppers are produced there. In the

TABLE 10.1. U and I factories whose operations were discontinued because of curly top

Factory	Year Commenced Operations	Year Ceased Operations	Eventual Disposition
Lehi, Utah	1891	1924	Dismantled
Nampa, Idaho	1906	1910	Moved to Spanish Fork, Utah
Elismore, Utah	1911	1928	Dismantled
Payson, Utah	1913	1925	Dismantled
Moroni, Utah	1917	1925	Moved to Toppenish, Wash.
Delta, Utah	1917	1924	Mixed to Belle Fourche, S.Dak.
N. Yakima, Wash.	1917	1918	Moved to Chinook, Mont.
Rigby, Idaho	1919	1924	Dismantled
Toppenish, Wash.	1919	1924	Moved to Bellingham, Wash.
Sunnyside, Wash.	1919	1919	Moved to Raymond, Alberta

SOURCE: From Arrington (4).

spring, they migrate for hundreds of miles northward within the intermountain region and occasionally in a northeasterly direction along the east side of the Continental Divide. In many sugarbeet-producing areas, severe curly top may occur as a result of long-distance migrations from the south, but in other areas such as Washington and Oregon the abundance of locally produced leafhoppers is wholly responsible.

Appreciable damage has developed on sugarbeets from curly top during years when warm, dry springs followed moderately cold winters. Maxson (129) reported that in Colorado the weed hosts of the leafhopper remained green unusually late in the spring of 1922, and since the leafhoppers were not forced to move until after mid-June, there was little spread of curly top to sugarbeets. However, a dry spring in 1924 caused the winter host plants to die early and leafhoppers were forced to move to sugarbeets

TABLE 10.2. Years of great leafhopper abundance, or other years when curly top was quite serious in various sugarbeet-producing states

State	Years of Abundance
Arizona	1939
California	1889, 1899, 1905, 1914, 1919, 1922, 1925, 1926, 1940, 1950, 1956, 1960, 1966
Colorado	1900, 1901, 1903, 1908, 1921, 1922, 1926, 1928–31, 1933, 1937, 1938, 1939, 1940, 1943, 1945, 1947, 1950–53
Idaho	1919, 1921, 1922, 1924, 1926, 1930, 1931, 1934, 1935, 1937, 1940, 1941, 1947, 1950, 1969
Kansas	1953
Oregon	1904
Texas	1906, 1930, 1931, 1938–40, 1943, 1945–53, 1959
Utah	1898, 1899, 1900, 1905, 1910, 1915, 1919, 1924, 1926, 1930, 1931, 1933, 1934, 1935, 1937, 1940, 1941
Washington	1918, 1919, 1923, 1924, 1934, 1935, 1960, 1966, 1967
New Mexico	1953

during the latter part of May. That year several thousands of acres of small sugarbeets were abandoned because of curly top damage before July 1.

Factors other than weather have also affected the abundance of beet leafhoppers during certain years, and these include changes in the kind and abundance of suitable alternate host plants. For example, Arrington (4) reported that the overexpansion of agriculture during World War I affected the sugarbeet industry in an unexpected way. The demand for food caused vast areas of submarginal land to be cleared of sagebrush and planted to wheat. At the end of the war, much of the land was abandoned and became infested with weeds on which the beet leafhopper propagated in great numbers. In this and similar man-made breeding grounds in Idaho and California, potentially destructive populations of leafhoppers were controlled during some years with insecticides applied by airplane. Douglass (32) reported that large numbers of beet leafhoppers and serious losses from curly top would probably have occurred in south-central Idaho in 1950 and in 1953 if the most productive of the spring breeding grounds had not been sprayed. Also in Idaho, the Bureau of Land Management, in cooperation with the Entomology Research Division of the USDA, reseeded thousands of acres of abandoned land with crested wheat grass which is not a host of the leafhopper.

A strain of sugarbeet partially resistant to the curly top virus was developed in 1929 by the USDA, in cooperation with several sugar companies and the University of California. From the U.S. 1 strain of sugarbeet, plant breeders developed strains that were more resistant to curly top and also somewhat resistant to feeding by the beet leafhopper.

Much has been done to alleviate the beet leafhopper-curly top problem on sugarbeets, but the leafhopper still is able to propagate in large numbers in favorable locations during some years. Curly top continues to suppress yields up to seven tons per acre in some years when large numbers of leafhoppers become infected with virulent strains of the virus. Some years ago, Douglass (32) warned that "the beet leafhopper weed-host complex has changed in many areas during the past thirty years and there is more than the bare possibility that this leafhopper is becoming acclimated to new areas. Therefore, we must be prepared for major changes over a longer period of years." Since 1950 the state of California by legislative commitment has continued an insect control program in a vast area of approximately 250 miles long by 20 miles wide in the breeding range of the beet leafhopper, although a control program by two sugarbeet companies was in effect from 1931 and was continued by the state in 1943. The leafhopper in 1950 attained epidemic population levels because of several years of low rainfall and an abundance of Russian thistle. The program as outlined by Green (52) indicates that since the yearly spraying operation in 1950 there have been no major outbreaks of the leafhopper and curly top, although losses have been sustained in local areas and particularly in areas where beets have been planted in or adjacent to the beet-breeding areas (The

California Sugar Beet—1969, pp. 28, 30). The incidence of more virulent strains of curly top have also complicated the picture in California (9).

The disease of sugarbeet called virus yellows was first positively identified in the United States by Coons and Kotila (29) in 1951. Bennett (8) has adequately covered the host range, distribution, epidemiology, and other aspects of this particular disease. It was not until the 1957 Kern County crop failure (93), with an average decrease of 10 tons per acre in roots and a 1 to 2.5 percentage point drop in sucrose, that the full significance of this virus disease became apparent. Curly top was also evident as was sugarbeet mosaic, and later it was found that the effects of these diseases on root yield were additive and that together they could cause severe losses.

In 1960 Duffus (35) reported a second yellowing virus, now known as beet western yellows, and this virus has been proved to be responsible for a yellowing condition and yield decreases in western United States. Investigations carried out by the USDA Agricultural Research Station at Salinas, California, in cooperation with the Beet Sugar Development Foundation, the California Beet Growers' Association, and the University of California, have resulted in the release of two monogerm hybrid varieties, US H9A and US H9B, which when grown under conditions of severe yellows infection yield approximately 20 percent more sugar than the widely grown US H7 variety. The use of these varieties, together with cultural control and the proper timing of certain systemic insecticides, has temporarily alleviated the threat of the yellowing viruses in California and in certain other areas. This is another example of the continued cooperation among all agencies in solving critical problems which menace the sugarbeet industry.

Mention should be made of the many circulars and books available to the sugarbeet grower and other interested persons, which cover the principal arthropod pests of the sugarbeet both in the United States and in other countries. The most outstanding examples of books covering the pests and associated natural enemies of any crop are the two by Maxson (128, 129) that cover the principal insect pests and diseases of sugarbeet in the United States and Canada. In the early 1900s, both Adams (1) and Harris (58) presented field manuals of sugarbeet production which included sections on insect pests. Leach (116) and Lange (100) cooperated with the Spreckels Sugar Company in producing a series of colored plates of the important diseases, insects, and nematodes of the beet crop in California, and this was followed in 1951 (102) by a section on sugarbeet insects in a book on sugarbeet technology. Insects of the beet seed crop are presented in a circular by Hills (77) and in one on other sugarbeet insects by Peay (132). One of the most outstanding publications available is one published in England on sugarbeet pests by Jones and Dunning (92), which not only gives diagrammatic representations of the type of injury caused by the more important sugarbeet pests but also attempts to set up economic thresholds for some of the insects. Even though the insect species are not always the same ones occurring in this country, the bulletin is very useful to sugarbeet growers and technologists in this country. Publications such as those by McKinney (123) and Pettit (136) are useful, since beet pests of specific areas in the United States are covered. The circular by Hull (85) on sugarbeet diseases is very useful, and sugarbeet pests and diseases are covered in some detail (often with adequate colored illustrations) by several excellent foreign publications. The two books by Lüdecke (117) and Lüdecke and Winner (118) are very useful as is a Belgian publication by Ernould and Van Steyvoort (40) and a publication by Gram and Bovien (51) on root crop pests in Denmark.

CROP ECOLOGY

It is essential to the control of insects and mites affecting sugarbeets that we understand the nature of the sugarbeet ecosystem, since the application of an incorrect control measure, such as the application of certain insecticides, may actually cause resurgences in the populations of certain insects or mites. In California, for example, the application of some insecticides has been observed to upset the natural parasite and predator balance, where insects actually cause more damage than if no control measures were applied. One of the difficulties with sugarbeets is that it is often not only difficult to determine the exact time to apply a control measure but the economic thresholds of most of the important pests are not understood. Many chemical or other control measures are used which act in an adverse fashion rather than assist in the control of the pest in question.

The overwintering of large acreages of sugarbeets in the central valley area of California, often 25,000 to 30,000 acres, has greatly complicated the pest and disease situation inasmuch as some beets may remain in the ground for as long as 12 to 15 months. This means that many insects and mites have a means of overwintering and carrying on to new crops and there is a ready source of virus inoculum present. Even though an attempt has been made to harvest beets in blocks to get away from adjacent new plantings and also to instigate beet-free periods, it is still difficult under these conditions to protect new spring plantings from sources of viruses.

Sugarbeet fields often sustain large populations of many insects and mites and their associated parasites and predators. Most of the insects and mites on sugarbeets are primarily associated with various weeds, particularly those in the genera *Chenopodium* and *Amaranthus* and their relatives. Weeds are the sources of many species of aphids, thrips, cutworms, armyworms, leaf miners, flea beetles, webworms, sugarbeet petiole borers, spider mites, leafhoppers, and many other harmful insects. In many cases, the sugarbeet is not the preferred host, but large acreages of sugarbeets make it possible for these insects to increase in large numbers.

The past history of the field is often important in predicting possible pest infestations, and this is particularly true with a number of the soil inhabitants such as symphylans, wireworms, springtails, and root maggots. In some cases, continuous growing of sugarbeets in individual fields may increase nematodes and the sugarbeet root aphid. In the Salinas Valley of California, the author has observed the sugarbeet as a source of large numbers of the vegetable leaf miner, *Liriomyza langei* Frick, with damage to adjacent lettuce crops. Border rows of lettuce are occasionally completely destroyed by this particular leaf-mining insect. In other situations, the large numbers of hymenopterous parasites of the leaf miner on sugarbeets seem to be beneficial in suppressing leaf miners in adjacent lettuce. In another instance in the Salinas Valley, the application of DDT to large acreages of sugarbeets apparently decreased the natural enemies of this leaf miner to the extent that a resurgence actually defoliated large acreages of sugarbeets.

The spacing of sugarbeets may influence the buildup of certain insects such as leafhoppers and aphids. The microclimate versus macroclimate in sugarbeet fields should receive some attention. Reed (137) found that in the interior valley area of California where temperatures may often reach 100° F or above there was an 18° F difference between ambient temperatures and temperatures taken in the crowns of the plants. The green peach aphid, for example, can oversummer to a limited extent in the centers of the plants, whereas ordinarily they cannot survive the summer temperatures. It can also oversummer in dense clumps of *Malva* spp. or other protected spots adjacent to beet plantings where temperatures are lower. Another interesting observation relative to the green peach aphid is that it is adversely affected by feeding on sugarbeet. The crop of this aphid is reported by Chang (22) to become so swollen that it is almost impossible for the insect to reproduce. Biotypes of this insect adapted to sugarbeet, however, can survive fairly well but may never reproduce to the extent that they can on other more favorable hosts. From this discussion, it is apparent that a complete knowledge of the sugarbeet agroecosystem is of extreme

FIG. 10.1. The Pacific Coast wireworm *(Limonius canus)* and damage to sugarbeet seedlings.

importance in working out a satisfactory pest management control program.

DESCRIPTIONS OF PESTS AND INJURY

From the standpoint of the grower and field technologist, several different classifications of insects and mites could be used, but in this chapter I will divide them into (1) those that are present at the time of planting, (2) pests of seedlings, (3) pests of older plants, (4) pests of seed production, (5) occasional pests, and (6) natural enemies. Some of the general pests such as cutworms attack sugarbeets in all stages of plant growth and can even be present when the crop is planted.

PESTS PRESENT AT TIME OF PLANTING

Wireworms, the larvae of click beetles (family Elateridae), frequently attack sugarbeets at the time of germination and also feed on the taproots of seedlings (Fig. 10.1) (90), often causing death of the plants. Most of the species causing damage in the western part of the United States are in the group of irrigated-land wireworms belonging to the genus *Limonius*. These include the sugarbeet wireworm, *L. californicus* (Mann.), *L. canus* LeConte, *L. infuscatus* Mots., and *L. subauratus* LeConte. In addition to these species, damage can also be caused by wireworms in the following genera: *Hemicrepidius, Anchastus, Aeolus, Conoderus, Ctenicera, Melanotus,* and *Cardiophorus*. The adults of wireworms deposit their eggs in the soil, in many cases in the spring of the year, and the larvae are active two to six years before pupating in the fall of the year and emerging as adults the following spring. Some species have a one-year life cycle and in such cases adults may emerge in the fall of the year and overwinter in the adult stage. As early as 1914, Graf (50) reported on work carried out on the sugarbeet wireworm in southern California from 1909 to 1912, and Stone (157) presented an excellent life history of the sugarbeet wireworm. A review of the literature in the western states up to about 1958 was presented by Lange et al. (114).

Wireworms are often very abundant following certain cropping systems, and usually control is necessary before sugarbeets can be planted following the discing of previously uncultivated land, grassland, or pasture land. Alfalfa grown under certain conditions decreases populations of wireworms but under other cultural situations may increase particular species. In California, chlorinated hydrocarbon insecticides have greatly reduced wireworms as a problem. The rather extensive use of seed treatment materials at the time of planting has greatly lessened their damage (105).

Larvae of the seed-corn maggot, *Hylemyia platura* (Meigen), occasionally feed on germinating sugarbeets, particularly in spots where excessive amounts of organic matter are incorporated into the soil and during cool weather where adequate moisture reaches the surface. In certain cases, the first generation is initiated in plant debris and the maggots present in the soil feed on germinating sugarbeets. In other instances, the impetus for egg deposition is the cultivation of the soil, and the first generation of maggots then go into the seed parts as they are germinating.

Several species of springtails (Collembola) occasionally cause injury. The group called seed springtails, *Onychiurus* spp., commonly attack germinating seeds and small seedlings in California. The garden springtail *Bourletiella hortensis* (Fitch), is another common species. Springtails are usually associated with low areas rich in organic matter or plowed-under debris.

The garden symphylan, *Scutigerella immaculata* (Newport), is almost a universal pest and is particularly damaging in certain areas rich in organic matter where they feed on the germinating seeds and also on the small seedlings. The small roots are continually pruned back, the plants become stunted, and in certain cases areas in fields are completely eliminated of plants.

The sugarbeet root aphid, *Pemphigus populivenae* Fitch, is a common pest in many parts of the West. Detailed life history studies (53) have shown that this species forms saclike galls on the leaves of *Populus trichocarpa* in California and Oregon and has as alternate hosts Swiss chard, spinach, *Chenopodium alba,* and *Amaranthus graecizans.* Contrary to some published records (99), this species has not been found to feed on the roots of docks, *Rumex* spp. (110). It is possible that *Pemphigus balsamifera* Williams, which forms leaf galls on *Populus angustifolia,* is a similar but distinct species in the Great Basin region and Canada. The primary host of the *Rumex*-dwelling species is not known. Although lettuce is commonly listed as a host of the sugarbeet root aphid, the investigations of Lange (110) would indicate that lettuce infestations are caused by a related species, *Pemphigus bursarius.* White, woolly masses of aphids on the roots of the plants are characteristic of this particular insect (Fig. 10.2). In California, several generations are possible during the season on the roots of beets or other host plants. In the fall of the year sexuparae leave beets and other hosts and fly to almost any poplar, give birth to male and female beakless sexuales which mate, and each female gives birth to a single large egg (Fig. 10.3). The eggs overwinter and in the spring hatch into stem mothers which crawl up to the new foliage and start leaf galls. The fundatrigeniae which emerge later from the galls fly directly to sugarbeets and other hosts and give rise to living young which start new infestations. If the eggs are deposited on nonhost poplars, no galls are produced. This species is quite capable of producing continuous anholocyclic generations on weed hosts or on sugarbeets without reference to the primary host (poplars). Reductions in yield from this aphid have been recorded by a number of workers, and in cases where culls overwinter in the soil or even in soil lacking beets, the aphid is often able to carry through to the next crop.

Fig. 10.2. Sugarbeet root aphid *(Pemphigus populivenae)* showing masses of white, woolly material on roots of sugarbeet.

Several species of white grubs (family Scarabaeidae) occasionally cause damage to the roots of sugarbeets, particularly in cases where grass or sod-land is plowed under prior to planting beets. In California and in Washington, the ten-lined June beetle, *Polyphylla decemlineata* (Say), has occasionally damaged beets. Maxson (129) recorded damage to the roots of sugarbeets in Illinois in 1900 by the rugose May beetle, *Lachnosterna rugosa* Melsheimer, and by *L. anxia* LeConte in Colorado. The carrot beetle, *Bothynus gibbosus* (De Geer), damaged sugarbeets in Nebraska in 1890.

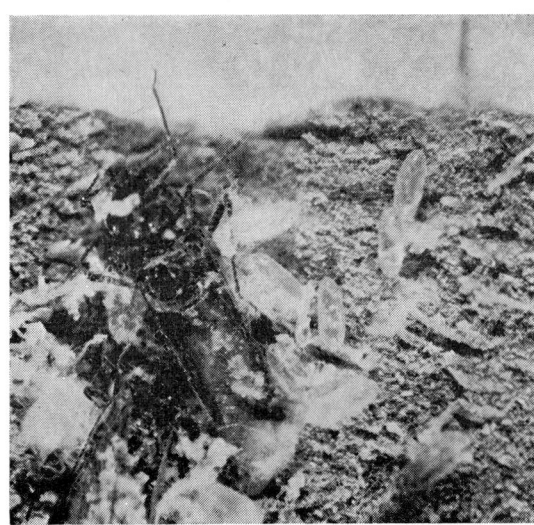

Fig. 10.3. Eggs of sugarbeet root aphid *(Pemphigus populivenae)* in crevices of cottonwood bark showing dead aphids (sexsuparae) that produced sexual forms.

Rootworms, the larvae of cucumber beetles, *Diabrotica* and *Acalymma* spp., occasionally feed on sugarbeet seedlings and are present in the soil at the time sugarbeets are planted. In other situations, they feed on the mature taproot, causing characteristic round circular pits. Sugarbeets do not seem to be a preferred host but are probably selected when other plants are not available. The adults are foliage feeders. The western striped cucumber beetle, *Acalymma vittata* (Fabricius), was collected many times from the foliage of sugarbeets in the Pacific Coast states from 1907 to 1910 but caused no serious injury (27). The western spotted cucumber beetle, *Diabrotica undecimpunctata undecimpunctata* Mannerheim, destroyed a field of sugarbeets in California in 1908.

Cutworms and armyworms belonging to the very large family Noctuidae are often severe pests of sugarbeets and many other field crops. It would be impossible in the space available to discuss all the different species, but a few of them will be taken up as examples of the group. In general, cutworms are more surface or subterranean feeders, cutting off the plants at the ground level, whereas armyworms often feed more on the foliage of the plants and may move into sugarbeet fields from adjacent crops. Some cutworms lay their eggs in groups on any substrate and others scatter them over the soil. Many armyworms deposit the eggs in masses on the plants, intermingled with scales from the bodies of the females. As early as 1905, Pettit (136) reported that many cutworms feed on the sugarbeet. The variegated cutworm, *Peridroma saucia* (Hübner), is one of the most widely distributed of the cutworms and has been known in this country since 1841. It attacked sugarbeets on the Atlantic coast as well as on the Pacific coast before 1901 (24, 26) and was especially destructive in Oregon, Washington, and British Columbia in 1900, when damage to several crops totaled $2,500,000. Several minor outbreaks have occurred since then, but a report that no outbreaks occurred in the Great Western Sugar Company's territory from 1913 to 1948 (129) seems fairly applicable to other sugarbeet areas as well.

The pale western cutworm, *Agrotis orthogonia* Morrison, is more restricted in distribution, but its great capability for causing damage makes it the most dangerous of the western species (129). Dry winters and springs must occur at least every 10 years for the pale western cutworm to maintain itself. Outbreaks occurred in northern Colorado in 1912 and 1921. Of less importance were the red-backed cutworm, *Euxoa ochrogaster* (Guenee), that attacked sugarbeets in Montana in 1929 and 1930, and the spotted cutworm, *Amathes c-nigrum* (Linnaeus), that attacked beets in Wisconsin before 1948 (129).

The armyworm, *Pseudaletia unipuncta* (Haworth), was reported to be always present and capable of building up very rapidly (129). It is ordinarily associated with grasses.

In California, the black cutworm, *Agrotis ipsilon* (Hufnagel), is usually an annual pest, cutting off seedling sugarbeets at the ground level, and is almost impossible to control in cloddy soil as it remains mostly underground.

INSECTS AND MITES AND THEIR CONTROL 299

FIG. 10.4. Larva and pupae of the rough-skinned cutworm *(Proxenus mindara)*.

In recent years, the rough-skinned cutworm, *Proxenus mindara* Barnes and McDunnough, has become an increasing pest on the roots of sugarbeets in many areas of California in addition to attacks on many other crops. The light tan to brown nondescript larvae (Fig. 10.4) are characteristically subterranean, doing most of their damage to the roots where typical feeding consists of black, gouged-out areas which often run together, making elongate darkened blemishes (Fig. 10.5), but they can also feed on foliage touching the ground. The damage may be superficial but in many cases is deeper and allows for the entrance of decay organisms which discolor the root tissues and may kill the beets completely (Fig. 10.6). Larvae often pupate at the bases of the plants in superficial cocoons or on the soil. The smoky brown moths (Fig. 10.7) have a wing expanse of about

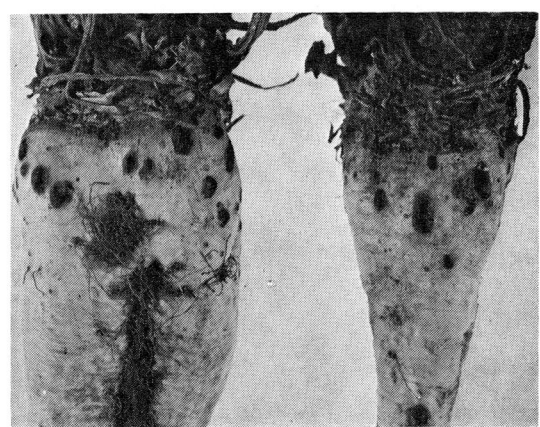

FIG. 10.5. Damage to sugarbeet roots caused by the rough-skinned cutworm *(Proxenus mindara)*.

Fig. 10.6. Cross section through roots of sugarbeet to show primary feeding damage by the rough-skinned cutworm and secondary invasion by decay organisms.

one inch, with white hind wings, and often stay hidden about the bases of the plants, flying only when disturbed. There are apparently several generations a year of this cutworm. The caterpillars may be confused with the sugarbeet armyworm but the skin granules on the surface of the larvae are characteristic of *Proxenus.*

The beet armyworm, *Spodoptera exigua* (Hübner), is a periodic pest in California, attacking the roots as well as the foliage of sugarbeets (Fig. 10.8) (17). It was observed attacking sugarbeets in Oregon and California as early as 1876 (25). In 1888, hundreds of acres of sugarbeets were too heavily damaged to be harvested in Colorado and a new sugar factory was forced to close at Eddy, New Mexico, because of the armyworms' depredations. It was very destructive to sugarbeets in Utah in 1908, and Essig and Hoskins (41) report it to be one of the commonest pests in the West. In California, this insect is controlled quite effectively in the field by natural epizootics of a polyhedral virus.

The western yellow-striped armyworm, *Prodenia praefica* Grote, is a cyclic pest in California. Severe attacks occurred in 1956 and 1957 (104). These armyworms are general feeders on many crops, and damage to sugarbeets often follows their migration from adjacent newly cut alfalfa. They feed readily on the foliage and migrate as an army to feed on the leaves, stems, and crowns of sugarbeets. Severe attacks often strip the leaves and only the leaf petioles remain intact. There are three generations a year and a partial fourth under central California conditions. The eggs are laid in feltlike masses on the leaves. Up to 1,000 eggs can be laid by individual females, with an average of over 500 eggs per mass. During these epidemics, a polyhedral virus disease destroys large numbers of larvae and the flaccid, disintegrating reddish brown caterpillars hang dead from the plants.

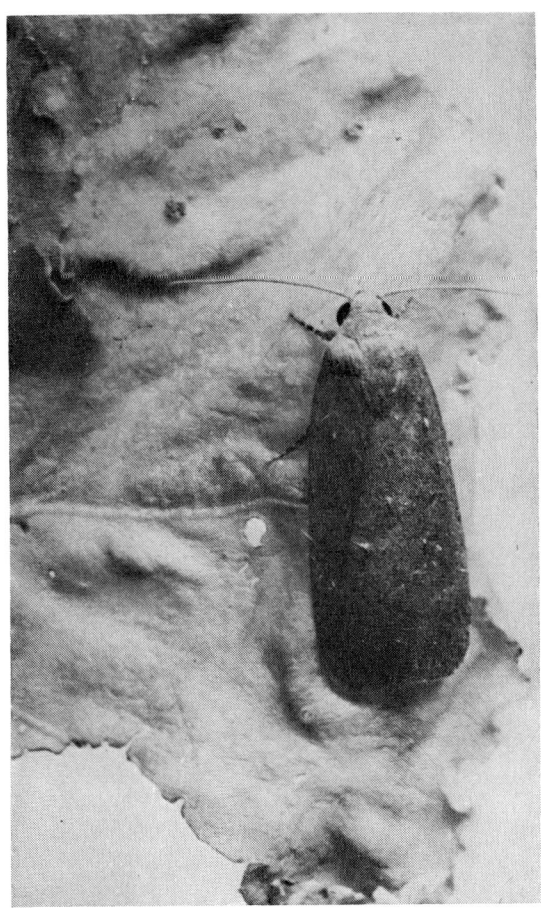

Fig. 10.7. Adult of the rough-skinned cutworm *(Proxenus mindara)* resting on sugarbeet leaf.

PESTS OF SEEDLINGS

Many of the cutworms present at the time of planting sugarbeets attack the seedlings and also feed on the roots or defoliate mature plants. In some cases, however, there are insects primarily associated with the young plants, and as the plants become older they become fairly resistant. In the case of the transmission of certain viruses by aphids or leafhoppers, the diseases are more severe if they are transmitted at the seedling stage.

A shot-hole appearance in the leaves of sugarbeets is characteristic of the adult feeding of several species of flea beetles. The larvae can also cause damage by feeding on the roots. Most of the species feed on weeds and move to sugarbeets. In California (100), *Hemiglyptus basalis* (Crotch), Halticus spp., and *Chaetocnaema* spp. have been reported damaging sugarbeets.

Fig. 10.8. Top and crown injury to young sugarbeet plant caused by larval feeding of the sugarbeet armyworm *(Spodoptera exigua)*.

Adults of the tuber flea beetle, *Epitrix tuberis* Gentner, damaged 3,000 acres of sugarbeets so seriouly in Montana in 1948 that they were either abandoned or replanted (144), and the same trouble was experienced in 1955 (134). This flea beetle was an occasional pest of sugarbeets in Washington from 1935 to 1950. The banded flea beetle, *Systena taeniata* Say, frequently damaged sugarbeets in the West and Southwest (26), particularly in Colorado, Idaho, Montana, South Dakota, Utah, and Wyoming (129). The pale-striped flea beetle, *Systena blanda* Melsheimer, was found first on sugarbeets in Michigan in 1886 and practically destroyed some beet crops there in 1899 (23) and 1900 (26). It damaged beets in Indiana in 1890 and 1901, New Jersey in 1891 and 1893, Colorado in 1900, and South Carolina in 1901. The three-spotted flea beetle, *Disonycha triangularis* (Say), was reported slightly injurious in Colorado, Idaho, Montana, Ohio, and Washington and very serious in South Dakota (129). The spinach flea beetle, *Disonycha xanthomelas* (Dalman), was reported in greater abundance in 1900 than in previous years (25) and damaged beets in 1900, 1902, and 1903 in the East. The red-headed flea beetle, *Systena frontalis* Fabricius, was first observed attacking sugarbeets in Nebraska in 1891 and injured the crop in New York in 1899 (25). The western black flea beetle, *Phyllotreta pusilla* Horn, damaged sugarbeets in Colorado in 1901 (26) and the Colorado cabbage flea beetle, *P. albionica* (Lec.), attacked sugarbeets in Idaho in 1931 (45).

The sugarbeet crown borer, *Hulstia undulatella* (Clemens), is a periodic pest in California, Colorado, Oregon, Utah, and Washington. Outbreaks of this insect were recorded as early as 1904 by Maxson (129). Damage was sustained in the Utah sugarbeet fields in 1921 where 30 to 50 percent of the plants were killed (61). It was first observed on sugarbeets east of the Rocky Mountains in Colorado in 1937 and outbreaks occurred again in 1939 and 1943. In 1949 Lange (101) reported a severe attack of this borer in the central valley area of California.

The crown borer attacks the crowns of many plants, including broccoli, pigweed, purslane, sour dock, and spinach. On sugarbeets they feed on the crowns of the young plants, on the petioles near the ground, or even on leaves that touch the soil surface. The larvae move back and forth inside characteristic silken tubes which are often two to six inches long and radiate out from the beet roots just under the surface of the soil. The caterpillars feed primarily upon the crown area and the feeding may be superficial (Fig. 10.9) or may actually cause a girdling of the roots. Partial girdling of the roots causes a weakened condition so that the plant often breaks off at the ground level. The feeding holes may also allow an avenue for the entrance of decay organisms. In 1949 the moths observed laid an average of 294 eggs singly on the petioles of the plants. Two generations were found, and at an average temperature of 76° F a complete life history was completed in 34 to 39 days. The occurrence of several hymenopterous parasites would indicate that parasitism may play a part in the periodic nature of the outbreaks.

Several species of webworms belonging to the lepidopterous family Pyralidae are defoliators of seedling beets. The beet webworm, *Loxostege sticticlis* (L.), causes the greatest damage. It damaged sugarbeets as early as 1869 in Utah and appeared there again in 1921. A belated application of 84,000 pounds of Paris green was applied to 31,000 acres of sugarbeets in Colorado in 1918, but damage was estimated at $264,500. In 1919 it damaged 90,666 acres of sugarbeets in Colorado, Nebraska, Wyoming, and Montana (129). The garden webworm, *Loxostege similis* (Guenee), has

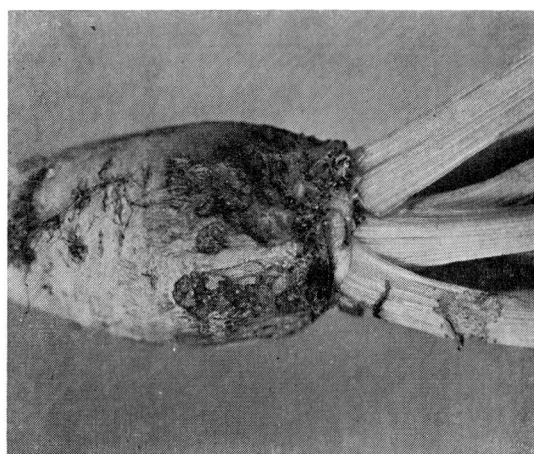

FIG. 10.9. Sugarbeet crown borer *(Hulstia undulatella)* damage to sugarbeet.

occasionally caused damage to a number of crops in the South and Middle West (26). It was the most serious pest of sugarbeets in Nebraska before 1891 (129) but there have been few reports of serious damage since. Sugarbeets grown for seed were damaged in Arizona in 1937 and 1938 and slight injury has been observed to sugarbeets in eastern Washington for many years. The alfalfa webworm, *Loxostege commixtalis* Walker, was extremely abundant on alfalfa in Colorado in 1914 and caused some damage to sugarbeets. A major outbreak occurred in the West in 1921 when reports of injury to beets were received from Colorado, Montana, Utah, and Wyoming.

We have already discussed cutworms and armyworms but it should be mentioned that these caterpillars are extremely injurious to seedling beets and in many cases are very difficult to control. In recent years, the sugarbeet armyworm, for example, has been causing subterranean damage, particularly in cloddy, rough soil where it can get to the crowns and roots of the seedling plants.

Several species of small, darkling ground beetles, particularly in the genus *Blapstinus* (family Tenebrionidae), damage seedling beets by cutting off the beets at the ground level. The larvae of these beetles are commonly called false wireworms and can feed on the roots. It is not uncommon for the adults to move in from the edges of fields and cause considerable injury.

The sugarbeet root maggot, *Tetanops myopaeformis* (von Roder), is a native fly that is present in many of the western states but has become a pest of sugarbeets only in parts of North Dakota, Montana, Utah, Idaho, and Colorado, where it damages seedling plants. The adult is about ¼ inch long and shiny black, with a bulb-shaped body, yellow legs, and two characteristic pale brown crossbands on each wing that extend from the costal cell to the fourth vein, slightly beyond the base of the discal cell. The larvae are maggotlike, tapering from the front to the back, and are slightly more than ½ inch long when mature. The puparium is about ⅖ inch long, brown, and slender.

The insects may overwinter as full-grown maggots, which change to puparia in early spring. The flies emerge from the soil in May and June and lay their eggs in clusters of 10 to 20 in the soil close to small beets. The eggs hatch in June and the maggots start feeding on either the small lateral roots or the taproot, but later they seem to concentrate on the taproot. Sometimes 20 to 60 maggots may feed on a single beet. The rasped-out area on the taproot is characteristic of injury caused by this insect. The injured part turns black, bleeds profusely, and the leaves wilt badly during warm days. In severe cases, infested plants die. There is probably one and a partial second generation annually. Damage is most severe and is not limited to beets in sandy soils.

The sugarbeet root maggot was collected in California in 1881 but does not seem to be a pest in this state. Hawley (60) reported an outbreak on beets in Utah, Idaho, and Colorado about 1920. It was also reported attacking beets in Montana and Wyoming (129) before 1948. In North Dakota, it caused $150,000 damage to sugarbeets in 1954 and $95,000 in 1955 (49).

As soon as beet seedlings are out of the ground, several species of aphids are attracted to them. In general, aphids cause two types of damage: (1) acute damage from direct feeding injury or the injection of toxins, and (2) the transmission of viruses such as beet yellows and sugarbeet mosaic. In a later discussion we will cover the role of aphids in virus transmission. In some cases, aphids may occur during the entire growth of the plants, and Maxson (129) reported that aphids even damage the seed crop in Colorado, New Mexico, and Utah. The green peach aphid, *Myzus persicae* (Sulzer), is often the commonest aphid on young plants and is also the most efficient vector of several types of yellowing viruses (8). The green peach aphid has several hundred host plants and usually sugarbeets are not the most preferred host, although biotypes may occur which readily select sugarbeets. The bean aphid, *Aphis fabae* Scopoli, often causes injury by the injection of toxins, causing stunting of the plants and reduced yields. Unlike in Europe where it is a common species, it is usually less abundant in the western states in field plantings and it is also a less efficient vector of the yellowing viruses. The potato aphid, *Macrosiphum euphorbiae* (Thomas), is another common aphid found on sugarbeets and ordinarily increases in abundance during the season in both California and in the Northwest. Woolly masses on the roots of seedling beets often indicate the early establishment of the sugarbeet root aphid, *Pemphigus populivenae* Fitch.

Although several species of the leafhopper family (Cicadellidae) occur on seedling sugarbeets, the most important species is the sugarbeet leafhopper *Circulifer tenellus* (Baker), inasmuch as it transmits the curly top virus. The southern garden leafhopper, *Empoasca solana* DeLong, is a common species damaging beets. In most cases, *Empoasca* spp. cause more damage to older beets. Investigations by Reynolds (139) in the Imperial Valley of California demonstrated that this leafhopper caused a wilting of beets and a significant reduction in yield at the time of harvest.

PESTS OF OLDER PLANTS

Several small flies of the genus *Liriomyza* are important pests of sugarbeet and several other crops in California. The vegetable leaf miner, *Liriomyza langei* Frick, is a very important species in the coastal areas of California where the larvae make winding mines inside the leaves and bore into the main veins and the petioles, causing a darkening and complete collapse of the tissues (Fig. 10.10). The adults also cause damage to seedling plants by their feeding and oviposition punctures. The adult flies use the ovipositors to puncture holes in plant tissues and feed upon the exuding droplet of sap and also to insert eggs into plant tissues. With heavy infestations, the mines run together, causing large blotch mines with extensive necrosis. This species has a history of damage to spinach, lettuce, cruciferous crops, celery, and beans in the coastal regions of California. In the interior valleys, this species is replaced by others (46).

Fig. 10.10. Whitening and discoloration of leaves of sugarbeet caused by larval mining of the vegetable leaf miner *(Liriomyza langei)* (leaves on right lightly attacked).

Larvae of the genus *Pegomya* produce large, blotch-type mines in the leaves. The beet leaf miner, *Pegomya betae* (Curtis), attacks sugarbeets in the West, and a similar fly, the spinach leaf miner, *P. hyoscyami* (Panzer), attacks the crop in the East. The two insects are very similar in appearance and habits, except for their alternate host plants. The adult of the *Pegomya* leaf miners of the sugarbeet are small, gray flies nearly ¼ inch long. These insects overwinter as puparia in the soil and the flies are among the first insects to emerge in the spring. Small, white, spindle-shaped eggs are laid in neat rows of 1 to 10 on the beet leaves, starting when the plants are in the 4-leaf stage. Eggs hatch in 4 to 5 days and the greenish white larvae enter the leaf mesophyll. Small larvae make narrow, serpentine mines but the nearly mature larvae produce large blotches ½ to 2 inches in diameter. Mature larvae leave the blotch mines and enter the soil where the puparia are formed. There are several generations annually. *Pegoyma* damage consists of pruning the small plants and thus retarding growth, although some plants may be killed outright. The leaf miners are attacked by many parasites and predators and these, apparently, control the insect during the summer. During some years, however, heavy damage may develop near harvest time.

Early western records of *P. hyoscyami* probably refer to *P. betae*. The *betae-hyoscyami* complex was reported in California as early as 1891 (129) and in New York in 1899 (47). Hawley (62) reported that the insect was especially annoying when beet tops were eaten as greens because the larvae often floated to the top of the dish. It was unusually abundant in Utah in 1921, 1933, and 1934; Montana, 1934 and 1959; Michigan, 1936; Indiana, 1941; Idaho, 1945, 1958, and 1959; Colorado, 1955; Ohio, 1958 and 1959; and Wisconsin, 1962.

Several mites of the genus *Tetranychus* are pests of sugarbeets in California, eastern Washington, northeastern Oregon, and southwestern Idaho. The two-spotted spider mite, *T. urticae* (Koch), is the commonest species in California, and Chittenden as early as 1903 (26) reported it in sugarbeet fields east of the Rocky Mountains. This mite has been particularly damaging in the San Joaquin and Sacramento valleys of California, and during 1964 and 1965 control was necessary. The whitish or yellowish discoloration of the injured foliage, together with stippling and webbing, are common symptoms of the mite. In the southern desert valley areas of California, a closely related species, the carmine spider mite, *T. telarius* (Linnaeus), is the common species (139). Yield decreases were reported by Reynolds (139) in the Imperial Valley and by control investigations by Lange in 1965 (110).

In addition to several species of armyworms and cutworms which have been discussed earlier, there are a number of other lepidopterous larvae which defoliate mature sugarbeets. In recent years, the celery leaf tier, *Oeobia rubigalis* (Guenée), has been harmful late in the season as the larvae web and roll the leaves and in many cases defoliate beets in the interior valleys of California. In the northern part of California, a related species, the false celery leaf tier, *O. profundalis* Packard, is the species involved. These moths commonly occur on weeds such as *Chenopodium* spp., and the small yellowish moths are often abundant in beet fields late in the year.

The zebra caterpillar, *Ceramica picta* (Harris), was found attacking sugarbeets in Massachusetts as early as 1841 (26) and has also been reported causing damage to the crop in Colorado, Utah, and Washington (128).

The so-called woolly bear or salt-marsh caterpillars are often abundant in the western states and in certain other localities. Early records show damage occurred from the salt-marsh caterpillar, *Estigemenae acrea* (Drury), in California and Oregon. A similar species, the yellow woolly bear, *Diacrisia virginica* Fabricius, defoliated 1,000 acres of sugarbeets in Colorado in 1909 (127) and later caused damage in South Dakota. In California, the salt-marsh caterpillar is one of the commonest species found in sugarbeets late in the season. They often migrate in large numbers across roadways to pupate along fence lines and other protected places. In some cases, they almost completely defoliate mature beets.

The silvering and black-spotting of the undersurfaces of sugarbeet leaves may be due to the feeding of bean thrips, *Caliothrips fasciatus*

(Pergands) (5). This species is associated in California with prickly lettuce and annual sowthistle which are preferred hosts. The sugarbeet thrip, *C. femoralis* Reuter, occurs on sugarbeets, primarily in greenhouses.

Several different kinds of Hemiptera or "true bugs" attack sugarbeets and of these the false chinch bug, *Nysius ericae* (Schilling), several species of lygus bugs, *Lygus* spp., and the Say stink bug, *Chlorochroa sayi* Stol, are the most important. The minute false chinch bug, *Nysius minutus* Uhl., has also been reported as an important pest. Although lygus bugs are very important pests of the sugarbeet seed crop, they can also cause injury to growing beets not only by their feeding, with resultant introduction of toxins, but by the permanent egg deposition scars left in the leaf veins and petioles.

A small ephydrid leaf miner, *Psilopa leucostoma* (Meigen), was found attacking sugarbeets in Washington in 1962 by Landis and has since been observed as abundant in other sugarbeet areas of the Northwest. This small brine fly mines the leaves of sugarbeets. The adult is a small black fly about 1/8 inch long, with yellow legs and a small rectangular smoky patch extending crosswise on the wings from the junction of the medial cross vein to the cubitus vein. The insect lays small, white, spindle-shaped eggs on the underside of the leaves. They differ from *Pegomya* eggs in that they are always laid singly and are much smaller in size. The eggs hatch in four to five days and the small larvae enter the leaf where they make long, twisting mines. The mines may be 1/4 to 1/2 inch long, and slightly wider at the distal than proximal ends. The full-grown, greenish white larva drops to the ground where it seeks cover under fallen leaves and pupates. There are several generations annually. Since it was first discovered infesting sugarbeets in 1962, the ephydrid leaf miner has spread into all sugarbeet production areas of the Northwest east of the Cascade Mountains. It can be more serious than the beet leaf miner because it continues to increase in abundance on beets from May through September.

Leafhoppers have already been discussed as seedling pests in the transmission of viruses. They can also occur in large numbers on mature beets (*Empoasca* spp. and others), causing stippling and desiccation of the foliage.

The beet petiole borer, *Cosmobaris americana* Casey, was found attacking sugarbeets in Ontario in 1935 and later in California on Swiss chard in 1943 (108); Colorado, 1944 (122); Oregon, 1965; and Idaho, 1966. In California, an outbreak on Union Island near Tracy, observed by Lange, damaged a field of 40 acres. The adult weevils insert the eggs into the leaf veins and petioles and the whitish larvae cause typical necrotic spots (Figs. 10.11, 10.12) similar to lygus bug injury. In heavy attacks, the petioles are irregularly swollen due to early larval feeding, and this can lead to complete necrosis of the petioles (Fig. 10.13).

Blister beetles (family Meloidae) were reported damaging sugarbeets as early as 1903 by Chittenden (26). Maxson (129) lists a number of species which have been particularly destructive to beets. Although some of these beetles are associated in their larval stages as predators on grasshopper egg

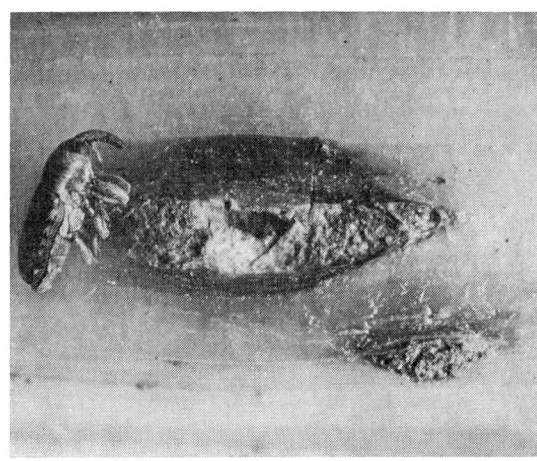

FIG. 10.11. Adult and egg puncture of the sugarbeet petiole borer *(Cosmobaris americana)*.

masses, the adults often feed on sugarbeets and other plants and are considered pests by sugarbeet growers. The black blister beetle, *Epicauta pennsylvanica* (De Geer), caused severe damage in Montana and South Dakota; the spotted blister beetle, *Epicauta maculata* (Say), caused damage in Colorado, Montana, Wyoming, South Dakota, and Nebraska; the striped blister beetle, *E. vittata* (Fabricius,) caused serious damage in Colorado in 1937; and the ash-gray blister beetle, *E. fabricii* (LeConte), caused severe damage to beets in Montana and South Dakota (26). Chittenden (26) also reported that the immaculate blister beetle, *M. immaculata* Say, injured beets in Kansas in 1897 and in Colorado in 1902; the Nuttall blister beetle, *Lytta nuttallii* Say, attacked beets at several locations before 1903; the spotted blister beetle, *M. albida* Say, destroyed a field of beets in a single day in the "Indian Territory"; the segmented blister beetle, *M. segmentata* Say, heavily damaged beets in Kansas in 1897; and the three-lined blister

FIG. 10.12. Eggs of the sugarbeet petiole borer *(Cosmobaris americana)* in sugarbeet petiole (epidermis removed to show egg).

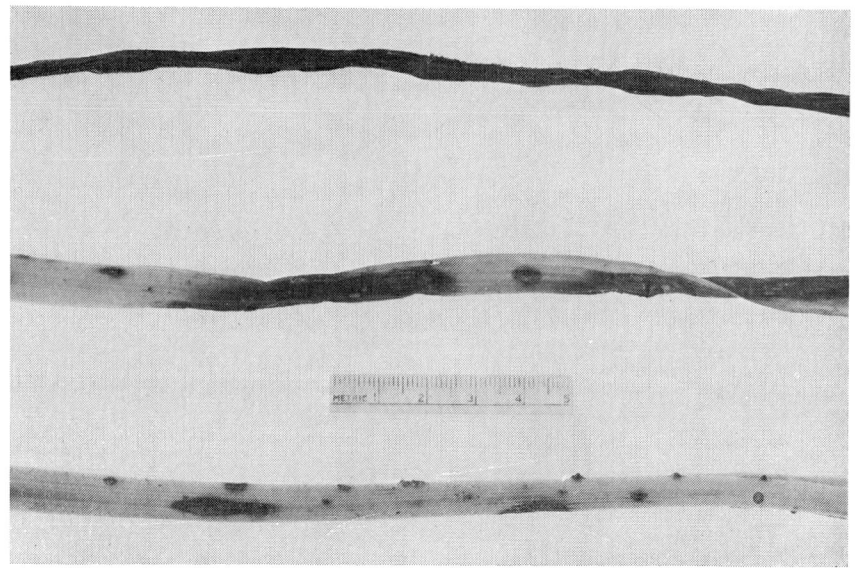

FIG. 10.13. Damage to sugarbeet petioles by the sugarbeet petiole borer *(Cosmobaris americana)* at Union Island, California (Sept. 5, 1962).

beetle, *E. lemniscata* Fabricius, caused complaints from beet growers in the southeastern states before 1903.

Many species of grasshoppers and crickets attack vegetation, and sugarbeets are not immune from their attacks. In a rather extensive account of grasshoppers as pests of sugarbeets in Utah, Hawley (62) recorded that the ravages of the Rocky Mountain grasshopper, *Melanoplus spretus* (Walsh), that occurred to crops in the midwestern states from 1875 to 1877 subsided before beet culture was introduced and that it has been a long time since a destructive outbreak occurred. However, he reported that the two-striped grasshopper, *Melanoplus bivittatus* (Say); the lesser migratory grasshopper, *M. atlanis* Riley; the red-legged grasshopper, *M. femurrubrum* (De Geer); the clear-winged grasshopper, *Camnula pellucida* (Scudder); and the Carolina grasshopper, *Dissosteira carolina* (Linnaeus) were most injurious to sugarbeets. The red-legged grasshopper was destructive to sugarbeets in Illinois (26), and Maxson (129) recorded it damaging sugarbeets in California, Colorado, Montana, Minnesota, Nebraska, and South Dakota. He also reported that the two-striped grasshopper injured sugarbeets in Colorado, California, Minnesota, Montana, Nebraska, North Dakota, and Wyoming; the clear-winged grasshopper damaged beets in Colorado and New Mexico; and the lesser migratory grasshopper damaged beets in Minnesota, Nebraska, and North Dakota. The Mormon cricket, *Anabrus simplex* Haldeman, is sometimes a pest of sugarbeets in Montana and Wyoming according to Hawley and Maxson (62, 129).

PESTS OF SEED PRODUCTION

Sugarbeets are planted during late summer for production of a seed crop the following summer in restricted parts of Arizona, southwestern Utah, Nevada, and western Oregon. The insects that attack a crop of sugarbeets grown for seed are little different from those that attack a crop of beets grown for sugar production in the same general area. However, since the seed crops are planted in late summer, the succulent plants are more attractive to some kinds of insects than the mature plants of a sugar crop. Also, the large, fibrous seedstalks and heavy production of developing seed are also more attractive to certain insects than the maturing leaves of the sugar crop.

Hills (77) adequately covered the arthropod pests of the sugarbeet seed crop and for this reason a complete review of the insects and mites involved is not considered necessary. Several investigators (18, 76, 82, 84, 96) have shown the relationships between certain Hemiptera and Homoptera and seed production. These investigators, and others, have shown that several species of lygus bugs, *Lygus hesperus* Knight, *L. elisus* VanDuzee, and *L. lineolaris* (Palisot De beauvois), are involved in plant deformation, dieback and reductions in seed yield and seed viability. Carlson (19, 20) has shown that the economic level for table beets is two to four bugs per sweep. He found that *L. hesperus* reduced the yield, increased the number of small seeds, reduced the weight of individual seeds, and had effects on the number of viable germs per multigerm seed. Hills et al. (78, 79, 81) have shown that the yellowing viruses, both separately and together with curly top, have adverse effects on yields of sugarbeet seed and that time of infection seems to be very important in that early infected beets had greater reductions in seed than those exposed at a later time.

In the autumn, the young sugarbeet plants of the seed crop are attacked by several kinds of lepidopterous larvae, including the beet armyworm, *Spodoptera exigua* (Hübner); the yellow-striped armyworm, *Prodenia orthogalli* Guenée; and the garden webworm, *Loxostege similalis* (Guenée). Cutworms may also be a problem during autumn in seed beet fields in the Southwest. The commonest cutworm in the fields is the granulated cutworm, *Feltia subterranea* (Fabricius). The salt-marsh caterpillar, *Estigmene acrea* (Drury), is also a common pest of seed beets in Arizona. In the spring, the larvae of the celery leaf tier, *Oeobia rubigalis* (Guenée), may attack sugarbeets in Arizona and Utah and cause some defoliation. A similar moth, *Platynota stultana* Walsingham, also infests seed beets both in autumn and in spring.

Two kinds of spider mites attack sugarbeets grown for seed. The two-spotted spider mite, *Tetranychus urticae* (Koch), is present and destructive through the seed production area, whereas *T. desertorum* Banks is common in the South.

The southern garden leafhopper, *Empoasca solana* DeLong, is often the commonest insect in seed beet fields in the Southwest in the winter and early spring. Large numbers of these insects can reduce the yield but apparently not the viability of the seed. The false chinch bug, *Nysius ericae* (Schilling), also may congregate in beet fields in large numbers but populations of as many as 500 adults per plant from bloom until harvest do not seem to reduce yields.

The green peach aphid, *Myzus persicae* (Sulzer), is one of the most destructive insects to beets grown for seed. Large numbers of aphids can cause primary damage but greatest damage arises from the transmission of harmful viruses, such as beet yellows and beet western yellows. The bean aphid, *Aphis fabae* Scopoli, infests the seed beets in some localities.

The beet leafhopper, *Circulifer tenellus* (Baker), also causes considerable indirect damage during some years when it infects with curly top large numbers of the overwintering plants.

OCCASIONAL PESTS

Many species of insects and mites infest sugarbeets occasionally and a few of these are mentioned as it is always possible that some of the obscure pests may become destructive in the future. One of the most interesting insects recorded recently is a scale insect, *Asterolecanium arabidis* Signoret, by Landis (98). The pit-making scale infests the lower leaves of sugarbeets and they are present on the shortened petioles or on the undersides of the tightly curled leaves. When the infested leaves drop later in the year, plants that have been infested with this scale can be identified by large, black leaf scars at the base of the plant crown. There are a number of other hosts, but broad-leaved cress, *Lepidium latifolium* L., may be the source of the infestation in Washington. At the present time, this scale is known from California, Oregon, and Washington, although it is apparently a species of European origin and was first collected in Connecticut.

The pavement ant, *Tetramorium caespitum* L., was first observed as a pest of sugarbeets in 1953 in California by Lange (106). These ants feed on the germinating seeds and on the primary roots just below the crowns of the seedling plants (Fig. 10.14). Their feeding may cause a complete girdling of the roots and subsequent death of the plants. The ants feed on the sap exuded from the destroyed root areas. Damage to sugarbeets seems to occur when beets follow several years of alfalfa, particularly late-planted beets which are in a suitable stage for attack at a time when the ants are active.

Two small, yellowish brown beetles that feed as adults and larvae on weeds received early attention on sugarbeets in the West (28). The larger sugarbeet leaf-beetle (alkali-beetle), *Monoxia puncticollis* Say, injured sugarbeets in Colorado in 1898 and 1902; Utah, 1909; and New Mexico, 1913 (28). The western beet leaf beetle, *M. consputa*, Lec., injured sugarbeets in Oregon in 1890 and 1891. To date they have not become major pests of the crop.

Two species of carrion beetles attack sugarbeets occasionally; the black

FIG. 10.14. Damage to sugarbeet seedlings by the pavement ant *(Tetramorium caespitum)*. Damage appears as blackened areas on the roots (undamaged on right).

carrion beetle, *Silpha opaca* L., and the spinach carrion beetle, *Silpha bituberosa* Lec., feed on young sugarbeets both as adult beetles and as larvae, causing damage to the margins of leaves. Chittenden reported *S. opaca* damaging sugarbeets in Colorado, Montana, South Dakota, and Wyoming.

The imbricated snout beetle, *Epicaerus imbricatus* (Say), causes occasional damage to sugarbeets. It was first observed feeding on beets in Illinois in 1900, but apparently caused little concern until it severely damaged sugarbeets in Nebraska in 1927 (129).

The clover-root mealy bug, *Dactylopius trifolii* Forbes, was reported to damage the roots of sugarbeets in Michigan in 1901 (26).

NATURAL ENEMIES AND PATHOGENS

In terms of some field crops, sugarbeets represent a rather long-term agroecosystem, and for this reason there are many natural enemies and pathogens in sugarbeet fields which frequently keep injurious pests under control.

In the Hemiptera, big-eyed bugs of the genus *Geocoris* are common predators in sugarbeet fields where they feed on aphids, leafhoppers, and other soft-bodied insects. In many cases they are associated with adjacent alfalfa fields and are both plant-feeders and predators. Minute pirate bugs are common predators in sugarbeet fields, feeding upon thrips, mites, and various eggs. A common pirate bug is *Orius tristicolor* (White), which has been observed feeding on mites in all stages. Damsel bugs, such as *Nabis ferus* (Linneaus), is a common inhabitant of sugarbeets, as is *Zelus socius* Uhler, which is called the leafhopper assassin. In the order Neuroptera, common inhabitants of beets include green lacewings in the family Chrysopidae and brown lacewings in the family Hemerobiidae. These insects feed both as larvae and adults on various soft-bodied insects and are particularly noted for their feeding habits on aphids. The twisted-winged parasites of the order Strepsiptera are found as parasites of leafhoppers. In the Diptera we find many species of syrphid flies (family Syrphidae), which in their larval stages feed on aphids and other soft-bodied insects. In many cases, aphids are completely controlled by these insects. In the central valley of California we find a number of species feeding on aphids, including *Syphus opinator* O.S., *Scaeva pyrastri* L., *Metasyrphus wiedmanni* John., *M. meadii* Jones, and *Eupeodes volucris* O.S. Many adults of parasitic flies belonging to the family Tachinidae are found on beets or associated with flowers of weeds growing in sugarbeet fields. The larvae are internal parasites of a number of caterpillars and certain other insects associated with sugarbeets. In the family Chloropidae, we find an interesting relationship with several species of *Thaumatomyia*, especially *T. glabra* (Meigen), which is an active predator on the beet root aphid on roots of beets. The adults are normally found resting on sugarbeet plants. In many cases the larvae of this fly completely decimate populations of root aphids. Many minute to moderately sized wasps belonging to the order Hymenoptera occur in sugarbeet fields and most are parasitoids on various insects.

Many parasites belonging to the families Eulophidae, Encyrtidae, Ichneumonidae, Braconidae, and Mymaridae are commonly found resting on the foliage of sugarbeet. The green peach aphid parasite, *Aphidius matricariae* Haliday, is now fairly well established in many localities in California, and another species, *Lysiphlebus testaceipes* (Cresson), occurs as an internal parasite of the green peach and bean aphid. A third species, *Diaeretiella rapae* M'Intosh, has also been found parasitizing the green peach aphid. Leaf miners of the genus *Liriomyza* are commonly parasitized by numerous hymenopterous parasites, particularly members of the genus *Solenotus* (111). The adults of these small parasites of leaf-mining flies sometimes completely cover sugarbeet foliage in the central part of California. Members of the genus *Polynema* are of particular interest because they parasitize eggs of the beet leafhopper. There are several predacious thrips in sugarbeet fields. One species, *Scolothrips sexmaculata* (Perg.), was observed feeding on spider mites in the San Joaquin Valley of California, along with the black hunter, *Leptothrips mali* (Fitch), and a species of *Aeolothrips*. Several species of ground beetles in the family Carabidae commonly occur in sugarbeet fields, feeding on caterpillars or on slugs.

Ladybird beetles of the family Coccinellidae include some of the commonest inhabitants of sugarbeets. One species, *Hippodamia convergens* Guer., migrates from the Sierra Nevada mountains each year and within a short period of time is instrumental in almost completely reducing populations of the green peach aphid. Other species occurring in California on beets include *H. sinuata* Muls., *H. parenthesis* (Say), and *H. quinquesignata* (complex) Kby. We also find *Ceratomegilla fuscilabris* (Muls.) and *Coccinella novemtata franciscana* Casey. Ladybird beetles feed both as larvae and as adults on aphids, eggs, and other soft-bodied insects and are fairly efficient predators.

Certain naturally occurring insect pathogens often are instrumental in reducing populations of insects. The entomogenous fungus, *Entomophthora aphidis* (Hoffm.), has been observed to completely reduce populations of the green peach aphid within a period of a few days. Fungi in this group also attack leafhopper caterpillars and other insects. In the group of viruses, the nuclear polyhedroses often reduce populations of armyworms and cutworms. In certain cases some of the granulosis viruses are also involved in controlling caterpillars.

A number of predacious mites reduce populations of spider mites. Two species that have been observed in the San Joaquin Valley of California are *Metaseiulus occidentalis* (Nesbitt) and *Amblyseius fallacis* Garman. These predacious mites often reduce populations of spider mites damaging sugarbeets.

VECTOR-VIRUS RELATIONSHIPS

In dealing with vector-virus relationships, it is absolutely necessary to understand epidemiology and other more pathological phases of the diseases. We also need added information on the life histories, ecology, population dynamics, economic levels, and flight patterns of the vectors, as well as much other background information, in order to carry out a successful control program.

The curly top virus is carried only by the sugarbeet leafhopper, *Circulifer tenellus* (Baker). Severe reductions in yield can occur to sugarbeets even with our resistant varieties, and in addition the disease can also affect spinach, tomatoes (western blight), peppers, beans, squash, cantaloupes, potatoes (green dwarf), and a number of other hosts. The work on the systematics of the group by Oman (130) made it possible for Young and Frazier (164) to revise the group and also search for parasites in the Mediterranean region. Adequate accounts of the symptoms and other aspects of curly top are covered in several articles (21, 33, 145, 146).

With the planting of sugarbeets closer to the breeding areas of the leafhopper and the appearance of more virulent strains of the virus (9), it is apparent that continual investigations are necessary to adequately control this disease. In addition to various mustards (*Brassica* spp.), the most important summer host is Russian thistle. The state of California since 1950 has attempted to initiate a large-scale Russian thistle eradication program.

Of the yellowing viruses, the most important are those causing beet yellows (BY) (Figs. 10.15, 10.16), beet western yellows (BWY), and beet mosaic (BM). The literature is extensive on the subject, but the effects of BY on sugarbeet growth and yield have been adequately documented by Bennett and co-workers (13, 14, 15), by detailed work in Great Britain by Hull (87, 88), and in California (86). Changes in the amino acid relationships in sugarbeets have been described in several articles by Fife (42, 44), including a satisfactory method of measuring the resistance of growing beets to BY virus by determining growth rate.

The more widespread BWY was described originally by Duffus (35) as radish yellows. Subsequent publications by the same worker (36, 38) established the economic significance of this virus and also several of the strains involved. BWY has a much wider distribution than BY and in several USDA publications it was reported that the effects of several of these viruses are additive. Duffus in 1963 (37) also determined the relative significance of each of these viruses on overwintering sugarbeets.

BM was described by Severin (148) and by Severin and Drake (149).

FIG. 10.15. Symptoms of beet yellows in a field in the Salinas Valley, California.

Fig. 10.16. Winged and wingless green peach aphids *(Myzus persicae)* on sugarbeet leaf. This is the chief vector of the yellowing viruses affecting sugarbeet.

Shepherd et al. (153, 155) determined losses due to BM and also described a necrotic disease caused by a virulent strain of BM. Bennett (10) defined the various isolates of the disease.

Although several species of aphids are involved in the transmission of these viruses to sugarbeets, the green peach aphid, *Myzus persicae* (Sulz.), is the main and most efficient vector. Lange and other workers at the University of California have found intraspecific variation in the ability of the green peach aphid to transmit sugarbeet virus yellows. It was shown that 2 of 14 biotypes of this aphid were unable to transmit the disease with any degree of efficiency and the ability to transmit the virus was inheritable, as selections could be made for more efficient transmission. Such a relationship did not occur with BM. Chang (22) believed that the differences were correlated with the ability of the different biotypes to reach the phloem for acquisition and transmission of the virus.

Reed (137) found that both biotic and abiotic factors affect the relative abundance of the green peach aphid and that temperatures were very important in determining reproduction rate of the green peach aphid. The sexual cycle of this aphid took place on peach and related trees in the central valley area of California. Certain strains produced sexual forms regularly in the fall of the year, even in the laboratory. The relation between the green peach aphid and peaches or other deciduous trees is apparently not as important in California as it is in certain other areas in the the Pacific Northwest. Many cruciferous overwintering weeds are a ready source of BWY. Weeds are very important in connection with the maintenance of large populations of the green peach aphid throughout the year (110) and weeds were also found important in England.

Beets are the chief source for BY. Kennedy and co-workers (94, 95) carried out fundamental work on the relationships between leaf age and the abundance of both the green peach aphid and the bean aphid. The investigations of Heie and Petersen (68) in Denmark are outstanding examples of careful investigations dealing with BY and green peach aphid. Flight of the green peach aphid is very important as shown by Dickson and Laird (30), who studied populations on desert beets. In the Imperial Valley of California most aphids fly in every year as none can survive the high temperatures during the summer.

Lange (107) and Lange and Hills (112) utilized yellow traps in determining the flight patterns of the green peach aphid in connection with the timing of insecticides. Seasonal trapping records in other areas, using yellow pan traps, were initiated about 1958. It was found in several areas that a 5- to 10-year summary of the flight pattern for each location could be used to determine planting dates and also obtain fundamental information on the differences in flight patterns in the different areas. A typical flight pattern for the Salinas Valley, for example, is shown in Figure 10.17, where a limited flight occurs in the spring of the year during April and May and a more extensive flight occurs from September through October. This pattern is completely reversed in the central valley area of California

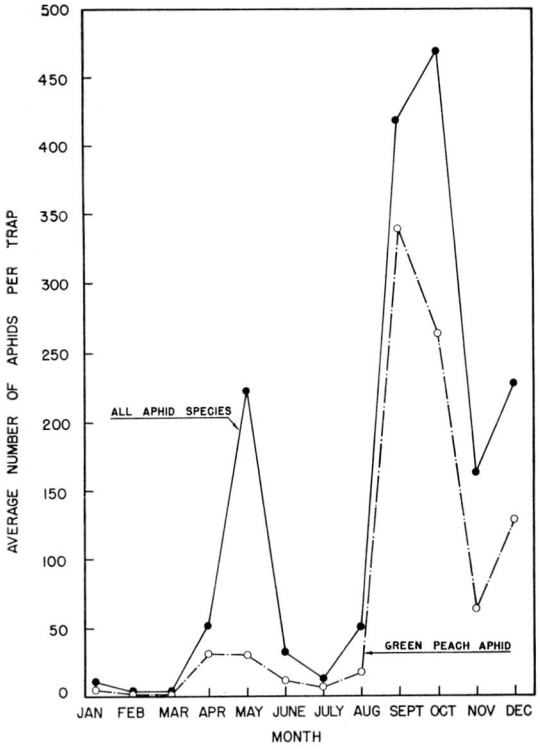

Fig. 10.17. Flight pattern of the green peach aphid compared with all aphid species in the Salinas Valley, California, based on yellow pan water traps (average of 4 traps).

(Table 10.3). The peak flights of the green peach aphid in Yolo County (1967–69) occurred in April and May, with a more limited flight in the fall of the year. A comparison of yields in two areas of California for the period 1955–68 shows that during years of extensive flights of the green peach aphid (1958, 1961, 1967) there was a decrease in tons per acre of 24 to 35 percent for an average decrease of 30 percent (Table 10.4). A significant decrease in sucrose was not always obtained.

Sampling methods for the green peach aphid on the plants are very important and an evaluation of methods has been described by several investigators, including Heathcote et al. (67) and Kershaw (97). A sequential sampling system for aphids on sugarbeets was developed by Sylvester and Cox (159). In California an economic population level of 0.25 to 0.50 apterous aphids per plant is used as an economic threshold. Differences in the transmission of BY virus by alates and apterous forms of green peach aphid were described by Heathcote and Cockbain (65). Lange and co-workers (110) have shown a definite antibiosis in the resistant sugarbeet hybrid US H9B relative to the rate of reproduction of the green peach aphid in the laboratory and field. In addition, by using the new methods developed by McLean and co-workers (124, 125), they have been able to use the aphids to determine the degree of resistance of individual plants

TABLE 10.3. Numbers of green peach aphids and total numbers of all aphids collected in 8 yellow pan aphid traps in Yolo County, California, 1967–69

Month	1967 GPA	1967 Total	1967 GPA %	1968 GPA	1968 Total	1968 GPA %	1969 GPA	1969 Total	1969 GPA %
Jan.	6	11	63	16	39	41	0	0	0
Feb.	2	10	20	6	7	85	0	0	0
Mar.	46	88	52	31	75	41	46	64	72
Apr.	87	115	76	585	1,181	49	2,421	3,217	75
May	40,809	54,087	75	30	680	4	4,358	6,976	63
June	1,223	3,607	34	3	59	5	3	204	2
July	6	38	16	49	393	12	47	344	14
Aug.	20	326	18	205	3,093	9	196	968	20
Sept.	627	1,726	36	179	446	40	75	152	49
Oct.	754	1,026	73	101	191	52	54	87	62
Nov.	1,574	2,237	70	18	85	21	104	155	67
Dec.	27	31	87	7	16	44	53	70	76

SOURCE: Lange (110).
NOTE: GPA = Green peach aphid. Total = Total numbers of all aphid species.

to BY. Due to the heterozygosity in resistant lines, this could prove very helpful to plant breeders. They also found that in US H9A and B, red hypocotyls and petioles are associated with susceptibility to BY.

In addition to the viruses on sugarbeets described by Shepherd et al. (154) and Sylvester and Severin (160), several new diseases or more unusual diseases should be mentioned. Bennett (11) described marble leaf of sugarbeets which is only juice transmissible. He also pointed out the leafhopper-

TABLE 10.4. Root yields per acre and percent sucrose in two California areas for the period 1955–68

Year	Area 1 Yield (T/A)	Area 1 Sucrose (%)	Area 2 Yield (T/A)	Area 2 Sucrose (%)
1955	19.86	14.93	19.14	14.88
1956	20.24	15.12	19.44	15.04
1957	23.81	13.46	23.60	13.25
1958	16.59	14.73	16.74	14.51
1959	25.43	15.38	25.09	15.35
1960	20.75	13.33	21.45	13.01
1961	16.48	13.06	16.95	12.41
1962	22.85	13.88	23.74	13.71
1963	18.57	14.45	18.29	14.61
1964	22.35	14.75	23.33	14.54
1965	20.45	14.94	20.90	14.74
1966	21.98	14.14	22.53	14.06
1967	14.92	13.86	15.03	14.09
1968	22.33	14.62	22.08	14.59

SOURCE: Jack Brickey, Spreckels Sugar Company, Woodland, Calif.
NOTE: Area 1—Liberty Island to Grimes, Calif. Area 2—Woodland, Calif.

transmitted virus causing yellow wilt, a serious disease in South America (12). Severin (147) described the alfalfa mosaic virus from sugarbeets, and western cucumber mosaic has been described by Severin and Freitag (150) and Shepherd et al. (152). Severin et al. (151) have described a viruslike disease caused by leafhopper toxins.

CONTROL PROCEDURES

Chemicals are still widely used in the control of insects and mites affecting sugarbeets at all stages of growth. The common formulations are sprays (as emulsible concentrates and wettable powders) applied to the soil or plants; seed treatments; granules applied at the time of planting in the furrows, side-dressed or applied topically; dusts; concentrates, as low volume formulations, or in liquid fertilizers, or sprayed on granular fertilizers, or applied in the soil by special injection equipment; baits; and soil fumigants. An attempt was made (Table 10.5) to give a list of currently suggested materials for the control of sugarbeet insects and mites (root crop). It is always necessary to check for additions, changes, or deletions to a compilation of this sort and to check for state deviations in uses.

Since the establishment of seed treatment using insecticides and an adequate fungicide in 1950 by Lange and Leach (103, 115), this method has been commonly used in some areas for the control of soil-inhabiting insects. The Holly Sugar Company[1] treats seed with lindane at the rate of 2.4 ounces of actual material per 100 pounds of seed together with 4 ounces of a Dexon-PCNB (35–35 percent mixture), together with 5 grams of Victoria green dye. Using an advanced seed treater, they can treat 1,000 pounds of seed per hour.

Seed and soil treatments for a number of sugarbeet insect pests have been reported by investigators. Seed treatment has been used for beet leafhopper control (80, 126) and for the control of aphids relative to virus yellows (63). Dorst (31) used both seed and soil treatments for the control of sugarbeet pests. Insecticides are commonly used for the control of the sugarbeet root maggot (2, 57, 91, 133), *Liriomyza* leaf miners (39, 163), and root aphids (55). In some work by Reed (138), spider mites were controlled by several materials but the best control was obtained with dicofol (Kelthane).

The beet webworm has been adequately controlled by chemicals (34, 54, 119, 131, 135). Hills, Lange, and co-workers, evaluating investigations on the control of beet yellows extending from 1962 through 1969 (69, 70, 71, 72, 73, 74, 75, 113), found that properly timed applications of oxydemetonmethyl (Metasystox-R) adequately controlled the green peach aphid and in most years gave substantial increases in root yield. It was found in their investigations that under most circumstances the granular systemic materials

---

1. Personal communication from Alex Lange.

TABLE 10.5. Suggested materials for the control of sugarbeet insects and mites (root crop)

Name	Tolerance (ppm)	Dosage (pounds actual per acre unless specified)	Limitations (time between application and harvest, days)
carbaryl	100 (tops)	2.0	14
carbophenothion	5	1.0	14
D-D mixture	nonfood use	600.0	7 days per each 10 gal applied before planting
demeton	5	0.5	30
Dexon	nonfood use	2.8 oz per 100 lb	seed treatment
diazinon	0.75 (roots) 10.0 (tops)	4.0	...
dichloropropenes mixture	nonfood use	253.0	14–21 before planting
dicofol	extended to 12-31-70†	1.2	21
disulfoton	2 (tops) 0.5 (roots)	1.0	30
dyfonate	0.1	4.0	...
endosulfan	extended to 12-31-70†	1.0	...
lindane	nonfood use	4.0 oz per 100 lb	seed treatment
malathion	extended to 12-31-70†	2.5	7
naled	0.5 (pending)	1.0	2 (pending)
oxydemetonmethyl	extended to 12-31-70†	0.5	90
parathion	extended to 12-31-70†	0.8	15
PCNB	nonfood use	2.8 oz per 100 lb	seed treatment
phorate	3.0 (tops) 0.3 (roots) 1.0 (dried pulp)	1.5 granular 1.0 spray	30
phosphamidon	nonfood use	1.0	30
sulfur	safe	50.0	...
trichlorfon	0.1 (roots) 12.0 (tops)	1.5	14

SOURCE: Calif. Agr. Exp. Sta. "Pest and disease programs for sugar beets," 1970, and USDA, Wash. Agr. Exp. Sta. "Chemical Insect Control Handbook," Wash. State Univ., Pullman, Wash., 1970. The mention of proprietary materials is not recommendation for the use of these materials or discrimination against other available materials for use in sugarbeets.

† See your agricultural extension agent or other agricultural authority for current status of this material.

applied at the time of planting did not give adequate protection through the entire season and in this area the sprays of Metasystox-R gave better control than other available systemic materials. Five-year trials, using the unregistered carbamate insecticide, aldiarb (Temik), gave excellent control of the green peach aphid, leaf miners, root aphids, and several other insects affecting sugarbeets and in most experiments gave increased yield benefits. Being more water soluble than some systemics, it appeared that this material was released more readily from side-dressed applications and became available.

An example of the selective action of insecticides, based on the results of a replicated experiment at Gonzales, Monterey County, in 1966, is given in Figure 10.18. Oxydemetonmethyl was applied March 21, April 11, and

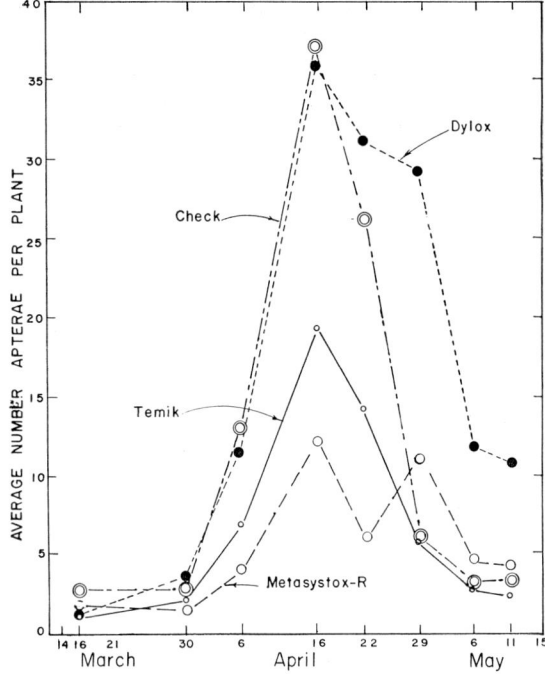

FIG. 10.18. Population trends of the number of apterous (wingless) aphids on sugarbeets showing selective nature of one of the insecticides.

May 11, using 8 ounces of actual material per acre. Aldicarb (Temik) was side-dressed March 24, using 2.12 pounds of actual material, and a second application was made April 26, using 2.0 pounds. Trichlorfon (Dylox) was applied March 21, April 11, and May 13. The results of the aphid counts (Fig. 10.18) demonstrate the selective action involved. The populations of apterae on plants treated with trichlorfon parallel those of the untreated check plants but increased in late April and May to remain higher than untreated plants. This chemical had an observed effect on some of the natural predators present. The granular systemic material (Temik) initially reduced the population a little slower than oxydemetonmethyl but ended up with lower populations toward the end of the observation period. Trichlorfon was selective in that it did not control the green peach aphid but was effective in controlling leaf miners (110).

The suppression of both curly top and BY in sugarbeets by controlling planting date and vector control was demonstrated in the work of Ritenour et al. (141). Control of yellowing viruses in the Pacific Northwest, which includes Washington, Idaho, and western Oregon, is commonly practiced through the application of phorate applied three times at the rate of one pound of actual material per acre for each application (3). Materials are applied either by ground or by air. In many cases, the airplane is used or preferred over ground application. The first two sprays consist of phorate. The first spray is applied as soon as the rows come up, and the second spray is applied two to three weeks later. The third application consists of the use of a topical 15 percent granular application. Time of planting applications or side-dressing applications in dry soil in the Pacific Northwest have not proved satisfactory because it is necessary to irrigate at the time of planting. In this area, the seedlings are not irrigated until after thinning, which delays the systemic action of the material. In certain other areas such as in Idaho, the application of systemics in the soil might prove more advantageous. Phorate also controls many of the other pests on sugarbeet except spider mites which come in late in the season and may cause extensive late damage.

In California tests using granular phorate and disulfoton at time of planting, side-dressed and topical applications resulted in positive increases in yield in only 3 of 18 experiments carried out during a three-year period (112, 158).

One difficulty with the use of certain chemicals for the control of sugarbeet insects is that the tops cannot be fed to livestock. The income per acre to the grower from beet tops may vary considerably, depending upon whether they are fed in the field for livestock feed or used as ensilage (59). The decision to use these chemicals by the grower will depend upon economics and area differences. The extent of chemical residues found in beets at the time of harvest (roots and tops) depends upon a number of factors, but residues should always be considered. The fate of aldrin, dieldrin, and endrin residues during the processing of raw sugarbeets is given in an article by Walker et al. (161). Some systemic insecticides concentrate more in the leaves than in the roots and it is important to determine these relationships.

As far as biological control is concerned, the many predators and parasites available in sugarbeet fields have already been mentioned, together with some mention of natural pathogens. At the present time, the University of California is introducing parasites for the control of the green peach aphid and certain other insects which attack sugarbeets, but in general very few special attempts have been made to control sugarbeet insects through biological control efforts. Bioenvironmental methods are commonly used by most growers and entomologists in the control of sugarbeet insects. These involve planting to avoid peaks in aphid flights, using weed control to do away with alternate hosts of many insects, leaving beet-free periods so that aphids cannot move directly from old sources of viruses to new plantings, and the use of many other manipulations to suppress infestations of insects and mites.

Resistant sugarbeet varieties have already been discussed to some extent. Both Harper (56) and Wallis and Gaskill (162) have investigated resistance to the sugarbeet root aphid, *Pemphigus populivenae* Fitch, and found some resistant types. The outstanding development of US H9A and US H9B yellows-resistant hybrids has provided substantial increases in yield in work carried out by McFarlane and co-workers (120, 121, 122). Russell (142, 143) has been working on aphid resistance in sugarbeets and also resistance of aphids to yellowing viruses relative to their resistance to organophosphorous insecticides.

OUTLOOK

Our goal for the future in the control of sugarbeet insects, mites, and other pests should involve the perfection of an integrated control program which takes into account all methods of control possible for the suppression of these pests. Smith and Reynolds (156) define integrated pest control as "a pest population management system that utilizes all suitable techniques in a compatible manner to reduce pest populations and maintain them at levels below those causing economic injury. Integrated control achieves this ideal by harmonizing techniques in an organized way by making the control practices compatible and by blending them into a multi-faceted flexible, evolving system."

Lange and co-workers (109, 112) have attempted to integrate several available control techniques in suppressing yellowing viruses in the interior valley of California. This technique involves several points: (1) the use of a resistant variety (US H9A and B); (2) a delay in the planting of beets to avoid peak populations in flight as determined by yellow pan water traps (and flight patterns derived from about 10 years of data in this area); and (3) the application of a systemic insecticide at the proper time to protect the plants during a critical period prior to the natural decrease in aphid populations caused by the oncome of hot summer temperatures and increased predator activity. It was found by experience that waiting too long to plant may result in decreased yields due to a loss of ideal growth conditions and increased nematode or insect pests.

An overall increase of 39 percent in production was obtained by Hills et al. (69) in a factorial-designed experiment in 1969. In this particular area, the only available material giving adequate protection was oxydemetonmethyl, applied two to five times at the rate of eight ounces of actual material per acre. Other granular materials have given protection under certain conditions, and in other cases sprays combined with granular materials have been effective.

The mechanization of all aspects of the production of sugarbeets is becoming more apparent (140) and it is necessary that attention be given to early seedling protection to insure an adequate stand. This is particularly important when planting to a stand. Studies such as those of Young (165), covering the eastern beet sugar industry, will play an important part in determining where beets can be grown in the future. There is great need for more work on the population dynamics of insects in sugarbeet fields, particularly in relation to the use of time tables, economic levels, and other detailed aspects of sugarbeet pests.

The use of cover crops (64) or trap crops and aluminum foil to trap or repel aphids and protect against viruses is worthy of more investigations.

Even though it is commonly considered that natural enemies of native insects cannot always be manipulated, it would be desirable in the future to attempt to at least conserve the natural enemies present in sugarbeet fields to utilize them to greater advantage. Finally, we need more selective insecticides and pathogens which can be used to control insects affecting sugarbeets without the disadvantages of detrimental effects on natural enemies, residues in roots and tops, pollution of the environment, and adverse effects upon wildlife or on man or animals. We need more work like that of Hurst (89), who found that he was able to forecast the abundance of the green peach aphid and BY in Great Britain by determining the mean average temperature for February. In the future we will have to make more and better judgments relative to forecasting populations of pests in sugarbeet fields and their resultant damage.

REFERENCES

1. Adams, R. L. Field Manual for Sugar Beet Growers, pp. 113–30. Beet Sugar Gazette Co., Chicago. 1913.
2. Allen, W. R., W. L. Askew, and K. Schreiber. Insecticidal control of the sugar beet root maggot and yield of sugar beets. J. Econ. Entomol. 54:178–81. 1961.
3. Anonymous. Virus yellows. U and I Cultivator 29:45–47. 1969.
4. Arrington, L. J. Beet sugar in the West, pp. 101–22. Univ. Washington Press, Seattle. 1966.
5. Bailey, S. F. Thrips of economic importance in California. Calif. Agr. Exp. Sta. Circ. 346. 1938.
6. Baker, C. F. New Homoptera received from New Mexico Agricultural Experiment Station 11. Psyche 7(suppl. 1):24–26. 1896.
7. Ball, E. D. The beet leaf-hopper *(Eutettix tenella)*. Utah Agr. Exp. Sta. Ann. Rept. 16 (1904/1905), p. 16. 1906.
8. Bennett, C. W. Sugar beet yellows disease in the United States. USDA Tech. Bull. 1218. 1960.
9. ———. Highly virulent strains of curly top virus in sugar beet in western United States. J. Am. Soc. Sugar Beet Technologists 12:515–20. 1963.
10. ———. Isolates of beet mosaic virus with different degrees of virulence. J. Am. Soc. Sugar Beet Technologists 13:27–32. 1964.
11. ———. Marble leaf of sugarbeets caused by a juice transmissible virus. J. Am. Soc. Sugar Beet Technologists 13:33–41. 1964.
12. Bennett, C. W., J. F. Hills, R. Ehrenfeld K., J. Valenzuela B., and C. Klein K. Yellow wilt of sugar beet. J. Am. Soc. Sugar Beet Technologists 14:480–510. 1967.
13. Bennett, C. W., and J. S. McFarlane. Damage produced by beet yellows and beet western yellows under greenhouse and field conditions. J. Am. Soc. Sugar Beet Technologists 14:619–36. 1967.
14. Bennett, C. W., C. Price, and G. E. Gillispie. Effects of virus yellows on yield and sucrose content of sugar beets in tests at Riverside, California. J. Am. Soc. Sugar Beet Technologists 8:236–40. 1954.
15. Bennett, C. W., C. Price, and J. S. McFarlane. Effect of virus yellows on sugar beet with a consideration of some of the factors in changes produced by the disease. J. Am. Soc. Sugar Beet Technologists 9:479–94. 1957.

16. Callenbach, J. A., W. L. Gojmerac, and D. B. Ogden. The sugar beet root maggot in North Dakota. J. Am. Soc. Sugar Beet Technologists 9:300–304. 1957.
17. Campbell, R. E. and V. Duran. Notes on the sugar-beet armyworm in California. Calif. Dept. Agr. Mon. Bull. 18:267–75. 1929.
18. Carlson, E. C. Effects of the green peach and bean aphids on table beet seed plants. Calif. Agr. 14:11–12. 1960.
19. ———. Lygus bug damage to table beet seed plants. Calif. Agr. 15:12–14. 1961.
20. ———. Lygus bug damage to table beet seed plants. J. Econ. Entomol. 54:117–19. 1961.
21. Carsner, E. and C. F. Stahl. Studies on curly-top disease of the sugar beet. Agr. Res. 28:297–319. 1924.
22. Chang, V. C. Intraspecific variation in the ability of the green peach aphid, *Myzus persicae* (Sulz.), to transmit sugar beet virus yellows. Ph.D. thesis. Univ. Calif., Davis. 1968.
23. Chittenden, F. H. Some insects injurious to garden crops. USDA Div. Entomol. Bull. N.S. 23. 1900.
24. ———. The fall armyworm and variegated cutworm. USDA Div. Entomol. Bull. N.S. 29. 1901.
25. ———. Some insects injurious to vegetable crops. USDA Div. Entomol. Bull. N.S. 33. 1902.
26. ———. Principal insect enemies of the sugar beet. USDA Div. Entomol. Bull. 43. 1903.
27. ———. Some insects injurious to truck crops. USDA Bur. Entomol. Bull. 82, pp. 67–75. 1912.
28. Chittenden, F. H. and H. O. Marsh. The beef leaf-beetle. USDA Bull. 892. 1920.
29. Coons, G. H. and J. E. Kotila. Virus yellows and its occurrence in the United States. Phytopathology 41:559. (Abstr.) 1951.
30. Dickson, R. C. and E. F. Laird. Green peach aphid populations on desert sugar beets. J. Econ. Entomol. 55:501–4. 1962.
31. Dorst, H. E. Insect control on sugar beets by seed and soil treatments. J. Am. Soc. Sugar Beet Technologists 13:649–53. 1965.
32. Douglass, J. R. Outbreaks of beet leafhoppers north and east of the permanent breeding areas. Proc. Am. Soc. Sugar Beet Technologists 8:186–93. 1954.
33. Douglass, J. R. and W. C. Cook. The beet leafhopper. USDA Circ. 942. 1954.
34. Douglass, J. R. and Van E. Romney. Comparative toxicity of some new insecticides to the webworm in Colorado, 1951. Proc. 7th General Meeting, Am. Soc. Sugar Beet Technologists, pp. 503–6. 1952.
35. Duffus, J. E. Radish yellows, a disease of radish, sugar beet, and other crops. Phytopathology 50:389–94. 1960.
36. ———. Economic significance of beet western yellows (radish yellows) on sugar beet. Phytopathology 51:605–7. 1961.
37. ———. Incidence of beet virus diseases in relation to overwintering beet fields. *Plant Disease Reporter* 47:428–31. 1963.
38. ———. Host relationships of beet western yellows virus strains. Phytopathology 54:736–38. 1964.
39. Duffus, J. E. and N. F. McCalley. Phorate and demeton for control of the pea leaf miner on sugarbeets. J. Econ. Entomol. 57:221–22. 1964.
40. Ernould, L. and L. Van Steyvoort. Atlas des ennemis et maladies de la betterave. Institut Belge pour l'Amelioration de la Betterave, Tirlemont, pp. 5–70. 1958.

41. Essig, E. O. and W. M. Hoskins. Insects and other pests attacking agricultural crops. Calif. Agr. Ext. Sta. Circ. 87, rev., pp. 27–30. 1944.
42. Fife, J. M. Changes in the concentration of amino acids in sugarbeet plants induced by virus yellows. J. Am. Soc. Sugar Beet Technologists 11:327–33. 1961.
43. ———. Growth rate of young sugar beet roots as a measure of resistance to virus yellows. J. Am. Soc. Sugar Beet Technologists 12:497–502. 1963.
44. ———. Changes in concentration of three amino acids, in mature leaves of sugarbeet plants, produced by mass selection from a population infected with beet yellows. J. Am. Soc. Sugar Beet Technologists 14:334–40. 1967.
45. Fox, D. E. Occurrences of the beet leafhopper and associated insects on secondary plant succession in southern Idaho. USDA Tech. Bull. 607. 1938.
46. Frick, Kenneth E. *Liriomyza langei*, a new species of leaf miner important in California. Pan-Pacific Entomologist 27:81–88. 1951.
47. Frost, S. W. A study of the leaf-mining diptera of North America. Cornell Univ. Agr. Exp. Sta. Memoir 78. 1924.
48. Gillette, C. P. and C. F. Baker. A preliminary list of the Hemiptera of Colorado. Colo. Agr. Exp. Sta. Bull. 31:100. 1895.
49. Gojmerac, W. L. and J. A. Callenbach. Sugar beet root maggot—one year's trials in the Red River Valley show that chemical control is feasible. N. Dak. Agr. Exp. Sta. Bull. 58:115–20. 1956.
50. Graf, John E. A preliminary report on the sugar-beet wireworm. USDA Bur. Entomol. Bull. 123. 1914.
51. Gram, E. and P. Bovien. Rodfrugternes sygdomme og skadelyr. Kgl. Danske Landhusholdningsselskab, Copenhagen, pp. 12–73. 1942.
52. Green, H. J. The beet leafhopper control program. Calif. Sugar Beet, pp. 49–50. 1962.
53. Grigarick, A. A. and W. H. Lange. Host relationships of the sugar-beet root aphid in California. J. Econ. Entomol. 55:760–64. 1962.
54. Hagan, A. F. Evaluation of thiodan and sevin for control of webworms in sugarbeets J. Econ. Entomol. 54:799. 1961.
55. Harper, A. M. Effect of insecticides on the sugar-beet root aphid, *Pemphigus betae*. J. Econ. Entomol. 54:1151–53. 1961.
56. ———. Varietal resistance of sugar beets to the sugar beet root aphid, *Pemphigus betae* Doane (Homoptera: Aphididae). Can. Entomol. 96:520–22. 1964.
57. Harper, A. M., C. E. Lille, and P. Bergen. Effect of insecticides on control of the sugar-beet root maggot. J. Econ. Entomol. 54:895–900. 1961.
58. Harris, F. S. Pests and diseases. *In* The Sugarbeet in America, pp. 184–204. Macmillan, New York. 1919.
59. Harris, L., D. C. Clanton, and M. A. Alexander. Sugar beet tops and modern sugar beet production. J. Am. Soc. Sugar Beet Technologists 13:432–47. 1965.
60. Hawley, I. M. The sugar-beet root-maggot *(Tetanops aldrichi* Hendel), a new pest of sugar-beets. J. Econ. Entomol. 15:388–91. 1922.
61. ———. Notes on the insect pests of Utah. J. Econ. Entomol. 16:377–79. 1923.
62. ———. The more important insects injurious to the sugar beet in Utah. Utah Agr. Exp. Sta. Circ. 54. 1925.
63. Heathcote, G. D. The use of menazon seed dressing to decrease spread of virus yellows in sugar-beet root crops. Ann. Appl. Biol. 62:113–18. 1968.
64. ———. Protection of sugar beet stecklings against aphids and viruses by cover crops and aluminum foil. Plant Pathol. 17:158–61. 1968.
65. Heathcote, G. D. and A. J. Cockbain. Transmission of beet yellows virus by alate and apterous aphids. Ann. Appl. Biol. 53:259–66. 1964.

66. Heathcote, G. D., R. A. Dunning, and M. D. Wolfe. Aphids on sugar beet and some weeds in England, and notes on weeds as a source of beet viruses. Plant Pathol. 14:1–10. 1965.
67. Heathcote, G. D., J. M. P. Palmer, and L. R. Taylor. Sampling for aphids by traps and by crop inspection. Ann. Appl. Biol. 63:155–66. 1969.
68. Heie, O. and B. Petersen. Investigations on *Myzus persicae* Sulz., *Aphis fabae* Scop., and virus yellows of beet *(Beta* virus 4) in Denmark, Danish Acad. Technolog. Sciences, Copenhagen, pp. 7–52, 16 maps. 1961.
69. Hills, F. J., W. H. Lange, and J. Kishiyama. Varietal resistance to yellows, vector control, and planting date as factors in the suppression of yellows and mosaic of sugar beet. Phytopathology 59:1728–31. 1969.
70. Hills, F. J., W. H. Lange, R. S. Loomis, H. L. Hall, and J. L. Reed. Sprays for aphid control increase sugar beet yields in Davis tests. Calif. Agr. 19:6–7. 1965.
71. Hills, F. J., W. H. Lange, R. S. Loomis, J. L. Reed, and D. H. Hall. Late plantings reduce yellows infection, improve beet yields and sugar production at Davis. Calif. Agr. 17:14–15. 1963.
72. ———. Late-planted sugar beets damaged by yellows viruses. Calif. Agr. 18:10–11. 1964.
73. Hills, F. J., W. H. Lange, J. L. Reed, D. H. Hall, and R. S. Loomis. Response of sugarbeet to date of planting and infection by yellows viruses in northern California. J. Am. Soc. Sugar Beet Technologists 12:210–15. 1962.
74. ———. Later planting dates in California save sugar beets from yellows virus damage. Calif. Agr. 16:5. 1962.
75. Hills, F. J., W. H. Lange, J. L. Reed, and R. S. Loomis. Aphid control and planting date for the control of yellows of sugar beets. J. Am. Soc. Sugar Beet Technologists 14:117–26. 1966.
76. Hills, O. A. Isolation-cage studies of certain hemipterous and homopterous insects on sugar beets grown for seed. J. Econ. Entomol. 34:756–60. 1941.
77. ———. Insects affecting sugar beets grown for seed. USDA Agr. Handbook 253. 1963.
78. Hills, O. A., C. W. Bennett, H. K. Jewell, C. L. Coudriet, and R. W. Brubaker. Effect of virus yellows on yield and quality of sugar beet seed. J. Econ. Entomol. 53:162–64. 1960.
79. Hills, O. A., D. L. Coudriet, C. W. Bennett, H. K. Jewell, and R. W. Brubaker. Effect of three insect-borne virus diseases on seed production. J. Econ. Entomol. 56:609–93. 1963.
80. Hills, O. A., F. H. Harries, and A. C. Valcance. Treatment of sugar beet seed with systemic insecticides for control of the beet leafhopper. J. Am. Soc. Sugar Beet Technologists 9:124–28. 1956.
81. Hills, O. A., H. K. Jewell, C. W. Bennett, and R. W. Brubaker. Effect of aphid-borne beet yellows and beet western yellows on sugar beet seed production under conditions of varying fertility. J. Am. Soc. Sugar Beet Technologists 14:168–73. 1966.
82. Hills, O. A., K. B. McKinney, and W. E. Peay. Lygus control in sugar beets grown for seed. Proc. 45th General Meeting, Am. Soc. Sugar Beet Technologists, pp. 298–318. 1946.
83. Hills, O. A., V. E. Romney, and K. B. McKinney. Effect of *Empoasca solana* on sugarbeets grown for seed. J. Econ. Entomol. 37:698–702. 1944.
84. Hills, O. A. and E. A. Taylor. Lygus damage to sugar beet seed in various stages of development. Proc. Am. Soc. Sugar Beet Technologists, pp. 481–87. 1950.
85. Hull, R. Sugar beet diseases. Gt. Brit. Min. Agr. Fisheries Food Tech. Bull. 142, pp. 36–37, figs. 12, 13, 16, 21, 36. 1949.

86. Hull, R. Sugar beet yellows in California. Calif. Sugar Beet, pp. 34–36, 52. 1967.
87. ———. The effect of infection with beet yellows virus on the growth of sugar beet. J. Am. Soc. Sugar Beet Technologists 15:192–99. 1968.
88. ———. Yellows and aphids in sugar beet. Rept. Rothamsted Exp. Sta., pp. 270–75. 1968.
89. Hurst, G. W. Forecasting the severity of sugar beet yellows. Plant Pathol. 14:47–53. 1965.
90. Jones, D. P. and F. G. W. Jones. Wireworms and the sugar-beet crop: field trials and observations. Ann. Appl. Biol. 34:562–74. 1947.
91. Jones, E. W., J. R. Douglass, C. P. Parrish, and Vernal Jensen. 1952 experiments on control of the sugar-beet root maggot. Proc. Am. Soc. Sugar Beet Technologists, pp. 490–96. 1952.
92. Jones, F. G. W. and R. A. Dunning. Sugar beet pests. Gt. Britain Min. Agr. Fisheries Food Tech. Bull. 162, rev. 1969.
93. Kendrick, J. The 1957 Kern County beet crop. Spreckles Sugar Beet Bull. 21:46. 1957.
94. Kennedy, J. S., A. Ibbotson, and C. O. Berth. The distribution of aphid infestation in relation to leaf age. I. *Myzus persicae* (Sulz.) and *Aphis fabae* Scop. on spindle trees and sugar-beet plants. Ann. Appl. Biol. 37:651–79. 1950.
95. ———. II. The progress of *Aphis fabae* Scop. infestations on sugar beets in pots. Ann. Appl. Biol. 37:680–96. 1950.
96. Kershaw, W. J. S. Aphid infestations in sugar-beet seed crops. Plant Pathol. 13:90–91. 1964.
97. ———. Aphid sampling in sugar beet. Plant Pathol. 13:101–7. 1964.
98. Landis, B. J. *Asterolecanium arabidis*, a scale attacking sugar beets. J. Econ. Entomol. 61:871–72. 1968.
99. Lange, W. H. Notes on the sugar-beet root aphid in California. J. Econ. Entomol. 32:727. 1939.
100. ———. Sugarbeet insects and nematodes and their control. Spreckels Sugar Beet Bull. 11:1–8. 1947.
101. ———. New sugar beet pest. Calif. Agr. 4:5, 14. 1950.
102. ———. Insect pests, nematodes, and their control. In R. A. McGinnis, Beet Sugar Technology, pp. 63–73. Reinhold, New York. 1951.
103. ———. Sugar beet pest control. Spreckels Sugar Beet Bull. 17:14–16. 1953.
104. ———. Another outbreak of the yellow-striped armyworm? Spreckels Sugar Beet Bull. 21:22–23. 1957.
105. ———. Seed treatment as a method of insect control. Ann. Rev. Entomol. 4:363–88. 1959.
106. ———. Pavement ant attacking sugar beets in California. J. Econ. Entomol. 54:1063–64. 1961.
107. ———. Sugar beet insect investigations. Calif. Sugar Beet, pp. 51–52. 1962.
108. ———. Two new insect pests of California sugar beets. Spreckels Sugar Beet Bull. 30:10–12. 1966.
109. ———. The integrated control approach to the suppression of yellows and mosaic of sugar beet. Calif. Sugar Beet, pp. 20, 36. 1969.
110. ———. Unpublished data. Calif. Agr. Exp. Sta. 1970.
111. Lange, W. H., A. A. Grigarick, and E. C. Carlson. Serpentine leaf miner damage. Calif. Agr. 11:3–5. 1957.
112. Lange, W. H. and F. J. Hills. Virus suppression on sugar beets by means of properly timed insecticides. Calif. Sugar Beet, pp. 49–51. 1966.
113. Lange, W. H., F. J. Hills, and J. Kishiyama. Early aphid control increases beet production. Calif. Agr. 21:14–15. 1967.
114. Lange, W. H., M. C. Lane, and M. W. Stone. A comparative appraisal of

wireworm and seed-corn maggot control in the western United States. Proc. 10th Intern. Congr. Entomol. 3:131–45. 1958.
115. Lange, W. H. and L. D. Leach. Insects and diseases controlled by seed treatment. Spreckels Sugar Beet Bull. 14:3, 8. 1950.
116. Leach, L. D. Sugar beet diseases. Spreckels Sugar Beet Bull. 9:1–8. 1945.
117. Lüdecke, H. Zuckerrübenbau, pp. 179–202. Paul Parey, Berlin. 1961.
118. Lüdecke, H. and C. Winner. Farbtafelatlas der Krankheiten und Schaedigungen der Zuckerruebe, pp. 55–80, 86 plates. DLG, Frankfurt am Main. 1959.
119. McDonald, S. Chemical control of the beet webworm on sugarbeets in southern Alberta. J. Econ. Entomol. 56:248–51. 1963.
120. McFarlane, J. S. and C. W. Bennett. Occurrence of yellows resistance in the sugar beet with an appraisal of the opportunities for developing resistant varieties. J. Am. Soc. Sugar Beet Technologists 12:503–14. 1963.
121. McFarlane, J. S. and I. O. Skoyen. New sugar beet varieties reduce losses from virus yellows. Calif. Agr. 22:14–15. 1968.
122. McFarlane, J. S., I. O. Skoyen, and R. T. Lewellen. Development of sugarbeet breeding lines and varieties resistant to yellows. J. Am. Soc. Sugar Beet Technologists 15:347–60. 1969.
123. McKinney, K. B. Common insects attacking sugar beets and vegetable crops in the Salt River Valley of Arizona. J. Econ. Entomol. 32:808–10. 1939.
124. McLean, D. L. and M. G. Kinsey. Identification of electrically recorded curve patterns associated with aphid salivation and ingestion. Nature 205: 1130–31. 1965.
125. McLean, D. L. and W. A. Weight, Jr. An electronic measuring system to record aphid salivation and ingestion. Ann. Entomol. Soc. Am. 61:180–85. 1968.
126. Malm, H. R. and R. E. Finkner. The use of systemic insecticides to reduce the incidence of curlytop virus disease in sugar-beets. J. Am. Soc. Sugar Beet Technologists 15:246–54. 1968.
127. Marsh, H. O. Papers on insects affecting vegetables. The sugarbeet webworm. USDA Bur. Entomol. Bull. 109, pp. 57–70. 1912.
128. Maxson, A. C. Principal insect enemies of the sugar beet. Great Western Sugar Company, Denver. 1920.
129. ———. Insects and diseases of the sugar beet. Beet Develop. Found., Fort Collins, Colo. 1948.
130. Oman, P. W. Notes on the beet leafhopper, *Circulifer tenellus* (Baker), and its relatives (Homoptera: Cicadellidae). J. Kansas Entomol. Soc. 21:10–14. 1948.
131. Peay, W. E. Laboratory screening tests of insecticides for control of the beet webworm. J. Am. Soc. Sugar Beet Technologists 13:645–48. 1965.
132. ———. Sugarbeet insects: how to control them. USDA Farmers' Bull. 2219. 1966.
133. Peay, W. E. and C. E. Stanger. Insecticide tests for control of the sugar beet root maggot in southern Idaho. J. Am. Soc. Sugar Beet Technologists 14:214–17. 1966.
134. Pepper, James H. Montana insect pests 1955–1956. Thirty-sixth Rept. of the State Entomologists, Mont. Agr. Exp. Sta. Bull. 526. 1956.
135. Pepper, James H. and E. Hastings. Life history and control of the sugarbeet webworm *Loxostege sticticalis* (L.). Mont. Agr. Exp. Sta. Bull. 389. 1941.
136. Pettit, Rufus H. Insects of the garden. Mich. Agr. Exp. Sta. Bull. 233. 1905.
137. Reed, J. L. The influence of certain physical factors on the relative abundance of the green peach aphid, *Myzus persicae* (Sulzer) (Homoptera: Aphididae). Ph.D. thesis. Univ. Calif., Davis. 1964.

138. Reed, J. L. The mite problem in northern California sugar beet fields. Spreckels Sugar Beet Bull. 29:20–21, 24. 1965.
139. Reynolds, H. T., R. C. Dickson, R. M. Hannibal, and E. F. Laird, Jr. Effects of the green peach aphid, southern garden leafhopper, and carmine mite populations upon yield of sugar beets in the Imperial Valley, California. J. Econ. Entomol. 60:1–7. 1967.
140. Ririe, D., H. L. Hall, F. J. Hills, M. Fireman, A. H. Lange, W. H. Lange, L. D. Leach, and L. J. Booker. Growing sugar beets without hand labor. Univ. Calif. Agr. Ext. Serv. AXT-188, pp. 11–13. 1965.
141. Ritenour, G., F. J. Hills, and W. H. Lange. Effect of planting date and vector control on the suppression of curlytop and yellows in sugarbeet. J. Am. Soc. Sugar Beet Technologists 16:78–84. 1969.
142. Russell, G. E. Preliminary studies in breeding for aphid-resistance in beets. 28th Winter Congr. I.I.R.B., Brussels. 1:117–25. 1965.
143. ———. Some effects of resistance to organophosphorus insecticides in *Myzus persicae* (Sulz.) on the transmission of beet yellowing viruses. Bull. Entomol. Res. 56:191–96. 1965.
144. Scalley, F. T. and Rowland M. Cannon. Flea beetle control in the Montana district. Proc. Am. Soc. Sugar Beet Technologists, pp. 491–93. 1950.
145. Schneider, C. L., A. M. Jafri, and A. M. Murphy. Greenhouse testing of sugarbeet for resistance to curly top. J. Am. Soc. Sugar Beet Technologists 14:727–34. 1967.
146. Severin, H. H. P. Curly top symptoms on the sugarbeet. Calif. Agr. Exp. Sta. Bull. 465. 1929.
147. ———. Symptoms of additional cucumber-mosaic viruses in sugar beets. Hilgardia 18:531–38. 1948.
148. ———. Sugar-beet mosaic tests. Calif. Agr. 3:6–14. 1949.
149. Severin, H. H. P. and R. M. Drake. Sugar-beet mosaic. Hilgardia 18:483–521. 1948.
150. Severin, H. H. P. and J. H. Freitag. Outbreak of western cucumber mosaic on sugar beets. Hilgardia 18:523–30. 1948.
151. Severin, H. H. P., F. D. Horn, and N. W. Frazier. Certain symptoms resembling those of curly top in aster yellows, induced by saliva on *Zerophloea vanduzeei*. Hilgardia 16:337–60. 1945.
152. Shepherd, R. J., D. H. Hall, and D. E. Purcifull. Occurrence of the alfalfa mosaic virus in sugarbeet in California. J. Am. Soc. Sugar Beet Technologists 13:374–77. 1964.
153. Shepherd, R. J., F. J. Hills, and D. H. Hall. Losses caused by beet mosaic virus in California grown sugar beets. J. Am. Soc. Sugar Beet Technologists 13:244–51. 1964.
154. Shepherd, R. J., W. H. Lange, F. J. Hills, and L. D. Leach. Important sugarbeet diseases in California. Calif. Sugar Beet, pp. 33–34, 38–40. 1963.
155. Shepherd, R. J., B. B. Till, and N. Schaad. A severe necrotic disease of sugar beet caused by a strain of the beet mosaic virus. J. Am. Soc. Sugar Beet Technologists 14:97–105. 1966.
156. Smith, R. F. and H. T. Reynolds. Principles, definitions and scope of integrated pest control. Proc. FAO Symposium on Integrated Pest Control 1:11–17. 1966.
157. Stone, M. W. Life history of the sugar-beet wireworm in southern California. USDA Tech. Bull. 744. 1941.
158. Sylvester, E. S., V. E. Burton, and G. V. Ferry. Aphid control trials on sugar beets and virus yellows disease in Kern County, California. J. Econ. Entomol. 54:758–61. 1961.
159. Sylvester, E. S. and E. L. Cox. Sequential plans for sampling aphids on sugar beets in Kern County, California. J. Econ. Entomol. 54:1080–84. 1961.

160. Sylvester, E. S. and H. H. P. Severin. Virus diseases of California sugar beets. Calif. Agr. Exp. Sta. Leaflet 46. 1955.
161. Walker, K. C., J. C. Maitlen, J. A. Onsager, D. M. Powell, L. I. Butler, A. E. Goodban, and R. M. McCready. The fate of aldrin, dieldrin, and endrin residues during the processing of raw sugarbeets. USDA, ARS, 33–107. 1965.
162. Wallis, R. L. and J. O. Gaskill. Sugar-beet root aphid resistance in sugar beet. J. Am. Soc. Sugar Beet Technologists 12:571–72. 1963.
163. Wilcox, J. and A. F. Howland. Control of the pea leaf miner in southern California. J. Econ. Entomol. 48:579–81. 1955.
164. Young, D. A., Jr., and N. W. Frazier. A study of the leafhopper genus *Circulifer* Zakhvatkin (Homoptera, Cicadellidae). Hilgardia 23:25–52. 1954.
165. Young, R. A. An economic study of the eastern beet sugar industry. Mich. Agr. Exp. Sta. Res. Bull. 9. 1965.

11

Nematodes and Their Control

JACK ALTMAN
Colorado State University
Fort Collins

IVAN THOMASON
University of California
Riverside

GENERAL MORPHOLOGY AND BIOLOGY	337
PLANT-PARASITIC NEMATODES ASSOCIATED WITH SUGARBEET CROPS	337
Sugarbeet Nematode, *Heterodera schachtii*	338
Root-knot Nematode, *Meloidogyne* spp.	347
Stem and Bulb Nematode, *Ditylenchus dipsaci*	353
Stubby-root Nematode, *Trichodorus* spp., and Needle Nematode, *Longidorus* spp.	355
Cobb's Root Galling Nematode, *Nacobbus aberrans*	360
GENERAL NEMATODE CONTROL	361
Early Planting	361
Rotations	363
Soil Fumigation	364
Resistant Varieties	367

AMONG the many pests and problems of sugarbeets, nematodes must surely stand as the most enduring if not the most important. They have been a major factor in beet production for over one hundred years. The sugarbeet cyst nematode, *Heterodera schachtii,* was a limiting factor in beet production in Europe in the second half of the nineteenth century and in the United States in the first half of the twentieth century, and as we shall see later, it has spread to other areas since then. Its presence in a field or beet district has dictated production practices and even the profitability of sugar production. In the past, its ravages were so severe in isolated areas that beet production ceased and processing facilities were closed.

In this chapter, the authors will provide a brief introduction to the general morphology and biology of nematodes. Consideration will then be given to the types of nematodes associated with sugarbeets, their distribution, and relative importance. The economically important, or potentially important, nematodes will then be discussed individually in more detail. Finally, general procedures used in controlling nematodes attacking sugarbeets will be discussed, along with implications that certain control procedures, for example, soil fumigation, might have on soil fertility and beet growth.

In some instances, plant-parasitic nematodes produce diagnostic disease symptoms in their hosts. The galls produced by the root-knot nematodes, *Meloidogyne* spp., and the enlargements caused by the stem and bulb nematode, *Ditylenchus dipsaci,* are such symptoms. Often, however, the symptoms of disease involving nematodes do not differ from symptoms of other root-debilitating agents. Determining nematode pathogenicity usually involves attempted applications of Koch's postulates. Koch's criteria were designed for bacterial pathogens of animals, and applied to bacterial and fungal pathogens of plants (21, 22). It is not surprising that there is difficulty in applying them to organisms so different from bacteria and fungi as nematodes.

Koch's first criterion, constant association with disease, is directly applicable to nematode parasites, with a minor reservation. More nematodes should not necessarily be expected around severely diseased plants than around healthy ones in the field. When a host plant becomes severely diseased as a result of nematodes and secondary organisms, it will not support as high a population of nematodes.

A prerequisite to determining association for nematodes is identification of the nematodes found in each instance of disease. Nematodes show much greater morphologic differentiation than bacteria or fungi, and identifications can usually be made on a morphologic basis by persons with nematological training. The morphology of nematodes associated with a diseased plant is also a clue to the nematode's possible relation to the plant. All the nematode parasites of higher plants have a characteristic feeding structure known as a stylet or spear.

In the case of nematodes, the criterion of constant association with disease should be extended to include, also, observation of the nematode species in question feeding on the suscept. Our present knowledge indicates that nematodes are not likely to be pathogenic to a plant unless they feed on it. Conversely, if a nematode species does feed on a plant it may well cause, or incite, disease.

GENERAL MORPHOLOGY AND BIOLOGY

All nematodes at some stage of their life are elongate worms, the body shape ranging from fusiform to nearly cylindrical, with rounded ends. Most nematodes are of this shape throughout their lives, but adult females of certain important plant parasites develop bodies which are pear-, lemon-, or kidney-shaped, or otherwise enlarged. Soil and plant-parasitic nematodes range in length from a minimum of about 0.25 mm to a maximum of about 10 mm. The great majority of species are < 2 mm long.

In their feeding habits, nematodes are saprophytic, predaceous, or plant parasitic. The saprophytes and predators feed on decaying organic matter and on the smaller forms of soil life, such as the bacteria, protozoa, and small nematodes. They complete their life cycles in or near decomposing organic matter. Plant-parasitic nematodes are found in rhizosphere soil, in plant roots, and in the foliage and seeds of plants. Mostly, however, they are limited to the rhizospheres and roots of their hosts. They are attracted to these areas by chemical stimuli, that is, by water-soluble metabolic products secreted by the host plant. The larvae of parasitic species may move randomly in the soil and persist therein for several weeks, but if they fail to migrate to a sphere of chemical stimuli and thence to a host rootlet, they will eventually die without completing their life cycle.

Extensive injury to the root systems of field crops and pasture plants can be caused by parasitic nematodes. This injury can greatly reduce efficiency of the root system to absorb water and essential nutrients and can lead to stunting, unthriftiness, and serious, though often unsuspected, loss of production.

PLANT-PARASITIC NEMATODES ASSOCIATED WITH SUGARBEET CROPS

Over the years, a number of species of plant-parasitic nematodes have been reported from soils in which sugarbeets were growing. However, in the authors' opinions, many of these associations are incidental and do not represent a definite pathogenic involvement. The nematodes listed below are believed to be important to beet production. They are listed in order of their importance in world beet production. The most important in the United States are *H. schachtii, Meloidogyne* spp., and *Nacobbus aberrans.*[1]

1. *Heterodera schachtii*—sugarbeet nematode—Tables 11.1, 11.2, 11.3, 11.4.
2. *Meloidogyne* spp.—root-knot nematode—Table 11.5.
3. *Ditylenchus dipsaci*—stem and bulb nematode—Table 11.6.
4. *Trichodorus* spp.—stubby-root nematode—Table 11.7.
5. *Longidorus* spp.—needle nematode—Table 11.7.
6. *Nacobbus aberrans*—Cobb's root-galling nematode, Nebraska root-galling disease (Schuster).

SUGARBEET NEMATODE, *Heterodera schachtii*

This nematode is one of the important pathogens of sugarbeets and is responsible for major reductions in yield. It is found almost everywhere beets are grown and can be introduced readily into areas where beets are being grown for the first time. If beets are grown continuously in infested soil, symptoms of sugarbeet nematode damage will become apparent within 5 years after initial introduction of the pathogen. Districts which have known infestations will be completely contaminated in 25 to 30 years, according to Jones and Dunning (20). Regarding new or virgin areas used for beet production, the primary source of infestation would appear to be cysts in soil moving with agricultural equipment. Seed can become infested during harvest operations when minute clods of soil containing cysts are picked up in the threshing procedure. There is evidence that infested seed served as a primary source of infection in new areas in the past. Passive distribution of larvae and cysts also occurs in irrigation water or contaminated soil. This can result in further distribution in a field or serve to initiate infestations in new areas. Larval movement in soil is restricted to several inches and serves primarily to bring the nematode in contact with host roots.

Newly infested districts follow a general pattern. Usually one or two fields are first observed in which nematodes have caused severe damage in areas several square rods in size, indicating that the pests probably have been present four years or more. By this time, they have been spread to neighboring fields, and a careful survey of the district will usually reveal additional infestations.

The dump-sampling technique has been used recently in several beet-growing areas for the early detection of sugarbeet cyst nematodes and studying their distribution. After the beets are delivered to the dump by truck, they are moved up a conveyer belt and then over a series of rollers that

---

1. For a detailed account of nematodes associated with sugarbeets in the United States, the reader is referred to the surveys of Caveness (10, 11), conducted for the Beet Sugar Foundation.

violently tumble the beets to remove clinging organic matter and soil. It is at this point that soil is collected below the rollers in a container having a cone-shaped screen top. A load of beets normally represents several passes across a field, and furthermore, it is precisely the soil intimately associated with the beet root that is most likely to contain cysts. After soil is obtained, it is processed by conventional cyst extraction processes. Experience in Utah and California has shown the dump-sampling technique to be efficient of time and labor and quite sensitive and valuable in detecting relatively low levels of infestation before they become of major economic importance.

Losses in sugarbeet yields due to this nematode normally range from 10 to 15 percent. In evaluating levels of nematodes where loss in yield may occur, the senior author, working in Colorado, has found that if the viable cyst population exceeds 10 cysts per 750 g (1 pint) of soil, economic losses can occur. In England, the level above which injury is expected to occur is 10 eggs per g of soil. With *H. schachtii*, sick patches appear in the crop at an average population density of about 10 eggs per g of soil, and a density of 100 eggs per g causes more or less complete crop failure (38).

Life History. Nematodes are observed in the field as females, which appear as small, white cysts (lemon-shaped bodies) clinging to the beet root (Fig. 11.1A). At maturity, the cysts will be a dark orange-brown color. Each cyst contains 100 to 600 eggs from which the wormlike larvae hatch (Figs. 11.1B, 11.1C, and 11.1D). These larvae are about 1/60 of an inch long and each has a stylet (Fig. 11.1C) with which it penetrates cell walls and makes its way into the beet root.

During the final or fourth molt (Fig. 11.2D), the males develop into slender, active eelworms which increase to three times in size, averaging about 1/20 of an inch in length. They then leave the roots and enter the soil in search of the females.

After maturation, the females are flask-shaped. As they increase in size, they break out of the root tissues and remain attached by their heads. At this stage they are fertilized by the males and soon develop into lemon-shaped bodies about the size of a pinhead. A gelatinous fluid is excreted, which collects in a mass about the posterior end and into which the females deposit a few of their eggs (Fig. 11.2C). These eggs soon hatch and the larvae find their way into the soil and then by serpentinelike movement to the beet roots.

In most temperate beet-growing areas, the nematode produces one to two generations per year. The life cycle (Fig. 11.2) is usually completed in 4 to 6 weeks, and in Colorado the nematode can complete a minimum of two full cycles each growing season. In some areas, such as the Imperial Valley of California, however, three to five generations occur (41). The females that have successfully invaded the host roots, and developed to maturity, die after a few weeks and change in color from white to dark brown. The eggs which have not been deposited externally remain within the cysts and the larvae contained in these eggs remain dormant during

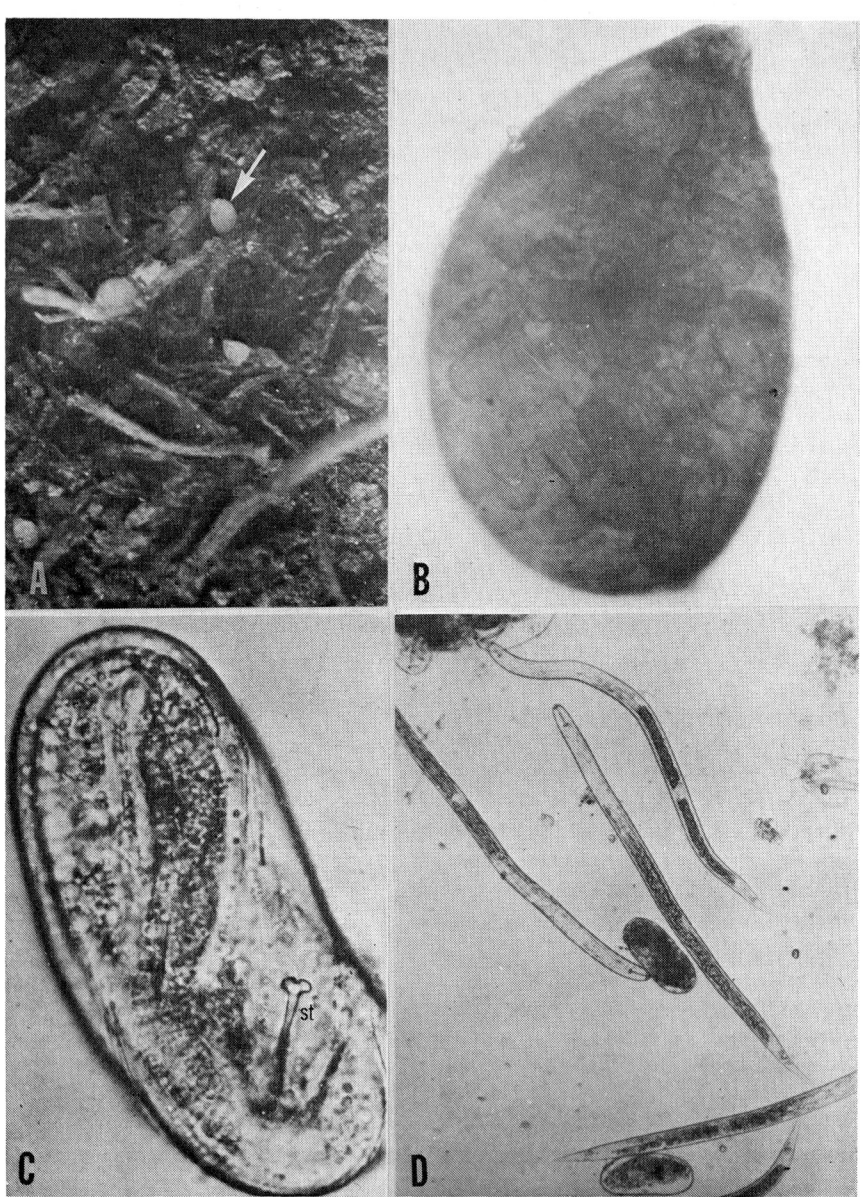

Fig. 11.1. *Heterodera schachtii*. *A*—mature females (white cyst stage) attached to sugarbeet roots by their heads; *B*—cyst containing numerous eggs; *C*—egg with second stage larva coiled inside, note the stylet *(st)*; *D*—eggs and second stage larvae.

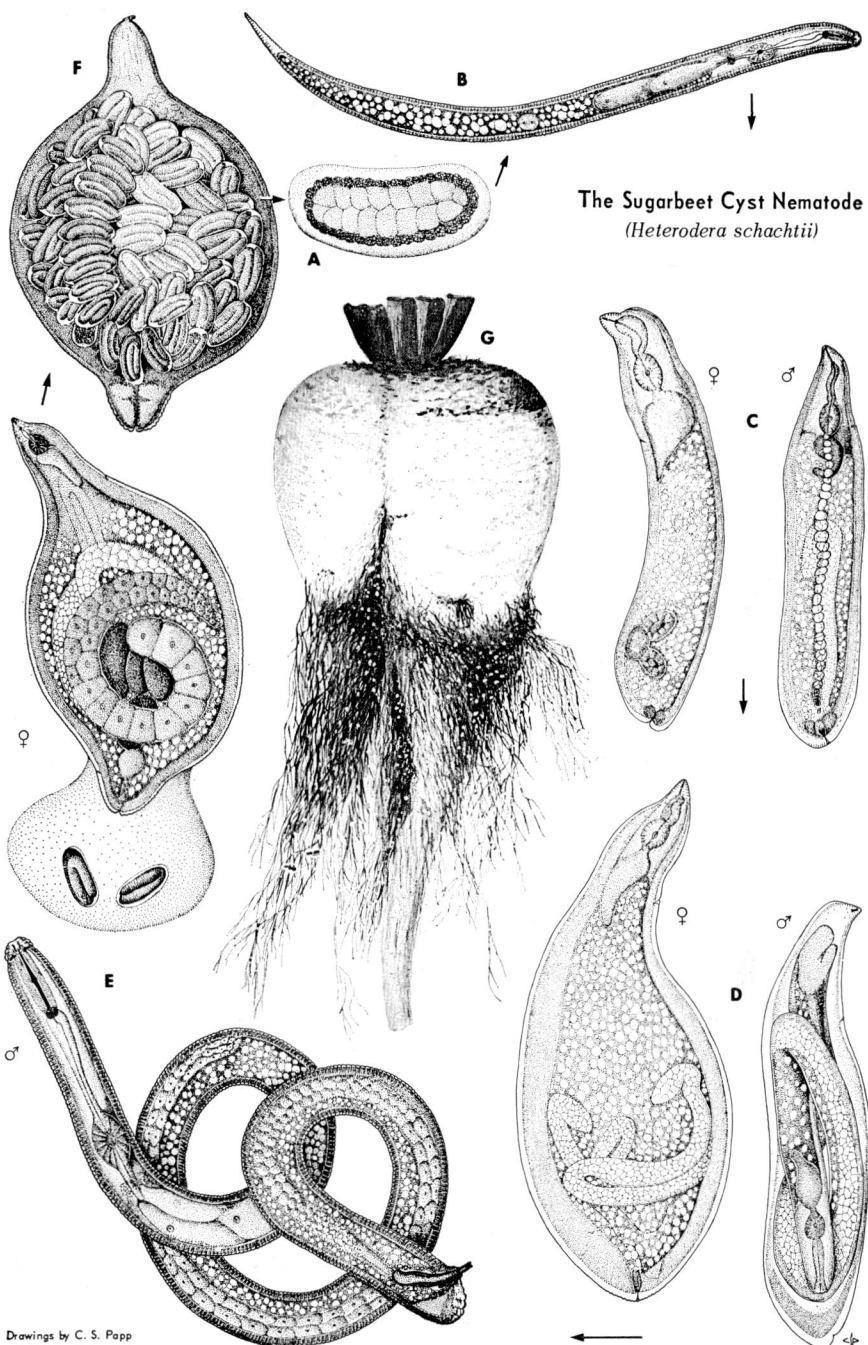

FIG. 11.2. Diagram illustrating the various stages in the life cycle of *Heterodera schachtii*. *A*—egg; *B*—second stage larva; *C*—third stage male and female; *D*—fourth stage male and female; *E*—fifth stage male and female; *F*—cyst; *G*—sugarbeet with white cyst stage on roots.

winter months or through other adverse conditions. If sugarbeets or other host plants (including weeds, such as wild mustard, lambsquarters, etc.) are grown in the field, most of the larvae hatch. In the absence of host plants, only a few of the larvae hatch each year, and these will die within a few months. Thorne (42) states that in the absence of a host, 50 percent of viable eggs hatch the first year, and hatch is reduced in subsequent years. Theoretically, the number of nematodes in the soil decreases with each passing year. However, survival of some nematodes is assured, since weed hosts exist in most fields. Eggs contained in some cysts will remain viable for periods as long as 6 to 10 years.

Ecological Relationships. Heterodera schachtii differs from the root-knot nematodes, *Meloidogyne* spp., in that it causes severe damage in all soil types, including sands, clays, and peat. Root knot is seldom a serious problem in fine-textured soils. *Heterodera schachtii* is also adapted to a wide range of soil pH, from the alkaline soils of the Imperial Valley of California and Colorado to the acid soils of Wisconsin and Michigan.

Heterodera schachtii has a rather narrow range of temperatures (55° to 82° F) at which growth and reproduction occur, with maximum development and reproduction occurring between 78° and 82° F. Above 82° F, development falls off very rapidly. This nematode is unique, however, in its capacity to survive adverse conditions. Eggs within mature cysts are capable of withstanding months of low temperatures in frozen soil and months of extreme temperatures (120°–140° F) in the surface soils of fallow fields in the Imperial Valley (41).

Field Symptoms. Nematode injury first appears in the field as small, distinct areas in which beet plants are severely stunted or killed. These areas are not likely to be large enough to attract attention for several years. By this time, many nematodes will have been scattered to start small infestations in other parts of a field. In severely infested areas, most of the young beets wilt and die shortly after thinning; those that do survive usually do not grow well. Infested parts of the field are particularly conspicuous on hot days, because the leaves of the affected beets wilt readily. Infested areas of sugarbeet fields are readily seen from the air and located for future reference by means of aerial photography (Fig. 11.3). Surviving beets in infested areas are usually small and stunted and have excessive hairlike roots (Fig. 11.4). Examination of the living roots will usually reveal the small lemon-shaped white bodies of the female nematodes (Fig. 11.1A).

In very dry fields, it is sometimes difficult to find the cysts on the beet roots except by careful examination. A soil test should be made on soil samples collected from suspected areas. Thorne (42) suggested placing one or two tablespoons of soil from composite samples in a glass of water and stirring. If cysts are present, they will immediately float to the surface and adhere to the edge of the glass, appearing as small, bright brown bodies that can be distinguished with the use of a hand lens from black weed seeds or

Fig. 11.3. Aerial view of sugarbeet cyst nematode damage to a beet field in the Imperial Valley of California.

other soil debris. The authors recommend having soil samples processed by commercial laboratories or a specialist associated with sugarbeet companies or experiment stations.

Relation of Nematode-infested Beets to Other Diseases. Beets infested with nematodes appear to be more susceptible to leaf spot and other diseases, which are sometimes blamed for the inferior growth, when actually the primary cause is nematode injury to the roots. In surveys conducted by the senior author in the Arkansas Valley, the presence of small areas of leaf spot in the fields was often found to be associated with nematodes, but because of the leaf spot, the nematode problem had been overlooked.

In Colorado, infections of sugarbeets by the fungus *Rhizoctonia solani* were more severe in fields infested with the beet nematode. This view is supported by Polychronopoulos et al. (28) who found that sugarbeet seed-

Fig. 11.4. Excessive lateral root production (whiskers) on sugarbeet incited by *Heterodera schachtii.*

lings infected by *H. schachtii* were more readily penetrated and colonized by mycelium of *R. solani*.

Distribution and Economic Loss from H. schachtii. The sugarbeet nematode is the most widely distributed of the cyst-forming *Heterodera* spp. It occurs in the United States, Canada, Australia, and the USSR, and is now common throughout Europe, including Sweden, Denmark, Holland, Belgium, France, and Germany, and in the countries to the east. Its damaging effects have been noted most recently in England. Table 11.1 presents some data accumulated on the distribution, occurrence, and economic loss caused by the sugarbeet nematode in the United States and throughout the rest of the world.

Control of H. schachtii. The primary control practice in the United States and foreign countries to permit economically profitable beet production in infested soil involves the use of crop rotations. Beets are grown no more frequently than every fourth or fifth year, and nonhost crops such as grains or hay crops are interspersed in the rotation. The senior author has observed that when conventional rotations are used, the greatest yield response occurs when the crop preceding the current beet crop has been a small grain or corn. This probably is related to the excellent level of control of weed hosts for *H. schachtii* that is obtained in grain and crop fields. These rotation practices will not eliminate the nematode but will reduce the populations of viable eggs and larvae to permit good beet growth.

As early as 1926, Thorne (42) pointed out the importance of early planting of sugarbeet to the establishment of vigorous plants in infested soil in cool temperature regions. The interrelationship of soil temperature to stand establishment, sugarbeet nematode activity, and crop injury was studied by Raski (29) and Raski and Johnson (30). They showed experimentally that plantings of sugarbeets made in severely infested fields prior to the time that average soil temperatures rose above 50° to 55° F would produce satisfactory yields (Table 11.2).

Furthermore, the move to fall fumigation of sugarbeet-nematode-infested soil may result in at least two favorable effects: the reduced direct phytotoxicity of the chemical applied shortly before planting and still persisting in the soil; and the advantage of the earliest possible planting date, which results in vigorous seedlings becoming established before nematodes are active.

In many areas, strict sanitation procedures are carried out in addition to the rotations to prevent the introduction and buildup of this nematode. Some countries have gone as far as to steam-clean planting and harvesting equipment. Most also dispose of tare dirt in areas that are not being used for beet production. In some areas, tare dirt samples have been taken at beet-receiving stations and processed for cyst nematodes as an aid in identifying infested fields prior to the time that serious injury has been observed. Although this precaution with regard to disposing of tare dirt has been stressed by many researchers, as recently as 1968 the senior author observed tare dirt being returned to the farm in Colorado (Fig. 11.5).

TABLE 11.1. World and United States distribution and estimated economic importance to sugarbeets of the sugarbeet cyst nematode, *Heterodera schachtii*

Country	Distribution	Percentage Economic Loss
Austria	Niederosterreich and Burgenland (98% infested); Stiermark and Karnten (75% infested); Oberosterreich not infested	± Unknown
Belgium	Less than 1% of crops affected; especially in central Belgium, sometimes in western Belgium	20–40% in crops affected
Denmark	Sporadic	Very little
England and Scotland	Fairly general, but especially in beet areas, principally peat soil and some sandy soils	25,000 A, approx., known infested. Beet only grown once in 4 or more. Very few A/annum show economic loss, say 25% on 10 A and 5% on 500 A
Finland	Known at five localities in southern Finland	< 20%
Germany	West Germany (all beet areas)	0–30% (indust. beets, 2–10%; seed beets, 30%)
Italy	Remarkable in the Po Valley	25–60%
Netherlands	General in sugarbeet- and cabbage-growing areas	10%
Sweden	Fairly common over the sugarbeet areas in southern Sweden, but in most cases, in low frequencies (100,000 A)	Estimated to vary between 0.5 and 3% of total yield
	About 10–15% of the sugarbeet area	In infested fields, < 1%
	Arlov, Sweden, 10–15% of the beet fields	1% in fields with nematode (as a mean)
Soviet Union (USSR)	Widely distributed in the USSR, especially in old sugarbeet regions (Ukraine, Moldavia)	Continuously decreasing — serious losses prior to World War II
United States		
Arizona	30,000 A	None
California coastal	12,000 A, 3,000 affected yearly	10–50% on affected fields
Sacramento	40,000 A, about 4,000 affected yearly	20–70% on affected fields
N. San Joaquin	25,000 A, 1,500 have some loss	20–70% on affected fields
S. San Joaquin	70,000 A, few small, known areas are recognized and beets not planted	None

TABLE 11.1. (cont.)

Country	Distribution	Percentage Economic Loss
Colorado		
eastern	Platteville, Greeley, Gilchrist, and Fort Morgan areas	10–50%
southern	Pueblo, Avondale, Rocky Ford, Otero areas	50%
western slope	2,030 A infested	3–10%
Idaho, Utah, and Oregon	Severe around Lewiston and Ogden, Utah, and in Franklin County, Idaho. Becoming widespread in lower Snake River Valley from Nampa to Weiser, Idaho, including North Malheur County of Oregon	On value of entire crop grown in Amalgamated Sugar Co. territory loss is 1.5–2%, counting cost of fumigation. Loss near 20% in Ogden and Cache Co., Utah, and in Franklin Co., Idaho area
Minnesota	No confirmed report	...
Montana		
southern	500 A infested	Trace
Wyoming		
central	3,270 A infested	7%
eastern	5,000 A infested	13%

Soil fumigation for the control of sugarbeet cyst nematodes has gained in popularity in recent years. Dichloropropene-type fumigants are injected into the soil in the fall or in the spring, at 20 to 22 gallons per acre, prior to the planting of the current year's crop. Several foreign countries, including Belgium, Italy, Austria, and England, have recently initiated experimental soil fumigation programs. Yield increases in the United States following soil fumigation treatment have averaged 5 to 7 tons per acre compared to beets grown in nonfumigated soil. This increase in yield has more than

TABLE 11.2. Average sugarbeet yields in date-of-planting tests and average daily minimum and maximum soil temperatures at 6-inch depth

Planting Date	Yield Tons/Acre*	Avg Min Temp (°F)	Avg Max Temp (°F)	Yield Tons/Acre†	Avg Min Temp (°F)	Avg Max Temp (°F)
Jan.	22.6	43	49	20.8‡	43	50
Feb.	18.6	47	55	18.1	48	55
Mar.	8.8	51	62	8.8	51	59
Apr.	6.1	53	62	8.3	56	65
LSD 0.05	5.1			4.2		
LSD 0.01	7.3			6.0		

SOURCE: Raski and Johnson (30).
* Harvest date—Oct. 27, 1955.
† Harvest date—Oct. 2, 1957.
‡ Planted Dec. 29, 1956, in dry soil, and first rains occurred Jan. 11, 1957.

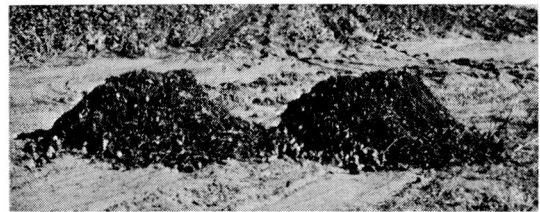

FIG. 11.5. Tare soil from a factory returned to a growers field, Colorado, 1968.

offset the cost of the chemical application and has permitted low-yielding beet acreage to return to average production (16 tons per acre). In some districts, yields have been doubled as a result of fumigation treatments (Tables 11.3 and 11.4).

ROOT-KNOT NEMATODE, *Meloidogyne* SPP.

Root-knot nematodes, *Meloidogyne* spp., are not as important in sugarbeet production in the United States, or in the world, as is the sugarbeet cyst nematode, *Heterodera schachtii*. However, in those areas where they do occur, they can be a very serious problem and in some cases result in a complete crop failure.

Distribution. In the United States, serious damage by root-knot nematodes occurs in growing areas south of the 40th parallel such as Arizona, Cali-

TABLE 11.3. Yield increases of sugarbeet roots and sugar following preplant soil fumigation with Telone in the Platteville area of Colorado in 1966

	Yield Tons/Acre	% Sugar	Lb Sugar/Acre
K Farm			
TELONE 10 gpa	20.25	16.0	6,482
Check	7.58	15.1	2,268
TELONE 15 gpa	21.93	14.7	6,469
Check	8.68	16.2	2,812
TELONE 20 gpa	24.41	15.6	7,614
Check	9.59	16.7	3,202
W Farm			
TELONE 12 gpa	27.33	17.6	9,628
Check	6.63	17.0	2,216
H Farm			
TELONE 20 gpa	16.06	16.4	5,268
Check	1.90	15.4	585

SOURCE: H. Lembright, Dow Chemical Co.
NOTE: The active ingredient in Telone is 1,3-dichloropropene. Other nematicides which contain 1,3-dichloropropene are D-D and Vidden D.

TABLE 11.4. Yield increases of sugarbeet roots and sugar following various crop rotations and preplant soil fumigation in the Platteville area of Colorado in 1967

	Fumigation Date	Planting Date	Yield Tons/Acre	% Sugar	Lb Sugar/Acre	Observations and Comments	Crop Rotation and Previous Beet Yields				
							66	65	64	63	62
Farm 305 Telone 20 gpa	3/24	4/17	15.70	14.0	4,290	severe hail, July 10	beet (11.7 tpa)	corn	beet	corn	alfalfa
Check			9.81	13.3	2,620						
Farm 306 Telone 18 gpa	3/15	3/27	16.28	17.0	5,540	4 severe hail storms; heavy flooding	beet	bean	cabbage	onion	bean
Check			7.57	17.7	2,680						
Farm 410 Telone 20 gpa	3/12	4/7	22.02	16.1	7,100	2 hail storms in June	onion	tomato	corn	beet (11.5 tpa)	bean
Check			13.78	17.2	4,760						
Farm 411 Telone 20 gpa	4/6	4/16	18.32	15.7	5,750	nematodes not considered severe	corn	tomato	bean	tomato	beet (poor)
Check			15.08	16.9	5,090						
Farm 510 Telone 20 gpa	3/23	3/29	16.10	16.7	5,380	4 hail storms; early repeated flooding	potato	beet (11.9 tpa)	potato	alfalfa	alfalfa
Check			12.32	15.9	3,900						
Farm 517 Telone 20 gpa	4/12	4/28	9.24	17.1	3,160	heavy Rhizoctonia	corn	beet (18 tpa)	beet (19.7 tpa)	wheat	bean
Check			5.72	17.1	1,960						
Farm 519 Telone 20 gpa	3/10	3/27	13.47	18.6	5,000	nitrogen deficiency	corn	potato	bean	alfalfa	alfalfa
Check			7.46	18.9	2,810						
Farm 527A Telone 20 gpa	4/1	4/17	18.33	17.3	6,311	10 yrs out of beets; severe nematodes '58	corn	corn	bean	bean	corn
Check			16.21	17.3	5,627						
Farm 529A Telone 20 gpa	4/17	4/24	19.90	14.8	5,893	nematodes not expected; 12 rows only, fumigated	bean	beet (19 tpa)	potato	alfalfa	alfalfa
Check			13.43	16.4	4,404						
Farm 529B Telone 20 gpa	4/17	4/24	21.80	15.6	6,811	5 yr out of beets; 12 rows fumigated	potato	corn	corn	beet (21 tpa)	...
Check			15.01	16.2	4,879						
Farm 503 Telone 20 gpa	11/66	3/10	21.70	17.9	7,769	3 hail storms; heavy nitrogen leaching	bean	beet (9.2 tpa)	bean	beet (11 tpa)	corn
Check			9.13	17.8	3,250						
Farm 509 Telone 15 gpa	3/14	3/22	18.38	16.8	6,142	4 hail storms; heavy flooding	bean	beet	corn	beet	fallow
Check			12.55	16.4	4,618						
Farm 532 Telone 20 gpa	...	...	21.91	18.4	8,062	no hail; dalapon and Pyrazon postemergence	bean	beet	bean	beet	potato
Check			4.85	16.9	1,632						

SOURCE: H. Lembright, Dow Chemical Co.

fornia, southern Colorado, and Texas. Damage is also confined, for the most part, to coarse-textured soils. Thousands of acres of beets in the southern San Joaquin Valley and the Sacramento Valley of California and the Safford area of Arizona are subject to serious damage if control procedures are not used (24, 37).

Most severe losses from root-knot nematode in the United States are attributed to *Meloidogyne incognita* and *Meloidogyne javanica*. These two species cannot survive out of doors in the more northern beet-growing areas of the United States and Europe. *Meloidogyne incognita*, however, has been reported to cause damage to beets in Italy (20) and in Israel. *Meloidogyne hapla* is widely distributed in both southern and northern beet-growing areas of the United States and will attack sugarbeet. It does not appear to be a major factor in production except in Japan. *Meloidogyne naasi*, reported by Franklin (16) to be a pathogen on field crops in England and Wales, has been reported recently (20) from beets in England and Belgium. This nematode also occurs in limited areas in California, Illinois, and Kansas, but has not been reported from sugarbeets.

The reader is referred to Table 11.5 for further details on the distribution and the extent of economic loss due to root-knot nematodes.

Field Symptoms. The extent of damage in the field is dependent on several factors, including the population level of the nematode and the soil temperature at planting time. If the population is high and soil temperatures are relatively warm, serious injury to seedlings may occur, resulting in stunting and even death of the plants. More often signs of injury are not seen until midseason, when plants show symptoms not unlike those attacked by sugarbeet cyst nematodes. Plants are weak, foliage is yellow and wilts readily on warm days. Severe infestations are characterized by a complete collapse of the foliage as illustrated in Figure 11.6.

Severe, early infestation is characterized by the formation of galls on the main taproot and lateral roots (Fig. 11.7). These galls often become very large. Mild or late infestation may result in galls on the lateral roots only. After making almost normal growth, severely infested roots may become completely decayed late in the growing season, due to secondary organisms that readily attack the galled roots. It is difficult in severe cases to keep the beets from rotting before the start of harvest, resulting in losses in tonnage, sugar, and purity. Late applications of irrigation water may accelerate the rotting and collapse of the sugarbeets.

Relationship of Nematode to Host. Root-knot nematodes survive as second stage larvae (Fig. 11.8C) and/or eggs (Fig. 11.8A) in the soil. They may also survive as egg masses in undecomposed root tissue from the previous crop. As the sugarbeet seedling develops, the second stage larvae penetrate the root tissue and become established in the cortical tissue. Feeding by the nematode initiates a series of host responses, culminating in the formation of galls and a number of giant cells (syncytia) within the tissue. It is from the giant cells that the nematode derives its nourishment.

TABLE 11.5. World and United States distribution and estimated economic importance to sugarbeets of root-knot nematodes, *Meloidogyne* spp.

Species and Country	Distribution	Percentage Economic Loss
Meloidogyne incognita and *M. javanica*		
England and Scotland	Very rare; has been recorded outdoors where glasshouse soil is dumped	Of no importance
Italy	*M. incognita* important, but not widespread	5–15%
United States		
Arizona	30,000 A, 10,000 subject to loss	25–50%
California		
coastal	12,000 A, 3,500 subject to loss	10–25%
Sacramento Valley	40,000 A, 15,000 subject to loss	25%
San Joaquin		
north	25,000 A, 8,000 subject to loss	25%
south	70,000 A, 25,000 subject to damage	10–50%
Idaho, Utah, Oregon area	General, but no heavy infestation in any area	Slight; negligible
M. hapla		
Great Britain	On light, sandy soils in eastern England	No loss as far as known
Japan	Hokkaido, the northernmost island in Japan, and Honshu (mainland)	14–31%
Netherlands	Some sandy areas and small gardens where no cereals are grown	1%
United States		
Colorado		
western slope	615 A infested	1%
Minnesota	Widespread	No one knows for sure, probably very little if any loss for state as a whole
Texas	Traces	...
Wyoming		
eastern	3,500 A infested	7%
M. naasi		
Belgium	New pest noted in 1964 in two fields in Belgium	In case of attack, 65%
Great Britain	Medium soils in western areas of Great Britain	No loss as far as known
Netherlands	Local	1%

During the life cycle of the nematode, it goes through four molts (Fig. 11.8). The first of these occurs in the egg. The second, third, and fourth occur in quick succession in the host tissue (Figs. 11.8D, 11.8E, 11.8F). At maturity the female is saccate and immobile (Fig. 11.8G) and can be seen as a pearly white body about the size of a pinhead, if the gall tissue is carefully teased apart. The male (Fig. 11.8G) remains elongate and slender.

Fig. 11.6. Complete collapse of foliage caused by severe infection with *Meloidogyne javanica*.

Mature females deposit numerous (50–1,000) eggs externally in a gelatinous matrix. The life cycle can be completed under ideal conditions in 20 to 25 days; therefore, four to five generations may occur in one growing season.

Damage by root-knot nematodes can be confused with the galling induced by *Naccobbus aberrans* (see page 360). Therefore, for positive diagnosis, soil and gall roots should be given to a specialist in nematology.

Crop rotation for control of root-knot nematodes, although still the most widely practiced method, is limited in effectiveness because of the wide host range of these pests (19). Small grains, combined with a summer fallow period, have given good control. Some reproduction of *Meloidogyne* spp. does occur on small grains if soil temperature is suitable, so these crops cannot be relied on for complete control. Maxson (26) reported that corn

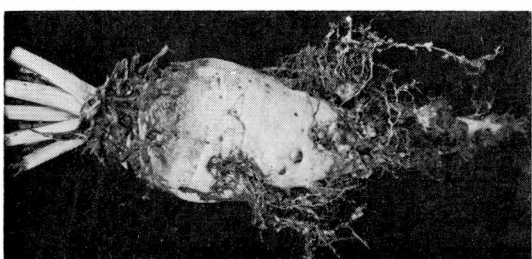

Fig. 11.7. Galls and root rot due to severe infection by *Meloidogyne javanica* and secondary organisms.

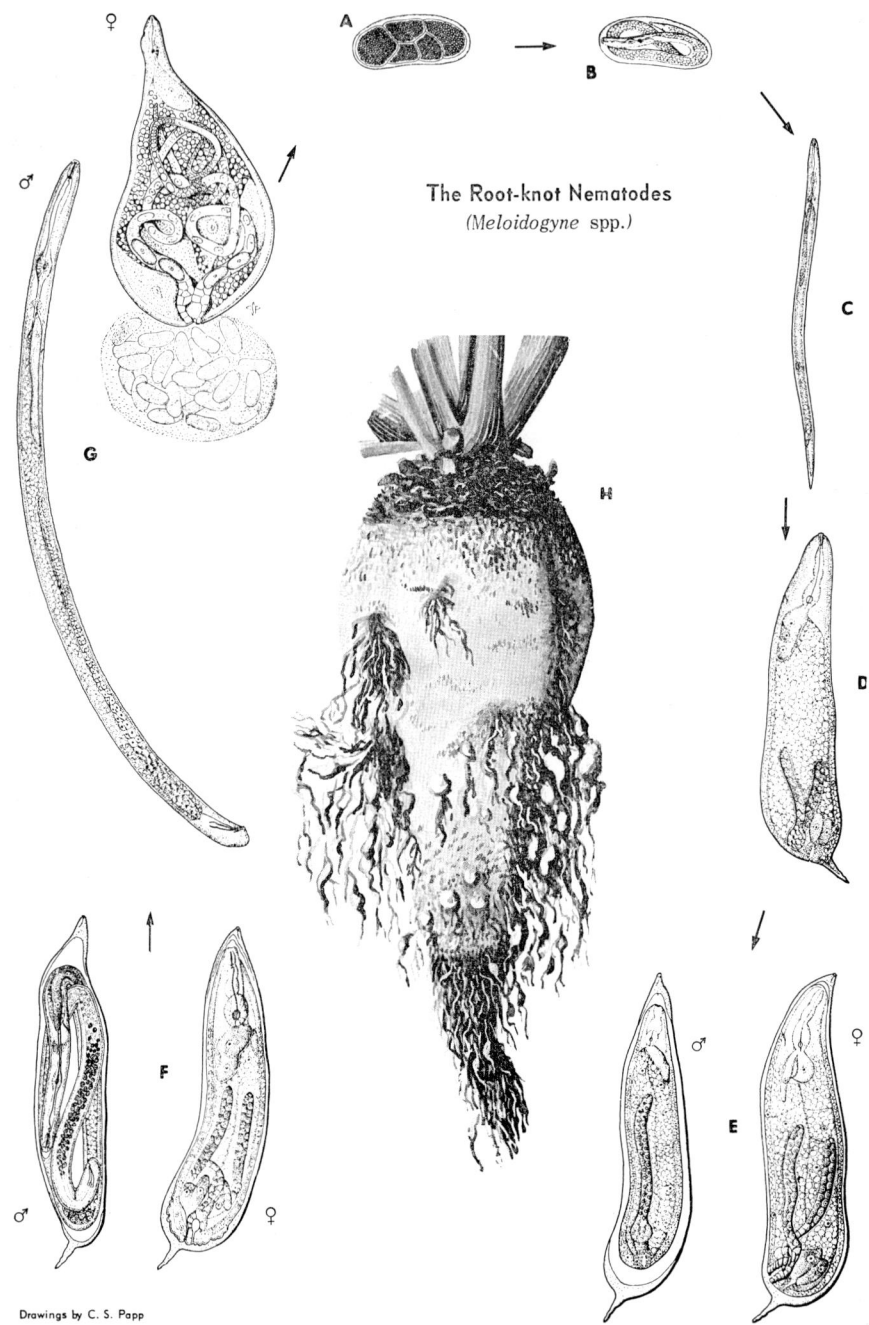

FIG. 11.8. Diagram illustrating the various stages in the life cycle of a root-knot nematode, *Meloidogyne* spp. *A, B*—egg; *C*—second stage larvae; *D*—spike-tail stage; *E*—fourth stage male and female; *F*—early fifth stage male and female; *G*—mature male and female; *H*—sugarbeet with galls caused by *Meloidogyne*.

was an effective crop for root-knot nematode control in northern Colorado. Good yields of beets were obtained following two years of corn.

Satisfactory crop rotations normally have to be developed for specific growing areas so that growing conditions, economics of crop production, susceptibility of available crops, and other similar factors can be taken into consideration.

Chemical control of root-knot nematodes is widely used and is usually successful because root-knot nematodes occur in coarse-textured soil. These soils are amenable to soil fumigation.

The 1,3-dichloropropene nematicides (D-D, Vidden D, and Telone) are most widely used. They can be applied as a broadcast treatment at 20 to 25 gallons per acre or as row treatments at the rate of 10 to 15 gallons per acre. The latter procedure is less costly and gives satisfactory results in most cases. In fields with serious infestations, dramatic yield responses have been obtained. Lear and Raski (24) report yield increases in excess of 20 tons per acre in Kern County, California, in plots where beets on nontreated soil were completely rotted by harvest time.

To date no root-knot-nematode-resistant sugarbeets have been developed.

STEM AND BULB NEMATODE, *Ditylenchus dipsaci*

Crown canker, caused by beet stem nematode, *Ditylenchus dipsaci*, is not known in the United States. *Ditylenchus dipsaci* is, however, a serious pest of sugarbeets in certain European countries, particularly Belgium, Germany, and Switzerland. *Ditylenchus dipsaci* is widely distributed in certain areas of the United States where sugarbeets are grown, for example, Arizona, California, and Utah, and it attacks alfalfa, garlic, and onions. Whether the lack of reports from sugarbeets is the result of ascribing injury to other pathogens or to the absence of pathogenic races is not known. Table 11.6 illustrates the known world distribution of *D. dipsaci* on sugarbeets, and the extent of losses incurred.

Symptomology and Field Injury. On seedlings, the normal symptoms are bloating and malformation of the petioles and midribs of the cotyledons, and rough leaves up to about the sixth pair (13, 20). These symptoms persist until the death of the leaf. Blindness as a result of invasion of the growing point occurs but is less common. Blind plants remain severely stunted, but auxilliary buds develop, leading to a multicrowned root.

The crown canker (Fig. 11.9A) seen in the fall begins among the leaf scars, usually in the form of raised, grayish pustules (14, 20). From here, the rot spreads outward and downward to form a continuous girdle around the crown. The canker, which is granular and slightly raised, may extend deeply into the tissue of the upper portion of the root (Fig. 11.9B). The rot may extend right through the shoulder so that the crown comes away when pulled. Cankers occur mostly above the soil level and multiple crowned beets can always be found in fields where stem nematode cankers

TABLE 11.6. Distribution of the stem and bulb nematode, *Ditylenchus dipsaci*, and its estimated economic importance to sugarbeets

Country	Distribution	Percentage Economic Loss
Italy	It is present	Dangerous only in some years and in small areas
East Germany	Southern and western	2–40%
Belgium	Whole country, but only after a wet and cold spring	Cannot give overall estimate; 5–10% in case of severe attack
Sweden	Single beets have been found attacked at two locations	None
West Germany	Southern and western	0–100%
Denmark	Occurs very seldom	...
Netherlands	Local and incidental	1%
Great Britain	Endemic. Occurs most commonly on heavy soils	5% on 5,000 A at most
Ireland	Local	May be as high as 20% of crop in some fields

occur. Large numbers of *D. dipsaci* are to be found in the relatively sound tissue at the advancing margins of the cankers. Serious rotting of the beet creates problems in topping and processing.

Disease severity is closely related to the climatic conditions prevailing during the development of the beet crop. *Ditylenchus dipsaci* and crown canker development are favored by moist, cool weather and conversely the disease is inhibited by warm, dry conditions.

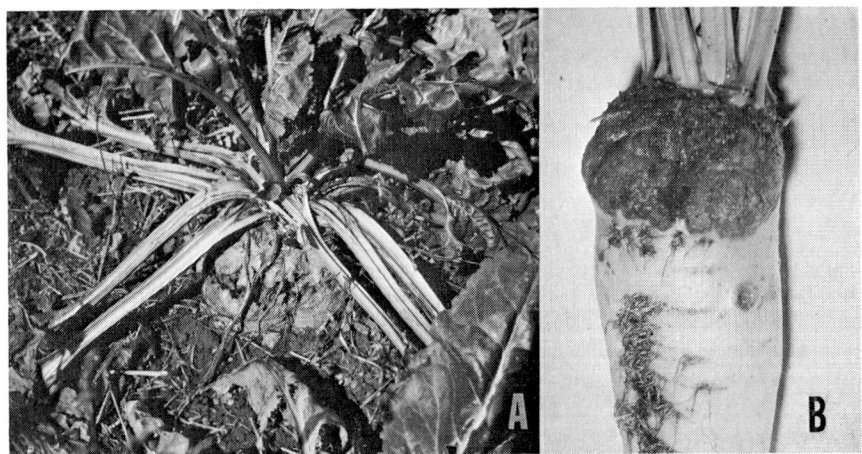

FIG. 11.9. Crown canker of sugarbeet caused by *Ditylenchus dipsaci*. A—field appearance of dry cracking and necrosis; B—raised pustulelike crown canker on mature beet.

Life Cycle. The exact source of the inoculum does not appear to be established. Apparently, the nematode can be carried on the seed (13), but this does not appear to be important. Survival of nematodes on weed hosts would appear to be a significant source. Sugarbeet is a host for the oats, rye, and onion race of *D. dipsaci* (13, 35). Consequently, damage to beets can be severe following infected onion or cereal crops. All stages of the nematode occur in the beet tissue. The various stages in the life cycle of *D. dipsaci* are shown in Figure 11.10. Survival of the nematode is better on loam to clay loam soils and it is on these types of soils that crown canker can be a problem.

Control. Sanitation, including weed control, and crop rotation to nonhost crops would appear to assist in control. Beets should not be grown after oats or onions that have been severely attacked by stem nematodes. In continental Europe, some degree of success in controlling the pests has been obtained with organophosphate nematicides. If *D. dipsaci* is detected early in the season, the British recommend early lifting of beets to avoid the risk of severe crown canker damage.

STUBBY-ROOT NEMATODE, *Trichodorus* SPP., AND NEEDLE NEMATODE, *Longidorus* SPP.

Specific experimental evidence of stubby-root nematodes, *Trichodorus* spp., and needle nematodes, *Longidorus* spp., causing damage to sugarbeets in the United States has not been reported. The stubby-root nematode, *Trichodorus christiei*, is widely distributed in the coarse-textured soils of central and southern California. A marked growth response to soil fumigation was obtained in a field in Fresno County in which *T. christiei* was the predominant nematode present. It has also been associated with stunted, poorly growing beets in other areas of the state. The world distribution of *Trichodorus* spp. and *Longidorus* spp. and the extent of economic loss are illustrated in Table 11.7.

In 1961, Kuiper and Loof (23) reported that *Trichodorus flevensis* was responsible for damage to sugarbeets in the reclaimed soils of polders in the former Lake Flevo in the Netherlands. Several species of *Trichodorus* and *Longidorus* are associated with Docking disorder of sugarbeet in England. According to Jones and Dunning (20), the disorder is named after the parish of Docking in West Norfolk where it has been studied since 1948. It is only recently that the relationship of ectoparasitic nematodes to the injury has been established.

Distribution and Economic Importance. Docking disorder is relatively unimportant in comparison to other nematode diseases of sugarbeet. However, in England, up to 20,000 acres are known to be affected annually. The extent of injury apparently fluctuates considerably from year to year, depending on rainfall and temperature. Severe damage is confined to sandy

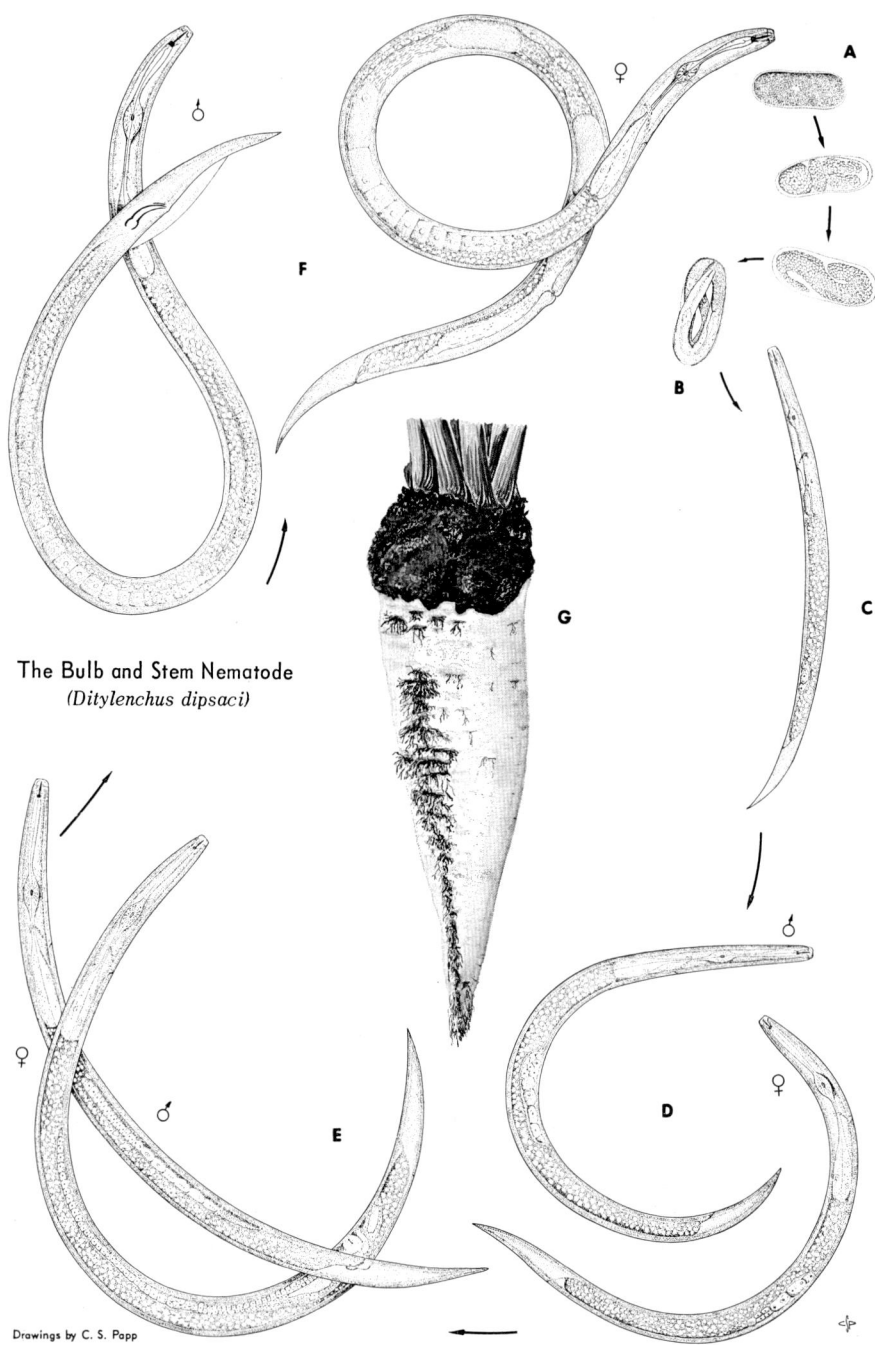

Fig. 11.10. Diagram illustrating the various stages in the life cycle of *Ditylenchus dipsaci*. *A*—egg; *B*—larva coiled in egg; *C*—second stage larva; *D*—third stage male and female; *E*—fourth stage male and female; *F*—mature fifth stage male and female; *G*—crown canker caused by *D. dipsaci* on sugarbeet.

TABLE 11.7. Known distribution of ectoparasitic nematodes, *Longidorus* spp. and *Trichodorus* spp., on sugarbeets and their estimated economic importance

Country and Genus	Distribution	Percentage Economic Loss
Great Britain (*Longidorus* and *Trichodorus*)	On all sandy soils (say 20,000 beets per annum). *Longidorus* also on peat soil	1969 worst year: 10,000 A up to 10%; 6,000 A 10–25%; 4,000 A >25% (probably a conservative estimate)
Netherlands (*Trichodorus*)	Local, only in marine sandy soils of certain texture, there rather widespread	2%
Sweden (*Trichodorus*)	Isolated in a few cases from beet fields	None
United States California (*Trichodorus*)	Discovered in light, sandy soils around Fresno	As high as 50–75% on several hundred A

soil, which is usually alkaline and of low organic matter content and poor structure (15). Since serious injury is dependent on large numbers of nematodes being present in the soil at planting time, the suitability of the previous crops as a host is important. Yields as low as four tons per acre have been taken from seriously affected fields.

Field Symptoms. Docking disorder is characterized by irregularly stunted plants which often have distorted, multiple taproots (fangy roots in the British terminology). The seedlings may emerge uniformly, but soon differences in growth appear in areas of irregular shape and extent. This usually occurs in the areas of coarse-textured soil. The plants may remain small in the early part of the season but often recover in midsummer, and top growth may look uniform. Roots, however, never reach normal size. Stunted plants may show leaf symptoms similar to plants suffering from lack of nitrogen and/or manganese.

Root Symptoms. Root symptoms may suggest which nematode is involved, but final confirmation is dependent on identifying the nematodes following their extraction from soil samples.

Seedlings attacked by stubby-root nematodes, *Trichodorus* spp., have stubby-ended lateral roots (Fig. 11.11A) that turn gray-brown and later black as they die and decay (15).

Seedlings from soil containing needle nematodes, *Longidorus* spp., are small and thin and many of the lateral roots are short, sometimes with swollen or darkened tips (Fig. 11.11B).

The interference with the normal taproot development leads to fangy root development in older plants.

FIG. 11.11. A—seedling showing root injury caused by a stubby-root nematode, *Trichodorus* spp. (After Dunning and Cooke, 1967.) B—seedling showing root injury incited by the needle nematode, *Longidorus elongatus*. (After Dunning and Cooke, 1967.)

Life Cycle. Both *Trichodorus* spp. (Fig. 11.12) and *Longidorus* spp. (Fig. 11.13) are ectoparasites that feed externally on root tissue. All stages of the nematode occur in the soil, including the egg stage. Under optimum conditions of temperature and moisture, *Trichodorus* spp. can produce a number of generations in one year and produce large populations. The life cycle of the *Longidorus* spp. is not clearly understood, but nematodes in this genus would appear to produce few generations per year and to be rather long lived.

Control. Soil fumigation with 1,3-dichloropropene-type fumigants has given good yield responses. Carbamate and organophosphate nematicides look promising.

Crop rotation may be difficult since the nematodes appear to have wide host ranges.

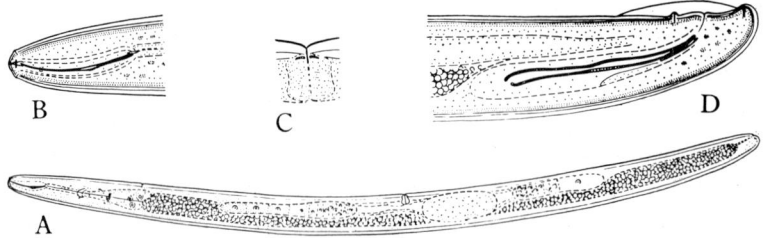

FIG. 11.12. Drawing of *Trichodorus christiei*. A—female; B—head with curved spear; C—vulva; and D—male tail. (After M. W. Allen, 1957.)

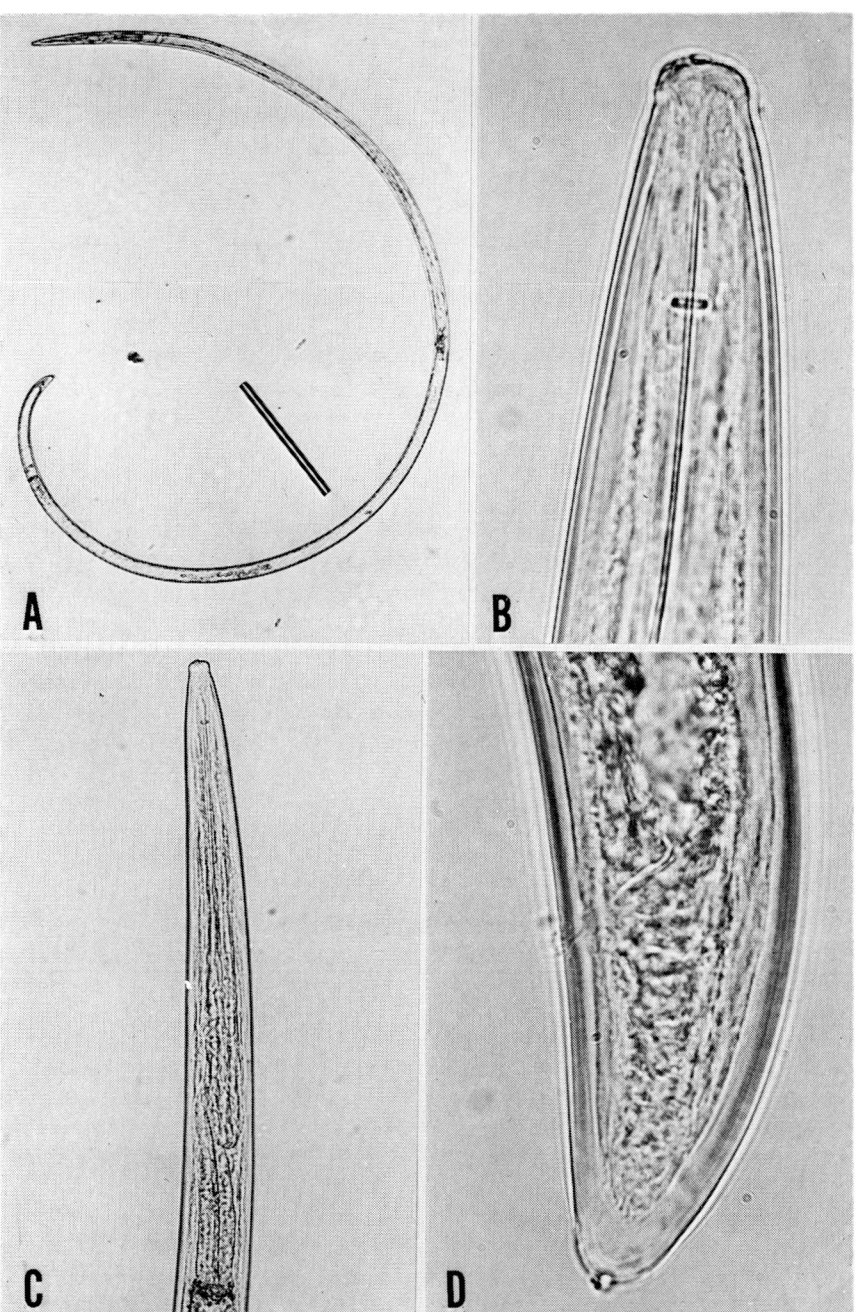

Fig. 11.13. *Longidorus africanus.* A—mature female; B—head showing tip of stylet and guide ring; C—anterior end showing esophagus; and D—female tail.

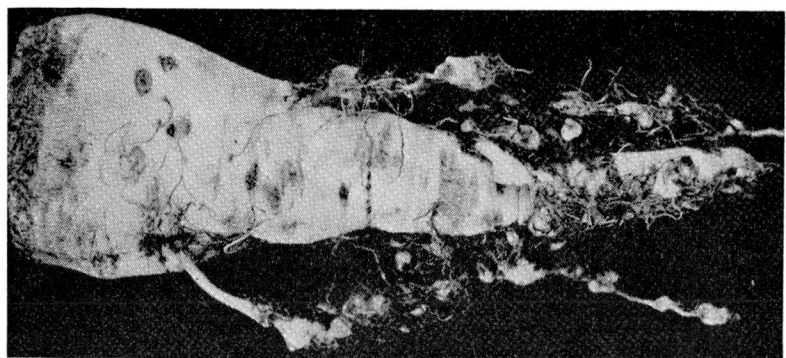

FIG. 11.14. Galls on the roots of sugarbeet incited by *Nacobbus aberrans*. (Courtesy M. W. Schuster.)

COBB'S ROOT GALLING NEMATODE, *Nacobbus aberrans*

Cobb's root galling nematode, *Nacobbus aberrans*,[2] is a serious pest of sugarbeet in a few states on the western slope of the Rocky Mountains. This nematode was first collected from beets by Dr. M. L. Schuster near Mitchell, Nebraska, in 1949. A survey in western Nebraska during 1953 and 1954 revealed that 32 percent of 125 beet fields were infested and in many of them severe losses occurred (43). It is reasonable to assume that for a number of years damage caused by *N. aberrans* was attributed to root-knot nematodes, *Meloidogyne* spp., since the symptoms on the beet roots are similar in appearance.

The nematode was described (44) and its distribution, host range, and pathology discussed for the first time in 1956 (34).

Distribution and Economic Importance. *Nacobbus aberrans* has been reported (10) from western Nebraska and Kansas, Wyoming, Montana, and eastern Colorado. It is apparently most serious in coarse-textured soils and losses due to this nematode can be in the range of 20 percent or higher.

Field Injury and Symptomology. In the field, complete loss of stand due to *N. aberrans* can occur. Often plants that remain never attain good growth. The wounds and necrosis in the cortical tissue of smaller branch roots and the taproot interfere with their function (33). Galls, similar to those produced by *Meloidogyne* spp., are produced on the main taproot and lateral roots (Fig. 11.14). They differ somewhat from the root-knot nematode galls in that *N. aberrans* characteristically stimulates from several to fifty lateral roots to form on the galls.

Positive identification of the causal agent, however, requires direct observation of the nematode by trained specialists. This nematode alone of

---

2. Originally described by Thorne and Schuster (44) as *Nacobbus batatiformis* and recently synonymized with *Nacobbus aberrans* by S. A. Sher (36).

the gall-forming nematodes stimulates starch formation in the roots. Schuster et al. (33), therefore, suggested that the iodine staining of a dissected gall could provide a quick method of identification. Young galls should be used since starch is depleted in older galls.

Life Cycle and Relationship to Host Tissue. Eggs and larval stages of *N. aberrans,* including immature fifth stage females (Fig. 11.15A, B), may be found in the soil. Larvae hatch from eggs and penetrate the cortical tissue of the beet roots. The nematode may remain within the root and continue its development or may return to the soil. However, after the fourth molt, the females usually become sedentary and a pronounced sexual dimorphism occurs. The male (Fig. 11.15C) remains eel-shaped while the female becomes swollen and saccate (Fig. 11.15D, E, F).

Penetration of the root does not appear to be confined to the root tip. The nematodes enter up and down the small roots. Hypertrophy of the epidermal and cortical cells occurs within a few days after penetration of the nematode. Necrosis of the tissue can also occur. Small galls are formed 7 to 10 days after infection as a result of the hypertrophy.

Continued activity of the nematode in the host tissue results in the development of syncytia (large specialized granular, deeply staining, multinucleate host cells) and the accumulation of starch in these cells. This nematode appears to be unique among nematodes in inducing the starch formation in host tissue.

Host Range. Hosts of *N. aberrans* include a number of plants in the Cactaceae, Chenopodiaceae, Cruciferae, and Zygophyllaceae. A few species in the Cucurbitaceae, Umbelliferae, Compositae, and Solanaceae are also hosts. Most plants in the Graminae, Liliaceae, Malvaceae, Iridaceae, Amaranthaceae, and Convolvulaceae are not hosts.

Control. Since this nematode is found for the most part in sandy soils, it is relatively easy to control by soil fumigation. The 1,3-dichloropropene-type nematicides (D-D, Telone, Vidden D) are recommended at 20 to 25 gallons per acre.

Crop rotation to nonhost crops is also effective. Four- to six-year rotations between sugarbeet crops, involving crops such as field corn, dry beans, alfalfa, and potatoes, have been effective.

GENERAL NEMATODE CONTROL

EARLY PLANTING

Planting as early as the season will permit is always advisable, especially in infested fields. The young beets become established and are better prepared to withstand the attacks of nematodes as well as insects and diseases

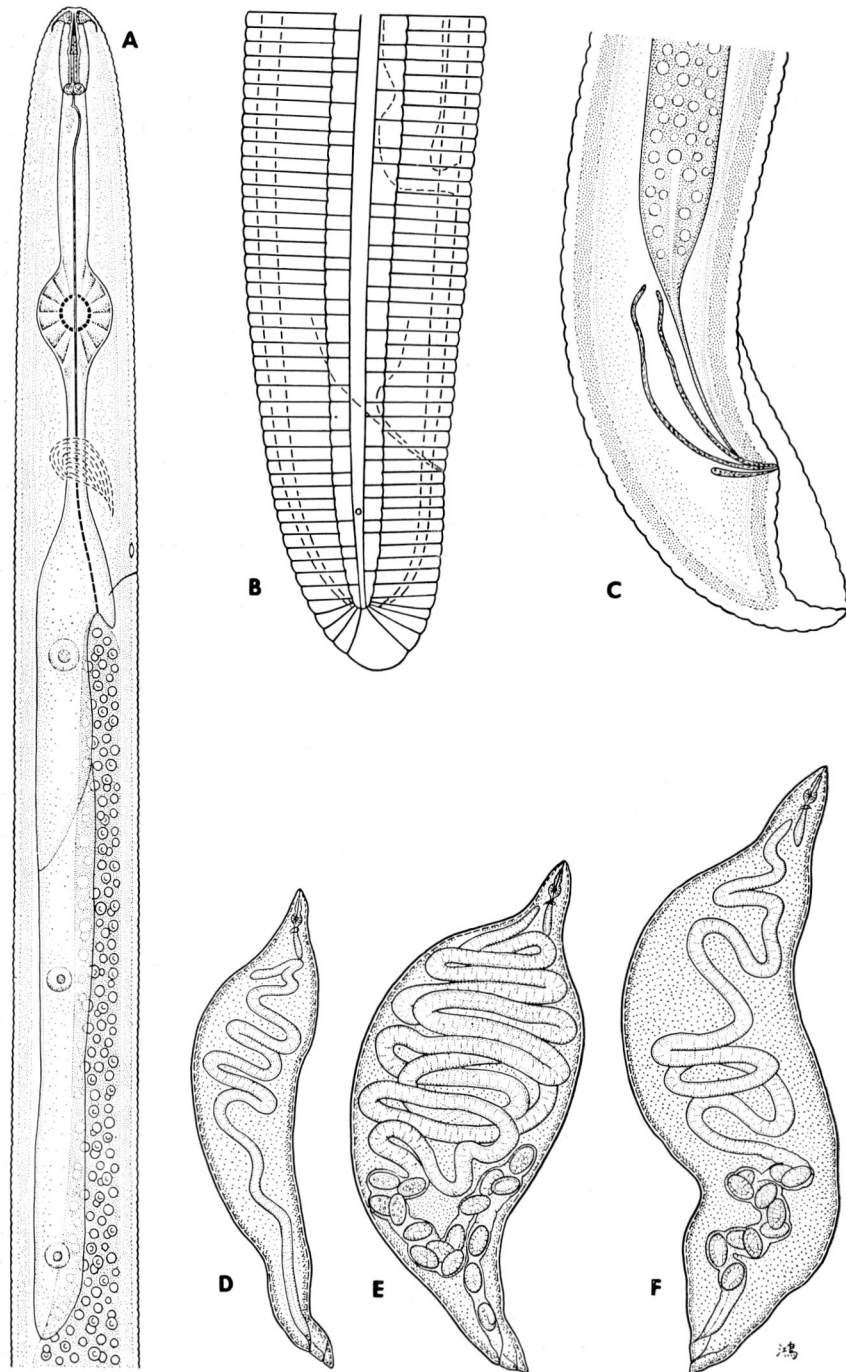

Fig. 11.15. Drawing of various life stages of *Nacobbus aberrans*. A—anterior end of fifth stage female; B—female tail; C—male tail; D, E, F—various stages in development of saccate fifth stage females. (Courtesy S. A. Sher.)

that may appear later. Even after a crop rotation, there may be enough holdover nematodes to check the growth of young beets unless the plants are growing well.

Raski (29) and Raski and Johnson (30) have shown the lower soil temperatures in the early growing season favor beet germination and growth but not cyst nematode hatching, migration, and root invasion in the coastal and interior valleys of California. They further reported that two to three months of beet growth when soil temperatures were below 70° F (21° C) resulted in significant yield increases on nematode-infested soil. In late-planted (May) beets on fumigated soil, yield increases averaged six tons, but in early-planted (January) beets, there was no significant difference between beets grown in fumigated or nonfumigated soils.

In the intermountain region, including Colorado, Wyoming, and Utah, beets are normally planted between March 15 and April 15. Planting delays due to spring fumigation do not normally affect yield or growth, and in the senior author's opinion, beets planted late in Colorado, due to spring fumigation, surpass by mid-June the rate of growth of beets planted earlier in nonfumigated soil.

ROTATIONS

If a field is planted to nonhost crops for a sufficient length of time, the number of nematodes in the soil will be reduced, and a satisfactory crop of beets can be grown. Since some nematodes still remain even after five or ten years, it is practically impossible to eliminate them entirely. In our western states and in Colorado in a few isolated instances where continuous beet cropping still occurs, conditions for the rapid increase in nematodes are ideal. Crop rotations, therefore, should be planned not with the idea of eliminating the nematodes but with the sole purpose of reducing their numbers, so that satisfactory crops of beets can be grown as often as possible. In planning a rotation and particularly in planning the length of a rotation, the following factors should be considered: (1) severity of the infestation; (2) crops adapted to the field; (3) market conditions; and (4) general fertility of the soil. In recent years, the use of three- to four-year crop rotations in Colorado has not reduced nematode infestations sufficiently to permit maximum beet production.

If nematodes are destroying one-fourth or more of the beet crop, the field is likely to require a rotation of four or five years. For this long rotation, alfalfa is most commonly used, but if desired, the rotation may include small grains, peas, potatoes, beans, tomatoes, and other suitable crops. A four- or five-year rotation usually reduces the infestation so that one profitable crop of beets can be grown. Experience has shown that a grain preceding a beet crop, and planted after a three-year alfalfa crop, gives excellent results. Under any system of rotation, the grower will need to increase the interval between beet crops if serious nematode injury becomes apparent.

When a grower gets a profitable yield of sugarbeets on nematode-infested soil after a rotation, he may believe that another crop the following year should be nearly as successful. This is not the case, since the few nematodes that remain after the rotation can increase to great numbers on the extensive root system of the first crop of beets, and thus severely retard the second crop.

In a survey completed in 1957 by Caveness (10) regarding rotations carried out in 273 fields, he found that a minimum rotation of four years with nonhost crops is required in order to produce a profitable sugarbeet crop. At that time, he pointed out that growers in California, Colorado, and Michigan were not practicing satisfactory crop rotations for minimum control of *H. schachtii*. Only 10 percent of the infested fields studied received the recommended four or more year rotation between susceptible crops.

When nematodes are present, it is all the more important to maintain adequate soil fertility. If beets do not make an immediate, thrifty growth from the beginning of the season, a comparatively small number of nematodes can check their growth and the yield may be seriously reduced.

SOIL FUMIGATION

In recent years, several thousand acres of land infested with the sugarbeet nematode have been fumigated successfully with commerical products containing dichloropropenes or dichloropropene-dichloropropane mixtures. Growers with limited acreage and with soils in a high state of fertility have found it economical to combine crop rotation and fumigation. Good results have been obtained on light, sandy soils having a moisture equivalent of 20 percent or less. However, heavy clay loams, if treated in the fall, also respond well to fumigation. Yield increases for beets planted in nematode-infested soil that has been fumigated have averaged 6 to 10 tons per acre for the past eight years (see Tables 11.3 and 11.4).

Applications of not less than 20 to 25 gallons per acre (broadcast treatment) of dichloropropene or dichloropropene-dichloropropane mixture are made under average conditions. It should be noted that this rate of fumigation will not kill all the nematodes. When properly applied, however, the fumigant kills a large percentage of the existing nematodes down to a depth of 14 to 16 inches. It permits young sugarbeet seedlings to become better established and to produce thriftier plants, with resultant significant yield increases.

A waiting period of 10 days to 2 weeks is recommended with spring fumigation. In some localities, farmers are reluctant to wait, since this delay between fumigation and planting is a critical period in relation to loss of soil moisture. Fall fumigation, on the other hand, permits planting just as early the following spring as the season permits. Research work in Colorado continues to show that fall fumigation is as effective as that done in the spring.

Soil fumigation is not a substitute for good farming and should not take the place of crop rotation. A good practice on infested land is to use a crop rotation and grow one crop of sugarbeets, then follow with several years of some nonhost crop, fumigate in the fall or spring, as conditions permit, and grow another crop of beets in the fourth or fifth year. One might assume that fields with unusually high fertility could produce several successive beet crops if fumigated each year. However, this practice is not generally recommended in Colorado, since it appears the nematode population may increase to a point where the usual rate of application may not give economic control. On the other hand, in many situations, fumigation may allow an extra crop of beets to be grown in a shorter rotation without too much of a drain on soil fertility if the nematode infestation is low.

Recent analyses of fumigated soil in Colorado (5) point out that a beneficial side effect resulting in dramatic growth responses is due in part to the nitrogen-releasing properties of the treatment.

Ninety soil samples from 45 fumigated fields had a 25 percent higher residual nitrogen content after beet harvest than comparable nontreated fields. It is possible that nitrogenous fertilizers could be reduced at least 25 percent in some fields when beets are grown in fumigated soil. This response has been observed in a range of soil types under Colorado conditions. Similar conclusions from work by Dunning and Cooke (15) in England support this contention of increased nitrogen availability.

Telone (1,3-dichloropropene) and D-D (1,3-dichloropropene, 1,2-dichloropropane) are chemicals which volatilize readily and move through the soil principally as a gas. Both chemicals have the ability to penetrate nematode cysts and larval body walls. Good control of nematodes is dependent upon soil conditions that provide for adequate but not excessive dispersion of the fumigant.

Factors affecting optimum benefit from use of these fumigants are:

1. Good soil structure to a depth of 18 to 24 inches—surface 8 to 12 inches should be in seedbed or planting condition.
2. Adequate air space for gaseous diffusion (+ 25 percent).
3. Optimum soil moisture (above permanent wilting point but below field capacity).
4. Soil temperature over 40° F (5° C) at 6-inch depth.
5. Low organic matter content—soil relatively free of crop residue.
6. Chemical should be applied about 10 to 14 inches deep for broadcast application, with chisels spaced 11 to 12 inches or 18 to 24 inches deep, with chisels spaced 22 to 36 inches.

If row or bed applications with single chisel are used, the shank should be placed 2 to 3 inches on either side of the seed row to avoid possible phytotoxicity from direct contact of chemical with germinating seed. The latter can be avoided with fall applications of fumigants or with a 10- to 14-day delay in planting after fumigation in spring.

Increased Growth Response Due to Soil Fumigation. Altman and Lawlor (8) reported the apparent utilization of chlorinated hydrocarbons, including D-D (a chlorinated propane-propylene mixture) and fractions of D-D, as carbon sources for certain soil bacteria. These compounds, in low concentrations, could also be utilized by the indigenous soil microflora. Waksman (45) and Alexander (1) made similar observations. They proposed that the indigenous microflora could then exert a direct or an indirect stimulatory effect on plants grown in the treated soil.

Effects of Soil Fumigation. Nitrification processes in soil were stressed in the late nineteenth century with the assumption that total plant growth was a reliable indication of increased microbial activity. Warington (48) showed that soil possessed the capability of converting ammonium-nitrogen to nitrate-nitrogen and that carbon bisulfide and other chemical fumigants stopped the process. That soil fumigation for the control of fungi and nematodes results in a temporary disruption of nitrification and an increase of ammonium-nitrogen has also been demonstrated (27, 39, 46, 47).

Tam (39) made one of the initial observations regarding nitrification inhibition following soil chemical treatment with halogenated hydrocarbons. This inhibition was shown to persist for 8 to 23 weeks, depending upon the concentration of chemical applied to the soil. Bollen et al. (9) have reported similar effects with soil insecticides.

Apart from inhibition of nitrification, most researchers have noted a stimulation of plant growth following soil treatment. Experiments reported by Altman (3, 4) and Altman and Tsue (7) indicate that there is an increase in ammonium, nitrate, and some amino acids following soil fumigation. Sugarbeets grown in fumigated soil showed increases in glutamic and aspartic acid in their roots (2). These nitrogenous compounds could account, in part, for stimulated plant growth. Increased growth response may also result from increased nitrogen fixation due to partial sterilization. Partial sterilization of soil is believed to render a greater amount of energy available to the nitrogen-fixing bacteria. The ammonia level has been shown (2) to increase after partial sterilization, and this may be the cause for the increased growth response. McKee (25) pointed out that many plants do, in fact, utilize ammonia more readily than nitrates. These beneficial results also occurred with pineapple (39, 40). The increased growth response may have been related to the greater availability of nitrogen or some nitrogenous component for plant growth.

In studies on increased plant vigor following soil fumigation with D-D (3, 7), there was an increase in ammonia for 7 weeks following soil treatment, and this was maintained at a level 25 percent greater than the control for the remaining 18 weeks. Comparisons with other physical and chemical soil treatments indicated that increased nitrogen could account for the stimulatory effect. The stimulatory effect was observed in the absence of plant pathogens and in nonamended fumigated soil. Chromatographic analyses of extracts from raw, treated soil showed an increase in glutamic and aspartic acids. This increase was detectable 2 weeks after treatment. In addition, a saprophytic bacterium was isolated from this treated soil which produced a stimulatory growth effect on sugarbeets in laboratory tests.

In field tests and in greenhouse tests, there has been a consistent slight depression of sucrose in sugarbeets grown in fumigated soil, but beet tonnage, and therefore net sucrose, was increased (6). Excessive nitrogen fertilization is known to depress sucrose percentage (17, 32). Since fumigation with D-D inhibits nitrification and can result in elevated soil nitrogen throughout the growing season, it is not surprising that there can be a slight reduction in sucrose percentage in beets grown on treated soil.

Altman and Tsue (7) suggested that one mechanism for stimulated seedling germination and increased growth response on fumigated soil was biological, and that the surviving soil microbial population may have the ability to make nitrogen more available for plant growth. The possibility also exists that the fumigant was broken down and absorbed into the host, thus resulting in the growth that was observed. Another possible explanation is that the fumigant or one of its by-products may have acted as an auxinlike or kininlike material and thus stimulated seed germination.

RESISTANT VARIETIES

Research to develop beet varieties resistant to *H. schachtii* is underway by the USDA and some sugarbeet companies. However, there are no commercial varieties with adequate resistance available at the present time. Although Golden (18) and Savitsky and Price (31) have reported a high degree of resistance to *H. schachtii* in several wild *Beta* spp., Doney and Whitney (12) report that interspecific breeding has been extremely difficult. Successful crosses have been made between tetraploid *Beta patellaris* (resistant parent) and tetraploid *Beta vulgaris*, but resistance to nematodes is linked with other undesirable characters. Efforts to find simply inherited resistance in cultivated sugarbeets have been unsuccessful.

Doney and Whitney (12) screened a number of cultivated lines of sugarbeet to determine if progress could be made within the *Beta vulgaris* species by selecting for different levels of quantitative resistance based on white female counts. They reported that differences between varieties were found in some tests, but these differences were not consistent from test to test. Selections did not have significantly fewer white females than their parents. There was a large test times variety interaction and test variance. Little genotypic variance for number of white females was observed. They concluded that there is either little or no resistance to the sugarbeet nematode in the cultivated sugarbeet; or the environmental variation involved in this method of selection is too great to detect small differences. This indicates that very little progress can be expected by selecting for a quantitative resistance based on this technique. Doney (personal communication) stated that previous reports claiming resistance to *H. schachtii* are not resistance but rather a tolerance or vigor factor appearing in the selections of the cultivated sugarbeet varieties.

REFERENCES

1. Alexander, M. Introduction to Soil Microbiology. John Wiley & Sons, Inc., New York. 1961.
2. Altman, J. Increase in nitrogenous compounds in soil following soil fumigation. Phytopathology 53:870 (Abstr.). 1963.
3. ———. The effect of chlorinated C_3 hydrocarbons on growth and amino acid production of indigenous soil bacteria. Phytopathology 54:886. 1964.
4. ———. Plant growth stimulation studies. Eleventh Conference on Control of Soil Fungi, Reno, Nevada, pp. 35–40. 1965.
5. ———. Side effects associated with the use of nematocides. Symposium on Nematode Control, pp. 39–44. Univ. Hawaii. 1966.
6. Altman, Jack and B. J. Fitzgerald. Late fall application of fumigants for the control of sugar beet nematodes, certain soil fungi and weeds. Plant Disease Reptr. 44(11):868–71. 1960.
7. Altman, Jack and King Mon Tsue. Changes in plant growth with chemicals used as soil fumigants. Plant Disease Reptr. 49(7):600–602. 1965.
8. Altman, Jack and Sue Lawlor. The effects of some chlorinated hydrocarbons on certain soil bacteria. J. Appl. Bacteriol. 29(2):260–65. 1966.
9. Bollen, W. G., H. E. Morrison, and H. H. Crowell. Effect of field and laboratory treatments with BHC and DDT on nitrogen transformations and soil respiration. J. Econ. Entomol. 47:307–12. 1954.
10. Caveness, F. E. A study of incidence of nematodes in sugar beet production. Beet Sugar Develop. Found. Rept., p. 109. 1957.
11. ———. The incidence of *Heterodera schachtii* soil population in various soil types. J. Am. Soc. Sugar Beet Technologists 10:177–80. 1958.
12. Doney, D. L. and E. D. Whitney. Screening sugarbeet for resistance to *Heterodera schachtii* Schm. J. Am. Soc. Sugar Beet Technologists 15:546–52. 1969.
13. Dunning, R. A. Beet stem eelworm. Nematologica 1(3):189–91. 1956.
14. ———. Stem eelworm invasion of seedling sugar beet and development of crown canker. Nematologica 2(suppl.):362–68. 1957.
15. Dunning, R. A. and D. A. Cooke. Docking disorder. Brit. Sugarbeet Rev. 36:23–29. 1967.
16. Franklin, Mary T. A root-knot nematode *Meloidogyne nassi*, n.sp. on field crops in England and Wales. Nematologica 11:79–86. 1965.
17. Gardner, R. and D. W. Robertson. The nitrogen requirements of sugar beets. Colo. Agr. Exp. Sta. Tech. Bull. 28, Ft. Collins. 1942.
18. Golden, A. M. Susceptibility of several *Beta* species to the sugar beet nematode *(Heterodera schachtii)* and root-knot nematode *(Meloidogyne* spp.). J. Am. Soc. Sugar Beet Technologists 10:444–47. 1959.
19. Hart, W. H. Nematodes injurious to sugar beets. Holly Agr. News 8(4):6, 7, 20, and 21. 1961.
20. Jones, F. G. W. and R. A. Dunning. Sugar beet pests. Gt. Brit. Min. Agr. Fisheries Food Tech. Bull. 162. (Pp. 63–86, eelworm pests.) 1969.
21. Koch, A. Untersuchungen uber die Ursachen der Rebenmudigkeit mit besonderer Berucksichtigung der Schwefelkohlenstoffbehandlung. Arb. Deut. Landw. Bes. 40:7–44. 1899.

22. ———. Uber die Wirkung von Ather und Schwefelkohoenstoff auf hohere und niedere Pflanzen. Antr. Bakteriol. Parasitenk., 2, Abt. 31:175–85. 1911.
23. Kuiper, K. and P. A. A. Loof. *Trichodorus flevenis* n.sp. (Nematoda: Enoplida), a plant nematode from new polder soil. Plziektenk. Dienst Wageningen 136:193–200. 1961.
24. Lear, Bert and D. J. Raski. Control by soil fumigation of root-knot nematodes affecting sugar beet production in California. Plant Disease Reptr. 42:861–64. 1958.
25. McKee, H. S. Nitrogen Metabolism in Plants. The Clarendon Press, Oxford. 1962.
26. Maxson, Asa C. Insects and diseases of the sugar beet. Beet Sugar Develop. Found. Rept. 1948.
27. Newhall, A. G. Disinfestation of soil by heat, flooding and fumigation. Botan. Rev. 21(4):189–250. 1955.
28. Polychronopoulos, A. G., B. R. Houston, and B. F. Lownsbery. Penetration and development of *Rhizoctonia solani* in sugar beet seedlings infected with *Heterodera schachtii*. Phytopathology 59:482–85. 1969.
29. Raski, D. J. Sugar beet nematode activity. Calif. Agr. 13(5):4, 14. 1959.
30. Raski, D. J. and R. T. Johnson. Temperature and activity of the sugar beet nematode as related to sugar beet production. Nematologica 4:136–41. 1959.
31. Savitsky, H. and Charles Price. Resistance to sugar beet nematode *(Heterodera schachtii)* in F_1 tetraploid hybrids between *Beta vulgaris* and *Beta patellaris*. J. Am. Soc. Sugar Beet Technologists 13:370–73. 1965.
32. Schmehl, W. R., R. Finkner, and J. Swink. Effects of nitrogen fertilization in yield and quality of sugar beets. J. Am. Soc. Sugar Beet Technologists 12:538–44. 1963.
33. Schuster, M. L., Robert Sandstedt, and Larry W. Estes. Host-parasite relations of *Nacobbus batatiformis* and the sugar beet and other hosts. J. Am. Soc. Sugar Beet Technologists 13:523–37. 1965.
34. Schuster, M. L. and Gerald Thorne. Distribution, relation to weeds, and histology of sugar beet root galls caused by *Nacobbus batatiformis* Thorne and Schuster. J. Am. Soc. Sugar Beet Technologists 9:193–97. 1956.
35. Seinhorst, J. W. Some aspects of the biology and ecology of stem eelworms. Nematologica 2(suppl.):355–61. 1957.
36. Sher, S. A. Revision of the genus *Nacobbus* Thorne and Allen, 1944 (Nematoda:Tylenchoidea). J. Nematol. 2:228–35. 1970.
37. Smith, G. D. and J. V. North. Ten points toward rot control. Holly Agr. News 17(1):16–17. 1969.
38. Southey, J. F. Plant nematology. Gt. Brit. Min. Agr. Fisheries Food Tech. Bull. 7, p. 97. 1965.
39. Tam, R. K. The comparative effects of a 50-50 mixture of 1,3-dichloropropene and 1,2-dichloropropane (DD mixture) and of chloropicrin on nitrification in soil and on the growth of the pineapple plant. Soil Sci. 59:191–205. 1945.
40. Tam, R. K. and H. E. Clark. Effect of chloropicrin and other soil disinfectants on the nitrogen nutrition of the pineapple plant. Soil Sci. 56:245–61. 1943.
41. Thomason, Ivan and Don Fife. The effect of temperature on development and survival of *Heterodera schachtii* Schm. Nematologica 7:139–45. 1962.
42. Thorne, Gerald. Control of sugar beet nematode by crop rotation. USDA Farmers' Bull. 1514. 1926.
43. ———. Principles of Nematology. McGraw-Hill Book Co., New York. 1961.
44. Thorne, Gerald and M. L. Schuster. *Nacobbus batatiformis* n.sp. (Nematoda; Tylenchidae) producing galls on the roots of sugar beets and other plants. Proc. Helminthol. Soc. Wash. D.C. 23:128–34. 1956.

45. Waksman, S. A. Principles of Soil Microbiology. Williams & Wilkins Co., Baltimore. 1932.
46. Waksman, S. A. and R. L. Starkey. Partial sterilization of soil, microbiological activities and soil fertility. Soil Sci. 16:137–56. 1923.
47. Warcup, J. H. Chemical and biological aspects of soil sterilization. Soils Fertilizers 20:1–5. 1957.
48. Warington, R. On nitrification. J. Chem. Soc. 33:44–51. 1878.

12

Factors Affecting Quality

J. T. ALEXANDER
Hawaiian Agronomics
Honolulu

QUALITY DEFINED 372

RIPENING OR QUALITY-IMPROVING FORCES . 374

PREPLANTING FACTORS 374
    Beet Varieties 374
    Land Suitability 375
    Nematode Infestation 375

GROWING-SEASON FACTORS 375
    Date of Planting and Length of Growing Season 375
    Beet Population 375
    Weed Control 375
    Nitrogen Supply 376
    Irrigation 376
    Disease 377
    Insects 377
    Hail 377
    Killing Frosts 378

PREHARVEST CONSIDERATIONS 378
    Nitrogen Availability 378
    Day and Night Temperatures . . . 379
    Rain 379

CONSIDERATIONS DURING HARVEST 380

SUMMARY 380

IT IS PROPER that a chapter pertaining solely to factors affecting quality be included in this book devoted primarily to the production of sugarbeets, since this characteristic is of major importance in the culture of sugarbeets. It is important to both growers and to processors.

During the past 15 years, the amount of sugar recovered per ton of beets purchased has declined significantly in all of the sugarbeet-growing areas of the United States (11). The most serious reductions have occurred in the warm climate areas of Arizona, California, Colorado, Kansas, and Texas.[1] There is general agreement that the decline is associated with a decrease in beet quality.

QUALITY DEFINED

To understand sugarbeet quality, certain components or groups of components of the beet root must be identified. Also, sugarbeet quality itself must be clearly and simply defined.

Root constituents may be divided into nine general groups: (1) water; (2) dry matter; (3) total soluble solids; (4) total insoluble solids; (5) sucrose; (6) nonsucrose, soluble solids; (7) soluble, nitrogenous organic compounds; (8) soluble, nitrogen-free, organic compounds; and (9) soluble, mineral matter (ash). Approximate relationships among these constituents and their relative concentration in the root are presented below in a diagram modified from Silin (12).

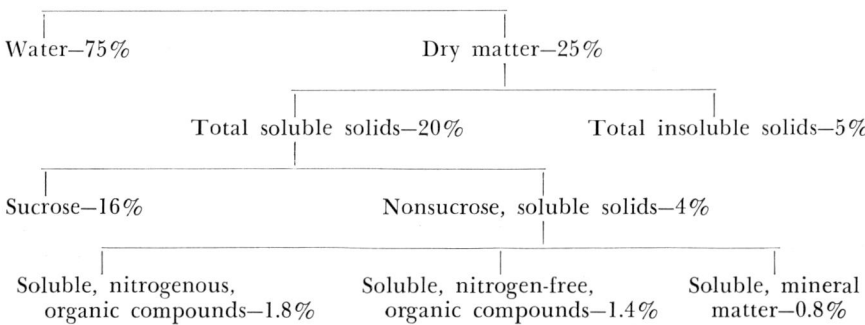

Chemical Composition of Sugarbeet Roots—Weight Percent

As indicated in the diagram, only 20 percent of the weight of the beet, the total soluble solids, makes up the material in the solution that is processed in the factory to produce refined beet sugar. Of course, the weight of the water in the beets must be handled and the 5 percent in-

---

1. J. T. Alexander, Relationship between nitrogen concentration in the petiole and root of sugarbeet with quality of beets in all Holly areas. Unpublished Holly Sugar Corp. Res. Rept. A112, 1964.

soluble solids must be removed; but the basic raw material, sugar, along with the impurities (nonsucrose, soluble solids) are found in the total soluble solids group. Sugarbeet quality is based upon this group of constituents.

Two criteria are generally used to define beet quality: sugar content and purity. Sugar content is the percentage, by weight, of sucrose found in the fresh beet root. Purity is the ratio of sucrose to total soluble solids expressed as a percentage.

Sucrose content of the beets is important to both the grower and the processor. The price per ton of beets paid to growers by processors in all beet-growing areas of the United States is determined by the sucrose content of the beets. Thus, to the grower, it determines the value of his crop. To the processor, it represents the potential basic raw material available in the beets he purchases. Other things being equal, the higher the sucrose content, the higher the potential recovery of crystallized sugar per ton of beets. A practical example might be cited. Analysis of extracted juice showed 20 percent total soluble solids and 17 percent sucrose. Sucrose was the main constituent of the soluble solids, comprising 85 percent ($.17/.20 \times 100$) of them; therefore, the purity of this solution was 85 percent. If the sucrose content of the total soluble solids was less, the purity of the solution would be proportionately lower. Since the purity value represents the percentage of the total soluble solids comprised by sucrose, then 100 minus the purity value is the percentage of solids other than sucrose in the solution. In the above case, 100 minus 85 equals 15 percent nonsucrose, soluble solids. These constituents include a wide variety of chemical compounds that have different effects upon the refining process.

Purity and the nonsucrose substances in the beets are of importance primarily to the processor. They do not directly affect the contractual amount of money paid for beets by the processor to the grower except in a few instances. They are generally indicative of the amount of the sucrose in the beet that can be recovered. However, if the nonsucrose substances occur in sufficient quantities to seriously affect the efficiency of refining sugar from beets, they potentially could reduce the amount of money paid for the beets.

The process by which sucrose is extracted from sugarbeets consists of diffusing the soluble solids from the finely shredded roots. This diffusion juice contains both the sucrose and the nonsucrose substances. Then in a series of chemical and physical steps these nonsucrose substances and water are separated from the sucrose, leaving the pure chemical sucrose or the ordinary sugar with which everyone is familiar.

Depending upon the operating characteristics of a factory, each pound of nonsucrose substance in the juice prevents 1.5 to 1.8 pounds of sucrose from crystallizing, and as a consequence, this amount of sugar is lost to molasses.

Among the nonsucrose substances that are found in the juice and that affect crystallization to the greatest extent are the soluble nitrogenous compounds and the nitrogen-free organic acids. Since these constituents are not stable at high temperatures, they may create processing difficulties before the actual crystallization stage is reached. The most important nitrogen-containing compounds found in the juice are in the form of ammonia, protein, amide, betaine, purine bases, and amino acids, while a partial list of nitrogen-free organic acids includes oxalic, succinic, malonic, glutaric, aconitic, glycolic, malic, citric, lactic, and tartaric (12, 11).

Both types of compounds, nitrogenous and nitrogen-free organic compounds, that contribute to impurity are products of metabolic activity, and their concentrations in the beet are influenced by any change of environment that affects the rate of growth of roots and tops. Rapid growth results in a decrease in sugar content and an increase in impurities. Conversely, a decline in growth rate in a healthy beet brings about an increase in sugar and a decrease in impurities. Practically all factors affecting beet quality do so through this mechanism (17).

RIPENING OR QUALITY-IMPROVING FORCES

The sugarbeet does not ripen or mature in the same manner as many other plants. In annual grains, for example, the plant produces seed and dies, and when the seed is sufficiently dry to store, it is referred to as ripe or mature. Sugarbeets are considered "ripe" or "mature" when the sugar content has reached a maximum for the conditions under which it is grown. Certain external forces prior to harvest can influence this process: (1) a deficiency of nitrogen; (2) moisture stress; (3) long, bright days; (4) maximum temperatures that do not exceed 85° F for extended daily periods, and (5) night temperatures near but not below freezing. The nitrogen deficiency, the moisture stress, and the cool nights slow root and top growth (metabolic activity) and permit sugar that normally would be consumed for growth to accumulate in the root. An intense and extensive period of sunlight results in increased sugar manufacture, and moderately warm days and cool nights reduce excessive respiration. The net effect is again an increase in sugar content and purity. Note, as the factors affecting quality are discussed in the following pages, how often they are associated with one or more of these five forces (4, 5, 6, 16, 17, 18).

The various factors and cultural practices affecting beet quality may be conveniently presented in chronological order as divisions of the growing season.

PREPLANTING FACTORS

BEET VARIETIES

Beet varieties differ in sugar content, purity, and adaptability to areas (7, 13, 21). Present varieties have been developed with the highest quality obtainable under the demands set by the local requirements for yield of roots, disease resistance, nonbolting tendency, and the monogerm characteristic. Improved sugar content and purity through variety improvement is difficult and time-consuming, but significant improvement can be anticipated for the future.

LAND SUITABILITY

Land suitability affects quality in that land unfit for beets because of high salt content, pH, texture, slope, drainage, or depth does not permit the normal, vigorous, uninterrupted early and midseason growth so necessary to exhaust the available nitrogen supply prior to harvest[2] (6, 16, 17).

NEMATODE INFESTATION

Nematode infestation may result in restricted growth, causing more nitrogen to be available to the crop than it can use during the season. Again quality is low as a result of excess nitrogen at harvest time.

GROWING-SEASON FACTORS

DATE OF PLANTING AND LENGTH OF GROWING SEASON

Date of planting and length of growing season may influence significantly harvested beet quality. The longer the period of healthy growth, the greater the chance the crop has to exhaust the soil nitrogen supply[3] (7, 16, 21). An ineffective growing season (characterized by a cold, late spring; a dry summer when water supply is short; an excessively hot summer; an early, wet fall; or a combination of these occurrences) usually results in disappointing beet quality. Yields of roots per acre also invariably suffer when the growing season is short or ineffective.

BEET POPULATION

Beet population and distribution influence quality (9). The competition for nutrients is reduced when stands are light and in-row gaps produce large beets more difficult to fill with sugar. Also field area without leaf canopy wastes sugar-producing sunlight. Here again, an excess of late-season nitrogen availability is the underlying cause of low beet quality.

WEED CONTROL

Weed control early in the season is important if beet quality is to be acceptable at harvest. The presence of weeds seriously retards the growth

---
2. Ibid.
3. Ibid.

of beets, which makes the season less effective (8). Weeds harbor foraging and disease-spreading insects and virus diseases. They also host soil pests such a nematodes (14). If not controlled, weeds late in the season shade the beets and reduce the quality of the crop at harvest by interfering with normal growth and sugar manufacture.

NITROGEN SUPPLY

Nitrogen supply has a profound influence on beet quality. If one controllable factor can be singled out as affecting beet quality to the greatest extent, it most certainly would be nitrogen fertilization. Nitrogen available to the beet crop comes mainly from three sources: (1) generated as a result of decomposing organic matter in the soil; (2) residual nitrogen applied to but not used by previous crops and remaining in the soil, and (3) nitrogen applied on or just prior to planting the beet crop. Too little nitrogen early in the season results in poor root yields, and it may reduce sugar content if top growth is severely reduced over an extended period. Nitrogen, if available in excessive quantities, encourages continued root growth at the expense of sugar content. Late nitrogen applications do not compensate for earlier inadequacies. They stimulate late-season growth, with the result that both sugar content and yield are low[4] (5, 7, 16, 17, 19).

Any practice or condition that provides beets with nitrogen late in the season just prior to harvest will lower quality. In this regard, previous crop and fertilizer programs influence beet quality. A small carry-over of nitrogen from the previous year may be just right to get the young beets off to a fast start. But if the carry-over is excessive, little can be done except to identify such fields in advance and avoid planting beets in them. On the other hand, following lightly fertilized crops such as barley or wheat with beets makes it quite easy for a nitrogen deficiency to develop prior to harvest (6). The decomposition of plant residues sometimes presents a problem as it may occur late in the season, with a release of nitrogen just prior to harvest (20).

Only experience, the use of indicator strips of varying levels of nitrogen application, and the use of tissue analysis can provide satisfactory guides to proper nitrogen management.

IRRIGATION

Irrigation affects the quality of beets in a number of ways. If fields are not irrigated immediately after planting, herbicide control may be lost, emergence may be delayed, and the length and effectiveness of the growing season may be reduced, producing light, irregular stands. Emergence may

---

4. Ibid.

be low due to poor germination or because fields stressed for moisture are more susceptible to frost damage. All these possible occurrences tend to lower quality through the physiological processes previously discussed, that is, retarded early growth due to moisture stress reduces early usage of nitrogen and provides a greater supply late in the season. Late irrigations keep the crop growing until harvest by leaching surface nitrogen into the active root zone (15). While this practice may be conducive to producing a maximum crop, it can also lower sugar percentage through hydration of the roots. Withholding water prior to harvest in rain-free areas tends to increase sugar percentage through dehydration[5] (5, 6).

DISEASE

Disease lowers quality because it interferes with normal beet growth and development and the formation of sugar (2, 10). A disease-induced reduction in growth often results in the presence of unused soil nitrogen at harvest. Root rot diseases destroy sugar in the process of root decomposition. Samples containing rotten beets test extremely low in sugar content.

INSECTS

Insects affect beet quality adversely by carrying and transmitting virus diseases. Perhaps the best known vector is the green peach aphid, which has been identified as the major disseminator of the virus complex of beet yellows, western yellows, and beet mosaic (2, 10). This complex when transmitted to a young beet crop interferes seriously with the growth and development of the beets and significantly retards the formation of sugar (2). Also, high concentrations of nitrogen are often found in beet roots infected with this disease complex[6] (8). These diseases act singly but their effect is less severe. Other insects spread virus diseases and affect quality and yield of beets in much the same way.

Defoliating insects reduce sugar-producing leaf area and lower sugar content. Insects such as the flea beetle, cutworm, and sugarbeet maggot that reduce stands effect a reduction in sugar content through reduced competition for nitrogen. Sucking insects may interfere with the movement of sugar from the leaf to the roots, thus reducing quality.

HAIL

Hail may cause serious reductions in sugar content when it occurs just prior to harvest. Leaf area is reduced and new leaf growth must come from

---
5. Ibid.
6. Ibid.

stored sugar (1). Early-season defoliations, either by hail or insects, affect quality very little, but yield of roots per acre may be significantly reduced.

KILLING FROSTS

A killing frost may bring about a marked lowering of the sugar content if subsequent temperatures are high enough to encourage regrowth. Such regrowth of tops occurs at the expense of the sugar reserve stored in the roots.

PREHARVEST CONSIDERATIONS

NITROGEN AVAILABILITY

Nitrogen availability just prior to harvest, as well as being the most important single factor influencing beet quality, is the factor that can most effectively be controlled. By personal experience during the period 1955 through 1964, using petiole analyses and the diphenylamine spot test on expressed juice from sugarbeet pulp, the strong inverse relationship between sugar content and the presence of nitrate nitrogen in the root at the time of harvest was confirmed. These investigations covered the states of Montana, Wyoming, Colorado, Texas, and California.

Burtch used the diphenylamine testing procedure and showed that the above average sugar percentage of beets grown in the southern San Joaquin Valley of California during the 1968 season was associated with low levels of nitrate nitrogen in the pulp (personal communication).

Hoff showed that *Cercospora betecola* and virus yellows infections widely distributed in California in 1958 were significantly correlated with low sugar content and purity but that the nitrate and sugar content relationship was even stronger and appeared to exert an effect independent of the disease complex (8). Fields were surveyed using the procedure of Ulrich et al. (19).

A simple correlation coefficient was computed for the relationship between nitrate nitrogen in beets at Hereford, Texas, during the 1969 harvest and sugar content. The computed value was —0.69. More than 24,000 samples were analyzed for nitrate nitrogen, using the Orion specific ion electrode. An almost identical correlation coefficient of —0.67 was found using the diphenylamine test.

The presence of nitrate nitrogen in the tissue depresses sugar percentage at low or high levels of sugar content. For example, the average sugar content of purchased beets from the Hereford, Texas, area was 10.1 percent—an all-time low. The beets from contracts having high nitrate nitrogen averaged only 8.3 percent sucrose, while those having low nitrate nitrogen averaged 12.4 percent sucrose. By comparison, beets testing low in nitrate nitrogen when harvested in the spring of 1970 in the Imperial Valley produced as high as 18.2 percent sucrose, while those testing high in nitrate nitrogen produced only an average of 14.6 percent sucrose. Other environmental factors, such as temperature, moisture, and sunlight, were responsible for the difference in the location averages.[7]

DAY AND NIGHT TEMPERATURES

Day and night temperatures preceding harvest influence quality almost as much as preharvest nitrogen deficiency (18). Low temperatures that tend to retard growth improve quality, while higher temperatures that are conducive to rapid growth prevent an increase in quality. Higher temperatures (above 100° F) retard growth but reduce quality. Efforts are made in some areas to use beneficial weather patterns for obtaining improved quality. For example, in the Pacific Northwest and in the beet-growing areas in and east of the Rocky Mountains, beets are harvested in a relatively short time, primarily in October. Essentially all potential growth has been accomplished by then and the lower temperatures at and prior to harvest have been beneficial to beet quality. Beets harvested in excess of factory capacity are piled on the ground and sliced as time permits in these areas.

In the Imperial Valley of California and the Salt River Valley of Arizona, on the other hand, beets are planted in September and harvested the following summer. In this case, an attempt is made to schedule the harvest to allow the best growth of the crop possible and yet complete the harvest before the adverse effects of the high summer temperatures have made the beets unsatisfactory for processing. In this area, no stockpiling of beets is done because temperatures at harvest time are so high that harvested beets begin to deteriorate almost immediately.

RAIN

In other areas of California, harvest begins in July and continues until the late fall rains halt field operations. Since only limited piling is feasible in these areas because of warm temperatures, the harvest is controlled to equal the slicing capacity of the factories. Beets left in the ground at the time when winter rains prevent further harvest are harvested the following spring as soon as weather permits.

Rain immediately preceding harvest in California, Texas, and other areas where high evaporation rates over a long, dry season deposit nitrate salts on the surface of the soil can abruptly and significantly reduce beet quality through the leaching of nitrate into the root zone (15). As pre-

---

7. Ibid.

viously stated, late rains or irrigation lower sugar percentage through simple hydration[8] (5).

CONSIDERATIONS DURING HARVEST

Beet storage conditions may result in significant sugar loss in areas where beets are held several months in storage piles. Where beet purchase agreements are based on sugar recovered or sugar in beets after storage, both grower and processor are vitally concerned. The amount of trash, tops, and dirt on the beets at the time of piling greatly influences their storability. Well-topped, clean beets lose less sugar in storage. Heating of the piles is reduced, and as a consequence respiration losses and rot are also reduced. Storage characteristics are affected by the nitrogen and moisture supply just prior to harvest. If the nitrogen and moisture supply have been low, the beets tend to have thickened walls and are less likely to chip and break when handled mechanically. They also tend to have a lower respiration rate and are less subject to rotting than beets that were growing rapidly as the result of an abundance of nitrogen and moisture at harvest (3).

SUMMARY

Quality has been defined and many factors affecting it have been discussed. As presented, most of the factors caused a depression of beet quality. To reverse this orientation, a positive approach must be taken. Thus, quality can be improved by growing healthy sugarbeets so that full advantage can be taken of any environment that favors sugar accumulation and an increase in purity late in the season and by providing a nitrogen control program that will permit a deficiency of this element to develop prior to harvest.

It is essential that certain practices be followed to promote the production of quality beets: a more careful selection of land; a closer adherence to the principles of a good rotation; lengthening the growing season through early planting; and increasing the effectiveness of the growing season by (1) leaving heavier, more uniform stands; (2) controlling weeds; (3) irrigating properly for germination (on time, lightly, frequently, and not too close to harvest); (4) controlling insects and disease; and (5) fertilizing correctly.

Carefully following these procedures will permit the climate environment of any one growing season to provide the best quality possible.

REFERENCES

1. Afanasiev, M. M. The effect of simulated hail injuries on yield and sugar content of beets. J. Am. Soc. Sugar Beet Technologists 13:225–37. 1964.

8. Ibid.

2. Bennett, C. W., Charles Price, and Glenn E. Gillespie. Effect of virus yellows on yield and sucrose content of sugar beets in tests at Riverside, California. Proc. Am. Soc. Sugar Beet Technologists 8:236–40. 1954.
3. Dexter, S. T., M. G. Frakes, and Grant Nichol. The effect of low, medium and high nitrogen fertilizer rates on the storage of sugar beet roots at high and low temperatures. J. Am. Soc. Sugar Beet Technologists 14:147–59. 1966.
4. Haddock, J. L. Sugar beet yield and quality. Utah Agr. Exp. Sta. Bull. 362. 1953.
5. ———. Yield, quality, and nutrient content of sugar beets as affected by irrigation regime and fertilizers. J. Am. Soc. Sugar Beet Technologists 10:344–55. 1959.
6. Haddock, J. L., P. B. Smith, A. R. Downie, J. T. Alexander, B. E. Easton, and Vernal Jensen. The influence of cultural practices on the quality of sugarbeets. J. Am. Soc. Sugar Beet Technologists 10:290–301. 1959.
7. Hills, F. J., L. M. Burtch, D. M. Holmberg, and A. Ulrich. Response of yield-type versus sugar-type sugar beet varieties to soil nitrogen levels and time of harvest. Proc. Am. Soc. Sugar Beet Technologists 8:64–70. 1954.
8. Hoff, John. Sugar Beet Quality Survey. Holly Agr. News (summer issue), pp. 17–19. 1959.
9. Loomis, R. S. and Albert Ulrich. Response of sugar beets to nitrogen depletion in relation to root size. J. Am. Soc. Sugar Beet Technologists 10:499–512. 1959.
10. McFarlane, J. S., C. W. Bennett, and A. S. Costa. Effect of virus yellows on the yield and sucrose percentage of the sugar beet at Salinas, California, in 1952. Proc. Am. Soc. Sugar Beet Technologists 8:215–18. 1954.
11. Rorabough, Guy O. Poor beet quality and its effect on the processor. Proc. Calif. Soc. Sugar Beet Technologists. Santa Barbara Meeting. 1961.
12. Silin, P. M. Technology of Beet-Sugar Production and Refining, pp. 43–47 and 75–80. (Israel Translation, 1964.) Clearinghouse, Springfield, Va. 1958.
13. Skuderna, A. W., C. W. Doxtator, Edward Swift, R. L. Bowman, and Arthur Deschamps. A study of varietal adaption with sugar beets, 1937 to 1941, inclusive. Proc. Am. Soc. Sugar Beet Technologists 3:349–55. 1942.
14. Steele, Arnold E. The host range of the sugar beet nematode, *Heterodera schachtii* Schmidt. J. Am. Soc. Sugar Beet Technologists 13:573–603. 1965.
15. Stout, Myron. Redistribution of nitrate in soils and its effects on sugar beet utilization. J. Am. Soc. Sugar Beet Technologists 13:68–80. 1964.
16. Ulrich, A. Growth and development of sugar beet plants at two nitrogen levels in a controlled temperature greenhouse. Proc. Am. Soc. Sugar Beet Technologists 8:325–38. 1954.
17. ———. Influence of night temperature and nitrogen nutrition on the growth, sucrose accumulation and leaf minerals of the sugar beet plant. Plant Physiol. 30:250–57. 1955.
18. ———. Variety climate interactions of sugar beet varieties in simulated climates. J. Am. Soc. Sugar Beet Technologists 11:376–87. 1961.
19. Ulrich, A., F. J. Hills, D. Ririe, Alan G. George, and M. D. Morse. Plant analysis a guide for sugar beet fertilization. Calif. Agr. Exp. Sta. Bull. 766. 1959.
20. Williams, W. A., D. Ririe, H. L. Hall, and F. J. Hills. Preliminary comparison of the effects of leguminous and non-leguminous green manures on sugar beet production. Proc. Am. Soc. Sugar Beet Technologists 8:90–94. 1954.
21. Woolley, Donald G. and W. H. Bennett. Glutamic acid content of sugar beets as influenced by soil moisture, nitrogen fertilization, variety and harvest date. J. Am. Soc. Sugar Beet Technologists 10:624–30. 1959.

13

Harvesting and Delivery

STEWART BASS
American Crystal Sugar Company
Denver, Colorado

P. B. SMITH
Great Western Sugar Company
Denver, Colorado

HARVESTING	384
Principles	384
Recovery of the Roots	384
Recovery of the Tops	385
Development of Mechanical Harvesting	386
RECEIVING	390
History and Development	390
Modern-day Receiving Equipment	392
Volume	393
Delivery Equipment	395
Sampling	396
Storage	397
Record Keeping	398
FUTURE	399

HARVESTING

PRINCIPLES

CONDITIONS during the sugarbeet harvest vary widely in the 23 states where the crop is grown. Climatic factors probably regulate the type and progress of harvest in much the same way as many other crops. Of chief importance in many sections of the country is the starting of harvest when the sugar content of the root is nearly optimum. It is necessary in the northernmost states to complete the harvest before freezing of the soil takes place, or actual loss of the crop may result. This means that there is an urgency about the operations in Montana, the Dakotas, Wyoming, Colorado, Nebraska, Minnesota, and other northern states.

In California it is possible to harvest in almost every month of the year, with the exception of northern California where it is generally too wet from mid-November to March. In the Imperial Valley the crop is planted in September instead of in the spring. Here harvest usually starts in April and extends through June. Hot temperatures necessitate immediate completion of the harvest, since the sugar stored in the root decreases as the temperatures climb. In the northern growing areas, the sugar stored in the root increases rapidly as winter approaches.

RECOVERY OF THE ROOTS

The fundamental idea is to harvest the roots free of clinging soil and to remove the foliage at the base of the lowest leaf scar. This is ideally the function of the present beet harvester. Mud, trash, rocks, and clods are supposed to be removed before the beet roots are delivered by the farmer to the processor. Additional screening equipment is provided by the receiving facilities of the sugar processor.

Under favorable soil and growing conditions, the beet root at maturity may extend six to eight feet deep (Fig. 13.1). Normally at harvest time, the beet foliage is about equal in weight to the root. This has publicized the statement, "Sugarbeets are two crops in one," which is credited to E. J. Maynard, noted animal nutritionist and former Dean of Utah State Agricultural College.

In hand-harvest methods before World War II, about 60 to 70 man-hours per acre were required to lift the roots with tops attached and hand-pile four to six rows together on each side of a sledded-out windrow (Fig. 13.2). Included in this time was the cutting off of the top by a special 16- to 18-inch-long hand knife. The tops were left on each side of the windrowed beet roots, while the beets were subsequently thrown by hand or fork into wagons or trucks. This was extremely hard work, particularly as the roots were generally forked off the wagon at the delivery station. Fortunately, this harvest operation is now completely mechanized.

Fig. 13.1. Deep rooted sugarbeet plant. Fine side roots may penetrate laterally in diameter of four to six feet.

RECOVERY OF THE TOPS

The dry substance in field-dried tops will average 10 to 12 percent of the root weight. In other words, there are about 200 to 240 pounds of dried tops for each ton of beet roots. During an eight-year trial, sugarbeet top silage averaged 1,107 pounds per ton of roots harvested. The greatest feed value from tops is secured by ensiling them as green and clean as possible.

Sugarbeet tops are very nutritious and palatable feed for cattle and sheep. Lionel Harris of the University of Nebraska's Scottsbluff Station states: "Pound per pound, beet top silage is equal to the feed value of the corn ensiled from a 100 bushel per acre corn crop." When dried tops are fed in fattening ration with grain and hay, each pound of tops is approximately equal to one-half pound of grain.

It should be emphasized that the concept of "two crops in the sugarbeet crop" is true. The production of sugar per acre at present yield averages about 5,500 pounds, and at the same time there is left as a byproduct an amount of stock feed about equal to that which would have

Fig. 13.2. Hand harvest of crop, requiring many field workers.

been grown on the same land if it had been devoted exclusively to the growing of feed crops.

DEVELOPMENT OF MECHANICAL HARVESTING

The beet crop was one of the first field crops to be completely mechanically harvested. It was accomplished in less than 10 years. This was made possible by the general use of row crop tractors just prior to World War II. The first field windrow loaders of hand-topped beets came in 1938.

The hand harvest of the beet crop required a great deal of hard labor. Based on a federal survey made in 1924–25, about 55 to 60 man-hours per acre were required for digging, topping, windrowing, and forking into a truck. With the advent of the war, there was a great shortage of workers for this type of agricultural labor.

The most rapid progress was made in the five-year period, 1945–49, largely due to promotion of machines by the Beet Sugar Development Foundation formed by the sugar companies. The proportion of the mechanical harvest progressed from 7 percent in 1945 to 52 percent in 1949 in the United States. By 1952 it had increased to essentially 100 percent. Probably the progress was most rapid in the intermountain states of Washington, Oregon, Idaho, and Utah, where, because of friable soil, digging problems were less than in the hard clay soils in some other sections of the country.

Machine harvesting of beets had some development about 1900. Scores of ideas were patented, including in 1872 a lifter wheel idea, which is used today. It was not until World War II approached, however, that any concentrated effort was made. In 1938 beet processing companies initiated a movement to energetically stimulate experimental harvesting machines. By 1941 four or five had been tried out. One of the first successful machines is shown in Figure 13.3. This was an idea of a farmer by the name of A. M. Jongeneel of Rio Vista, California, who together with E. F. Blackwelder of the Blackwelder Engineering Works, Rio Vista, California, constructed several one- and two-row machines. The basic feature was a six-foot diameter wheel that had closely spaced four-inch-long slightly curved spikes on which the beet roots were impaled. As the wheel moved at ground speed up the rows, the roots came to a point at the top of the machine where large discs separated the tops from the roots, which fell into a chain conveyor that loaded them directly in a truck traveling alongside. The crown and leaves were scraped off the surface of the wheel by chisels located between the rows of spikes. No effort was made in these first machines to save the tops. The Blackwelder machine was particularly adapted to dry soil conditions, but not too successful in wet soil.

Perhaps even earlier in appearance was the Scott-Urschel single-row harvester developed in Columbus, Ohio. An agricultural engineer of the USDA, Ernest M. Mervine, demonstrated the machine as early as 1936 in Fort Collins, Colorado. The beets were lifted from the soil by a single stinger, which expelled the root sufficiently so that the tops could be seized by chains (or later, V-Belts) pressing together and carrying the entire plant to the top of the machine. Here, the crown was separated by two concave discs and dropped onto a rod conveyor that put the roots into a truck alongside. The machine did a very acceptable job and harvested 35 to 50 tons a day in favorable conditions.

Thirty-two John Deere one-row harvesters were in operation in 1943 in the producing area east of the Rocky Mountains. This was a two-unit operation, with the first unit topping 10 or 12 rows and conveying both tops and roots into windrows (Fig. 13.4). The roots were then picked up

FIG. 13.3. Blackwelder beet harvester lifting two-row beds in the Imperial Valley of California.

FIG. 13.4. Model 54-A John Deere single-row harvester.

by a second machine that elevated them into a truck. By 1948 a total of 2,600 model 54 and 54A John Deere single-row harvesters were available.

In 1943 the International Harvester Company got into the beet-harvesting field with several experimental machines. The first machine was a two-unit operation, with the beet roots being topped in the ground ahead of the lifting, and elevated into a hydraulic hopper alongside. The hopper held about 1,200 pounds of beets which were tipped into a rod conveyor-loader, so hand sorting of clods and trash could take place.

Later IHC models were of single-unit design, with a rod chain in the bottom of the trailer cart for elevating roots into a truck at the end of the field. A sorting belt was built above the cart where one or more people could ride and sort beets from trash and clods. It was estimated that by 1949 there were 2,700 machines of this type being used by farmers.

One of the big difficulties in machine harvest was eliminating clods and doing an acceptable job of topping. The Marbeet harvester solved this with digger blades and a three-foot diameter spiked wheel (Fig. 13.5). This is still a popular harvester, especially where there are hard soil conditions that might produce clods.

Another answer to a mechanical separation of clods and roots was a double rod-chain device. The chains were arranged in a V shape, with a longer chain supplied with short spikes that held the roots to the end of the smooth rod chain when they fell off into another elevator. It was never employed to any great extent commercially.

One of the early power-driven toppers had a chain track which pressed down on the crowns of the two rows of beets, floating over them at 1½ times ground speed to prevent pushing high-crowned beets over as they were being cut off by single discs. Rubber-tired finder wheels used today in a six-row top harvester do a more adjustable topping job at speeds of

Fig. 13.5. Puller points and lower part of three-foot diameter wheel, which impinges roots and tops for elevating to topping discs of Marbeet Midget harvester, thus eliminating clods.

less than 1½ miles per hour. This type of top recovery machine is generally used in conjunction with two-row and three-row harvesters. The large-capacity beet top harvesters are very popular in areas where there is livestock feeding. Commonly, the six-row windrows are picked up by windrow field hay choppers for putting into silos about one week after the roots are harvested. This reduces the moisture content and makes for excellent feed high in protein (Fig. 13.6).

Today's sugarbeet harvesters, after 25 years of improvements, now comprise machines of one-row to six-row sizes. They have as much as 20 to 24 acres daily potential capacity in favorable soil conditions. Some of these giant combines have harvested upward of 500 tons in a 10-hour day.

Some of the latest beet harvesters are equipped with grab roll screens to remove clods, rocks, and trash, plus row finders and hydraulic power. Most American beet harvesters are essentially "all weather" machines capable of operating in all except the muddiest conditions. They are as dependable as almost any agricultural equipment, well suited to the

Fig. 13.6. Modern six-row beet top harvester, front view.

various producing areas. From a recovery of crop standpoint, today's beet harvesters are one of the most efficient pieces of harvesting equipment used in modern agriculture.

Field harvesting today in some areas is conducted 24 hours a day. Two or more working crews are used in shifts for the harvester operation and truck driving. This allows peak efficiency from the harvesters and trucks. It helps the grower remove his crop in as short a time as possible, and at a time when the crop has reached its peak potential under normal weather patterns for the areas involved.

RECEIVING

HISTORY AND DEVELOPMENT

In the early years of the beet industry, harvest was indeed an arduous task done entirely by man and beast. Horse-drawn lifters, hand topping, and hand loading were all employed. The loaded horse-drawn wagons were taken to the sugarbeet factory or to a railroad receiving station where they were hand shoveled or forked off the wagon into piles or into rail cars. It was not long until the processors devised easier means of unloading the wagons. "High line" stations allowed the loaded wagon to ascend a long ramp which at the unloading position put the wagon above the height of the railroad car. The beets were then tipped out of the side dumping wagon by means of a winch and cables attached to one side of the wagon (Fig. 13.7). The beets then slid into the rail car (and once in awhile also the wagon and teams).

As the windrow loaders appeared in the grower's field so did piling devices appear at the receiving stations. Hand forking of beets began to disappear. Horse-drawn wagons were soon replaced by trucks. More outside railroad receiving stations began to appear and so it was possible to grow beets at a greater distance from the factories. As machines began

Fig. 13.7. Early day "high line" beet-receiving station, showing team and wagon that just dumped load of beets into early-day beet railroad car.

their takeover of the field harvest, deliveries began to speed up. This necessitated faster receiving station equipment. The receiving equipment also had to be designed to help clean the dirt and field trash left in the beet loads in order not to reduce the slicing efficiency of the factories and allow the beets to remain in storage piles until the factory could slice them.

The early pioneer beet receiving equipment manufacturers soon designed a piler which was movable and capable of cleaning and piling beets in large storage piles. Since receiving equipment was first manufactured, there have been primarily only two companies that have supplied the beet industry with receiving equipment in the United States. These pioneer manufacturers, Ogden Iron Works and Silver Engineering (now C, F & I Engineers), have grown with the industry and today furnish the various processors with huge, high-speed and efficient beet pilers and loaders.

On the early pilers and direct rail receiving stations, the grower's trucks or tractor-drawn carts which then hauled two to four net tons, drove onto a platform adjacent to a hopper. Cables on an "A" frame tipped the truck or cart bed, allowing the beets to tumble into the hopper where a belt system elevated the beets to a cleaning screen (Fig. 13.8). From there the beets were dumped by a short belt system or slid directly into a railroad car for shipment to the factory for processing. Where a piler was employed, the beets came off the cleaning screen onto a belt on a long swinging boom which moved slowly in a semicircle, piling the beets. The piler was portable, allowing the forming of long storage piles. These beets were later reloaded into rail cars, usually by means of cranes, for direct processing as the factory was ready for them.

Today's modern receiving equipment is truly a great advancement from the first pilers and receiving stations; however, in principle they remain much the same. Their purpose is not only to receive beets but to clean and either load them directly in railroad cars or to pile the beets for later delivery to the factory.

With today's large acreage grown for each factory, it is necessary to

FIG. 13.8. Side-dumping truck being unloaded in the 1940s. Beets were tipped into receiving hopper, elevated to cleaning screen, and placed on storage pile.

stockpile beets in much larger quantities in most areas than in the early days of the industry. In California and Arizona, however, the warm climate allows very limited storage. Harvest is started in mid-to-late summer and continues until rains in the late fall or early winter force its stoppage. Only sufficient beets are harvested daily to keep the factories slicing at capacity. However, as the weather cools in the late fall in northern California growing areas, limited stockpiling is done to bridge possible short harvest stoppage due to rain. This allows continuous factory operations until such time as the weather forces complete stoppage of harvest until spring. Generally only a week or ten days' slice is stored ahead without fear of spoilage from the warm California temperatures.

In California and Arizona, beets are generally planted over a six-month period and they are also generally harvested during six months of the year. Due to the more seasonal weather experienced in all other areas of the industry, planting and harvesting are done in a much shorter period of time. In these areas, stockpiling is necessary and possible. Without stockpiling, it would not be possible to harvest and slice the large volume of beets grown before weather elements would stop the harvest. Planting is done in a two- to three-month period in the spring and harvest is started generally in mid-September and completed in mid-November, with the bulk of all harvest being done in October.

In all areas, harvest is timed to allow the maximum growth and maturity possible to the beets yet still assure adequate time to harvest the crop with existing equipment before the weather conditions in each respective area force the stoppage of harvest completely.

MODERN-DAY RECEIVING EQUIPMENT

Today's modern beet receiving equipment is capable of handling large quantities of beets either directly into railroad cars or into storage piles. In either case, the equipment must be capable of mechanically handling a grower's truckload of beets every few minutes. It must clean as much of the excess dirt, trash, and foreign material as possible from the beet. The grower's harvester has topped, dug, and cleaned the beet prior to loading on the trucks in the field; however, the larger cleaning devices on the receiving stations do a necessary second job of cleaning before the beet is delivered to the factory or placed in storage piles. Over the years the processor's receiving stations have employed several types of cleaning screens. The earlier screens were made of sloping one-inch pipe, separated to allow the dirt to fall through. Later developments provided for rod conveyors that bounced the beet along before discharge to the rail car. More elaborate beds of revolving star wheels replaced the rod chain. Soon more refined cleaning beds such as the Rienks rolls and Molnau screen followed and are still used and doing a fair job of cleaning. The most popular screen today, however, is known as the "grab roll" screen. It is a slanting bed of scrolled and smooth rolls. These rolls are perpendicular to the flow of beets and are turning at different speeds on every other roll. The scrolled roll turns faster than the rubber-covered smooth roll. The beet is bounced and rolled in a scrubbing fashion, with the dirt, trash, mud balls, and foreign material being sucked through the rolls into a collection hopper for return to the grower's truck immediately after the load has been dumped and cleaned. In some areas the dirt cleaned from the load is weighed separately on the receiving equipment for net weight determination but is hauled by separate trucks to a disposal area. This is done to avoid spread of disease and nematodes.

VOLUME

Today it is not uncommon for a direct loader which places the truckloads of beets directly into rail cars to load as many as 60 cars, each containing an average of 50 tons of beets, in a single day's operation. There are a few direct stations, particularly in California, at which the beets are loaded either into railroad cars or into large over-the-road transport trucks for delivery to the factory.

Today's modern high-speed pilers commonly unload and pile 500 to 600 truckloads of beets in a single day's operation. This means 5,000 to 6,000 tons of beets are piled by one piler in a day. It does the complete operation of unloading, cleaning, and piling the beets at a rate of 250 tons an hour or just over 4 tons per minute. The longer boom length on the newer high-speed super pilers makes a pile that is 210 feet wide at the base and 22 feet high. It is common for a modern piler to form a pile of 70,000 to 90,000 tons of beets. If loaded in rail cars, the pile would form a train of 1,800 beet hopper cars—a train just over 20 miles long. It would form a line of beet trucks $42\frac{1}{2}$ miles long. These piles will run 1,200 feet to 1,600 feet in length—truly a mountain of beets (Fig. 13.9).

The volume of tons to be harvested in each factory area has increased to where, in many areas, men and equipment in both the grower's field and the receiving station necessarily find it efficient to operate more than just daylight hours. Usually two shifts of 8 to 12 hours are used. In this method, peak efficiency is acquired from harvesters, trucks, and receiving stations. By placing ample lighting on the harvesters and receiving stations, after-dark operations become safe and relatively the same as daylight hours.

Receiving stations are strategically located in the factory and growing areas, so each grower has a relatively short distance to haul his beets. Most factory grounds have space to store large quantities of beets and generally have several pilers at that location (Fig. 13.10). In addition to movable pilers, some factories have receiving stations with elaborate belt or conveyor systems to pile or store in "bins" or "pens" tonnage similar to that which a piler would make. Many outside pile storage grounds now operate

Fig. 13.9. Country storage pile over a quarter of a mile long, formed by modern piler in foreground. It is 160 feet wide, 22 feet high, and contains over 70,000 tons. It will be reloaded into railroad cars for later processing at factory.

with more than one piler or with both a piler and a direct station. Direct stations are primarily used to supply the daily slice capacity of the factory during the harvest period. Piling stations are designed to supply the factory slice after harvest is over. In some areas direct stations load more beets daily than the factory slice. This is accomplished by unloading the rail cars with cranes or by other means and placing the beets in temporary

Fig. 13.10. Huge storage piles of sugarbeets can be seen in factory yards of northern factory. Five local yard pilers formed piles in background. Narrow, smoother piles in foreground were unloaded from rail cars. Nearly one-half million tons of beets are in picture.

storage piles at the factory ground for slicing at a later time. This releases the rail car to return to the station to be reloaded along with the empty cars acquired daily from direct unloading into the factory for the continuous slicing operation.

DELIVERY EQUIPMENT

Growers first delivered their beets to the receiving stations with teams and wagons that were hand loaded with forks. Multiteam and tractor-drawn trains were used by some of the larger acreage growers. The average tonnage hauled per wagon (two to three tons) is small compared to today's standards.

Growers' trucks now average 8 to 10 net tons hauled per load in most areas. More and more tandem axle trucks are used each year. These trucks haul 10 to 14 net tons per load. In California, large truck and trailer units are commonly used, and haul 20 to 25 net tons per load. Generally, the larger the truck used, the greater the distance the beets can economically be hauled.

The trucks are either self-end dumping trucks or the truck beet box is tilted to dump to the side by means of powered cables at the receiving station. The commoner and modern-end dumping trucks drive either parallel or across the receiving hopper of a piler. As the truck is in position to dump, a hydraulic gate opens over a conveyor belt, allowing the beets to be moved quickly to the cleaning screens and to be placed in the storage pile (Fig. 13.11). At direct stations, the commonest means of unloading the truck is for it to self-dump into an open receiving hopper with a conveyor belt system to move the beets over cleaning screens and into rail cars. With present-day equipment, a 12-ton net load of beets on a tandem axle truck can on the average be unloaded in under two minutes. On a normal day of receiving at several factories in the northern areas, where five or six pilers are located at a factory yard, as many as 3,000 truckloads of beets are received in a single twenty-four-hour period. At the peak of harvest in all areas, a steady stream of growers' trucks go to and from the receiving stations.

FIG. 13.11. Grower's end-dumping truck is shown dumping load of beets onto cross conveyor. Note hydraulic gate that opened after truck drove over parallel end-dumping platform. Belts convey beets over cleaning screens, and long, swinging boom piles beets.

When the harvest is completed, the factory slice is supplied from the storage piles located in the outside growing areas and those piled in the factory yards. In the past, the outside storage piles were reloaded into railroad cars daily at a rate equal to the factories' slicing capacities. Cranes were used to reload the beets. Now however, most piles are loaded with rapid-moving, rubber-tired, front-end loaders. These specially designed loaders have 10- to 14-yard capacity buckets with long reaching arms that can load into high gondola rail cars (Fig. 13.12). In less than 10 minutes they can load a 50-ton rail car with beets from the storage pile paralleling the rail line.

In recent years, an increasing number of tons have been hauled from country storage piles to the factory by large over-the-road transport semi-trucks. These trucks haul as many as 25 tons of beets a trip from piles as far as 80 miles away from the factory. The trucked-in storage piles are, however, generally closer to the factory. A front-end loader and enough trucks are used to supply the factory slice on an around-the-clock basis. The trucks generally dump directly into the factory supply hopper by means of hydraulic tilt dumps or bottom dumps on the trucks. Many of the outside storage piles to be hauled by the over-the-road trucks are not located adjacent to a railroad as has been the common history of country storage piles. Most factory yard storage piles are reloaded into trucks and hauled the short distance to the receiving wet hoppers of the factory. The railroad hopper cars used to transport the beets are unloaded by bottom-dump gates on the car that is placed over the receiving wet hopper of the factory.

SAMPLING

At the time the grower dumps his load of beets at the receiving station, a sample is taken from his load for tare analysis. The sample may also be used for sugar analysis where individual grower sugar determination is made. The sample is caught downstream from the cleaning screen at the receiving station. It is usually about a 25-pound sample, acquired by means of quickly swinging a bucket device either by hand or mechanically into and out of the flow of beets. In most cases, the sample is transported to a central laboratory where the analysis is made. The sample is weighed, washed or brushed to remove all dirt from the beet, top trimmed to remove excess crown and leaf material that does not contain purchasable beet according to the beet contract, and then weighed. The percentage of deductible material is arrived at by the weights and this in turn applied to the grower's gross load weight to acquire a net tonnage on which he is paid. In areas where individual sugar is determined for payment purposes, a pulp sample is taken from the load sample and through chemical analysis a percent sugar is determined for each load sampled. In a few areas, the grower's payment is based on the average percent sugar of the factory for

Fig. 13.12. Huge 12-yard special beet bucket on front-end loader dumps load of beets taken from nearby storage pile into rail car for movement to factory for slicing.

the campaign. In this case, only dirt and crown tare are taken from a grower load sample.

STORAGE

Sugarbeets are a perishable product and will not keep in storage piles unless the temperatures of the beets in the piles stay relatively cool. Piling in most areas is delayed until late fall when the soil and air temperatures have cooled. Care is taken to be sure beets going into storage piles are clean, well topped, and free of trash and excess dirt. This allows the beet pile to have the best possible air circulation. Dirt and trash tend to clog the air movement in a pile and trap the heat that is trying to escape to the top. Thus spoilage results at that point, causing a "hot spot." The hot spot must be removed or it will spread, causing further damage. In some areas, ventilation tubes are placed under the pile and air is forced through the tubes by means of fans. Thermometers are inserted in the piles when they are first formed and read daily to be sure the pile is cooling properly and to discover any sudden rises in temperatures that might indicate a hot spot forming. Generally the piles will cool to a safe range in a few days. The normal cool weather and cool winds experienced in late fall at piling time help reduce soil, beet, and storage pile temperatures.

Although the storage piles are reduced daily by the factory slice, many of them may remain until late February or mid-March. In the coldest areas, the piles are penetrated by frost 20 to 30 feet deep and remain frozen until removed for slice. If too large a percentage of the beets entering the factory is frozen, slice rate is reduced. Should weather, as it does in some piling areas, be moderate, allowing the frozen beets on the outside surface to thaw, then those beets will spoil quite rapidly.

To prevent too deep a freeze in the pile and to minimize the surface freezing and thawing, pile covering has been used in recent years. Many methods and types of material have been used with from fair to good success; however, conditions vary from area to area and within an area from year to year so that it is difficult to be sure of a workable system. Generally, a thin polyethylene sheet material is spread over the side of the pile from the toe of the pile to near the top. This is held down by ropes, old tires, and stakes to keep the wind from lifting it off the pile. In some areas, a straw and asphalt emulsion is blown on the poly sheets to form an insulation blanket, thereby reducing the freeze penetration. Experimental work has been done in completely sealing the pile, trying to keep the oxygen content at a very low level. Numerous other experiments have and are being done with the hope of reducing spoilage and breakdown of the beet during long storage periods.

RECORD KEEPING

The increased volume of beets at harvest forced many new and faster methods of keeping records of the beets delivered by each grower. Elaborate data-processing systems have been and are being installed and up-dated by most processors today. These have become vital in order to keep up with the tremendous volume of loads delivered in a short period of time in almost all areas. Information is now generally transmitted rapidly from the beet receiving areas to a central data-control center where records are compiled overnight and relayed back to the receiving areas. These records generally contain more complete information for grower and company than ever experienced before. Centralization of tare and sugar analysis allows adaptation of newer and more elaborate equipment, including data processing.

FIG. 13.13. Up to two tons of beets per minute can be topped, dug, cleaned, and loaded in continuous operation by today's modern beet harvesters.

FUTURE

New and increased cooperative effort within the industry is being directed to better methods of beet storage. Because the sugarbeet is a perishable crop, it has been the main limiting factor in the number of days a factory can operate. This in turn limits by area the number of acres to be harvested. Plant breeders, chemists, and engineers are all looking for ways to economically keep a sugarbeet for a longer period of time after harvest without losing the potential extractable sugar within it. Improved methods of transporting and handling beets have contributed to the growth of acres to harvest in many areas and the trend continues. New and improved field harvesting equipment is constantly being introduced (Fig. 13.13). Larger and faster receiving equipment is appearing in all areas.

Looking back over the development of sugarbeet harvest clearly shows the great advancements that have been made; however, a look to the future indicates that the advancements made to date may be nothing to what lies ahead.

14

Variety Development

J. S. MC FARLANE
Crops Research Division, ARS, USDA
Salinas, California

HISTORY	402
Early Selection Work in Europe	402
Early Variety Improvement in the United States	403
BREEDING METHODS	404
Mass Selection	404
Progeny Testing and Mother-line Breeding	405
Polycross Method and Recurrent Selection	406
Paired-plant Crosses	407
Inbreeding and Hybridization	407
Polyploidy and Triploid Hybrids	411
Interspecific Hybridization	413
RECENT PROGRESS IN BREEDING	416
Breeding for Disease Resistance	416
Nematode Resistance	423
Bolting Resistance	424
Monogerm Seed	425
Improved Processing Quality	427
COOPERATION IN BREEDING RESEARCH	428
FUTURE	430
Combined Resistance to Disease	430
Insect Resistance	431
Improved Production and Quality	431
Breeding for Better Beet Storage	431
Breeding Methods and Basic Genetic Studies	432
Other Breeding Objectives	433

THE BEET has provided food for man and animal since ancient times, but the plant's potential as a source of sugar was not realized until the middle of the eighteenth century. In the short span of 180 years, plant breeders have transformed the fodder beet, a plant used primarily for livestock feed, into an efficient producer of sugar. This newly developed crop provides nearly 45 percent of the world's sugar. The breeders responsible for this remarkable transformation are known, and the written records of their accomplishments are available.

HISTORY

EARLY SELECTION WORK IN EUROPE

In 1747 a German chemist, Andreas Marggraf, found that members of the beet family contained a sugar identical to that found in sugarcane (22). Marggraf envisioned the importance of this discovery, but there is no record that he made serious attempts to produce sugar or to improve the sugar-producing ability of the beet.

This was left to his student, Franz Karl Achard, who is universally recognized as the father of the beet sugar industry. Achard (1) not only developed processing methods for the production of sugar from the beet root but also made improvements in the sugar-producing ability of the plant. His selections were made from the forage beet, or Runkelrübe, which was grown for livestock feed by the farmers around Magdeburg, Germany. Achard's records show that the forage beet of that time was highly heterogeneous and contained roots of various sizes, shapes, and colors. He determined that roots with white skins, white flesh, and a conical shape were richest in sugar and in "pure, sweet juice." He also observed that the best plants were those with crowns that did not grow out of the ground. Achard selected roots with these characteristics for his seed stocks. This seed was made available to Freiherr von Koppy who had erected a sugar factory at Krayn, Germany. The seed was increased and additional selections were made. From the combined efforts of Achard and Koppy came the White Silesian beet which has been referred to as the "mother of all the sugarbeets of the world."

The White Silesian beet was highly variable for sucrose content. Reports in the literature differ but indicate a factory production of around 6 percent sucrose. The early factories were inefficient, and the actual sucrose percentage of the White Silesian beet was probably 3 to 4 percent higher than the factory production. The root yield and conformation of the White Silesian beet were similar to those of the present-day sugarbeet. This beet was used as a source of sugar in Germany and France from about 1810 to 1850.

The early records of the sugarbeet industry indicate that the first notable improvement over the White Silesian beet was made about 1850 by F. Knauer near Halle, Germany. The new variety was named "Imperial" and had a sucrose content of about 13 to 14 percent. This variety was used in both Germany and France.

The lack of an effective method for measuring the sucrose content of the root made sugarbeet improvement especially difficult during the first half of the nineteenth century. The polariscope had not been developed and a crude salt-bath flotation method was used to select roots with the highest density. Louis de Vilmorin, member of the well-known Vilmorin seed establishment, made notable contributions to sugarbeet breeding methods during the middle of the nineteenth century. In 1852 he introduced the silver-ingot method of determining density. This development enabled the breeder to select on the basis of the specific gravity of the juice rather than that of the entire root. Juice was extracted from the root by rasping, and a silver ingot of known weight and volume was weighed when immersed in the juice. He also experimented with the newly developed polariscope, and by 1860 it had become the accepted instrument for determining the quality of individual roots.

Louis de Vilmorin is best known for his discovery of the progeny system of breeding. He and his son used this system to improve the sugar-producing ability of the beet. Superior mother roots were selected from existing varieties, and the progenies of these mother roots were evaluated. Only those progenies with superior performance were maintained for seed production and further breeding. Through the use of this breeding system and the new laboratory techniques, the Vilmorins made rapid progress in sugarbeet quality improvement. Their contributions have been recorded in a book (42) written by Jacques L. de Vilmorin, a grandson of Louis de Vilmorin.

The sugarbeet industry spread rapidly in Europe after the mid-nineteenth century. Large quantities of seed were required, and commercial seed-producing firms were organized to meet the needs of the expanding industry. Varietal improvement was closely associated with seed production and was handled almost entirely by the seed firms. A few of the publicly supported institutions such as the Swedish Breeding Station at Svalöf, the Sugar Beet Institute at Kiev, USSR, and the Stazione Sperimentale di Bieticultura in Italy conducted genetic and plant breeding research; but they were exceptions. The commercial firms all used modifications and extensions of the Vilmorin progeny breeding system.

EARLY VARIETY IMPROVEMENT IN THE UNITED STATES

The beet sugar industry in the United States dates from 1870 with the establishment of the first successful factory at Alvarado, California. Most of the seed to provide beets for this and other factories built during the next 60 years came from European seed firms. Variety names such as Kleinwanzleben and Vilmorin were common during the early years of the industry. These European varieties were capable of good performance so long as conditions were similar to those found in Europe. Unfortunately, growing conditions were frequently not the same. Not only was the climate different but disease problems arose that either did not occur in Europe or were of minor importance. To make matters worse, the imported seed was often of poor quality. Problems associated with seed contributed to the failure of many of the first American factories.

The importance of adapted seed of high quality soon became apparent, and during the latter part of the nineteenth century, numerous tests were made to determine which of the European varieties were best adapted. The earliest recorded attempt at seed production and varietal improvement was made in 1890 by the U.S. Department of Agriculture at Schuyler, Nebraska. The Nebraska climate proved unsatisfactory for seed production, but the seed that was produced proved superior to most foreign-grown seed.

Some of the first selection work was done in South Dakota under the direction of James H. Shepard (39). In 1907 selections were made from a group of 26 sugarbeet and stock beet varieties on the basis of size, shape, and sucrose content. These selections were propagated and reselected in an attempt to breed uniform strains of beets with good yields and high sucrose. Frederick J. Pritchard (32) also described selections from European varieties that were made at about the same time at Fairfield, Washington, and Madison, Wisconsin.

The records of these early breeders indicate that they selected strains of sugarbeet that were superior to the commercial varieties imported from Europe. However, very little use was made of these selections because seed could be obtained easily from Europe at a reasonable price. Also, the idea prevailed that the high cost of labor and the unfavorable climate would not permit the successful development of a seed industry in the United States.

During World War I, the continental European ports were blockaded and the importation of seed became extremely difficult. Desperate attempts were made to produce domestic seed, but these reproductions were primarily of European varieties. By the end of the war, the sugar companies were able to produce most of their seed requirements, but costs were high and the yields very low. As soon as trade restrictions were removed, the companies discontinued seed growing and returned to the importation of European seed.

BREEDING METHODS

MASS SELECTION

Mass selection, the simplest of plant breeding methods, consists of picking superior plants from a mixed population and producing seed from these plants as a group. This is the method used by Achard, Koppy, Knauer, and others to improve the sugar production and uniformity of the first sugarbeets. These early breeders were able to bring about uniformity in shape, color, and other physical characteristics of the beet, but were only partially successful in materially increasing the sucrose-producing ability of their selections. This was caused in part by the lack of an accurate method for measuring sucrose content.

Breeders in the United States have effectively used mass selection to improve many sugarbeet characters. All of our curly-top-resistant varieties developed before 1950 were obtained by this method. The breeders simply made repeated mass selections until an acceptable level of resistance was incorporated. Eventually, a point was reached beyond which additional gains in resistance were insignificant. Mass selection has also been used in developing resistance to downy mildew, virus yellows, black root, and bolting. The method has not been effective in improving *Cercospora* leaf spot resistance.

Even though mass selection is a crude plant breeding technique that rarely produces the maximum possible gain, the method has been used to some extent by most sugarbeet breeders and has contributed to the development of our present varieties.

PROGENY TESTING AND MOTHER-LINE BREEDING

The pioneering work of Vilmorin with progeny testing provided the sugarbeet breeder with an effective method of variety improvement. This breeding principle was widely adopted in Europe, and most sugarbeet variety improvement prior to World War II was accomplished by a combination of progeny testing and mother-line breeding. The method was also used in some of the early breeding work in the United States.

The sugarbeet is a cross-pollinated crop that responds well to selection. Plants within an open-pollinated variety differ from one another, and the progenies of these individual plants segregate for a large number of characters. The breeder usually selects a group of roots (mother beets) on the basis of morphological characteristics and polariscope tests. He obtains seed from these roots and usually attempts to encourage self-fertilization by separating the plants during seed production. A progeny is grown from a portion of the seed from each plant, and the progeny performances are compared. If the performance of a progeny is superior, the reserve seed from the original mother beet is saved, and additional roots may be selected from within the progeny.

The reserve seed of the outstanding mothers is planted, and a seed increase of each line is produced. In Europe, the breeder plants the seed in the early fall and produces small roots (stecklings), which are siloed or placed in cold storage over the winter. The best stecklings are replanted in the spring for seed production. These larger quantities of seed are then used for evaluation tests, and inferior lines are further eliminated. The best lines may then be pooled to produce an elite seed stock for the production of commercial seed. Additional mother root selections are ordinarily made each year and their progenies tested. Seed is increased from the best lines and becomes an elite stock for use in commercial seed production.

Various modifications of the progeny testing and line breeding method are used. Some breeders employ various types of combining ability tests to determine which combination of lines performs well together.

POLYCROSS METHOD AND RECURRENT SELECTION

Sugarbeet breeders have found that many of the breeding methods developed for other crops work well with sugarbeets. The polycross, a method widely used with alfalfa, has been used successfully by several breeders. This method involves the selection of superior beets and interpollination of this selected group by natural means, followed by tests of the outcrossed progenies of each beet. Usually 25 or more beets are selected from a heterozygous population and placed in cold storage to induce bolting. The following spring, these beets are planted in an isolated plot and allowed to pollinate at random. In environments favorable for self-pollination, a branch of the inflorescence is bagged, and selfed seed is obtained. In areas where selfing does not occur, the breeder can maintain the genotype by dividing the selected beets longitudinally. Half of the beet is thermally induced for seed production and the second half is maintained as a clone without thermal induction. The clones can also be established from seedstalk cuttings if desired.

Outcrossed seed from each of the beets is planted in a replicated test to determine yield, sucrose percentage, and other desirable characters. Ten or more of the original beets with superior combining ability, as determined from the replicated test, are chosen to produce a new variety or to start another cycle of selection. When selfed seed of the original beets is available, an increase can be made directly from a composite of the seed from the best beets. When clones are used to maintain the original beets, the best clones can be thermally induced and a seed increase made the following year.

The polycross test can be extended to additional cycles of selection and the method is then known as recurrent selection. With recurrent selection, the selected beets are usually crossed in all possible combinations rather than allowing them to interpollinate at random.

In the first cycle of recurrent selection, superior beets are selected from the increase of outstanding beets identified from the polycross or similar test. If possible, selfed seed is usually obtained from these superior beets. This seed is grown in plant-to-row progenies in an environment favorable for thermal induction, and the progenies are crossed in all possible combinations. When a relatively few beets are involved, the crosses often can be made directly on the flowering branches of the selected beets without resorting to selfing. The hybrid seed from these crosses is composited and forms a bulk population to start another cycle of selection. The recurrent selection process may be continued as long as improvement occurs for the character being selected. The method is particularly adapted for quantitative characters such as sucrose content.

PAIRED-PLANT CROSSES

Problems associated with the production of selfed seed and maintenance of selected genotypes have discouraged many breeders from extensive use of the polycross progeny test and recurrent selection. Some breeders have adopted a method of paired-plant crosses to maintain a high frequency of favorable genes during the development of high-performing varieties (17). Several modifications of the paired-plant cross method are possible. A simple procedure involves crossing plants from a heterogeneous variety in pairs, and evaluating the performance of the individual paired crosses in a replicated yield trial. Short rows of the paired crosses are also planted in a nursery to produce mother beets. From the yield trials, approximately 25 percent of the highest performing paired crosses are identified. Selections are made from the mother-beet rows representing these superior crosses. Seed from the selected mother roots is bulked to form the next population or synthetic variety. This method utilizes the additive genetic variance present in sugarbeet varieties.

The method can be modified to utilize specific combining ability. To do this, the paired crosses obtained at random are crossed with a male-sterile tester. From yield trials of these male-sterile hybrids, the breeder can identify the best paired-plant crosses and they can be bulked to form a broad-based pollinator. A commercial hybrid is then produced by crossing the male sterile with the broad-based pollinator. To produce specific hybrids, seed from individual paired crosses could be increased for use as commercial pollinators.

INBREEDING AND HYBRIDIZATION

A major part of the sugarbeet acreage in the United States is planted to hybrid varieties. Breeders have known for many years (40) that hybrids between inbred lines often produce a higher root yield than do open-pollinated varieties. Without some means of emasculating the small sugarbeet flowers, there is no way of controlling pollination and producing commercial hybrids.

Cytoplasmic Male Sterility. A practical method of wide-scale emasculation became available with the discovery of cytoplasmic male sterility (CMS) by Dr. F. V. Owen (26) in 1945. This type of sterility is not determined by a gene carried on a chromosome but is carried in the cytoplasm of the cell (Fig. 14.1). CMS is inherited maternally and is transmitted from the mother to all the offspring. To produce large quantities of CMS seed, the breeder can simply plant alternate strips of CMS and pollen-fertile beets of the same breeding line. Pollen is wind-borne to the male-sterile plants, and all the seed produced on the CMS strips is theoretically male sterile. A hybrid is produced by planting alternate strips of the CMS and a second unrelated pollen-fertile line.

Unfortunately, the production of commercial hybrid seed has proved

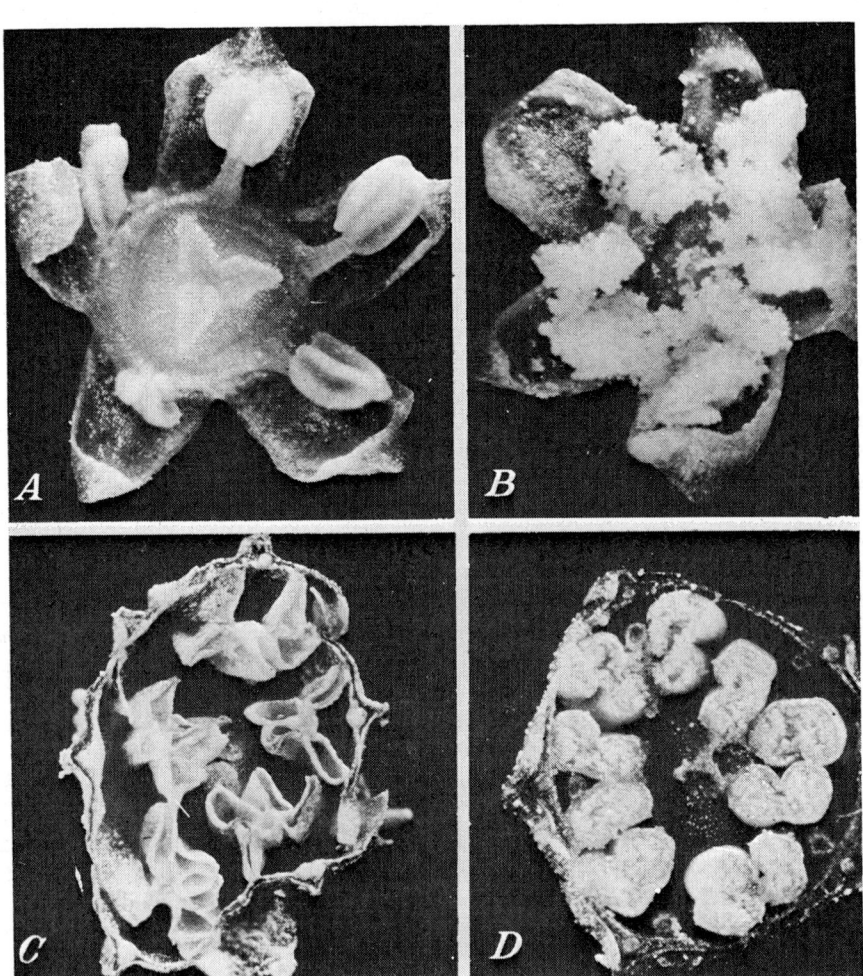

Fig. 14.1. *A*—Cytoplasmic male-sterile sugarbeet flower. *B*—Normal flower immediately after dehiscing of anthers. *C*—Cross section of male-sterile flower. *D*—Cross section of normal flower. (Photo by F. V. Owen.)

somewhat more complicated than has just been indicated. Rarely are all the offspring of a cross between a CMS plant and a pollen-fertile plant entirely male sterile unless the pollen-fertile parent has been specially selected. Usually, a few plants produce enough pollen to be a serious contamination hazard in the production of hybrid seed. The problem is associated with the genetic makeup of the male plant which furnishes pollen to fertilize the CMS flowers.

Most sugarbeet plants carry genes which suppress the expression of CMS. Dr. Owen postulated that two genes, X and Z, were responsible for this suppression. Assuming two types of cytoplasm, S for male sterile and N for normal, he obtained experimental evidence for the following genotypes: Sxxzz are completely male sterile; SXxzz and SxxZz are semi-male sterile, usually without viable pollen; and SXxZz are semi-male sterile, usually with some viable pollen and sometimes indistinguishable from the normal hermaphrodite.

When crosses are made between plants with the genotypes Sxxzz and Nxxzz, all the offspring are male sterile. Dr. Owen identified pollen-fertile plants with the Nxxzz genotype as Type O plants, and this designation has now been universally accepted by sugarbeet breeders. The production of Type O lines proved to be a tedious, time-consuming task and delayed the development of high-performing hybrid varieties. In order to identify Type O lines, the breeder has found it necessary to examine the progenies of hundreds of hand-pollinated crosses between a CMS line and pollen-fertile plants. Many breeders have used an annual CMS line, and progenies of the crosses can then be made to flower without resorting to thermal induction. Usually less than 5 percent of the plants in a non-selected open-pollinated population proved to be Type O.

Male sterility is introduced into a breeding line by crossing a CMS plant with a Type O selection and then backcrossing to the Type O parent. After a few backcrosses, a CMS line is available that is essentially the male-sterile equivalent of the Type O hermaphrodite. The sugarbeet is a prolific seed producer, and sufficient CMS seed can be produced from a pound of breeder's seed to enable the seedsman to start commercial hybrid seed production after three years of increase.

Cytoplasmic male sterility provides the plant breeder with a most useful tool. A means is provided of utilizing both inbred lines and heterosis in variety development. A high level of disease resistance can frequently be obtained in inbreds, and the resistance can be transferred to the hybrid. The introduction of productive monogerm varieties was hastened because the breeder could use a multigerm pollinator and needed only to develop the monogerm CMS parent. In recent years, European breeders have developed productive triploid varieties by crossing a monogerm CMS line with a tetraploid pollinator. Without CMS, commercial triploid hybrids would not be possible.

Inbred Lines. Before productive hybrids can be developed, a great number of inbred lines must be produced and tested. The sugarbeet is normally highly self-sterile and does not readily lend itself to selfing. Some breeders have incorporated a self-fertile gene discovered by Owen (25) and have produced inbreds that are highly self-fertile. Other breeders have induced self-pollination of otherwise self-incompatible sugarbeet plants by planting and bagging the plants at high altitudes. The number of selfings vary, but usually three or more are made, depending on the genetic makeup of the material. During the inbreeding process, selections are made for desired characters such as disease resistance, bolting resistance, processing quality, vigor, and good seed set. The Type O characteristic must also be introduced. Some breeders wait until the inbred development is well advanced before selecting for Type O. Other breeders have developed a series of Type O lines in self-sterile populations and then pooled these lines. Inbreds developed from this heterozygous pooled population are Type O or approach this condition.

To determine which inbreds will perform well in commerical hybrids, test crosses must be made and evaluated in field trials. Most breeders first make crosses with a heterozygous topcross parent. Oldemeyer et al. (24) developed a method of using a red beet that was obtained from Germany as the topcross parent. Hybrids with the red beet can be readily identified in the seedling stage, and those plants that are not hybrids are selectively thinned from the topcross performance test. Other breeders develop the CMS equivalent of a broad-based commercial variety, and use this line as the topcross parent. Both procedures have given satisfactory results and have enabled the breeder to eliminate inbred lines with poor general combining ability. Lines selected on the basis of topcross tests may then be further tested in single crosses. These crosses can be made through the use of CMS or by isolating several green hypocotyl lines with a red hypocotyl line. Hybrids will have red hypocotyls, and plants that are not hybrids can be selectively thinned from the test plot.

Hybrids. Several types of hybrids have been produced and grown commercially. The commonest has been the three-way topcross hybrid. An F_1 hybrid is produced between a CMS equivalent of one inbred and a second pollen-fertile inbred. By using Type O parents, the offspring of the F_1 hybrid will be entirely male sterile. This F_1 hybrid is used as the seed-bearing parent and is pollinated with a high-performing, open-pollinated line that possesses disease resistance and other required characteristics. Most breeders use a multigerm pollinator with a monogerm seed-bearing parent. The pollinator can then be mixed in a ratio of 1:20, or less, with the CMS for planting in the seed field. Sib crosses between pollinator plants will be multigerm and can be screened largely from the commerical seed.

Three-way topcross hybrids have often produced sugar yields from 5 to 20 percent better than open-pollinated varieties that were used prior to their development. They have also provided a convenient means of combining the monogerm character, disease resistance, bolting resistance, and good quality. Topcross hybrids usually have wide adaptability because the pollinator is open-pollinated.

Some breeders have produced three-way hybrids in which all three parents have been inbred. The pollinator is usually a multigerm inbred with good vigor and pollen production. This type of hybrid can be expected to perform very well in a narrow environmental range but may do poorly outside this environment. These closely bred hybrids need to be thoroughly evaluated, and their distribution needs close supervision.

Four-way hybrids, $(A \times B) \times (C \times D)$, have also been produced by utilizing both genic and cytoplasmic male sterility (27). Genic male sterility *aa* is introduced into inbred C; and by making appropriate crosses, a population of C can be produced that will segregate 50 percent fertile *Aa* and 50 percent genic male-sterile *aa*. A fully male-sterile population is obtained by roguing the *Aa* plants before the flowers open. By planting alternate strips of the segregating C line and the pollen-fertile D inbred, hybrid seed is produced on the *aa* plants of C. The progeny of this cross will be fully pollen-fertile *Aa*. It will be used as the pollinator to produce the four-way hybrid. The $A \times B$ seed-bearing parent is produced through the use of CMS.

A pollen-restorer line has recently been isolated by Theurer and Ryser (41) which can be used in the production of four-way hybrids. When a cross is made between a CMS line and a pollen-restorer line, all or nearly all of the plants from the cross are fertile. To produce $(A \times B) \times (C \times D)$, the breeder would cross a CMS C line with a pollen-restorer D line. This $C \times D$ cross would then be used as the pollinator in a four-way hybrid. This method would eliminate the tedious roguing job that is required when genic male sterility is used to produce the $C \times D$ pollinator.

Single-cross hybrids have been used widely with corn, but until recently, the sugarbeet industry has made little use of them. Inbreds tend to be weak, produce low seed yields, and are often difficult to propagate in the seed field. Single-cross hybrids are usually very uniform but are less widely adapted than topcross and four-way hybrids. For these reasons, most sugarbeet breeders have been hesitant to release them for commercial production.

Results by Nielson et al. (23) provide evidence that the problems with single-cross hybrids can be overcome, and that their performance is often superior to that of either three-way or double-cross hybrids. Limited commercial experience with single-cross hybrids has been encouraging, and an increase in their use is anticipated.

POLYPLOIDY AND TRIPLOID HYBRIDS

The sugarbeet plant is normally diploid (2n) and each cell contains 18 chromosomes. By soaking germinating seeds in a weak solution of colchicine, the normal process of cell division is disrupted, and plants can be selected in which each cell contains twice the normal number of chromosomes. These plants are tetraploid (4n) and the chromosomes occur in four sets of nine. Tetraploids can frequently be reproduced and true-breeding lines can be established. The performance of tetraploid lines is usually inferior to that of diploid lines and no commercial tetraploid varieties have been produced.

When a cross is made between diploid and tetraploid sugarbeet plants, a hybrid is produced with three sets of chromosomes. Triploid hybrids are often superior in some characteristics to either the diploid or tetraploid parents. The root yield of the triploid is frequently higher and the sugar yield may also be higher. The bolting resistance of the triploid is also usually superior to that of the diploid parents.

The triploid hybrid is sterile because normal reproduction depends on the pairing of like chromosomes, and sets of three cannot pair. Prior to the adoption of male sterility as a breeding tool, there was no way to commercially produce a completely triploid hybrid. European breeders devised a method of growing a mixture of tetraploid and diploid plants in the seed field, and this mixture yields a product that is partially triploid. The highest proportion of triploid seed is obtained from a mixture of approximately 80 percent tetraploid and 20 percent diploid plants. Natural hybridization occurs among these plants to produce a mixture of triploid, tetraploid, and diploid seed. Usually 50 to 55 percent of the seed is triploid. Varying proportions of the remainder are diploid and tetraploid, with tetraploid seed predominating. A variety with unequal ploidy or mixed ploidy is properly called an ansiploid variety, but the term "polyploid" has been adopted and is in general use in most countries.

This breeding method has been widely used in Europe, but few breeders have produced polyploids in the United States. In Europe, polyploid varieties have frequently been more productive than the open-pollinated diploid varieties. This is especially true in southern Europe. Tests in the United States have often shown good results with polyploid varieties so long as disease is not a factor. The susceptibility of the European polyploids to our major diseases has precluded their use in most parts of the United States.

Hybrid varieties that are completely triploid can be produced by utilizing cytoplasmic male sterility. A monogerm CMS is usually used in combination with a tetraploid pollinator. The same general seed production methods can be used with diploid hybrid seed. Less pollen is usually produced by tetraploid plants, and pollen production is more subject to variation in environmental conditions. High humidity is especially detrimental to pollen production and movement. The seedsman needs to give closer attention to the location and care of triploid seed fields than to diploid seed fields.

The relative merits of diploid and triploid hybrids have not been entirely determined. Results of European breeders show that monogerm triploid hybrids are frequently superior to their diploid open-pollinated varieties and to the diploid hybrids they have tested. Based on these results, the European sugarbeet industry is greatly interested in triploid hybrids, and they will soon be widely used in many countries.

VARIETY DEVELOPMENT 413

Most breeders in this country have not demonstrated a clear-cut advantage for triploid hybrids compared with our present diploid hybrids. Also, in this country, the germination of triploid monogerm seed tends to be inferior to that of diploid seed. For these reasons, few American breeders are recommending triploid varieties at this time. Developmental work is continuing and additional information will soon be available on the comparative productivity of diploid and triploid hybrids.

INTERSPECIFIC HYBRIDIZATION

The cultivated beet has several wild relatives that could serve as a source of valuable genes for the breeder. These wild species differ greatly in their plant type, root characteristics, disease resistance, and in their ability to hybridize with the sugarbeet. Dr. Coons (8) reviewed taxonomic studies with the genus *Beta* and classified the species into four sections (Table 14.1).

Vulgares. This section of the genus includes the cultivated beets and their close relatives. Taxonomists differ in their treatment of this section, and the species listed by Coons are treated differently by some other investigators. All Vulgares beets are diploid, cross readily with the cultivated beet, and produce fertile hybrids.

Many workers are of the opinion that all cultivated beets originated from *B. maritima.* This species possesses characters that are highly desirable when transferred to the sugarbeet. A notable example is *Cercospora* leaf spot resistance. O. Munerati discovered leaf spot resistance in biotypes of *maritima* growing along the mouth of the Po River in Italy, and transferred a high level of resistance to *B. vulgaris.* Some other biotypes possess resistance to curly top.

Segregating populations from crosses between the cultivated beet and the various species of the Vulgares section are extremely variable with respect to root shape, foliar type, seed size, percent dry matter in root, sucrose percentage, bolting resistance, and other characters. Varieties or breeding lines of immediate value are not to be expected from hybridiza-

TABLE 14.1. Species of *Beta*

I. *vulgares*	II. *corollinae*	III. *nanae*	IV. *patellares*
vulgaris L.	*macrorhiza* Stev.	*nana* Boiss. et Held.	*patellaris* Moq.
maritima L.	*trigyna* Wald. et Kit.		*procumbens* Chr. Sm.
macrocarpa Guss.	*foliosa* Hausskn.		*webbiana* Moq.
patula Ait.	*lomatogoma* Fisch. et Mey		
atriplicifolia Rouy	*corolliflora* Zoss.*		

SOURCE: Table from Coons (8).
* Some investigators consider *B. corolliflora* (4n) a separate species from the closely related *B. trigyna* (6n). Dr. Coons listed only *B. trigyna*.

tions of this kind. They may, however, enable the breeder to introduce valuable characters if followed by rigid selection and backcrossing.

Patellares. This section includes three clearly defined species characterized by a viny growth habit and a hard-coated monogerm seed. The section is a potential source of desirable genes for sugarbeet improvement. Each species is resistant to the sugarbeet nematode, *Cercospora* leaf spot, and curly top. Nematode resistance is of particular interest because there is no other known source of high-level resistance to this important pest.

The Patellares section is distantly related to the Vulgares section and transfer of genes is difficult. Crosses have been made between sugarbeet and each of the species, but the chromosomes are not homologous. This means that genes can rarely, if ever, be transferred from these species to sugarbeet through normal crossing and selection procedures. The incorporation of a desired character requires a transfer of chromatin material from the Patellares section. The chromatin must carry the desired gene or possibly two or more closely linked genes. This chromatin transfer is called a translocation. If resistance genes occur on more than one chromosome, additional translocations are required.

There are several types of translocations. One of the commonest is the reciprocal translocation in which a section of one chromosome is exchanged for a section of some other chromosome. To transfer a desired character such as nematode resistance, the investigator must find the Patellares chromosome with the resistance gene or genes, and include it as an extra chromosome in an otherwise normal sugarbeet cell. To obtain translocations, flowering plants with the extra chromosome are often treated with X rays. Pollen from these irradiated plants is used to fertilize the flowers of normal sugarbeet plants.

A portion of the plants from these crosses may carry translocations. To identify the desired translocations, large numbers of these plants need to be screened for resistance and examined for cytological behavior. The translocated section of the chromosome should contain the gene or genes for resistance and as few undesirable genes as possible. If a single reciprocal translocation is to be successful, the desired gene or genes should be located near the end of the chromosome. The end segment of the Patellares chromosome will be exchanged for the end segment of the sugarbeet chromosome. After the translocation is completed, the Patellares chromosome bearing the short translocated section of the sugarbeet chromosome is lost. The sugarbeet chromosome bearing the short section of the Patellares chromosome will then function normally in cell division. The loss of the sugarbeet section may or may not be deleterious, depending on the genes that occurred on the lost section.

When the gene or genes for resistance lie some distance from the end of the chromosome, translocation plants bearing the resistance genes can be expected to suffer from the loss of several sugarbeet genes and addition of alien genes. Sometimes a second translocation can be induced to retain the resistance genes from Patellares, and retrieve most of the sugarbeet genes that have been lost in the first translocation.

The ideal solution is an intercalary translocation. This involves the insertion of a very short section of the Patellares chromosome into a sugarbeet chromosome without any loss of sugarbeet chromatin. The inserted section must carry the resistance genes and no deleterious genes from Patellares. Unfortunately, intercalary translocations are rare, because two breaks must be obtained close together in the chromosome arm.

Work is underway in both the United States and Europe to transfer nematode resistance from the Patellares section to the cultivated beet. Dr. Helen Savitsky (33) has crossed sugarbeet with species of the Patellares section. She has backcrossed to sugarbeet and isolated nematode-resistant plants with an extra chromosome from Patellares. This extra chromosome carries the gene or genes for resistance.

This most significant achievement required hundreds of crosses, nematode resistance tests, and cytological examinations. Unfortunately, a plant with an extra chromosome is an abnormal plant and will not breed true. A small section of the alien chromosome must be transferred to a sugarbeet chromosome by techniques similar to those we have just described.

These procedures are complicated and will require much painstaking work. Encouragement is provided by the success of Sears (38) in transferring leaf-rust resistance of wheat from the genus *Aegilops* to common wheat in the genus *Triticum*. He made this transfer with an intercalary translocation.

Nematode resistance is needed in most sugarbeet production areas and would be of great economic importance. Even though transfer of resistance through interspecific hybridization will be difficult and the outcome somewhat uncertain, the potential benefits are sufficient to justify these special efforts.

Corollinae. This section includes species with distinct floral characteristics marked by a corollalike perianth. Taxonomic treatment of the section is not entirely clear and the list of the species differs among investigators. Like Patellares, this section is distantly related to the cultivated beet, and the transfer of desirable genes is difficult. Investigators have made crosses between each of the species and the cultivated beet. These hybrids are sterile, and there is little or no homology between chromosomes.

Characters of potential value in the Corollinae section include curly top resistance, monogerm seed, and apomixis. *Beta corolliflora* is highly resistant or immune to the most virulent strains of the curly top virus. This is a tetraploid species that closely resembles the hexaploid *B. trigyna*. Dr. Savitsky (34) has crossed sugarbeet with *corolliflora* and then backcrossed to sugarbeet. She selected second backcross plants that were immune or highly resistant to curly top. These resistant plants carried 2 to 5 *corol-*

liflora chromosomes in addition to the diploid set of *vulgaris* chromosomes. Chromatin transfers will probably be required to incorporate this resistance into a sugarbeet plant with the normal set of 18 chromosomes.

Some attempts have been made to transfer the monogerm seed character of *B. lomatogona* to sugarbeet, but the lack of chromosome homology is a problem with these crosses. A pentaploid (5n) form of *B. trigyna* has been found to be apomitic. This could be a useful character if introduced into the sugarbeet.

Nanae. This section is represented by a single species, *B. nana*. Some investigators include the species in the Corollinae section. The plants are very small and seldom produce a leaf rosette more than 4 inches across. No hybrids have been reported with sugarbeet.

RECENT PROGRESS IN BREEDING

Problems encountered in seed production during World War I, and our inability to compete with the European seed suppliers following the war, further delayed the establishment of sound breeding programs in this country. This situation might have continued indefinitely had not recurrent outbreaks of disease occurred in many of our production areas. Curly top was especially serious in the western states and caused the abandonment of several factories. In the eastern production areas, sporadic epidemics of *Cercospora* leaf spot frequently caused severe crop losses. The continued success and expansion of the sugarbeet industry was in jeopardy, unless some means could be found to control these diseases or unless resistant varieties were developed. The need for resistant varieties prompted the establishment of breeding programs by the U.S. Department of Agriculture, the sugar companies, and state agricultural experiment stations.

BREEDING FOR DISEASE RESISTANCE

Curly Top. Curly top, caused by a leafhopper-transmitted virus, occurs on sugarbeet throughout the western United States and is especially prevalent west of the Rocky Mountains. The disease is extremely damaging, and when susceptible varieties are infected in the seedling stage, the plants are often killed or are so severely damaged that a profitable crop cannot be produced. The first serious attempt to select for resistance was begun by Dr. Eubanks Carsner of the U.S. Department of Agriculture in 1918. Dr. Katherine Esau, who was employed first by the Spreckels Sugar Company and later by the University of California, carried on a selection program from 1919 to 1929. Other sugar company breeders also made selections during this period. Most of these early selections were from commercial fields sown to European sugarbeet varieties. Roots of the most promising of these selections were assembled by the U.S. Department of Agriculture at Twin Falls, Idaho, in 1929, and planted together in a block for seed production. The seed from this group planting constituted the original seed supply of the first curly-top-resistant variety and was given the official designation of US 1 (5).

Tests with this new variety showed that a moderate improvement in resistance had been made, but severe damage still occurred under heavy disease exposure. Some deterioration in sucrose percentage and bolting resistance had taken place during the selection process. Selection work was continued to improve the curly top resistance, root yield, and sucrose content. A. M. Murphy of the U.S. Department of Agriculture, Twin Falls, Idaho, developed a dependable method of producing uniform, heavy exposures to the curly top virus. This was done in a field adjacent to the desert breeding grounds of the leafhopper vector of the virus. Curly-top-infected beet roots from the previous season were transplanted into strips of a susceptible variety that had been early sown at about 100-foot intervals across the nursery. Migrating leafhoppers from the desert fed on the diseased plants and infected the susceptible variety. Reproduction occurred on the susceptible plants, and high populations of viruliferous leafhoppers developed. These hoppers moved to breeding material which was sown between the infected strips in May or June. In most years, uniformly heavy infection occurred in the late-sown beets and reliable selections could be made.

Varieties with progressively higher resistance were developed by the mass selection method. Eventually, successive selections in open-pollinated lines showed little or no gain in resistance, and in recent years, the greatest gains have occurred in selections from inbred populations accompanied by progeny testing. This selection work has been done both by the U.S. Department of Agriculture and the sugar companies. These resistant inbreds have been used as components in hybrid varieties. The varieties now available to the growers have sufficient resistance to withstand the curly top attacks that occur in most seasons. Improvements in yields that have occurred in southern Idaho, an area subject to frequent curly top epidemics, are illustrated in Figure 14.2.

The contribution of curly top resistance to the sugarbeet industry is also well illustrated by results from Washington, a state that abandoned sugarbeet production in the 1920s because of curly top. Resistant varieties are now used exclusively in Washington and recently produced an average 5-year yield of 23.8 tons on 53,600 acres.

The curly top virus is a complex of strains that vary in virulence. Highly virulent strains are widely distributed throughout western United States and are capable of causing severe injury to our present varieties. Fortunately, the more virulent strains apparently kill the desert host plants that harbor the leafhoppers throughout the winter, and are thereby largely eliminated (14). A low percentage of the leafhoppers that migrate back to the sugarbeet fields in the spring carry virulent strains and infect scattered beet plants. Strains of curly top virus do not offer cross protection

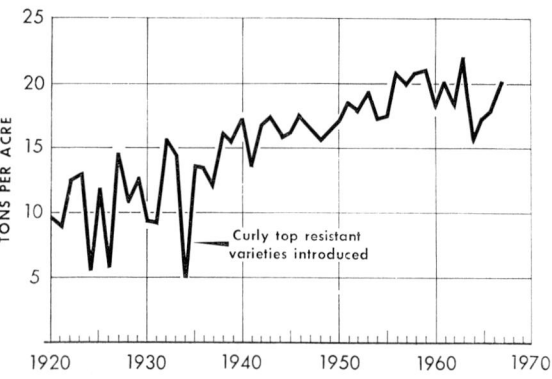

FIG. 14.2. Average sugarbeet yields in southern Idaho, 1920–67. (Data from Amalgamated Sugar Company.)

against each other, so there is no barrier to infection with the virulent strains. By the end of the growing season, the virulent strains are widespread in the beet field. If these fields or scattered plants are maintained throughout the winter, a source of the most virulent strains is available to infect spring-sown beets. Breeding lines with good resistance to these virulent strains have been selected in the greenhouse. This resistance will be incorporated into the inbred components of hybrid varieties.

Cercospora *Leaf Spot.* Leaf spot caused by the fungus *Cercospora beticola* Sacc. destroys the foliage of the beet plants and causes a reduction in root yield and sucrose percentage. Most sugarbeet production areas east of the Rocky Mountains are subject to sporadic epidemics of leaf spot. Losses may also occur in California, especially in the northern Sacramento Valley. Some selection work was started in the early 1920s by the Great Western Sugar Company and American Crystal Sugar Company (formerly American Beet Sugar Company) to find resistance to leaf spot. A breeding program was started by the U.S. Department of Agriculture in 1925 at Rocky Ford, Colorado, with emphasis on leaf spot resistance. Dr. G. H. Coons, Dewey Stewart, and their colleagues had available some 200 strains of sugarbeet developed at Fort Collins, Colorado, by W. W. Tracy, Jr. From these 200 lines, representing a wide range of morphological types, 14 were found to possess some resistance.

Attempts to select individual plants within a variety or line were not successful. No immune or highly resistant plants were found, and the individual plants were difficult to classify. The 14 Tracy lines that showed some resistance had been selfed two or more generations. The results showed that the greatest progress could be made by inbreeding selected plants and identifying the resistant lines through progeny testing. The procedure centered on inbreeding to isolate resistant genotypes and testing the inbreds under conditions of drastic leaf spot exposure.

After many years of inbreeding and intensive selection, inbreds with good resistance were obtained. A series of hybrids were produced to find resistant inbreds with good combining ability. The first resistant variety, US 217, was produced in 1936 from five inbred lines (7). Seed was produced by pairing (a × b), (b × c), and (d × e) in three isolated fields. The seed from the three fields was mixed in approximately equal proportions to supply stock seed for commercial seed production. Additional varieties were produced by pooling other hybrids between inbred lines. Even though some selfing occurred with this breeding procedure, the varieties yielded significantly better than the standard nonresistant varieties when exposed to leaf spot.

With the discovery and introduction of cytoplasmic male sterility into sugarbeet breeding, true hybrids could be produced between leaf-spot-resistant lines. Hybrid varieties are widely used in eastern United States, but at the present time, only the male parent is leaf spot resistant. Male-sterile parents with leaf spot resistance have been developed and are being increased for use in commercial hybrids.

Leaf-spot-resistant varieties have also been developed by the American sugar companies and by European breeders. The Italian varieties, Cesena, Mezzano 71, and Rovigo 581, have proved valuable sources of resistance. These varieties trace back to the breeding work started by O. Munerati in 1910 at the Station Sperimentale di Bieticultura, Rovigo, Italy. He crossed the cultivated sugarbeet with *Beta maritima* collected along the estuaries of the Po River. From these hybrids, he developed lines high in leaf spot resistance and sucrose percentage. These lines were turned over to breeders at the commercial seed stations at Cesena and Mezzano, Italy. These breeders used the Munerati lines to develop commercial varieties bearing the names of the respective stations. Tests in the United States showed these Italian varieties to be high in both leaf spot resistance and sucrose content. The Great Western Sugar Company used Cesena in the development of G.W. 304 and G.W. 359. Mezzano 71 was used as a source of leaf spot resistance in US 201, a highly leaf-spot-resistant line developed by the U.S. Department of Agriculture.

Some information is available on the nature of resistance to leaf spot. Harrison et al. (15) determined that a phenolic compound (3-hydroxytyramine) is related in some way to leaf spot resistance. Oxidized 3-hydroxytyramine is toxic to *C. beticola* and resistant varieties contain more of this phenolic compound than do the susceptible types. Hecker et al. (16) found that the quantity of 3-hydroxytyramine in sugarbeet leaves appears to be conditioned by four or more genes.

Black Root. A common cause of poor sugarbeet stands is a complex of seedling diseases known as black root. This name is used because the dead or dying plants are black in appearance. Fungi primarily responsible for serious injury include *Pythium* species, *Rhizoctonia solani* Kuehn, *Phoma beta* (Oud.) Frank, and *Aphanomyces cochlioides* Drechs. Each of these organisms is capable of killing seedlings before emergence or during the first 2 weeks following emergence. Plants that escape this acute form of the disease may be subject to a chronic form of black root. Plants attacked by this form of the disease are not killed, but the lateral roots and the terminal part of the taproots are destroyed. Plants so affected grow very slowly and may never reach a marketable size. The chronic form is caused by *A. cochlioides*. The fungi responsible for the acute form may be partially controlled by seed treatment, whereas fungicides have provided little or no control of the chronic form.

Differences in resistance were observed in 1940 and 1941 by Dr. G. H. Coons in variety trials in the Great Lakes region. An inbred, US 216, and hybrids involving this inbred remained vigorous, whereas other varieties were stunted by black root. This observation furnished evidence that genetic factors for black root resistance existed, and a program of field selection was started. Selections were made, in heavily black-root-damaged fields, by Henderson and Bockstahler (18) and by Doxtator and Downie (10). These early efforts led to the development of US 400, US 401, and American Crystal No. 3. These open-pollinated, multigerm varieties had moderate resistance and yielded well when black root epidemics were not too severe. A higher level of resistance was needed in many beet-growing districts.

Field selecting and testing were often found to give unsatisfactory results because of low disease intensity, the lack of uniformity, or the masking effect of environmental and nutritional factors (6). In recent years, most of the selection work has been done in the greenhouse for resistance to *A. cochlioides*. Methods of producing inoculum and for selection in the seedling stage have been developed by Dr. G. E. Coe and Dr. C. L. Schneider (6, 37). Sugarbeet seedlings are grown in 6-inch saucers and inoculated with zoospores of the pathogen when 7 to 10 days old. Some plants are relatively vigorous and recover rapidly from the infection, but the more susceptible plants are killed. Plants showing the least damage are transplanted to 6-inch pots and are again inoculated with approximately 1 million zoospores per pot to confirm resistance. The seed is produced from the surviving plants and the progenies tested in the greenhouse. Through the use of this technique, Coe and Schneider have selected lines that are substantially more resistant than the present commercial varieties. Their work suggests that the inheritance of resistance is complex and is influenced by many genetic factors.

Virus Yellows. Yellowing diseases caused by viruses occur in many sugarbeet-producing areas of western United States. Losses have been severe in areas where sugarbeets, or other susceptible plants, are present during the entire year. Two yellowing diseases, beet yellows and beet western yellows, are common on beet (2, 11). They are caused by two distinct viruses, beet yellows virus and beet western yellows virus. The green peach aphid, *Myzus persicae* (Sulzer), is the principal vector of both viruses. Beet yellows is the more damaging of the two diseases and may cause crop losses of 20 to 40 percent. Losses from beet western yellows range from about 10 to 20 percent. When both viruses are present, the losses are additive.

A breeding program to find resistance to yellows has been underway at the U.S. Agricultural Research Station, Salinas, California, since 1955. The breeding of yellows-resistant varieties has proved to be more difficult than many other breeding problems with sugarbeet. A survey of more than 350 different varieties and breeding lines failed to uncover a source of high resistance. Yield losses in this material ranged from 10 to 65 percent, and most commercial varieties showed yield losses ranging between 20 and 40 percent. A survey was also made of several wild species of *Beta* in hopes of finding a source of high resistance or immunity. None of these species proved to be resistant, and it is unlikely that they will be of value in the yellows resistance breeding program.

The identification of resistant plants has also presented some problems. The correlation between root yield reduction and top symptoms is low. This means that the breeder cannot depend on the severity of yellows as a measure of resistance. At present, the only way to determine the level of resistance in a variety or breeding line is to compare the root yield and sucrose content of healthy and yellows-infected plants. This is a costly and time-consuming operation and limits the amount of material that can be evaluated.

Selections are made from field plantings that are inoculated with a combination of beet yellows virus and beet western yellows virus when the plants are about 7 weeks old (21). Plants that show severe yellows symptoms are rogued. At harvest, large, well-shaped roots are selected from plants that show the least number of dead leaves. An additional selection is made on the basis of sucrose content. Seed increases are made from these selections and the degree of resistance determined by comparing the performance of inoculated and noninoculated plots of each selection.

Improvements in resistance have been obtained in both self-sterile and self-fertile sugarbeet lines. The open-pollinated US 75 variety proved heterozygous for yellows resistance and has been used extensively in the breeding program. Root yield losses from yellows averaged 20 percent for a seventh successive selection from US 75 compared with 41 percent for the parent (21). Under severe yellows, the selection produced an average 53 percent higher root yield and was 0.8 percentage points higher in sucrose than US 75. Under light yellows infection, the performance of the selection was similar to that of the parent.

This yellows-resistant selection from US 75 has been used as the pollen parent in the commercial hybrid varieties US H9A and US H9B (21). These hybrids have produced 20 percent more sugar per acre than the standard check variety in widespread tests under conditions of moderate to severe yellows. These yield increases have been obtained without sacrificing sucrose content.

Emphasis is currently being placed on the development of resistant monogerm inbreds and male-sterile lines that can be used as seed-bearing parents in combination with the resistant multigerm pollinator.

Downy Mildew. Downy mildew, caused by the fungus *Peronospora farinosa* (Fr.) Fr., occurs primarily in the coastal valleys of California and in the seed production areas of western Oregon. Low temperature and high humidity requirements for conidial germination and infection are responsible for the limited distribution of the disease. Losses are expressed as a reduction in root size, in sucrose content, and in purity of the beet.

Breeding for resistance was started by Dr. L. D. Leach of the University of California. The work was continued by Dr. J. S. McFarlane and by the Spreckels Sugar Company. Open-pollinated selections with moderate resistance, and inbred lines with high resistance, were developed (20). Varieties used in California prior to 1954 either lacked mildew resistance or had only fair to moderate resistance. By 1956 the susceptible varieties had been largely eliminated and replaced with moderately resistant varieties. Later, when hybrid varieties were developed, moderate mildew resistance was introduced into the inbred and male-sterile parents.

Losses from downy mildew have been of minor importance in the coastal valleys of California since 1954. Sufficient infection to cause economic loss has occurred in isolated areas in some years, but there is no record of wide-scale infection. The principal change in sugarbeet production has been in varieties. The level of resistance has been raised, and the rapidity with which the disease develops has undoubtedly been reduced. This means that the number of conidia available for infection is probably much lower, particularly during the late spring months. Downy mildew continues to cause economic losses to the seed crop in western Oregon. However, the losses are confined almost entirely to varieties with little or no resistance to mildew.

Even though downy mildew has not been a serious problem during recent years in the beet-growing areas, the potential for infection and economic loss still exists. As new varieties are developed, breeders should strive to maintain at least the present level of resistance. The possible appearance of new strains of *P. farinosa* should not be overlooked. The California varieties are not considered resistant in England, and the differential behavior may be caused by strain differences in the pathogen. Selections have been made that are much more resistant than our present varieties. If needed, this resistance could be incorporated into adapted varieties.

Rhizoctonia. Root and crown rot, caused by *Rhizoctonia solani* Kuehn, is a serious problem in most sugarbeet-producing areas of the United States. The fungus occurs in all agricultural soils and is capable of persisting indefinitely. Sugarbeet plants attacked in the seedling stage may be killed by the fungus or only cankered. Many of the cankered plants develop crown and root rots. Yield reduction occurs because of poor stands and damage from rots. The rots also cause losses in the storage piles and problems in processing.

Numerous attempts have been made to select for resistance, but progress has been hampered by the erratic behavior of the disease. Naturally occurring soil inoculum is undependable and gives unsatisfactory results. John O. Gaskill (13) reports that exposure by this method has resulted in negligible to complete loss of stands. He obtained his best results by placing inoculum either in the center of the foliar rosette or in a semicircle 1½ inches from the taproot, and about 1 inch below the soil surface. Both methods proved too severe if inoculation was performed sooner than 3 weeks after thinning.

By making several cycles of mass selection under artificial disease exposure, Gaskill (13) was able to substantially raise the level of *Rhizoctonia* resistance. Two of the most promising selections have been released by the U.S. Department of Agriculture as FC 701 and FC 702. These selections show a marked improvement in resistance but may still be damaged by the disease. The resistance is apparently relatively ineffective when the plants are small.

A large number of strains or biotypes of *Rhizoctonia* are known to occur. These strains differ in their pathogenicity on sugarbeet. FC 701 and FC 702 were selected primarily against one pathogenic strain and may not necessarily be resistant to a wide range of biotypes. These two lines are considered of value primarily as sources of genes for *Rhizoctonia* resistance. Even though they are not suited for use as commercial varieties, the lines represent a significant contribution to a most difficult breeding problem.

NEMATODE RESISTANCE

The sugarbeet nematode *Heterodera schachtii* Schmidt is one of the most serious pests of the sugarbeet. The first nematode infestation in the United States was observed in Utah, about 1895; and since then, the pest has become established in many sugarbeet-growing areas of the country. Heavy infestation causes poor stands and the production of small, stunted plants. Fields that are heavily infested will not produce an economic crop of beets.

Nematode populations can be reduced through crop rotation or by soil fumigation. To be effective, a crop rotation program should include sugarbeet or other susceptible crops no more than one year in five. Careful attention must also be given to the control of nematode-susceptible weeds in those years when crops other than beets are grown. A rotation program that limits the production of beets to one year out of five drastically reduces the sugarbeet acreage in any given area. This loss of acreage has been a contributing factor in the closing of some sugar factories. Soil fumigation will effectively control nematodes on light, sandy soils but has not been successful on heavy, clay soils.

A serious need exists for a variety with nematode resistance. Growers and plant breeders have frequently observed healthy-appearing plants in fields that are heavily infested with nematodes. The occurrence of these vigorous plants has suggested that resistant segregates exist, and repeated selections have been made. Invariably, the offspring of these resistant-appearing plants have proved to be highly susceptible.

In recent years, breeding programs have been started in both the United States and Europe to specifically breed for nematode resistance. Methods of inoculating sugarbeet seedlings in the greenhouse with large numbers of either nematode cysts or larvae have been developed (9, 31). Charles Price (31) screened thousands of plants from several varieties and selected plants that possessed the least number of nematode cysts. Progenies of these selections showed some improvement in resistance, but opportunities for finding a high level of resistance within *Beta vulgaris* do not appear promising.

Surveys of the wild relatives of the sugarbeet have revealed that the wild species *B. patellaris* Moq., *B. procumbens* Chr. Sm., and *B. webbiana* Moq. are highly resistant. These species will cross with sugarbeet, and the F_1 hybrids are resistant (35). Dr. Helen Savitsky of Salinas, California, has backcrossed to the sugarbeet and selected segregates with a very high level of resistance, but these plants are abnormal and possess an additional chromosome from the wild species. Attempts are being made at Salinas and in Europe to break this extra chromosome and to transfer a small section bearing the genes for resistance to one of the sugarbeet chromosomes.

BOLTING RESISTANCE

The sugarbeet is a biennial plant, and bolting (seedstalk production) ordinarily is not a serious problem when the crop is planted in the spring. In parts of California and Arizona, sugarbeets are grown as a winter crop. Planting is started in September in the Salt River Valley of Arizona and the Imperial Valley of California for harvest the following spring. In the coastal valleys much of the planting is done between December and February. Varieties that have been developed for spring planting usually bolt severely when planted in the fall or winter months.

The initiation of seedstalks and flowering of biennial beets is brought about mainly by the cumulative effect of low temperature exposure followed or accompanied by the effect of long photoperiods (28). The relative importance of thermal induction and photoperiodic induction is influenced by the genetic constitution of the beet. There is also some evidence that the optimum temperature for thermal induction varies from one genotype to another.

The first varieties developed in the United States tended to be susceptible to bolting and sometimes produced an objectionable number of seedstalks even in spring plantings. An exception was the US 15 variety that was selected from the old European variety R and G Pioneer. This variety was widely used for winter plantings in California until replaced with varieties combining good bolting resistance with curly top resistance.

Selection for bolting resistance is done in such areas as the coastal valleys of California, which have a mild winter climate. Varieties or breeding lines heterozygous for bolting resistance are planted in the late summer or early fall and allowed to grow through the winter. Frequently, 75 percent or more of the plants of a susceptible variety will bolt the following summer. Roots of the nonbolting plants are placed in a thermal-induction chamber and seed is produced the following year. By repeating this process two or three times, a line can ordinarily be developed with a high level of bolting resistance. Little difficulty has been experienced with declining quality or other undesirable characteristics, provided the selected populations are of sufficient size and adequate tests are made with each selection.

The greatest gains in bolting resistance have been obtained in inbred lines through evaluation of the progenies of single-plant selections. Inbred lines have been selected that differ greatly in their bolting behavior in different environments. Some lines respond more sharply to a long photoperiod and others to a long period of thermal induction.

Selections for bolting resistance can also be made under controlled conditions in refrigerated thermal-induction chambers, in low temperature greenhouses, or in a combination of the two. John O. Gaskill (12) worked out a method of inducing reproductive development by prolonged exposure to low temperatures and continuous artificial light. By modifying Gaskill's method, Dr. R. T. Johnson (19) was able to select for bolting resistance in an unheated greenhouse at Spreckels, California.

MONOGERM SEED

In nature, the seeds of the beet plant are usually produced in a cluster containing two or more germs that have fused into a multigerm seedball. When this seedball is planted, a clump of seedlings that requires handwork for singling emerges. To reduce handwork, seed-processing methods that enabled the seedsman to reduce the size and germ number of the seedball were devised.

Processed seed was only a partial answer to the multiple seedling problem, and a need existed for a single-germed (monogerm) seed (Fig. 14.3). A search for a sugarbeet plant that bore entirely monogerm seed was started more than 60 years ago. The search proved discouraging, and success was not achieved until the late 1930s, when Dr. V. F. Savitsky and his colleague, M. G. Bordonos, found a monogerm plant at the Sugar Beet Institute in Kiev, Russia. Dr. Savitsky immigrated to the United States in 1947 and continued his genetic research with sugarbeets at the U.S. Sugarbeet Laboratory at Salt Lake City. From his work in Russia, Dr. Savitsky had learned the botanical characteristics of the monogerm sugarbeet plant and was successful in finding a few plants with predominantly monogerm seed in a commercial seed field in Oregon (36). Two of these plants, designated SLC 101 and SLC 107, proved to be true genetic monogerms. Seed from SLC 101 was distributed to sugarbeet breeders in the United States,

Fig. 14.3. Single sugarbeet seedling from monogerm seed compared with a clump of seedlings from multigerm seed. (Photo by I. O. Skoyen.)

Canada, and Europe. Dr. Savitsky found that monogermness is determined primarily by a single recessive gene which he identified as *m*. In segregating populations, some double-germ seeds occur on *mm* plants, and he concluded that this is caused by genes that modify the action of the basic *m* gene. These modifying genes have not been identified.

The SLC 101 monogerm was a weak, self-fertile inbred that lacked disease resistance but had a moderate degree of bolting resistance. Two general methods were used to transfer the *m* gene to adapted varieties. One group of breeders made crosses between their most productive varieties and SLC 101. Self-sterile *mm* types were selected from segregating populations and then a series of backcrosses were made to the adapted variety. After a few years, the monogerm equivalent of the adapted variety was developed. A second group concentrated on the development of monogerm inbreds. This was accomplished by crossing SLC 101 with desirable multigerm inbreds or open-pollinated lines and by selecting vigorous self-fertile *mm* segregates in the F_2 and F_3 generations. Usually one or more backcrosses were made to the multigerm parents. Male-sterile monogerm lines were also produced concurrently with the inbreds. The monogerm inbreds and male steriles were often weak, but when crossed, many of the F_1 hybrids were vigorous and produced a good seed set. These monogerm, male-sterile F_1 hybrids were used as the seed-bearing parents in conjunction with multigerm pollinators. The seed produced on the male-sterile monogerm plant is monogerm even though the pollinator is multigerm. By choosing pollinator lines with a high level of either curly top or leaf spot resistance, the breeder was able to develop commercially acceptable monogerm hybrid varieties within a few years after the discovery of the monogerm character. Inbreds with a high level of disease and bolting resistance have been developed, and monogerm varieties with greater productivity than the older multigerm varieties are now available.

IMPROVED PROCESSING QUALITY

The amount of sugar that can be produced from a ton of sugarbeets is determined not only by the sucrose content of the beets but also by the presence of nonsugar substances that interfere with the recovery of sucrose from the raw product. The measure of the combined effect of these nonsugars is referred to as juice purity.

In this country, most of the selection work has been for higher sucrose content, and until recently, little emphasis has been placed on breeding for improved juice purity. In general, these selections have resulted in the so-called "sugar type" varieties that are relatively high in sucrose content but low in yield of roots and sugar per acre. These high-sugar varieties have been used very little commercially because of the low yields.

The results of this selection work suggested that sucrose content is conditioned by additive factors or genes without any expression of dominance or heterosis. Studies by Powers et al. (30) with F_1 hybrids and inbred lines have provided evidence which strongly suggests the existence of dominance and even heterosis for sucrose content. Even though root yield and sucrose content are undoubtedly closely associated, there is evidence that some improvement can be made in the two characters simultaneously. These results suggest the possibility of developing varieties with higher sugar yields accompanied by higher sucrose content.

High juice purity is a much more complicated character to breed for than is high sucrose content. The accurate measurement of juice purity is a laborious process, particularly with individual beet roots. Furthermore, the processor is interested primarily in the purity of the second carbonation (thin juice) rather than that of the raw juice. Brown and Serro (3) have developed a laboratory method of determining thin juice purities that can be used with single beets. This method measures the total impurity components in the thin juice and provides reliable information on the processing characteristics of the individual beet or breeding line.

The nonsugars are made up of a large number of components, and the thin juice purity does not provide information on either the qualitative or quantitative composition of the individual components. Rather than select against the nonsugars as a group, many breeders feel that more rapid progress can be made by selecting for low concentrations of a few of the more important impurity components. Carruthers and Oldfield (4) reported that 70 percent of the thin juice nonsugars are potassium and and sodium salts, amino acids, and betaine. Potassium, sodium, and amino nitrogen can be easily measured spectrophotometrically on the raw juice at the same time that sucrose determinations are made. Several breeders in both Europe and the United States routinely measure these three components in their breeding programs. Sodium and potassium have proved to be heritable constituents and can be changed through breeding (30). As yet, there is little evidence that amino acid content can be shifted materially by genotypes (29). The quantity of betaine in the beet root differs among varieties, but the laboratory measurement of this nonsugar is more difficult than for sodium, potassium, and amino nitrogen.

Research in the United States suggests that some other impurities may be just as important as sodium, potassium, amino nitrogen, and betaine. One of these, raffinose, tends to build up in storage piles and is especially troublesome in some areas. This impurity has been reported to be heritable.

In recent years, there has been much concern about processing quality in relation to nitrogenous fertilization. We have evidence that the increased use of nitrogen has contributed to the deterioration in sucrose content and juice purity. A question exists regarding the feasibility of developing varieties capable of utilizing higher applications of nitrogen for increased root yields without seriously affecting processing quality. A study by Powers et al. (30) showed that sugarbeet populations differ in their ability to produce a higher sucrose content when fertilized with increased applications of nitrogen. Payne et al. (29) found that sodium and potassium increased in each of 12 sugarbeet populations with increased nitrogen application, but at different rates in different populations. They found little evidence that selections could be made that would overcome the production of high amino nitrogen levels caused by high nitrogen fertilization. On the other hand, the quantity of betaine was affected more by variety than by nitrogen fertilization.

The results of these studies are not conclusive, but they do indicate the possibility of developing varieties capable of overcoming a portion of the deleterious effects associated with increased nitrogen application. However, improved varieties will probably only partially overcome these effects, and the indiscriminate use of nitrogen will always need to be avoided.

COOPERATION IN BREEDING RESEARCH

Sugarbeet breeding and genetic research in the United States is a cooperative undertaking of the sugarbeet industry, the U.S. Department of Agriculture, and the state agricultural experiment stations. Close cooperation has existed between industry and government ever since the first domestically developed variety became available.

When breeder's seed of the first curly-top-resistant variety was released in 1931, nearly all of the nation's sugarbeet seed was imported. A domestic organization with qualifications to increase seed of the new variety did not exist. The sugar companies in the curly top area banded together and formed the Curly Top Resistance Breeding Committee. This organization was cooperative with the U.S. Department of Agriculture and sponsored increases of US 1 and other curly-top-resistant releases. Later the committee expanded its activities and provided financial support to the government's curly-top-resistance breeding work.

In 1947 the scope of the committee's activities was again extended and the number of participating companies increased. Dr. and Mrs. V. F. Savitsky, prominent Russian sugarbeet geneticists, left the Sugar Beet Institute at Kiev during World War II to seek employment in either Europe or America. The Curly Top Resistance Breeding Committee was able to employ the Savitskys. They were stationed at the U.S. Sugarbeet Laboratory in Salt Lake City. These able scientists soon had a productive research program underway, and in 1948 Dr. V. F. Savitsky discovered monogerm seed in an American variety.

The activities of the Curly Top Resistance Breeding Committee were absorbed by the Beet Sugar Development Foundation in 1953. The foundation had been organized during World War II to work on sugarbeet mechanization and represented all the major sugarbeet companies in the United States. The support of U.S. Department of Agriculture sugarbeet research was expanded to include projects at each of its major sugarbeet laboratories and field stations. These projects are concerned with studies on diseases, bolting, nematodes, interspecific hybridization, breeding methods, sugarbeet quality, and the development of basic breeding lines. The industry support of these projects has steadily increased, and through the foundation the companies now contribute more than $100,000 each year.

Much of the sugarbeet variety development work in the United States is done by the sugar companies. In developing these varieties, the company breeders utilize breeding lines and methods developed by the U.S. Department of Agriculture. Each year the government breeders make promising lines available to the foundation members. These are often inbred or male-sterile lines that possess disease resistance and other needed characters. The company breeders can either use these lines directly as components of hybrid varieties or can incorporate desired characters into their own breeding lines.

A few companies do not have breeding programs and have provided additional funds to the U.S. Department of Agriculture for commercial variety development. The Union Sugar Division of Consolidated Foods Incorporated has supported variety work in California. The Farmers and Manufacturers Beet Sugar Association, an affiliation of both processors and growers, supports variety development for Michigan and Ohio. Varieties developed in these programs are available to anyone who wishes to use them.

In recent years, some grower organizations have also provided financial support to public agency research. The California Beet Growers Association supports research on virus yellows and problems associated with variety development in the Imperial Valley. The Red River Valley Sugarbeet Growers Association contributes to the state and federal program at Fargo, North Dakota. This research is directed toward the solution of pathological and physiological problems related to beet storage, and it contributes indirectly to breeding programs at other locations.

The joint efforts of industry, the federal government, and the state experiment stations have been most effective in accelerating the development of high-performing varieties. Without the financial support provided by industry, the scope of sugarbeet breeding research would have to be substantially reduced.

The cooperative effort has served to keep industry leaders and research workers in close contact with the public agency breeding programs. These frequent meetings have not only made industry aware of research progress but have also provided a stimulus to the research workers. Industry leaders responsible for seed propagation have been able to observe the performance of breeding lines during the early stages of their development. Lines of particular promise or interest to industry can be increased and tested in hybrid combinations over a wide range of conditions. Lines that perform well in these cooperative tests can then be used immediately as components in hybrid varieties.

This cooperative approach to research has speeded the development of monogerm seed, disease resistance, bolting resistance, and hybrid varieties. These accomplishments have demonstrated how the industry and the public agencies can effectively work together to solve research problems and increase the productivity of our food crops.

FUTURE

Plant breeders and geneticists have made notable improvements in the sugarbeet during the past half century but opportunities exist for even greater achievement. Many of the remaining problems are difficult and will require the combined efforts of well-trained scientists in many fields of biological science.

COMBINED RESISTANCE TO DISEASE

Progress has been made in developing varieties resistant to many of our most damaging diseases. A still higher level of resistance to the individual diseases would be desirable and combined resistance is needed. With the expansion of the sugarbeet industry to new areas in Texas and New Mexico, an especially urgent need exists for combined resistance to curly top and *Cercospora* leaf spot. The increased use of sprinkler irrigation in many western areas has tended to create a favorable environment for leaf spot, and combined resistance would also be useful in these areas.

Several pathogens are responsible for the damping-off of young seedlings. Some resistance has been found to *Aphanomyces cochlioides* but little progress has been made with seedling diseases caused by *Phoma betae, Rhizoctonia solani,* and *Pythium* species. The search for resistance to these pathogens needs to be accelerated and resistance to the individual pathogens should eventually be combined.

Sclerotium root rot caused by *Sclerotium rolfsii* is a serious problem in some beet-producing areas and some evidence of resistance has been observed. Resistance to this root rot should be developed and combined with resistance to other diseases. Varieties for the coastal valleys of California need combined resistance to downy mildew, rust, virus yellows, curly top, and bolting.

INSECT RESISTANCE

Little work has been done on insect resistance in sugarbeet. Differences in resistance to the green peach aphid *(Myzus persicae)* and the bean aphid *(Aphis fabae)* have been observed. These aphids are vectors of virus yellows, and there may be a possibility of reducing yellows losses by selecting for aphid resistance.

Sugarbeet breeding lines have shown variation in resistance to spider mites. These insects cause damage in some areas and an effort should be made to develop resistant varieties. No evidence of resistance to root-feeding insects such as maggots and wireworms has been reported.

IMPROVED PRODUCTION AND QUALITY

In the future, large genetic gains in productivity will probably require different breeding approaches. Some physiologists have suggested that higher sugar yields could be obtained by increasing photosynthesis. One approach would be to alter leaf canopies to expose greater leaf areas to sunlight. Genetic studies now underway should provide a better understanding of the inheritance of root yield and sucrose content. In general, a lower root yield has accompanied an improvement in sucrose content. Research by Dr. LeRoy Powers suggests that the breeder may be able to improve sucrose content and at the same time maintain root yield. This possibility deserves further study. There is undoubtedly a limit to the amount of sucrose that can be stored in the beet root without interfering with physiological processes.

Much research remains to be done on beet quality. We need more information on the relative importance of the various nonsugars that interfere with sugar extraction. The breeder can then study the heritability of these nonsugars and their relationship to one another. This information will help him breed varieties with better processing quality.

BREEDING FOR BETTER BEET STORAGE

Large quantities of beets are stored in piles to await processing. In some parts of the country, a portion of these beets may remain in the storage piles four months or longer. Sugar losses during these long storage periods are very high.

Major causes of these losses are respiration and molds or rots. The sugarbeet root is living tissue and continues to respire after harvest. Sucrose is oxidized and one of the end products is energy in the form of heat. This heat accelerates the breakdown of the beets. The molds and rots not only destroy beet tissue but also cause the conversion of sucrose into invert sugars.

Storage losses can be most effectively reduced through proper control of the storage environment. The breeder may be able to develop varieties that will also help reduce these losses. There is some evidence that varietal differences occur in the respiratory activity of harvested beets. Sound, healthy beets store better than those with rots. We need to know more about the identification of rots that occur when the beets enter storage and of the molds that develop during storage. The breeder may be able to develop resistance to some of these rots and molds.

BREEDING METHODS AND BASIC GENETIC STUDIES

A number of breeding methods are used in the United States and Europe. In this country, breeders have emphasized the development of hybrid varieties at the diploid level, whereas in Europe the greatest emphasis has been on polyploids. By combining the two breeding methods, the breeder can produce triploid varieties. The relative merits of diploid and triploid hybrids have not been resolved and additional research is required. Cytoplasmic male sterility is required for the production of hybrid varieties. The breeder needs additional basic information on male sterility and its inheritance.

Relatively little is known about the inheritance of root yield, sucrose content, disease resistance, and many other characters. Only a start has been made in working out the linkage groups for sugarbeet. This knowledge would be most useful to the breeder. Information is needed on the nature of disease resistance. If the breeder could select for or against some specific chemical entity or physiological process rather than for freedom from disease symptoms, progress would frequently be more rapid.

Interspecific hybridization offers a great challenge. Desired characters can be transferred from species in the Vulgares section by ordinary crossing and selection procedures. The transfer of characters from the Corollinae and Patellares sections will probably require chromatin transfers. Emphasis needs to be placed on these wide species crosses to obtain nematode resistance and other characters not readily available within *B. vulgaris*.

Sugarbeet geneticists may also wish to investigate other approaches to plant breeding. Mutation breeding has been proposed by several geneticists and has been used with some crops. Mutations can be produced by either chemicals or irradiation and provide additional opportunities to obtain disease resistance and other desired characters.

OTHER BREEDING OBJECTIVES

Much progress has been made in the development of selective herbicides for sugarbeets. Research by the Great Western Sugar Company has shown that the breeder can select for tolerance to herbicide damage. Several years would be required to develop tolerant varieties and the breeder would need assurance that the herbicide would still be in use when variety development was completed.

Highly productive monogerm varieties are available, but improvements are still needed in seed size and shape. Germination is frequently a problem and there is evidence that differences occur among varieties for this character. Monogerm seed capable of producing seedlings with greater vigor under a wide range of growing conditions would be desirable. Twin seedlings are sometimes produced from a single embryo. The breeder can select lines that are relatively free of this twinning characteristic. Eventually, the cytogeneticist may wish to investigate the transfer of the monogerm seed characteristics of some of the wild species to the sugarbeet.

Bolting-resistant varieties have been developed but additional resistance would be useful in many areas. More information is needed on the physiological factors responsible for bolting and seed production. There is evidence that breeding lines respond differently to the effects of temperature and light. The breeder may be able to develop varieties that bolt readily in the environment of the seed-growing area, and yet are highly bolting resistant in the sugar production areas.

Yellow wilt, a South American virus disease, offers a potential threat to our western sugarbeet industry. This disease is more destructive than either curly top or virus yellows, and resistant varieties have not been developed. Neither the virus nor the leafhopper vector occur in the United States, but the danger of their being introduced is great. The U.S. Department of Agriculture, the Beet Sugar Development Foundation, and the Industria Azucarera Nacional S.A. are participating in a cooperative research project in Chile. The purpose of the project is to search for sources of yellow wilt resistance and to make selections. A survey of a large number of varieties and breeding lines has thus far failed to uncover a good source of resistance.

Soil salinity is a problem in the Imperial Valley and some other western areas. The sugarbeet is salt tolerant after the plants are well established but losses can occur during germination and when the plants are small. Attempts are being made to breed for tolerance during this early growth period.

The growing season is short in some parts of the United States and we need varieties capable of forming a harvestable root in a short period of time. Resistance to subfreezing temperatures during the seedling stage would also be a valuable character.

This listing of unsolved problems serves to remind us that the sugarbeet is one of our newer food crops. Although much has been accomplished since Franz Achard made his first selections from the fodder beet, we still have a great opportunity to increase the productivity of the crop.

REFERENCES

1. Achard, F. C. Anleitung zum Anbau der zur Zuckerfabrication anwendbar Runkelruben und zur vortheilhaften Gewinnung des Zuckers aus denselben. 1803. (Reprinted in Ostwald's Klassiker der exakten Wissenschaften 159. Engelmann, Leipzig. 1907.)
2. Bennett, C. W. Sugar beet yellows disease in the United States. USDA Tech. Bull. 1218. 1960.
3. Brown, Robert J. and R. F. Serro. A method for determination of thin juice purity from individual mother beets. Proc. Am. Soc. Sugar Beet Technologists 8(2):274–78. 1954.
4. Carruthers, A. and J. F. T. Oldfield. Methods for the assessment of beet quality. III. Summation of non-sugars. Intern. Sugar J. 63:137–39. 1961.
5. Carsner, Eubanks. Curly-top resistance in sugar beets and tests of the resistant variety U.S. No. 1. USDA Tech. Bull. 360. 1933.
6. Coe, Gerald E. and C. L. Schneider. Selecting sugar beet seedlings for resistance to *Aphonomyces cochlioides*. J. Am. Soc. Sugar Beet Technologists 14:164–67. 1966.
7. Coons, G. H. Improvement of the sugar beet. USDA Yearbook Agr., pp. 625–56. U.S. Govt. Printing Office, Washington. 1936.
8. ———. The wild species of *Beta*. Proc. Am. Soc. Sugar Beet Technologists 8(2):142–47. 1954.
9. Doney, D. L. and E. D. Whitney. Screening sugarbeet for resistance to *Heterodera schachtii* Schm. J. Am. Soc. Sugar Beet Technologists 15:546–52. 1969.
10. Doxtator, C. W. and A. R. Downie. Progress in breeding sugar beets for resistance to *Aphanomyces* root rot. Proc. Am. Soc. Sugar Beet Technologists 5:130–36. 1948.
11. Duffus, James E. Economic significance of beet western yellows (radish yellows) on sugar beet. Phytopathology 51:605–7. 1961.
12. Gaskill, John O. Induction of reproductive development in sugar beets by photothermal treatment of young seedlings. Proc. Am. Soc. Sugar Beet Technologists 7:112–20. 1952.
13. ———. Breeding for *Rhizoctonia* resistance in sugarbeet. J. Am. Soc. Sugar Beet Technologists 15:107–79. 1968.
14. Giddings, N. J. Combination and separation of curly-top virus strains. Proc. Am. Soc. Sugar Beet Technologists 6:502–7. 1950.
15. Harrison, Merle, M. G. Payne, and J. O. Gaskill. Some chemical aspects of resistance to *Cercospora* leaf spot in sugar beets. J. Am. Soc. Sugar Beet Technologists 11:457–68. 1961.
16. Hecker, R. J., G. W. Maag, and M. G. Payne. Inheritance of 3-hydroxytyramine in sugarbeet: a phenolic compound associated with *Cercospora* leaf spot resistance. J. Am. Soc. Sugar Beet Technologists 16:52–63. 1970.
17. Helmerick, R. H., R. E. Finkner, and C. W. Doxtator. Paired-plant crosses in sugar beets. J. Am. Soc. Sugar Beet Technologists 13:548–54. 1965.
18. Henderson, R. W. and H. W. Bockstahler. Reaction of sugar beet strains to *Aphanomyces cochlioides*. Proc. Am. Soc. Sugar Beet Technologists 4:237–45. 1946.
19. Johnson, Russell T. A rapid method of making a non-bolting selection in sugar beets. Proc. Am. Soc. Sugar Beet Technologists 8:84–87. 1954.
20. McFarlane, J. S. Breeding for resistance to downy mildew in sugar beets. Proc. Am. Soc. Sugar Beet Technologists 7:415–20. 1952.

21. McFarlane, J. S., I. O. Skoyen, and R. T. Lewellen. Development of sugarbeet breeding lines and varieties resistant to yellows. J. Am. Soc. Sugar Beet Technologists 15:347–60. 1969.
22. Marggraf, A. S. Chymische Versuche, einen wahren Zucker aus verschiedener Pflanzen, die in unseren Ländern wachsen, zu ziehen. 1747. *In* Chymischen Schriften, Theil 2, pp. 70–85. Berlin. 1767. (Reprinted in Ostwald's Klassiker der exakten Wissenschaften 159. Engelmann, Leipzig. 1907.)
23. Nielson, Kent, D. Quayle, and S. Bergeson. Comparison of sugarbeet single cross hybrids with double cross hybrids for seed and root characteristics. J. Am. Soc. Sugar Beet Technologists 15:726–30. 1970.
24. Oldemeyer, R. K., W. H. Davis, H. L. Bush, and A. W. Erichsen. The evaluation of and the use of the top-cross test as a method of selecting inbred lines of sugarbeets for general combining ability. J. Am. Soc. Sugar Beet Technologists 15:49–60. 1968.
25. Owen, F. V. Inheritance of cross- and self-sterility and self-fertility in *Beta vulgaris*. J. Agr. Res. 64:679–98. 1942.
26. ———. Cytoplasmically inherited male-sterility in sugar beets. J. Agr. Res. 71:423–40. 1945.
27. ———. Hybrid sugar beets made by utilizing both cytoplasmic and Mendelian male sterility. Proc. Am. Soc. Sugar Beet Technologists 8:64. 1954.
28. Owen, F. V., E. Carsner, and M. Stout. Photothermal induction of flowering in sugar beets. J. Agr. Res. 61:101–24. 1940.
29. Payne, Merle G., R. J. Hecker, and G. W. Maag. Relation of certain amino acids to other impurity and quality characteristics of sugarbeet. J. Am. Soc. Sugar Beet Technologists 15:562–94. 1969.
30. Powers, LeRoy, R. E. Finkner, G. E. Rush, R. R. Wood, and D. F. Peterson. Genetic improvement of processing quality in sugarbeets. J. Am. Soc. Sugar Beet Technologists 10:578–93. 1959.
31. Price, Charles. Breeding sugar beets for resistance to the cyst nematode *Heterodera schachtii*. J. Am. Soc. Sugar Beet Technologists 13:397–405. 1965.
32. Pritchard, Frederick J. Some recent investigations in sugar-beet breeding. Botan. Gaz. 62:425–65. 1916.
33. Savitsky, Helen. Viable diploid, triploid and tetraploid hybrids between *Beta vulgaris* and species of the section Patellares. J. Am. Soc. Sugar Beet Technologists 11:215–35. 1960.
34. ———. Meiosis in hybrids between *Beta vulgaris* L. and *Beta corolliflora* Zoss. and transmission of resistance to curly top. Proc. 12th Intern. Congr. Genetics 1:180 (Abstr.). 1968.
35. Savitsky, Helen, and C. Price. Resistance to the sugar beet nematode *(Heterodera schachtii)* in F_1 tetraploid hybrids between *Beta vulgaris* and *Beta patellaris*. J. Am. Soc. Sugar Beet Technologists 13:370–73. 1965.
36. Savitsky, V. F. Monogerm sugar beets in the United States. Proc. Am. Soc. Sugar Beet Technologists 6:156–59. 1950.
37. Schneider, C. L. Methods of inoculating sugar beets with *Aphanomyces cochlioides* Drechs. Proc. Am. Soc. Sugar Beet Technologists 8:247–51. 1954.
38. Sears, E. R. The transfer of leaf-rust resistance from *Aegilops umbellulata* to wheat. *In* Genetics in Plant Breeding, pp. 1–22. Brookhaven Symposium in Biol. 9. Brookhaven National Laboratory, Upton, N.Y. 1956.
39. Shepard, James H. Sugar beet culture in South Dakota. S.Dak. Agr. Exp. Sta. Bull. 142:163–84. 1913.
40. Stewart, Dewey, J. O. Gaskill, and G. H. Coons. Heterosis in sugar beet single crosses. Proc. Am. Soc. Sugar Beet Technologists 4:210–22. 1946.
41. Theurer, J. C. and G. K. Ryser. Double-cross sugar-beet hybrids utilizing pollen restorers. Crop Sci. 9:610–12. 1969.
42. Vilmorin, J. L. de. L'Hérédité chez la Betterave Cultivaée. Gauthiers-Villars, Paris. 1923.

15

Seed Production

SAM C. CAMPBELL
West Coast Beet Seed Company
Salem, Oregon

A. A. MAST
Western Seed Production Corporation
Phoenix, Arizona

HISTORY	438
ORGANIZATION	440
SCOPE	440
TYPES OF SEED PRODUCTION	441
REPRODUCTIVE DEVELOPMENT	442
CLIMATE AND SOIL	442
AGRONOMIC PRACTICES	443
GROWTH HABIT	444
HARVESTING, CLEANING, AND PROCESSING	445
PLANTING METHODS USED IN PRODUCTION OF HYBRID SEED	446
MAINTENANCE OF VARIETAL PURITY	447
FERTILIZER PRACTICES	448
WEED CONTROL PRACTICES	448
INSECTS AND DISEASES	448
SEED GERMINATION	449
LIMITATIONS AND FUTURE	449

HISTORY

PRIOR to World War I, all sugarbeet seed planted for sugar production in the United States was imported from Europe. During that war, the American beet industry experienced extreme difficulty in obtaining adequate supplies of seed, and of necessity, embarked on a seed-producing enterprise of its own, using the "steckling," or transplanting, method as practiced in Europe. This method of seed production, which is still practiced on a large scale in Europe, involves the growing and digging of roots in one season, siloing or storing in root cellars over winter, and transplanting to the field in the spring. During the five-year period 1916 through 1920, the American industry produced over five million pounds of seed annually by this method, but because of production problems and high labor costs they turned again to European supplies after the war.

Our present seed production enterprise had its beginning at the New Mexico Agricultural Experiment Station at Las Cruces, New Mexico. In the early 1920s, plantings of sugarbeets were being made at the station each month of the year to determine the best planting date for the production of a root crop. It was found that sugarbeets planted in the late summer or early fall survived the comparatively mild winters in southern New Mexico. With the advent of spring growing weather, the beet plants sent up seedstalks and produced a seed crop in the summer. These observations led immediately to an intensive research program, conducted by the Division of Sugar Plant Investigations and the New Mexico Agricultural Experiment Station, to determine the possibilities of growing seed by leaving the plants in the ground over winter. The results of the experimental work led to the conclusion that seed could be produced economically by this new method. Thus was born the "overwintering" method of producing sugarbeet seed as now practiced throughout the North American continent.

The first commercial planting by the overwintering method was made near Las Cruces, New Mexico, in the fall of 1927 for the American Beet Sugar Company. A good yield of excellent seed was harvested the following year. Encouraged by the initial success of this venture, other sugar companies, including Great Western Sugar Company, Amalgamated Sugar Company, and Holly Sugar Corporation, began trial plantings in the area. Experimental work was also started in southern Utah and southern California and it soon became evident that several areas of the Southwest offered promise of seed production.

The pioneering efforts of these sugar companies, in close cooperation with the Division of Sugar Plant Investigations and the New Mexico Agricultural Experiment Station, established the foundation for the domestic seed enterprise. Added impetus was provided when the U.S. Department of Agriculture developed a new strain of sugarbeets resistant to curly top disease, which had ravaged certain sugar-producing areas in the West. Elite seed of this new variety, U.S. No. 1, was made available for increase in 1931. Seed was apportioned among several sugar companies in whose areas curly top disease was most serious. Plantings of approximately five acres each were made by Amalgamated Sugar Company in New Mexico, Utah-Idaho Sugar Company at St. George, Utah, and Spreckels Sugar Company near Hemet, California. Production from these plantings amounted to approximately 23,000 pounds of stock seed which became available for large-scale commercial plantings in 1932. In addition, the sugar companies began production of new seed varieties of their own which were considered better adapted to their beet-growing areas than varieties then in use.

Agricultural statistics for 1932 show that a total of 62 acres of sugarbeet seed from fall plantings was harvested in the states of New Mexico, Utah, and California. Production amounted to approximately 131,000 pounds. As more companies recognized the advantages of a domestic seed enterprise, there followed a spectacular increase in production. By 1935 a total of 1,692 acres was harvested, including the first trial planting of 10 acres in the Salt River Valley of Arizona. By 1938 acreage was being contracted in the states of Nevada, Washington, and Oregon. In that year the harvested acreage had increased to 7,854 acres and production exceeded 13 million pounds. For the first time, domestic seed production exceeded imports, and by the advent of World War II, the American beet sugar industry was no longer dependent on foreign seed (Fig. 15.1).

FIG. 15.1. Solid planting of sugarbeet plants for seed production. Notice height of seedstalks.

ORGANIZATION

In 1937 the four major companies involved in the development of the seed enterprise in New Mexico and Arizona formed Western Seed Production Corporation, with headquarters at Phoenix, Arizona. These companies included American Crystal Sugar Company, Amalgamated Sugar Company, Great Western Sugar Company, and Holly Sugar Corporation. The development of the seed industry in California and later in Oregon created interest among other sugar companies in forming a second seed-growing agency. In 1940 West Coast Beet Seed Company was formed at Salem, Oregon, as a subsidiary of eight major beet sugar companies. In addition to the four companies involved in the Arizona corporation, the new organization included Spreckels Sugar Company, Union Sugar Company, Utah-Idaho Sugar Company, and Farmers and Manufacturers Beet Sugar Association. The Utah-Idaho Sugar Company, while a member of West Coast Beet Seed Company, had organized its own seed-producing agency, with operations at St. George, Utah.

The membership of these organizations today is essentially the same as it was at their inception.

The seed-producing organizations are charged with the production of seed as required by its member companies. Orders are placed for the quantities of seed desired and the seed agencies contract with farmers for the production.

SCOPE

A review of agricultural statistics reveals that the seed requirements of the industry during the period up to the mid-1940s ranged from 16 to 20 million pounds annually. By 1941 domestic production had increased to 18 million pounds, which was sufficient to meet the industry requirements. During and after World War II, production was maintained at a level above domestic requirements, and large quantities of seed were exported to countries whose production facilities had been disrupted due to the war.

As European countries reestablished their seed-producing areas, imports from the United States were sharply curtailed. During the same period, the American industry had developed seed-processing techniques which led to a significant reduction in planting rates in sugarbeet fields. These two factors, primarily, were responsible for reducing annual seed requirements to about 10.5 million pounds which, today, satisfy the needs of the industry.

The drastic reduction in seed requirements brought about the abandonment of many seed-growing localities in several states. As farmers gained experience in seed production, average seed yields increased and caused a further decline in acreage requirements. At the present time, the only areas used for seed production are those immediately adjacent to the headquarters of the seed-growing agencies.

In Arizona the bulk of the seed is produced in the Salt River Valley, an area comprising approximately 1,300 square miles. The Utah acreage is limited to an area near St. George in Washington County. In Oregon, most of the seed is produced in the Willamette Valley in northwestern Oregon. The growing area in the Willamette Valley covers approximately 2,500 square miles.

Seed yields in Arizona and Utah usually run 3,000 pounds or more per acre. Yields in Oregon are generally lower, averaging about 2,400 pounds per acre.

Yield statistics for the three areas during the past ten years are shown in Table 15.1.

TYPES OF SEED PRODUCTION

From its beginning until the early 1950s, seed production consisted entirely of open-pollinated multigerm lines. The development of male-sterile lines in the mid-1950s led immediately to the production of large quantities of multigerm hybrid seed. These new hybrid varieties generally were superior to the open-pollinated varieties and quickly replaced them in most sugar-producing areas.

The discovery of a monogerm seed plant by Dr. V. F. Savitsky in 1948 created intense interest among plant breeders in the industry. During the early 1950s, seed breeders, while engaged in the development of multigerm hybrid varieties, placed major emphasis on the development of both open-pollinated and male-sterile monogerm lines. Commercial production of monogerm seed was initiated in 1956, and by 1960 nearly 50 percent of the total seed crop consisted of monogerm varieties. During the past 10 years, monogerm hybrids have gradually replaced multigerm hybrids, so that today, nearly 100 percent of all commercial seed produced consists of monogerm hybrid varieties.

The seed produced in Arizona and Utah is limited, primarily, to commercial seed. In Oregon the production consists of large quantities of commercial seed and also probably 90 percent of the foundation seed used in all the seed-growing areas. In addition, many dozens of experimental hybrids, used by the sugar companies in variety-testing trials, are also produced.

TABLE 15.1. Average harvested acreage, yield per acre, and sugarbeet seed production per year for the United States during the past ten years

	Avg. Acres Harvested	Avg. Yield per Acre	Avg. Total Production
		lb	lb
Arizona	1,566	3,116	4,880,000
Oregon	2,092	2,404	5,029,000
Utah	200	3,263	653,000
	3,858	8,783	10,562,000

REPRODUCTIVE DEVELOPMENT

In the production of seed, if a variety is to retain all of its characteristics, it is essential that it be planted under conditions which help insure complete and full reproduction. The sugarbeet plant is a biennial, and photothermal induction is necessary to bring about complete reproductive development. The plant must be exposed to a period of cool temperatures, followed by one in which the daylight hours are long. The reproductive process is controlled by genetic factors that determine bolting tendency, or seedstalk formation, in the plant. Bolting tendency varies widely among sugarbeet varieties. In the trade, seed varieties are usually referred to as either "easy bolting" or "nonbolting."

In the case of easy bolting varieties, comparatively light cold exposure is adequate to bring the plants to seed. The nonbolting varieties require longer periods of cold temperatures to insure complete reproduction.

CLIMATE AND SOIL

A wide range of climatic conditions exist in the present seed-growing areas. In Arizona and Utah the winters are comparatively mild and the summers long and hot. Precipitation averages slightly over 7 inches in Arizona and 7.9 inches in Utah. The winters in Oregon are longer and cooler and the summers usually mild. Precipitation averages 40 inches, most of which falls during November through March. Snowfall occurs only occasionally in the Utah area but never in Arizona. The Oregon area is more subject to snow, usually 6 to 10 inches annually. Normally, the lowest winter temperatures in Oregon fall in the range of 15° to 20° F, while in Utah and Arizona the low range is usually 25° to 30° F. Winter injury to beets due to freezing conditions never occurs in Arizona and only occasionally in Utah. In Oregon a number of seed crops have been partially lost due to excessive cold conditions.

Because of the mild winter climate in Arizona and Utah, only seed varieties of easy to moderate bolting tendency are grown. The seed produced in these areas is delivered to factory districts which practice spring planting and fall harvesting for sugar production.

In Oregon the longer and colder winters favor the induction of bolting-resistant seed types, and much of the seed produced is used in the sugarbeet areas of California where the crop is planted during the fall and winter months. Easy bolting varieties also are grown in Oregon and the seed produced is utilized for spring plantings in certain beet areas.

Soil types utilized for seed production in Arizona and Utah are predominantly silty clay loams. In Oregon most of the seed is produced on sandy loam soils and the balance on heavier loam types. Good drainage is an important consideration in selecting seed fields in Oregon, due to heavy fall and winter rains.

AGRONOMIC PRACTICES

In all areas of seed production, fields are carefully selected for growing of the crop. Consideration is given to the grower's crop rotation practices, soil fertility, weed potential, irrigation facilities, and equipment.

Planting date is especially important in the production of the crop. It varies slightly in each growing area but the primary consideration is to plant as early as is necessary for optimum root size for overwintering and for reproductive development.

The seed crop is planted with whole (unprocessed) seed. Seeding rates in Arizona and Oregon approximate 10 to 12 pounds per acre, while in Utah the rate is 15 to 18 pounds.

Irrigation at planting time is necessary in all areas of production. Furrow irrigation is practiced in Arizona and Utah, but in Oregon the uneven topography of the land necessitates the sprinkler method of irrigation.

In Arizona the crop is planted during the first two weeks of September. It usually follows early harvested barley in the rotation, and seedbed preparation begins with plowing down stubble in early June, irrigating and fallowing through the summer. Fertilizer is applied and worked into the ground prior to planting. The beets are planted on 40-inch beds, 2 rows per bed, and irrigated immediately for germination.

In Utah seedbed preparation starts with plowing down alfalfa cover crops. The land is irrigated and fallowed through the early summer for weed control. In early August, fertilizer is applied and the crop is planted in late August and early September. Planting is on beds with row spacings 14 to 20 inches, depending on each grower's planting and cultivating equipment.

In Oregon, beets are planted during the period August 1 to about September 15. The land selected may be fallowed, or it may be land from which other early-maturing crops have been harvested in time to permit proper soil preparation. After preparation of the seedbed and application of fertilizer, the land is irrigated. Immediately after irrigation and subsequent drying of the topsoil, the land is worked lightly and planted. The crop is planted on level land, as opposed to bed plantings in the other areas, in rows 24 inches wide. Depth of planting varies, depending on moisture, but most of the seed is planted $3/4$ to 1 inch deep.

The comparatively thick stands of beets resulting from the heavy seeding rates are similar to those obtained for root production prior to the use of processed seed. The plants are left unthinned, primarily for the purpose of providing heavy foliage cover to suppress weed growth. In Arizona and Utah the shading effect of heavy foliage during the winter months helps to maintain cooler soil and beet crown temperatures which enhance bolting. In Oregon heavy foliage affords some protection of the beet root from winter injury.

The seed crop is cultivated as soon as possible after the beets emerge and the rows are clearly marked. Cultivations are continued through the fall growing season as soil conditions permit.

In Arizona and Utah frequent irrigations are required throughout the fall growing season. In Oregon irrigations usually are required during August and September but are no longer necessary after fall rains begin.

GROWTH HABIT

As winter approaches, the seed crop becomes partially dormant in Arizona and Utah and almost completely dormant in Oregon. Under normal winter conditions, the beet tops will frost down, depending on the amount of cold weather experienced in the particular seed area. In Oregon, because of its colder temperatures, the older foliage growth usually freezes down completely, leaving only the short, green leaf growth in the crown of the plant. In the other growing areas, loss of leaves is less severe, due to the milder temperatures.

Root size, before dormancy, is dependent on a number of factors. The most important of these include length of growing season, care of the growing crop, soil fertility, and moisture. In Arizona, under normal growing conditions, most plants will attain a root size of $3/4$ to $1\frac{1}{2}$ inches at the crown. In Utah average plant size is slightly less than in Arizona. Under Oregon conditions, most plants will attain a root size of $3/8$ to $3/4$ inch at the crown, which is considered near optimum size for overwintering.

Sustained periods of temperatures down to 20° F are common in Oregon but it is unusual to lose beets to winter injury in this temperature range except the very large plants. Very small plants are sometimes lifted from the ground due to freezing and thawing action. Temperatures under 10° F, with no snow cover, have caused partial or complete loss of seed crops.

Dormancy of the seed crop occurs for a period of two to three months, depending on locality. In Arizona and Utah growth usually resumes in early February. In Oregon no appreciable growth is apparent until early March. Spring temperatures climb rapidly in Arizona and Utah and enhance crop development in these areas. In Oregon the crop develops more slowly, due to cooler spring temperatures and cloudy, wet conditions.

Bolting of the Arizona crop usually commences in late March, and in Utah, about 10 days later. The difference in crop development in these two areas generally is maintained throughout the balance of the growing season. In Oregon the easy bolting varieties begin seedstalk formation in late April. The bolting-resistant varieties develop more slowly and are usually 10 days behind the easy bolting types.

The growth of a seed plant after it begins to bolt is phenomenal. Usually, within six to seven weeks, the seedstalks have reached a height of six to nine feet, with heavy lateral branching from the main stalk.

In all seed areas, the first blooms appear five to six weeks from the time seedstalks begin to form. The full-bloom stage of development is reached approximately two weeks later. The crop, thereafter, begins to mature and is ready for harvest about six weeks after the full-bloom stage.

As the seed crop approaches maturity, it usually takes on a straw color. The coloring is not as pronounced in monogerm seed types, particularly bolting-resistant monogerms, as in the multigerm varieties. Definite changes in color of the seedstalk and foliage, and the amount of shattering of early maturing seed, are factors to look for in determining the proper time for harvesting the crop. In the bolting-resistant varieties, maturity is more difficult to determine. These types remain quite green in color for longer periods and closer inspection of the crop is desirable. In such instances, maturity can be determined by dissecting average seedballs from representative branches on several plants to observe the stage of development of the endosperm. The seedballs may be quite green in appearance and high in moisture content, but if the true seed is in a moderate to hard dough stage, the crop is ready to harvest.

HARVESTING, CLEANING, AND PROCESSING

In Arizona the crop usually is ready for harvest about June 15 and in Utah about June 25. In Oregon harvesting of the easy bolting varieties begins about August 1 and the later-maturing, bolting-resistant varieties about August 10. The seed is cut with either factory-made or specially constructed windrowers, with vertical sickle attachments to permit cutting through the tangled mass of branches. The crop is swathed into three- to six-row windrows, depending on the size combines to be used in threshing. Drying time between the cutting and threshing operation varies by seed area. In Arizona and Utah the crop usually dries sufficiently to thresh in 7 to 10 days, while in Oregon 10 to 14 days are required. The threshing operation is accomplished in all areas with large, self-propelled combines which can thresh five to seven acres per day (Fig. 15.2).

FIG. 15.2. Combine harvesting of cut, windrowed sugar-beet seedstalks.

The seed is delivered to the seed company warehouse and cleaning plants for storage and final cleaning. The cleaning operation consists of passing the seed through air separators and other equipment to remove trash and to size the seed in accordance with contract specifications. The grower is paid on the basis of the "cleaned seed" delivered under his contract, and which meets the requirements of germination and purity.

After the cleaned seed weight is determined, much of the seed is partially processed at the seed warehouses. The processing operation consists of passing the seed through equipment designed to polish the seed, or remove by grinding action a portion of the rough, corky material surrounding the true seed. The purpose of the operation is to reduce the weight of the whole seed and to make the product more uniform in size.

A partial processing operation, which is practiced at the seed warehouses in Arizona and Oregon, reduces whole seed weight from 20 to 25 percent. The seed then is delivered to the sugar companies who further process the seed to their own specifications and prepare it for distribution to their sugarbeet growers. In Utah the processing operation is completed entirely at the seed company's plant.

PLANTING METHODS USED IN PRODUCTION OF HYBRID SEED

Two methods of planting are employed in the production of commercial seed hybrids. One predominant method is to plant a mixture of the monogerm male sterile and the multigerm pollinator wherein 5 to 10 percent of the mixture is pollinator. The pollinator seed usually is larger than the monogerm, and caution must be exercised in planting such mixtures to obtain uniform distribution of pollinator seed in the row.

The other method of growing hybrid seed is to plant alternate strips of the male-sterile line and the pollinator. Two blank rows between the strips are left so that harvest can be accomplished without mixing the two lines. Usually, the ratio of pollinator rows to male sterile is 4 to 12, or 4 to 16, depending on the pollen-producing ability of the pollinator. The two blank rows between the strips, or a space of six feet, usually is sufficient to keep the two lines separated. If some plants on the outside rows of each strip fall against each other, it is necessary, before harvesting, to put back the lodged seedstalks by hand to confine them to the strips in which they were planted. Where the pollinator seed is not to be used, it is eliminated by discing down or tilling about three weeks before harvest of the male-sterile line (Fig. 15.3).

The strip planting method is used in instances where the pollinator may be a weak pollen producer, or its growth habit such that it will not adequately compete in a mixed planting with more vigorous male-sterile lines.

Individual parents in hybrid seed combinations are critically examined for a period of two years or more prior to planting for commercial produc-

FIG. 15.3. Four-row strips are pollinator rows, while sixteen-row strips are male-sterile seed-bearing parent line.

tion. Each year, variety observation plots are planted in the seed-growing areas to assess such factors as growth habit, vigor, pollen production, male sterility, and bolting tendency of many seed lines.

MAINTENANCE OF VARIETAL PURITY

In the production of seed, the primary concern is to produce seed as free of contamination as possible. It is well known that wind-borne pollen will carry long distances, and in planting a seed field, consideration must be given to its location in relation to other seed fields or sources of pollen. In the selection of fields, efforts are made to separate plantings of related seed types by distances of one mile or more. Fields of unrelated seed types are usually separated by two miles or more, taking into consideration terrain and prevailing winds. In the production of commercial seed, which is not again increased, a slight admixture of foreign pollen is not considered critical. In the production of elite, or stock seed, which will be increased, caution must be exercised to prevent contamination by stray pollen and by any other source.

To avoid contamination of seed by mechanical means, close supervision is necessary in the planting, harvesting, and cleaning of seed of different varieties.

FERTILIZER PRACTICES

The overwintered seed crop uses two distinct seasons of growth and requires fertilizer treatments in each season. In Arizona the standard practice is to apply 50 pounds of nitrogen and 60 pounds of phosphate prior to planting. An additional 100 pounds to 150 pounds of nitrogen are side-dressed during the fall. In the spring, 200 pounds to 300 pounds of nitrogen are side-dressed or applied to the crop through the irrigation water.

In Utah the fall treatment consists of applying 240 pounds of nitrogen and 125 pounds of phosphate prior to planting. In the spring the crop is side-dressed with 60 pounds of nitrogen.

In Oregon the standard fall treatment is 75 pounds of nitrogen, 100 pounds of phosphate, 50 pounds of sulfur, and 5 pounds of boron. In the spring, 160 pounds of nitrogen are applied, usually in two applications. Additional nitrogen may be applied on an individual farm basis.

WEED CONTROL PRACTICES

In Arizona no effective herbicides have been found suitable, but extensive tests of a number of compounds are underway. As a rule, spring cultivation of the seed crop is not feasible because of heavy foliage growth. Winter weeds, if present, are chopped back just prior to bolting. Rapid growth of the bolting plants soon shades most of the weeds to a point where they are largely eliminated.

In Utah cultivation in the early fall, followed by hand hoeing, is the principal method of weed control. Herbicides have been used on a small scale in the area.

In Oregon the major herbicides in use include Eptam, Roneet, IPC, and Endothal. Eptam and Roneet are applied as a preplant treatment at the rate of 3 pounds to 4 pounds of active ingredient per acre. Both herbicides generally provide good control of grasses and most broadleaf plants during the fall growing season.

IPC is a very effective herbicide for the control of grasses and common chickweed. It is applied at the rate of 6 pounds of active ingredient per acre, usually in late fall. Endothal is used as a contact spray for specific weeds. It is applied during the winter months when the beet crop is dormant. An application rate of 4 pounds of active ingredient per acre gives good control of broadleaf weeds and retards growth of grasses.

A number of other compounds are being tested through cooperative work with Oregon State University.

INSECTS AND DISEASES

In Arizona high temperatures and accompanying low rainfall provide ideal conditions for many injurious insects. Those present in the fall which require control are the beet leafhopper, the vector for curly top disease, and the beet armyworm. The green peach aphid, vector for virus yellows disease, infests seed fields in the late fall and must be controlled to minimize the spread of that disease. At the time of flowering, lygus bug control is mandatory. Usually, two to three treatments are required to protect the crop against damage to germination. Spider mites also are a serious problem in many fields and must be controlled to prevent desiccation and resulting reduction in seed yield.

The most commonly used insecticides are the organic phosphate compounds which do not present soil residue problems.

In Utah infestation of insects during the fall are rare. Occasionally, the beet armyworm infests seed fields, requiring control measures. In the spring the area is subject to infestation of black bean aphids, green peach aphids, Says green plant bugs, and lygus bugs. These insects, especially the plant bugs and lygus bugs, require several treatments for their control.

In Oregon lygus bugs are considered the most harmful insects attacking seed beets. Control measures are a standard practice. Other insects which occasionally infest fields are the green peach aphid, spider mite, leaf cutworm, and spittlebug. Control measures are taken as required.

Diseases most prevalent in Oregon are *Ramularia* and *Phoma* leaf spot. These diseases occur during the late fall, but as the beet leaves wither away during the winter, the infection is greatly reduced. In the more susceptible seed varieties, leaf spot appears again in early summer. In some instances it will almost completely defoliate fields by harvest time. Control measures have not been effective.

Western yellows virus occurs occasionally in the seed area, but as yet it is not considered damaging to the crop. Other diseases of minor importance are downy mildew and mosaic.

SEED GERMINATION

Seed germination probably is one of the more critical problems associated with seed production. During the years of multigerm seed production, no particular germination problem was encountered, unless the seed was damaged by insects. When monogerm seed came into production, it was obvious that germination would be reduced, due to the nature of the single germ seed type.

Generally, commercial multigerm seed germinated in the range of 85 to 95 percent in all seed areas. By comparison, most monogerm seed lots germinate from 70 to 85 percent. As the industry continues to reduce planting rates, or plant to a stand, it becomes more important that every seed planted should germinate.

LIMITATIONS AND FUTURE

Sugarbeet seed production is very closely tied to the requirements of the beet sugar industry. Such requirements will be influenced by the acreage of beets planted and the amount of seed planted per acre. There appears to be a continuing trend toward lighter seeding rates, and in some instances, planting to a stand. Should this procedure become a general practice, seed requirements will decrease still further.

REFERENCES

1. Overpeck, J. C., W. B. Morrow, A. A. Mast. History of the Sugarbeet Seed Industry in Arizona and New Mexico. Western Seed Production Corp., Phoenix, Ariz. 1955.
2. Tolman, B. Sugarbeet seed production in southern Utah, with special reference to factors affecting yield and reproductive development. USDA Tech. Bull. 845, 1943.
3. Tolman, B. and C. H. Smith. Sugar beet seed growing in Utah. Utah Agr. Ext. Bull. 118, 1943.

16

Economics of Production

BION TOLMAN
Utah-Idaho Sugar Company
Salt Lake City, Utah

CHARLES WHIPPLE
Utah-Idaho Sugar Company
Salt Lake City, Utah

Costs Involved in Production 452

Effect of Sugar Content on Price per Ton 454

Effect of Yield on Costs and Returns . 454

Trends in Acreage and Yields 456

Trends in Man-Hour Requirements . . 457

Beets as a Primary or Secondary Crop in a Cropping System 458

THROUGH the years, sugarbeets have earned the reputation of being the contract crop from which the return was sufficiently stable that it could be counted on to buy the land, buy the machinery, pay the taxes, and provide a living for the farm family.

In recent years, young men have found it possible to specialize in sugarbeet production and become "professional beet growers," renting land from landowners who wanted beet production in their crop rotation program. These young men have been able to acquire a full line of beet machinery and gradually expand farming operations to the point where they are involved in the production of several hundred acres of sugarbeets.

COSTS INVOLVED IN PRODUCTION

Successful beet farmers soon learn that profits are determined by three principal factors: control of production costs, yield level, and sucrose content.

Production costs vary rather widely from farm to farm and area to area. Size of farming operation is an important factor. Sugarbeet production should be on a scale that will efficiently utilize the machinery needed for production of the crop.

For example, a one-row harvester will harvest the beets on 80 acres, and a two-row harvester will harvest beets on 150 to 160 acres. A farmer who has 40 acres of beets should not own a two-row harvester unless he intends to do some custom work for a neighbor so that the harvester can be utilized efficiently. This same principle applies to planting, cultivating, and mechanical thinning equipment. The balance can generally be achieved by doing custom work or by hiring some custom work.

Weed control is one of the most costly operations in row crop production. This cost varies from less than $20 per acre to as much as $100 per acre. Economical weed control is largely a farm management problem and involves recognition of the fact that weed control must be practiced in all crops.

Whenever one sets himself to the task of approximating total crop production costs, he is immediately confronted with making decisions as to the approach to the problem. One approach is to calculate machinery costs, including maintenance, interest, depreciation, operating costs, labor, etc., plus fixed overhead costs, such as water cost, interest on land investment, taxes, interest on operating capital, and management fees.

Another approach is to base all operations other than fixed overhead costs on a custom-rate basis. The custom rate includes interest, depreciation, operating expense, and labor costs for the various operations. All approaches, including survey data, have been considered in the production costs presented in this chapter. Sources of cost data considered are listed under references 2 to 11.

In Table 16.1 the various items of expense are shown down the left

TABLE 16.1. Comparative beet production costs of various areas

Item or Material	Comparative Production Areas				
	1	2	3	4	5
Preplanting					
Plowing	$ 6.00	$ 6.00	$ 6.00	$ 6.00	$ 5.00
Discing and harrowing	7.00	7.00	4.00	4.00	4.00
Leveling or floating	4.00	4.00	4.00	4.00	4.00
Fertilizing	30.00	32.00	20.00	20.00	18.00
Subtotal	$ 47.00	$ 49.00	$ 34.00	$ 34.00	$ 31.00
Planting					
Seed	$ 3.00	$ 3.00	$ 3.00	$ 3.00	$ 3.00
Planting	3.00	3.00	3.00	3.00	3.00
Herbicide	7.50	7.50	7.50	7.50	7.50
Subtotal	$ 13.50	$ 13.50	$ 13.50	$ 13.50	$ 13.50
Growing the crop					
Thinning	$ 20.00	$ 20.00	$ 16.25	$ 16.25	$ 15.00
Hoeing or weeding	20.00	20.00	10.00	10.00	10.00
Cultivating, etc.	16.00	16.00	12.00	12.00	9.00
Irrigation labor	10.00	10.00	8.00	8.00	6.00
Insect and disease control	7.50	7.50	2.00	2.00	2.00
Subtotal	$ 73.50	$ 73.50	$ 48.25	$ 48.25	$ 42.00
Harvest costs					
Harvesting @ $1.75/ton	$ 40.25	$ 40.25	$ 31.50	$ 31.50	$ 24.50
Hauling @ $1.25/ton	28.75	28.75	22.50	22.50	17.50
Subtotal	$ 69.00	$ 69.00	$ 54.00	$ 54.00	$ 42.00
Overhead					
Taxes	$ 5.00	$ 5.00	$ 4.00	$ 4.00	$ 4.00
Water cost	12.00	12.00	4.00	4.00	12.00
Maintenance	4.00	4.00	4.00	4.00	4.00
Interest on operating	8.00	8.00	5.00	5.00	5.00
Interest on Investment*	30.00	30.00	25.00	25.00	25.00
Subtotal	$ 59.00	$ 59.00	$ 42.00	$ 42.00	$ 50.00
Total production cost	$262.00	$264.00	$191.75	$191.75	$178.50
Yield per acre	23 tons	23 tons	18 tons	18 tons	14 tons
Average sugar content	14.5	15.5	15.0	16.0	16.5
Price per ton @ $8.50 net†	$ 14.89	$ 16.27	$ 15.58	$ 16.96	$ 17.65
Gross return per acre	$342.47	$374.16	$280.42	$305.23	$247.04
Net return per acre	$ 80.47	$110.16	$ 88.67	$113.48	$ 68.54

* Interest charged on land investment only. Interest on other investments is included in custom charges.
† See Table 16.2.

side of the table. Five production areas are shown. Some have similar yield levels but varying sucrose percentage levels. In general, the high production areas are long-season areas and they have higher costs for fertilizer and weed control. Higher-yielding areas have higher per acre harvest and hauling costs inasmuch as these are generally on a tonnage basis.

Water costs vary considerably, depending on type of irrigation system and water charges, and when pumping is involved, on the depth of water lift, cost of power, etc.

It is evident from that data on price per ton that sugar content plays an important part in the price paid per ton and that within limits increased sugar content can compensate for lower yield.

EFFECT OF SUGAR CONTENT ON PRICE PER TON

The data in Table 16.2 indicate in more detail the effect of sugar content on the price received per ton. At the present time, an $8.50 net selling price for sugar is a reasonable level to use in comparing the return for beets at various levels of sugar content. Note that the price varies from $11.44 per ton for beets of 12 percent sugar content to $20.40 per ton for beets of 18.5 percent sugar content. While sucrose content is affected by weather, pests, and diseases, production practices such as irrigation, fertilizer usage, and harvest procedures have a large influence on sucrose level and these latter factors can usually be controlled by the farmer.

EFFECT OF YIELD ON COSTS AND RETURNS

Generally speaking, yield level is the most important factor in determining the net profit from sugarbeet production.

TABLE 16.2. Effect of net selling price of sugar and sucrose percentage of beets on the price paid farmers per ton of beets

Sucrose Percentage of Beets	Net Selling Price of Sugar per Cwt						Government Compliance Payment
	8.30	8.40	8.50	8.60	8.70	8.80	
12.0	$11.168	$11.307	$11.445	$11.561	$11.678	$11.793	$1.753
12.5	11.840	11.987	12.134	12.257	12.380	12.502	1.826
13.0	12.511	12.667	12.823	12.953	13.083	13.212	1.899
13.5	13.182	13.347	13.512	13.649	13.786	13.922	1.972
14.0	13.854	14.027	14.201	14.345	14.488	14.631	2.045
14.5	14.525	14.708	14.890	15.040	15.191	15.341	2.118
15.0	15.197	15.388	15.579	15.736	15.893	16.050	2.192
15.5	15.868	16.068	16.268	16.432	16.596	16.760	2.265
16.0	16.539	16.748	16.957	17.128	17.299	17.470	2.338
16.5	17.211	17.428	17.646	17.824	18.001	18.179	2.411
17.0	17.882	18.109	18.335	18.519	18.704	18.889	2.484
17.5	18.554	18.789	19.024	19.215	19.406	19.598	2.557
18.0	19.225	19.469	19.713	19.911	20.109	20.308	2.630
18.5	19.896	20.149	20.402	20.607	20.812	21.018	2.703

NOTE: Prices per ton shown in the above table include the government compliance payment shown in right-hand column. Prices are representative of a typical beet contract but will vary some from area to area.

ECONOMICS OF PRODUCTION

Table 16.3 indicates production cost in a typical beet-producing area, with yields varying from 10 tons to 30 tons per acre. Note that in this same area it cost $174.30 per acre to produce a 10-ton crop, resulting in a deficit of $11.62 per acre, and $251.25 per acre to produce a 30-ton crop, resulting in a net profit of $236.79 per acre.

These figures assume a constant price for sugarbeets of $16.27 per ton. This might be questionable with so large a range in yield level. However, studies have shown that both high yield and high sugar content crops are possible.

TABLE 16.3. Comparative production costs and returns of various yield levels of sugarbeets based on contract labor rates in a given area where costs are comparable

Item or Material	Yield Levels—Tons/Acre				
	10	15	20	25	30
	(dollars)	(dollars)	(dollars)	(dollars)	(dollars)
Preplanting					
Plowing	6.00	6.00	6.00	6.00	6.00
Discing and harrowing	4.00	4.00	4.00	4.00	4.00
Leveling or floating	4.00	4.00	4.00	4.00	4.00
Fertilizing	20.00	20.00	25.00	30.00	35.00
Subtotal	34.00	34.00	39.00	44.00	49.00
Planting					
Seed	3.00	3.00	3.00	3.00	3.00
Planting	3.00	3.00	3.00	3.00	3.00
Herbicide	7.50	7.50	7.50	7.50	7.50
Subtotal	13.50	13.50	13.50	13.50	13.50
Growing the crop					
Thinning*	15.00	15.00	15.00	15.00	15.00
Hoeing or weeding*	10.00	10.00	10.00	10.00	10.00
Cultivating, etc.	9.00	9.00	9.00	9.00	9.00
Irrigation labor	6.00	6.00	6.00	6.00	6.00
Insect and disease control	2.00	2.00	2.00	2.00	2.00
Subtotal	42.00	42.00	42.00	42.00	42.00
Harvest costs					
Harvesting @ $1.75/ton	17.50	26.25	35.00	43.75	52.50
Hauling @ $1.25/ton	12.50	18.75	25.00	31.25	37.50
Subtotal	30.00	45.00	60.00	75.00	90.00
Overhead					
Taxes	4.00	4.00	4.00	4.00	4.00
Water cost	12.00	12.00	12.00	12.00	12.00
Maintenance	4.00	4.00	4.00	4.00	4.00
Interest on operating	4.80	5.40	5.75	6.25	6.75
Interest on investment	30.00	30.00	30.00	30.00	30.00
Subtotal	54.80	55.40	55.75	56.25	56.75
Total production cost	174.30	189.90	210.25	230.75	251.25
Expected gross per acre†	162.68	244.02	325.36	406.70	488.04
Indicated net	(11.62)	54.12	115.11	175.95	236.79

* Some areas average higher hand labor costs than those shown.
† Based on a price of $16.27 per ton. See Table 16.2 for 15.5 sugar and $8.50 net.

So far, the sugarbeet crop return has only considered the price paid for the beet roots as delivered for processing. The additional value of the tops for feed, either when pastured from the windrow or when fed as beet top silage, has already been pointed out in Chapter 13.

TRENDS IN ACREAGE AND YIELDS

During the past 20 years there have been significant changes in sugarbeet production (1). This includes a marked decrease in the number of farms producing beets, accompanied by a marked increase in the average acreage of beets per farm. For example, in 1946 there were 41,229 farms producing beets as compared to 19,847 in 1968. During this same period the average acreage of beets per farm increased from 19.8 acres to 71.0 acres. In 1946 the total harvested beet acreage was 818,005. By 1968 this had increased to 1,410,000 acres. Total tons of beets produced in 1946 were 10,863,000, and in 1968, 25,400,000. The average yield of beets per acre in 1946 was 13.28 tons, and in 1968 was 18.01 tons. These data are shown in graph form in Figure 16.1.

FIG. 16.1. An index of sugarbeet and beet sugar production in the United States for the years 1946 to 1968. The year 1946 was used as the base 100 index. (Source: USDA, ASCS Statistical Bulletin 244, vol. 2 (rev.) Feb. 1969; and USDA, ASCS Sugar Reports, Sept. 1969.)

TRENDS IN MAN-HOUR REQUIREMENTS

From 1946 to 1968 there have been significant changes in labor requirements per acre and per ton, and in the per-hour earnings of the beet laborer (1). This is illustrated in Figure 16.2, using index figures to compare man-hours per acre, man-hours per ton, labor cost per ton, and earnings per hour of fieldworkers. Man-hours per acre decreased from 88.32 hours in 1946 to 45.02 hours in 1968. This is even more striking when a comparison is made using man-hours per ton. In 1946 it was 6.64 hours per ton as compared to 2.50 hours per ton in 1968. The impressive reduction in man-hour requirement has resulted from the use of monogerm seed at low seeding rates and in precision planters, and the extensive use of herbicides for weed control. During this same period, the earnings per hour of the fieldworker more than doubled, increasing from 75 cents per hour to 1.67 dollars per hour, which has resulted in a rather modest decrease in the labor cost per ton of beets produced. However, trends of the past two years indicate that wages are increasing faster than man-hours are decreasing, resulting in an upward trend in labor cost per ton of beets. This trend can and no doubt will be reversed by further decrease in labor requirement

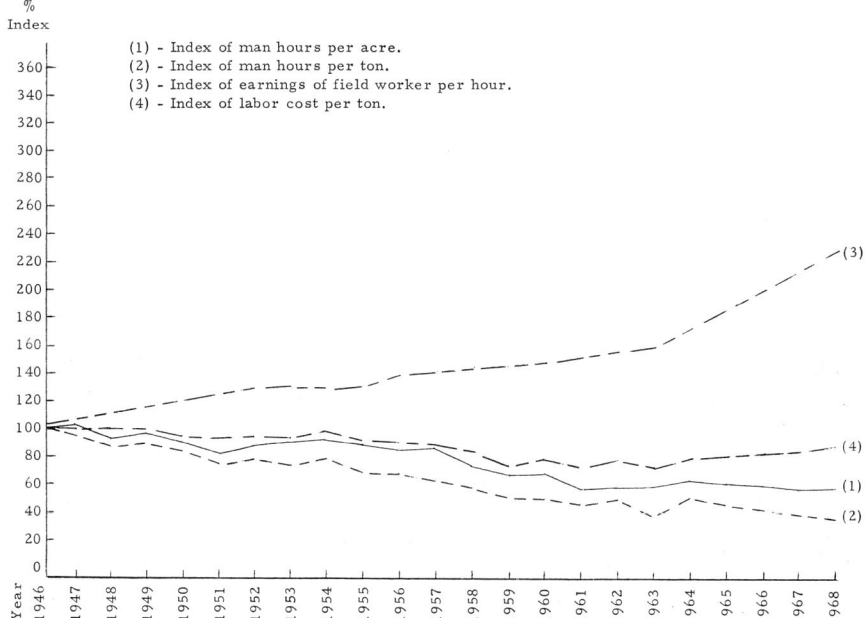

FIG. 16.2. An index of labor requirements in the production of sugarbeets, including the average hourly earnings per worker for the years 1946 to 1968 inclusive. (Source: USDA, ASCS Statistical Bulletin 244, vol. 2 (rev.) Feb. 1969; and USDA, ASCS Sugar Reports, Sept. 1969.)

through the use of electronic thinners, which were used in limited number during the 1969 season. While these were not fully successful, they did well enough to indicate they will be a major factor in decreasing the use of fieldworkers during the spring period. It now appears that as herbicides and electronic thinners are improved, the sugarbeet farmer will not be more dependent on the fieldworker for spring operations than for fall operations.

BEETS AS A PRIMARY OR SECONDARY CROP IN A CROPPING SYSTEM

The term cropping system suggests a crop rotation program in contrast to continuous cropping of a single crop. Crop rotation for control of weeds, insects, nematodes, and diseases has already been discussed in preceding chapters. Let us restate some of the general facts relating to the importance of crop rotation and the economic facts as they relate to selection of crops to be grown.

We live in an age of specialization and in many cases one-crop farming would be the most profitable if it were possible to ignore the dangers of speculation, pests, diseases, and weeds. However, farmers have run into disaster by excessive cropping to one crop. The sugarbeet industry, both farmer and processor alike, learned many years ago that disastrous results can follow continuous cropping of sugarbeets. Sugarbeet nematode buildup in the soil increases production costs, decreases yields, and makes it necessary to either fumigate the soil at a cost of $35 to $40 per acre or lengthen rotations so that beets will be grown on the same piece of ground only one year out of five or six years.

It is generally agreed that if the land is not infested with nematodes, buildup of this pest can be prevented by not raising beets more often than one year in three and that the cropping program can be managed so there will never be two successive years of beets on the same field.

Growers are having similar experiences with potatoes. Some farmers and some entire areas have had to abandon potato production due to the buildup of pests and diseases resulting from continuous cropping.

In mint-growing areas, mint growers are learning that wilt and rootknot nematode severely damage mint crops if mint is grown for longer than a two- to three-year period in the same field.

These examples suffice to illustrate the fact that the wise farmer who is looking for stability is following a carefully planned crop rotation.

Crop rotations will vary in different areas, depending on the crops available. In one area, a short rotation consisting of grain, potatoes, and sugarbeets may be most profitable. In other areas, grain, mint, and sugarbeets might be the best combination.

Production costs and possible returns from various crops are helpful information to have in planning the crops to be included in a rotation. Production costs for various crops in selected areas are presented in Table 16.4. Costs may vary somewhat from those shown if farming prac-

TABLE 16.4. Comparative crop production costs, gross receipts, and net returns of various crops in selected areas with varying yield levels

Crops Compared	Areas		
	Area 1	Area 2	Area 3
Sugarbeets			
Yield per acre	23 tons	18 tons	14 tons
Gross return	$374.16 @ $16.27	$298.74 @ $16.60	$247.04 @ $17.65
Cost per acre	$262.00	$191.75	$178.50
Net return	$112.16 per acre	$106.99 per acre	$ 68.54 per acre
Potatoes			
Yield per acre	380 cwt	250 cwt	200 cwt
Gross return	$513.00 @ $1.35 cwt	$337.50 @ $1.35 cwt	$300.00 @ $1.50 cwt
Cost per acre	$385.00	$262.00	$257.50
Net return	$128.00 per acre	$ 75.50 per acre	$ 42.50 per acre
Tomatoes			
Yield per acre	20 tons	18 tons	...
Gross return	$540.00 @ $27.00	$486.00 @ $27.00	...
Cost per acre	$345.00	$330.00	...
Net return	$195.00 per acre	$156.00 per acre	
Sweet corn			
Yield per acre	7 tons	5.5 tons	4.5 tons
Gross return	$175.00 @ $25.00	$137.50 @ $25.00	$112.50 @ $25.00
Cost per acre	$139.00	$110.00	$105.00
Net return	$ 36.00 per acre	$ 27.50 per acre	$ 7.50 per acre
Alfalfa hay*			
Yield per acre	8 tons	5.5 tons	4.0 tons
Gross return	$200.00 @ $25.00	$137.50 @ $25.00	$100.00 @ $25.00
Cost per acre	$160.00	$120.00	$108.50
Net return	$ 40.00 per acre	$ 17.50 per acre	$—8.50 per acre
Wheat			
Yield per acre	100 bushels	80 bushels	70 bushels
Gross return	$150.00 @ $1.50	$120.00 @ $1.50	$105.00 @ $1.50
Cost per acre	$115.00	$105.00	$100.00
Net return	$ 35.00 per acre	$ 15.00 per acre	$ 5.00 per acre
Mint†			
Yield per acre	80 pounds of oil	70 pounds of oil	60 pounds of oil
Gross return	$360.00 @ $4.50	$315.00 @ $4.50	$270.00 @ $4.50
Cost per acre	$309.00	$300.00	$290.00
Net return	$ 51.00 per acre	$ 15.00 per acre	$—20.00 per acre
Corn silage			
Yield per acre	20 tons	18 tons	15 tons
Gross return	$160.00 @ $8.00	$144.00 @ $8.00	$120.00 @ $8.00
Cost per acre	$126.00	$122.00	$116.00
Net return	$ 34.00 per acre	$ 22.00 per acre	$ 4.00 per acre

NOTE: Production costs are based on custom labor rates plus overhead costs. It should be recognized that some crops are speculative and many factors may alter returns per acre.

* Cost of establishing the alfalfa stand has been amortized over a 3-year period.

† Preplanting costs and cost of roots and planting have been amortized over a 3-year period.

tices vary from those on which these costs were based. However, production costs and crop returns are comparable as based on farming practices and current crop prices in the various areas. Costs for each crop were developed using contract labor and machine costs similar to those shown in detail in Table 16.1 for sugarbeets. Consequently, they are comparable and anyone using a similar approach can develop costs for any crop in any specific area of production.

The price indicated per ton for sugarbeets is based on the average sugar content for each of the various districts and the estimated net selling price of sugar.

In areas with individual sugar content payments, growers' payments may be either higher or lower than the price shown, dependent on whether their sugar content is above or below average.

Crop returns from other crops are based on average yields and current prices. It is recognized that prices for crops may in some cases be either higher or lower than those used.

Inasmuch as beets need to be grown in rotation, it has not been our purpose to specifically compare beet returns against the level of return from companion crops. Farmers should choose companion crops in a rotation that will accomplish the purposes of crop rotation and at the same time produce the highest consistent net crop returns for the area in which their farming enterprise is located.

REFERENCES

1. ASCS-USDA. Sugar Statistics and Related Data, vol. 2. Washington, D.C., 1968.
2. Bureau of Reclamation. Yearly Crop Summary Sheets. Ephrata, Wash. 1960–68.
3. Dahl, Merlyn M. Rates Paid for Custom Work in South Dakota. S.Dak. Agr. Exp. Sta. Circ. 663. 1968.
4. ———. Machinery Costs—Own, Lease or Custom Hire. S.Dak. Agr. Exp. Sta. Circ. 664. 1968.
5. Doran, Samuel M. and Wm. Foeppel. Sugar Beets Estimated per Acre Costs and Returns. Columbia Basin, Wash., E.M. 2754, Prosser, Wash. Exp. Sta. 1967.
6. Kennedy, V. D. Probable Cost of Production. Upper Snake Area, Extension Management Specialist, Univ. of Idaho, Moscow. 1968.
7. Morrison, Ernest M. Production Costs and Net Return for Sugar Beet Production in Utah, 1945–1963. (Mimeo.) Utah State Univ., Logan. 1964.
8. Tolman, Bion. Crops for the best rotation. U and I Cultivator 28:2–6. 1968.
9. ———. Sound cropping program. U and I Cultivator 27:24–28. 1967.
10. Withers, R. V. Sugarbeet production costs. Idaho Agr. Res. Progr. Rept. 119. Univ. of Idaho, Moscow. 1966.
11. Zuroske, C. H., I. D. Brawson, and J. P. Swanson. Sugarbeet production costs by acreage groups. Wash. State Univ. Circ. 470. 1966.

Index

Achard, Franz Karl, 5, 17, 402, 405, 434
Adams, R. L., 292
Adams, S. N., 150, 157
Alabama, 184
Albinism, 282
Alexander, M., 366
Alfalfa, 123, 132, 295, 300
  mosaic virus, 321
Altman, J., 366, 367
Aluminum, 177, 179, 180
Alvarado, California, 6, 403
Amalgamated Sugar Company, 10, 438, 439, 440
American Beet Sugar Company, 438
American Crystal Sugar Company, 9, 418
  hybrids developed by, 420
  and seed production, 440
Aminization, process of, 112
Amino acids, 114
Ammonification, process of, 112
Ammonium and compounds of, 112–13, 119, 120, 122, 123, 146, 180, 181, 182, 183, 366
Ammonium citrate method, 145
Ammonium nitrate-potassium chloride fertilizer treatment, 160
Anhydrous ammonia, 120, 122, 183
Anions, 182–84
Aphanomyces infection, 224, 265, 266
Aphids, 236–48 *passim*, 253, 256, 258, 259, 288, 294, 296, 301, 305, 312, 314, 420, 449
  control of, 321–24
  green peach and relation to virus diseases, 318–21
Argentina, 225
  diseases common to sugarbeets of, 232, 233, 239, 249, 251, 252
  end of sugarbeet industry in (1941), 249
Arizona, 10, 59, 225, 349, 372
  diseases common to sugarbeets of, 231, 233, 240, 248, 254, 270, 276, 278
  pests commonly attacking sugarbeets of, 304, 311, 448
  seed production in, 440–48 *passim*
Arizona, University of, 66

Armyworms, 298, 300, 304, 311
Arp, 103
Arrington, L. J., 289, 291
Atomic absorption test, 174
ATP (adenosine triphosphate), 139
Atrazine, 103, 104
Australia, 265, 275, 344

Bacterial diseases. *See* Fungus and bacterial diseases and specific diseases
Baker, C. F., 289
Ball, E. D., 289
Band-placement (of phosphorus), 147–48
Bedding, practice of, 54–55, 266
Beet crinkle. *See* Krauselkrankheit
Beetles, 304, 312–13, 315
  blister, 308–10
  carrot, 297
  cucumber, 298
  flea, 302
  June, 297
  May, 297
Beet mosaic, 242–45, 316–17, 318, 449
  causal agent of, 245
  control of, 245
  economic importance of, 243
  history and distribution of, 242–43
  and host range, 243
  symptoms of, 243–44
  transmission of, 244–45
Beet rust, 278
Beet Sugar Development Foundation, 12, 292, 386, 429, 433
Beet Sugar Society of Philadelphia, 5
"Beet Sugar Story," 6
Beet western yellows, 71, 239–42, 312, 316, 449
  causal agent of, 242
  control of, 242, 420–21
  economic importance of, 239–40
  history and distribution of, 239
  and host range, 240
  symptoms of, 241
  transmission of, 241

Beet yellows, 233–39, 312, 316, 320
  causal agent of, 237–38
  control of, 238–39, 321, 323, 420–21
  economic importance of, 234
  history and distribution of, 233–34
  and host range, 234–35
  symptoms of, 235–37
Belgium, problem of nematodes in sugarbeets of, 344, 349, 353
Benlate, 277
Bennett, C. W., 292, 316, 318, 320
Benson-Calvin carbon fixation mechanism, 43
Benzadox, 75, 84, 87, 98, 315
Black hunter, 315
Black root, 419–20
Blackwelder, E. F., 387
Black wood vessel diseases, 273
Blaney-Criddle formula, 204
Blight. *See* Leaf spot
Boawn, L. C., 123, 130
Bockstahler, H. W., 420
Bollen, 366
Bolting, 424–25, 431, 433, 442, 444–45
Bordonos, M. G., 425
Borers
  beet petiole, 308
  crown, 302–3
Boron, 172, 180, 182, 183–84, 186
  deficiencies of, 173, 178, 185–86
  and fungus control, 266
  and seed production, 448
Bovien, P., 293
Brandenburg, E., 267
Brandes, E. W., 5
Bray method, 152
Brazil, 232, 233
Breeding, 402–34
  and cooperation in research on, 428–30
  future of, 430–34
  history of, 402–4
  methods of, 404–16
    interspecific hybridization, 413–16
    mass selection, 404–5
    paired-plant crosses, 407–11
    polycross and recurrent selection, 406–7
    polyploidy and triploid hybrids, 411–13
    progeny testing and mother-line breeding, 405–6
  recent progress in, 416–28
Brimhall, Phil B., 71
British Columbia, 278, 298
Broadbent, F. E., 113
Bromine, 172
Brown, A. L., 154, 156
Brown, Robert J., 427
Buckeye Sugars, Inc., 10
Bugs, 314
  clover-root mealy, 313
  false chinch, 308, 312
  green, 449

lygus, 308, 449
Say stink, 308
C–A, 178
Calcium and compounds of, 122, 172, 177, 179, 183, 184, 186
  deficiences of, 173, 174, 178–79, 180, 185
  metaphosphate, 145, 146, 148
  and relation to potassium, 158, 159, 162
  and soil acidity, 177
California, 9, 17, 59, 190, 355, 372, 384
  diseases common to sugarbeets of, 226, 231–35 *passim*, 237–40 *passim*, 253, 254, 257, 260, 266, 269, 276, 277, 278, 280, 281, 422
  pests commonly attacking sugarbeets of, 289, 291–92, 296–300 *passim*, 302, 303, 305, 307, 308, 310, 312, 314, 315, 318
  seed production in, 438, 440, 442
  soil deficiencies in, 181, 182, 184
California Agricultural Experiment Station, 11
  Bulletin 766 of, 174
California Beet Growers Association, Ltd., 11, 292, 429
California, University of, 291, 292, 318, 324, 416, 422
Campaigns, length of, 14
Canada, 256, 292, 296, 344
Canopy architecture, of sugar beet communities, 45–46
Carbon, 102, 103, 172
  dioxide, 38, 40–44 *passim*, 112, 172
Carlson, E. C., 311
Carruthers, A., 427
Carsner, Eubanks, 416
Caterpillars, 307
  salt-marsh, 311
Cations, 177–82. *See also* specifications
Caveness, F. E., 364
Central Valley, California, 59
Chang, V. C., 294, 318
Chepil, W. S., 80
Chile, 224, 225, 233, 234, 249, 433
  diseases common to sugarbeets of, 232, 234, 239, 240, 249, 251, 252–53, 261, 265
Chimaera, definition of, 282
Chiseling, practice of, 52–53
Chittenden, F. H., 289, 307, 308, 309, 313
Chlorine, 172, 183, 186
Chlorophyll, 42, 179, 282
Climate, 50. *See also* Temperature
  effect of on sugarbeet production, 23, 24, 27–29, 192
  and seed production, 442
Cobalt, 172
Cockbain, A. J., 319
Codex Alimentarious Commission, 104
Coe, G. E., 420
Colorado, 9, 190, 360, 372, 384
  diseases common to sugarbeets of, 240, 242, 254, 271, 273, 277

pests commonly attacking sugarbeets of, 290, 298, 300, 302–5 *passim*, 307–10 *passim*, 312, 313
Connecticut, 312
Consolidated Foods, Inc., 429
Cooke, D. A., 365
Cook, R. L., 155
Coons, G. H., 292, 418, 420
Copper, 172, 183, 186
  deficiencies of, 174, 179–80
Cormany, C. E., 75
Corn use of as sweetener, 16
Cotyledons, emergence of, 20–21
Cox, E. L., 319
CP-52223, 75, 86
Crickets, 310
Crop competition, and weed control, 74
Crop rotation, 14, 50–51, 150
  and disease control, 264, 276
  economics of, 458
  and nematode control, 344, 351, 358, 361, 363–64
  and weed control, 74, 80–81
Cuba, 7
Cucumber mosaic virus, 244, 246–48, 321
  causal agent of, 248
  control of, 248
  economic importance of, 246
  history and distribution of, 246
  symptoms of, 246–47
  transmission of, 247–48
Cultivation machine, 73–74
Cunern, Silesia, 5
Curly top, 13, 71–72, 225, 226–33, 239
  Argentine, 232–33
  Brazilian, 233
  causal agent of, 230–31
  control of, 231–32, 323
  development of plants resistant to, 416–18, 424, 428, 430, 431, 439
  economic importance of, 226–27
  history and distribution of, 226
  and host range, 227
  symptoms of, 227–29
  transmission of, 229, 288, 289–92, 315–16
Curly Top Resistance Breeding Committee, 428, 429
Cutworms, 298–300, 301, 304, 311, 449
Cyclamates, 15
Cycloate and combinations of, 75, 82, 83, 86, 91, 92, 97, 98, 99

Dalapon and combination of, 75, 83, 95, 97, 98, 99, 102
Davis, J. F., 155
Dawson, J. H., 71, 98–99
D–D, 365, 366
DDT, 294
Deming, G. W., 75
Denmark, 344
Deoxyribose nucleic acid. *See* DNA
Deseret Manufacturing Company, 6
Dexon, 265, 321

Diallate and combinations of, 75, 83, 84, 85, 91, 97, 102
Diazinon, 99
Dicamba, 103, 104
Dickson, R. C., 318
Dicofol, 321
Diphenylamine test, 128, 133, 378
Diseases, and their control, 13–14, 186, 224–82, 377, 380. *See also* specific diseases
  fungus and bacterial diseases, 260–82
  virus diseases, 225–60
DNA, 138, 139
Docking disorder, 72, 355, 357
Dodder, 247, 251, 257, 281–82
Doneen, L. D., 124
Doney, D. L., 367
Dorph-Petersen, K., 155
Dorst, 321
Dotzenko, A. D., 103
Douglas, J. R., 291
  quoted, 291
Downie, A. R., 420
Doxtator, C. W., 420
Drake, R. M., 316
Draycott, A. P., 158, 162
Dry rot canker, 118
Duffus, J. E., 239, 242, 292, 316
Duncan's model, 45
Dunning, R. A., 292, 355, 365
Durrant, M. J., 158, 162
Dyer, E. H., 6, 9, 17

Eddy, New Mexico, 300
Eddy transfer, process of, 40, 42
Ehrenfeld, Roberto K., 252
El-Sheikh, A. M., 156, 172
Endothal and combinations of, 75, 83, 84, 85, 86, 105, 448
England. *See* Great Britain
Enzymes, 138, 154–55, 173, 179, 180, 183
EPTC, 75, 82, 83, 84, 91, 92, 93, 98, 102, 104
Erie, L. J., 202, 215
Ernould, L., 293
Esau, Katherine, 416
Essig, E. O., 300
Europe, 5, 151, 224, 226, 336, 412, 438
  diseases common to sugarbeets in, 233, 234, 236, 238, 240, 273, 280, 281
Evans, H. J., 155
Evapotranspiration, process of, 190, 191, 192–93, 196–206, 215–16, 220
  peak rates of, 203–4
  rate of in fall-planted beets, 196, 202–3, 204
  rate of in spring-planted beets, 196–202
  seasonal rates of, 204, 205

Family 41 yellows, 249–60
Farmers and Manufacturers Beet Sugar Association, 11, 429, 440
Ferry, G. V., 215

Fertilizers, 51, 54, 99–100, 148, 448, *See also* specific fertilizers
Firebaugh, California, 246
Flies, 305–7, 314
Florida, 16
Food and Drug Administration, 104
Fort Collins, Colorado, 418
Foster, L., quoted, 208
France, 6, 12, 242, 275, 344, 402, 403
Franklin, Mary T., 349
Franzoy, C. E., 206
Frazier, N. W., 315
Frederick William III. King of Prussia, 5
Freitag, J. H., 321
French, O. F., 215
Fumigants, 364-67
Fungicides, 265, 277, 321. *See also* specific fungicides
Fungus and bacterial diseases, 260–82. *See also* specific diseases
 of foliage, 275–81
 of roots, 265–71, 273–75
 of seedlings, 260–65
 of vascular structure, 271–73
Fusarium yellows, 271–73

Gaskill, J. O., 324, 423, 425
Germany, 12, 256, 275, 344, 353, 402–3
Gillette, C. P., 289
Glycolysis, process of, 139
Golden, A. M., 367
Goode, J. E., 212
Gowing, D. P., 97
Graf, John E., 295
Gram, E., 293
Grasshoppers, 310
Gray, Reed A., 91
Great Britain, 12
 diseases common to sugarbeets in, 233, 234, 240, 258, 265, 269, 277
 nematodes affecting sugarbeets in, 344, 349, 355, 365
 use of phosphate fertilizer in, 147, 151
Great Western Sugar Company, 10, 298, 418, 438
 hybrids developed by, 419
 and research activities, 433
 and seed production, 440
Grigsby, B. H., 75
Gypsum, 179, 182, 183

Haddock, Jay L., 130
Hail, 377–78
Harmer, P. M., 156
Harper, A. M., 324
Harris, S. F., 292
Harris, Lionel, quoted, 385
Harrison, Merle, 419
Harvesting, 384–90. *See also* Receiving and delivery
 development of mechanical, 386–90
 principles of, 384

 and recovery of the roots, 384
 and recovery of the tops, 385–86
Hawaii, Republic of, 7
Hawley, I. M., 289, 307, 310
Heathcoate, G. D., 319
Hecker, R. J., 419
Heie, O., 318
Henderson, R. W., 420
Herbicides, 51, 55, 63, 433, 448, 458. *See also* specific herbicides
 activity and selectivity of, 81–86
  organic chemical groups, 81–84
 cautions in use of, 100–102
 combinations of, 95–100
 dissipation of, 102–4
 factors affecting foliar-applied, 93–95, 98, 99
 factors affecting soil-applied, 90–93, 98, 99
 history of use of, 74–75
 principles of use of, 86–90
  classification, 86
  formulation and rate expressions, 86–87
 and rate calculations, 86–87
 registration of, 104–5
 research and development in, 105–7
 suggested dosages useful on sugarbeets grown from roots, 82
Herron, G. M., 204
Hewitt, E. J., 154
Hills, F. J., 154, 163, 174, 215, 292
Hills, O. A., 311, 318, 321, 325
Hoff, John, 378
Holly Sugar Corporation, 321, 438, 440
Hoskins, W. M., 300
Howe, O. W., 219
Hull, R., 139, 269, 293, 316
Hurst, G. W., 326
Husseini, K. K., 157, 160, 162
Hybrids, 407–16, 417, 419, 420, 421, 424, 428, 438
 and development of cytoplasmic male sterility (CMS), 407–12 *passim*, 419
 polyploidy and triploid, 409, 411–13, 432
 US 1, 417, 428, 438
 US H9A and B, 292, 319, 320, 324, 421
Hydrogen, 179
Hydrolysis, process of, 99, 102
Hypocotyl, growth of, 21

Idaho, 9, 190, 273
 pests common to sugarbeets in, 289, 291, 302, 304, 307, 308
Illinois, 226, 310, 313
Immobilization, process of, 112–13
Imperial Valley, California, 4, 59, 121, 130, 234, 240, 305, 307, 379, 384, 424, 429, 433
India, 5
Indiana, 177, 255, 302, 307

Industria Azucarera Nacional S.A., 433
Insecticides, 99–100, 106, 231, 321–25 *passim*. *See also* specific insecticides
Insects and mites, 51, 288–326, 377
    control procedures, 17, 300, 321–25, 431
    literature available on, 292–93
    and crop ecology, 293–95
    history of destruction by, 288–93
    as occasional pests, 312–13
    on older plants, 305–11
    outlook for control of, 325–26
    as pests of seed production, 311–12, 448–49
    those present at planting time, 295–305
International Harvester Company, 388
International Sugar Agreement, 8–9
IPC, 85, 102, 448
Ireland, 224
    diseases common to sugarbeets of, 240, 259, 260, 265, 269, 277
Iron, 172, 183
    deficiencies of, 174, 180, 186
    and soil acidity, 177
Irrigation and water management, 54, 55, 122, 185, 186, 190–220, 376–77, 380, 453–54
    advantages of, 190–191
    and effect of on nitrogen-soil moisture interaction, 215
    and effect of on yields, 211–14
    and fungus problems, 269
    furrow, 53, 64, 120, 443
    intervals of, 214–15
    literature on irrigation practices, 208–10
    methods of, 208–11, 215–19
    and relation to evapotranspiration, 206
    and relation to waterholding capacity of soil, 207
Israel, 268, 349
Italy, 349, 419
    diseases common to sugarbeets in, 265, 268, 275
    and Stazione Sperimentale di Bieticultura, 403

James, D. W., 116, 131, 159, 160
Japan, 12, 234, 268
Jensen, M. E., 206, 219
John Deere one-row harvesters, 387, 388
Johnson, R. T., 344, 425
Jones-Costigan Act (1934), 7
Jones, F. G. W., 292, 355, 387

Kansas, 254, 309, 360, 372
Kaudy, J. C., 163
Kaufman, D. D., 108
Kearney, Philip C., 102
Kennedy, J. S., 318
Kern County, California, 117, 234, 239, 292, 353
Kershaw, W. J. S., 319

King, L. J., 81
Kleinwanzleben variety, 404
Knauer, F., 403, 405
Koch, A., 336
Koppy, Freiherr von, 402, 405
Korea, 268
Kotila, J. E., 292
Krantz, B. A., 124
Krauselkrankheit, 256–57
Krayn, Germany, 402
Krionderis, Dennis J., 99

Labor, economics of, 457–58
Lacebug, 256, 257
Lacewings, 314
Laird, E. F., 318
Landis, B. J., 308, 312
Lange, W. H., 292, 296, 308, 312, 318, 319, 321, 325
Lawlor, Sue, 366
Leaching, process of, 191, 217–18, 220
Leach, L. D., 292, 321, 422
Leafhopper, 226, 229, 231, 233, 256, 289
    areas of and years of abundance, 290
    control of, 301, 305, 308, 311–12, 314, 315–16, 321, 417, 433, 449
    and curly top, 289–92, 416
Leaf miners, 305, 306, 308, 314, 321, 322
Leaf spot, 13, 224, 275–78, 281, 449
    breeding for resistance to, 416, 418–19, 430
Lear, Bert, 353
Legg, J. O., 131
Lehi, Utah, 289
Lehr, J. J., 161
Lightning, plant damage by, 282
Limestone, 179, 180, 186
Lipimeters, 193
Loof, P. A. A., 355
Loomis, R. S., 120, 121, 127, 128, 215
"Losses in Agriculture," 13
Louisiana, 6, 224
Lüdecke, H., 293

McBirney, S. W., 74
McDonnell, P. M., 160
McElroy, W. D., 154
McFarlane, J., 278, 324
McKay, R., 269
McKee, H. S., 366
MacKenzie, 130
McKinney, K. B., 293
McLean, D. L., 319
Macroclimate, 38
Macronutrients. *See* Nitrogen, Phosphorus, and Potassium
Magnesium, 172, 177, 179, 184, 186
    deficiencies of, 173, 174, 179, 185
    and relation to potassium, 158, 159, 162
    and soil acidity, 177

Maine Agricultural Experiment Station, 11–12
Maine Sugar Industries, Inc., 10
Manganese, 172, 179, 180–81, 183
  deficiencies of, 173, 174, 185, 186
  and irrigation water, 186
  and soil acidity, 177
Manure, green, 123–24, 132
Marbeet harvester, 388
Marble leaf, 258–59
Marggraf, Andreas, 5, 17, 402
  quoted, 5
Maryland, 226
Massachusetts, 184, 307
Maxson, A. C., 138, 139, 290, 292, 297, 302, 305, 308, 310, 351
Maynard, E. J., quoted, 384
Mech, S. J., 198
Mediterranean areas, 226, 229
Mendota, California, 246
Mercury compounds, 265
Mervine, Ernest M., 387
Mesa, Arizona, 66
Mexico, 226
Michigan, 9–10, 177, 179, 180, 184, 190, 255
  pests commonly attacking sugarbeets in, 302, 307, 313
Michigan Sugar Company, 10
Micronutrients, 173–86. *See also* specific micronutrients
Middleton, J. E., 198
Mildew, 278–81, 422, 431, 449
Minnesota, 310, 384
Mites, 307, 315
  spider, 311, 321, 449
Molnau screen, 392
Molybdenum, 172, 173, 182, 183, 186
Monitor Sugar Company, 10
Montana, 180, 184, 265, 271, 360, 384
  pests commonly attacking sugarbeets in, 298, 302, 303, 304, 307, 309, 310, 313
Mormon Church, 6
Mormon cricket, 310
Muller, K. A., 160
Munerati, O., 419
Munson, R. D., 159
Murphy, A. M., 417

NAD (nicotinamide-adenine-dinucleotide), 139
NADP (nicotinamide-adenine-dinucleotide phosphate), 139
Nagarajah, Sellappah, 180
Nakashima, T., 113
Nampa, Idaho, 289
Nason, A., 154
National Joint Commission on Fertilizer Application, 147
National Sugarbeet Growers Federation, 11
Nebraska, 181, 190, 360, 384
  diseases common to sugarbeets in, 226, 254, 269, 271, 273
  pests commonly attacking sugarbeets in, 297, 302, 304, 309, 313
Nelson, R. T., 75, 85
Nelson, W. L., 135
Nematodes, 17, 72, 118, 258, 336–68, 375, 458
  Cobb's root galling, 360–61
  development of varieties resistant to, 367–68, 423–24
  general control of, 361–68
    early planting, 361, 363
    resistant varieties, 367–68
    rotation, 363–64
    soil fumigation, 364–67
  general morphology and biology of, 337
  root-knot, 118, 336, 347–53
    distribution, 347, 349
    field symptoms, 349
    relationship to host, 349–53
  stem and bulb, 336, 353–55
    control of, 355
    life cycle, 355
    symptomology and field injury, 353–54
  stubby-root and needle, 355–58
    control of, 358
    distribution and economic importance, 355, 357
    field symptoms, 357
    life cycle, 358
    root symptoms, 357
  sugarbeet cyst, 13–14, 59, 72, 336, 338–47
    control of, 344, 346–47
    distribution and economic loss from, 344
    and ecological relationships, 342
    field symptoms, 342–43
    life history, 339–42
    and relation to nematode-infested beets and other diseases, 343–44
Netherlands, the, 177, 258, 273, 344
Nevins, D. J., 120
New Guinea, 5
New Jersey, 278, 302
New Mexico
  diseases common to sugarbeets in, 254, 270, 271, 278
  pests commonly attacking sugarbeets in, 305, 310, 312
  seed production in, 430, 438, 440
New York, 10, 180, 184, 278, 302, 307
New York Sugar Industries, Inc., 10
Nickle, H. G., 198
Nielsen, Kent, 411
Nitralin, 75, 92
Nitrogen and nitrates, 23, 33–35, 38, 102, 112–34, 140–43 *passim*, 145, 146, 148, 150, 172, 180, 182, 183, 186, 219, 265, 266, 374, 375–80 *passim*
  deficiency of, 30–33, 113–14, 126–28
  fertilization with for optimum produc-

INDEX 467

tion, 114–30
   amount of nitrogen required, 128–30
   application of nitrogen, 119–21
   and future developments, 133–34
   guides to fertilization with, 130–31
   in the plant, 113–14
   and relation to phosphorus, 149
   and relation to potassium, 157, 159, 160
   in the soil, 112–13
   source of, 121–24
Nivens, D. J., 127
Northampton, Massachusetts, 6
North Dakota, 304, 310, 384
Northern Ohio Sugar Company, 10
Nucleoproteins, synthesis of, 138, 139

Ogden Iron Works, 391
Ohio, 177, 265, 302, 307
Ohki, Kenneth, 183
Oklahoma, 254
Oldemeyer, R. K., 410
Oldfield, J. F. T., 427
Olsen, S. R., 146
Oman, P. W., 315
Ontario, Canada, 255, 308
Oregon, 10, 184, 225
   diseases common to sugarbeets in, 240, 258, 269, 277, 278, 281, 422
   pests commonly attacking sugarbeets in, 289, 290, 296, 298, 300, 302, 307, 308, 312
   seed production in, 440–46 *passim*, 448
Oregon State University, 448
Owens, F. V., 408, 409, 410
Oxnard brothers, 9

Parasites, 314–15
Pathogens, in sugarbeet seedling diseases, 261–65
   control measures for, 264–65
   identification of, 262–64
Pavement ant, 312
Payne, Merele G., 428
PCNB (pentachloronitrobenzene), 265
Peay, W. E., 292
Pebulate and combinations of, 75, 82, 91, 93, 102
Penman equation, 206
Pennsylvania, 184
Percolation, definition of deep, 191, 192
Pesticides, 13, 17. *See also* specific pesticides
Peterson, B., 318
Peterson, H. B., 146
Pettit, Rufus H., 293, 298
Phenmedipham, 75, 85, 94, 98, 99, 100, 105
Philippine Islands, 7
Phosphorus and phosphates, 138–53, 163, 164, 181, 185, 186, 265. *See also* Superphosphates

   determining the need for, 151–53
   and effect of as fertilizer on crop residue, 149–50
   fertilization by, 143–49
   and future consideration, 153
   physiological role of, 138–39
   residual value of, 150–53
   and soil acidity, 177
   uptake of, 139–43
Photometry, flame, 174
Photosynthates, 45, 47
Photosynthesis, process of, 21, 23, 24, 29–30, 38, 39, 42, 43, 45–47, 81, 94, 139, 179, 180, 184, 190
Phytin, 138
Planting. *See also* Seedbed, preparation of
   early and nematode control, 361, 363
   of hybrid seeds, 446–47
   practice of, 56–60
      planter selection and maintenance, 56–58
      planting dates, 58–59
Plowdown, practice of, 147
Plowing, and weed control, 72–73. *See also* Seedbed, preparation of
Poland, 256
Potash, 179
Potassium, 138, 142, 153–65, 177, 186
   fertilization, 164–65
   outlook for, 165
   physiological role of, 153–56
      and relation to sodium, 155–56, 163–64, 184
   predicting need of, 163–64
   root and sucrose yield interactions with other nutrients, 159–62
      magnesium and calcium, 162
      nitrogen, 160–61
      sodium, 156–57, 161–62
   uptake, 156–59
      critical level and seasonal pattern, 156–57
      distribution of potassium in the soil, 159
      plant and total uptake, 159
      relation to magnesium and calcium, 158–59
      relation to nitrogen, 157
      relation to sodium, 157–58
Potassium chloride fertilizer, 157, 160
Powers, Le Roy, 427, 428, 431
Price, Charles, 367, 424
Pritchard, Frederick J., 404
Pruitt, W. O., 201, 204
Puerto Rico, 7
Pyrazon and combinations of, 75, 83–86 *passim*, 91, 92, 94, 95, 97, 98, 99, 101–4 *passim*

Radiation, levels of, 34–35, 38, 39
   solar, 42
   thermal, 39, 40

Rain, 191, 379–80
Raski, D. J., 344, 353
Receiving and delivery operations, 390–99
 equipment
  delivery, 395–96
  receiving, 392–93
 history and development of, 390–92
 sampling, 396–97
 storage, 17, 397–98
 record keeping, 398
Red River Valley, 4, 11, 190
Red River Valley Sugarbeet Growers Association, 11, 12, 429
Reed, J. L., 294, 318, 321
Reeve, Perc A., 75
Respiration, process of, 24, 45, 139, 180
Reynolds, H. T., 305, 307
 quoted, 325
Ribose nucleic acid. *See* RNA
Rienks rolls, 382
Rio Colorado Valley, Argentina, 249
Rio Negro Valley, Argentina, 242, 249
Ririe, D., 123
Ritenour, G., 323
RNA, 138, 139
Robb, C. N., 206
Robbins, W. W., 74
Rocky Ford, Colorado, 418
Root rot
 charcoal, 270
 common scab, 273–74
 cotton, 270, 271
 crown gall, 274
 Phoma, 224, 265, 266–67
 Rhizoctonia, 224, 266, 422–23
 Sclerotium, 118, 224, 268–69, 430–31
 violet, 269–70
 wet, 270–71
Root worms, 298
Rosette, 257
Rübenkraüsel. *See* Krauselkrankheit
Rubidium, 172
Russell, G. E., 324
Russian thistle, 72, 229, 232, 291, 316

Saccharin, 15
Sacramento Valley, California
 diseases common to sugarbeets in, 234, 240, 269, 349, 418
 pests commonly attacking sugarbeets in, 307
Salinas, California, 424
 U.S. Agricultural Research Station at, 421
Salinas Valley, California, 294
 diseases common to sugarbeets in, 234, 238, 240, 278
Salter, P. J, 212
Salt River Valley, Arizona, 234, 246, 424, 439, 441
San Joaquin Valley, California, 59, 349, 378

diseases common to sugarbeets in, 231, 232, 234, 240, 248
pests commonly attacking sugarbeets in, 307, 315
Savitsky, Helen, 367, 424, 428–29
Savitsky, V. F., 425, 428–29, 441
Savoy, 255–56, 257
Scale, 312
Schmehl, W. R., 146
Schneider, C. L., 420
Schuster, M. L., 361
Schuyler, Nebraska, 404
Schweizer, E. E., 98
Scott-Urschel single-row harvester, 387
Seedbed, preparation of, 50–56
 early stages, 51–54
 field selection, 50–51
 final stages, 54–55
 in humid areas, 55–56
 and weed control, 72–74, 80–81
Seed-corn maggot, 295–96
Seeds, sugarbeet, 10, 11, 20–21
 monogerm, 425–26, 433, 441, 445
 production of, 438–50
  and agronomic practices, 443–44
  and climate and soil, 442
  and fertilizer practices, 447–48
  germination, 449
  and growth habits, 444–45
  and harvesting, cleaning, and processing, 445–46
  history of, 438–39
  and insects and diseases, 448–49
  limitation and future of, 449–50
  and maintenance of varietal purity, 447–48
  organizations involved in, 440
  planting methods for hybrids, 446–47
  and reproductive development, 442
  scope of, 441
 types of, 441
Serro, R. F., 427
Severin, H. H. P., 316, 320, 321
Sevin, 99
Shepard, James H., 404
Shepherd, R. J., 318, 320, 321
Singh, B., 172
Smith, P. B., quoted, 72
Smith, R. F., 325
Sodium, 74, 172, 177, 178, 182
 and interaction with potassium, 155–64 *passim*, 184
 test for deficiency of, 174
Sodium bicarbonate test, 152
Soil, 38–40 *passim*, 182
 acidity of, 174, 176–77, 178–79, 180
 and herbicides, 93, 100, 102–3
 management of moisture reservoir of, 191, 206–15
 and seed production, 442
Sorger, G. J., 155
Sorghum, as a source of sugar, 15–16, 17
South Carolina, 302

INDEX

South Dakota, 184, 271, 384
  pests commonly attacking sugarbeets in, 302, 307, 309, 310, 313
Spain, 268, 275
Spittlebug, 449
Spreckels, Claus, 9
Spreckels Sugar Company, 9, 292, 416, 422, 439
Standard Sugar Refining Company, 6
Stanford, G., 131
Steenbjerg, F., 155
Stewart, D., 271, 418
Subirrigation, 54
Subsoil, preparation of, 52–53
Sucrose, concentration and levels of, 5, 14–15, 15–16, 21, 23, 51, 66–67, 128, 139, 150, 319, 367, 373, 407, 418, 427, 453, 454, 455
  effect of diseases on, 234, 242, 255
  effect of temperature on, 25–35 *passim*
  importance of nitrogen to, 35, 36, 113, 116, 117, 120–23 *passim*, 126, 127
  and potassium, 159–64
  and soil moisture, 190, 215
Sugar, 5, 7, 15, 16, 372
  consumption of, 4
  regulation of, 7–9, 12
Sugar Acts, 6, 7, 8, 9, 12–13, 16
  Amendments of 1965, 7
Sugarbeet production
  economics of, 452–60
    cost and prices, 452–56
    trends in acreages and yields, 456
    trends in man-hour requirements, 457–58
  federal controls on, 12–13
Sugarbeet-root maggot, 304, 321
Sugarbeets. *See also* specific topic
  development of as a food resource, 4–17
  development of varieties of, 16, 402–34
  diseases of and their control, 224–82
  ecology of plants and communities, 20–47, 293–95
    environment-community relationships, 38–47
    environment-plant relationships, 24–38
    growth and development, 20–24, 30, 43, 194–96
  factors affecting quality of, 372–80
  harvesting and delivery of, 384–99
  insects attacking and their control, 288–326
  irrigation and water management of, 190–220
  nematodes affecting and their control, 336–68
  nutrition of
    nitrogen, 112–34
    phosphorus, 138–53
    potassium, 153–65
    secondary and micronutrients, 172–86
  production of seed for, 438–50
  seedbed preparation for and planting and thinning of, 50–67
  weeds affecting and their control, 71–107
Sugarcane, 4, 5, 16
Sulfur and sulfates, 172, 182–83, 186, 448
Sun and sunlight, 25, 33–35, 39, 43, 45. *See also* Radiation
Superphosphates, 145–46, 148, 182, 183
Sweden, 12, 264, 344
  Breeding Station of, 403
Switzerland, 353
Sylvester, E. S., 319, 320

Taylor, John, 6
TCA, 75, 82–83, 85, 91, 97, 98, 102
Temperature, 40, 379
  effect of on sugarbeet growth and development, 23–27, 30–33
  and herbicidal activity, 95
  root, 25–26
  soil, 38
Texas, 184, 270, 372, 430
Thiabendazole, 277
Thimet, 99
Thinning, 60–67, 72, 120, 219
  plant population, 63–67
  stand reduction, 60–63
  and weed control, 81, 105, 107
Thiocarbamates, 92, 93, 103. *See also* EPTC
Thorne, Gerald, 342, 344
Thrips, 307–8, 315
Tinker, P. B. H., 158, 160, 162, 164
Tisdale, S. L., 130
Tobacco rattle, 258
Tomato black ring, 258
Tracy, W. W., Jr., 418
Transpiration, process of, 24, 40–42, 191. *See also* Evapotranspiration
Triazine compounds, 51
Trifluralin and compounds of, 51, 91, 92, 102, 103, 104
Tsue, King, 366, 367
Turkey, 226, 229, 239, 265
Twin Falls, Idaho, 417
2,4-D, 74–75
Tyler, K. B., 119

Ulrich, A., 127, 128, 151, 152, 154, 156, 157, 163, 174, 180, 183, 378
Union of Soviet Socialist Republics, 344
  Sugar Beet Institute, 403, 425, 429
Union Sugar Company, 440
United Nations (FOA), 104
United States, 11
  controls on beet sugar growth and production in, 6–9
  diseases common to sugarbeets in, 225, 226, 233, 236, 237, 239, 249, 256
  low price of sugar in, 12

United States *(continued)*
  manufacture of planters in, 56, 57
  pests commonly attacking sugarbeets in, 288–93
  regional nutritional problems in, 185–86
  sugarbeet industry in, 5–6, 9–12
  sugar requirements of, 8, 13, 16–17
  use of phosphate fertilizers in, 138, 147, 151
  variety development in, 403–4
United States Department of Agriculture, 12, 15, 17, 291
  and control of sugarbeet production, 7–8
  and disease control, 277, 281, 292
  and herbicide registration, 104
  and pest control, 288, 291
  publications of,
    Farmers' Bulletins (392, 1645, 1867, 1903), 208–9
    Handbook 60, 218
    Yearbook of Agriculture (1955), 209–10
  and variety development, 404, 416–19 *passim*, 423, 428, 429, 433, 438
United States Salinity Laboratory, 218
United States Sugarbeet Laboratory, 425
United States Weather Bureau, 206
Urea, 122
Uruguay, 232, 233
Utah, 6, 225, 423
  diseases common to sugarbeets in, 240, 254, 266, 281
  pests commonly attacking sugarbeets in, 302–5 *passim*, 307, 310, 311, 312
  seed production in, 438, 441–46 *passim*, 448
Utah-Idaho Sugar Company, 10, 289, 440

Vanadium, 172–73
Van Steyvoort, L., 293
Verticillium wilt, 273
Vilmorin, Jacques L. de, 403
Vilmorin, Louis de, 403, 405
Vilmorin variety, 404
Virus diseases. *See* Diseases, virus
Virus yellows, 13, 17, 292, 431. *See also* Beet yellows and Beet western yellows

Waksman, S. A., 366
Walker, A. C., 116
Walker, K. C., 324
Wallace, T., 139
Wallis, R. L., 324
Ware, Lewis S., 6
Warington, R., 366
Washington, 181, 182, 184, 190
  diseases common to sugarbeets in, 240, 269, 273, 277, 278, 281
  pests commonly attacking sugarbeets in, 289, 290, 297, 298, 302, 304, 307, 308, 311, 312
Wasps, 314
Water, 38, 50, 52, 53. *See also* Irrigation and water management
  and herbicide dissipation, 102–3
Watson, D. J., 45
Watsonville, California, 289
Weather. *See also* Temperature
  effect of, 38–39, 50, 52
  and planting dates, 58–59
  and relation to foliar-applied herbicides, 95, 98
Weatherspoon, D. M., 98
Webworms, 303–4, 321
Weeds, 51, 52, 380
  biology of, 80–81
  and competitive emergence of, 81
  control of, 17, 54, 55, 62, 64, 72–75, 105–7, 375–76, 452. *See also* Herbicides
  distribution of, 76, 80
  morphology and physiology of, 85–86
  problem of, 71–72
  and seed production, 448
  varieties of commonly infesting sugarbeet fields in United States, 75–76
West Coast Beet Seed Company, 440
Whitefly. *See* Leafhopper
Whitegrubs, 297
White Pigeon, Michigan, 6, 9, 10
White Silesian sugarbeet, 5, 402–3
Whitney, E. D., 367
Wilcox, L. V., 218
Willamette Valley, 441
Williams, W. A., 123, 124
Wilson, R. H., 155
Wind, effect of, 40, 42, 50
Winner, C., 293
Wireworm, 13, 295
Wisconsin, 184, 226, 298, 307
Wit, C. T. de, 157
Worker, G. F., 121
Wort, D. J., 172
Wyoming, 190, 271, 360, 384
  pests commonly attacking sugarbeets in, 302, 303, 304, 309, 310, 313, 360

Yellow net, 253–54
Yellow splotch, 260
Yellow vein, 254–55
Yellow wilt, 225, 249–53, 321, 433
  causal agent, 252
  control of, 252–53
  economic importance of, 249
  history and distribution of, 249
  and host range, 250
  symptoms of, 250–51
  transmission of, 251–52
Young, D. A., Jr., 315
Young, R. A., 325

Zinc, deficiencies of, 172, 173, 174, 180, 181–82, 183, 185, 186